Handbook of Research in Mobile Business:
Technical, Methodological, and Social Perspectives

Volume II
Chapters 31–63

Bhuvan Unhelkar
University of Western Sydney, Australia

IDEA GROUP REFERENCE
Hershey · London · Melbourne · Singapore

Acquisitions Editor:	Michelle Potter
Development Editor:	Kristin Roth
Senior Managing Editor:	Amanda Appicello
Managing Editor:	Jennifer Neidig
Copy Editor:	Maria Boyer
Typesetter:	Diane Huskinson
Support Staff:	Renée Davies
Cover Design:	Lisa Tosheff
Printed at:	Yurchak Printing Inc.

Published in the United States of America by
Idea Group Reference (an imprint of Idea Group Inc.)
701 E. Chocolate Avenue, Suite 200
Hershey PA 17033
Tel: 717-533-8845
Fax: 717-533-8661
E-mail: cust@idea-group.com
Web site: http://www.idea-group-ref.com

and in the United Kingdom by
Idea Group Reference (an imprint of Idea Group Inc.)
3 Henrietta Street
Covent Garden
London WC2E 8LU
Tel: 44 20 7240 0856
Fax: 44 20 7379 0609
Web site: http://www.eurospanonline.com

Library of Congress Cataloging-in-Publication Data

Handbook of research in mobile business : technical, methodological and social perspectives / Bhuvan Unhelkar, editor.
       p. cm.
    Summary: "This reference book brings together various perspectives on the usage and application of mobile technologies and networks in global business"--Provided by publisher.
    Includes bibliographical references and index.
    ISBN 1-59140-817-2 (hardcover) -- ISBN 1-59140-818-0 (ebook)
    1. Mobile commerce. 2. Mobile communication systems--Economic aspects. I. Unhelkar, Bhuvan.
    HF5548.34.H36 2006
    658'.05--dc22
                            2005032111

British Cataloguing in Publication Data
A Cataloguing in Publication record for this book is available from the British Library.

All work contributed to this book is new, previously-unpublished material. The views expressed in this book are those of the authors, but not necessarily of the publisher.

Editorial Advisory Board

List of Contributors

Table of Contents

Section III
Technical

Section IV
Network

Section V
Security

Section VI
Strategy

Section VII
Application

Section VIII
Method

Section IX
Customer

Section X
Social

Detailed Table of Contents

Section I
Location

Chapter I

Mobile computing leads us into a new and fascinating journey into location-based services (LBSs) that was not feasible with land-based Internet connectivity. This is the dynamic creation of service offerings based on a location. However, the location itself can change from time to time. Thus LBSs form the crux of what is specific to mobile technologies, as discussed in this extremely well-researched and well-written chapter by Vyas and Yoon. The authors have rightfully argued that the recent rise in the level of comfort and demand to access various types of information using mobile devices can be attributed to the advancements in location-based services, which in turn are engendering new passion in mobile services utilizing users' location information. Such spatio-temporal information processing entails the need for a dynamic middleware that accurately identifies changing user location and attaches dependent content in real time without putting extra burden on users. This chapter succinctly describes the creation of a distributed infrastructure that is capable of supporting such scalable content dissemination. The Location-Aware Intelligent Agent System (LIA) offered by the authors is a conceptual framework in integration with Publish/Subscribe middleware to comprehensively address dynamic content dissemination (based on PUSH and PULL strategies) and related issues.

Chapter II

As already stated, location-based services are expected to play an integral role in the mobile commerce domain. This ability of mobility in creating dynamic and location-based services opens up opportunities for mobile network operators and service providers to add value and create additional revenue streams through a variety of personalized services based on location of individual wireless users. This chapter makes a crucial and substantial contribution to the strategic thinking in this area of location-based services. The issues and challenges discussed in this chapter include ownership of networks and their use by network operators and third parties, privacy concerns of consumers, and the corresponding business models for these services. The major areas covered by this chapter include an overview of location-based wireless services and their related technologies, a critical examination of the LBS value chain, and the strategic implications of location-based services for network operators and service providers.

Chapter III

This chapter discusses the primary features of location services in cellular networks and mechanisms to implement them. The authors start with an excellent overview of the most important location-based services, followed by the main location techniques (including their constraints and mechanisms to overcome them) that facilitate the provision of these services. The solutions proposed in this chapter have been used by local regulatory bodies in their official recommendations. Finally, this chapter also reviews the location architectures standardized for use in the main cellular networks and presents the concept of location middleware as a natural addition to these architectures.

Chapter IV

An expressive and comprehensive service description is vital when offering Web services. This is so because the discovery relies on the ability to match a user's need accurately to a service description. Ontologies are a flexible and powerful method to describe services. In this chapter the authors demonstrate how ontologies can be used to improve service discovery considerably in a mobile context by offering location-based information. This discussion on ontologies is followed by an example ontology, and the authors explain how to integrate such an ontology into existing technologies, thereby providing an effective way to describe location-based services.

Section II
Health

Chapter V

The healthcare domain stands to gain immensely by the incorporation of mobile information and communication technologies (MICTs)—as is demonstrated succinctly by the authors in this chapter. Researched by a medical doctor, this case study-based chapter delves deeper into the use of different MICT devices by doctors in specific hospital settings. While some doctors easily adopt MICT devices and find them a helpful tool, others encounter problems with their usage and, as a result, stop using the devices. This chapter identifies five factors influencing the uptake of MICTs in clinical work practices and proposes a framework for analyzing their interactions with the aim of increasing its uptake in medicine.

Computing and IT support within intensive care units (ICUs) has traditionally focused on monitoring the patients and delivering corresponding alarms to care providers within a hospital setting. However, many intensive care unit admissions are via intra- and inter-healthcare facility transfer, requiring receiving care providers to have access to patient information prior to the patient's arrival. The author discusses the opportunities that exist for mobile gadgets, such as personal digital assistants (PDAs), to substantially increase the efficiency and effectiveness of processes surrounding healthcare in the ICUs. This chapter provides invaluable reading and discussion on transcending beyond the current use of mobile devices in hospitals, which is restricted to mere personal information management and static medical applications, and takes the readers into the deployment of mobile-enabled solutions with overall considerations including privacy, cost, security, and standards.

This chapter reports on a study to research and evaluate the use of latest generation wireless devices—typically personal digital assistant (PDA) devices—by clinical staff at the large Westmead Hospital located in the west of Sydney, Australia. Currently, medical reports in this and other hospitals are primarily recorded on paper supported by personal computers at nursing stations. However, there is very little or no access to medical reports and decision-making tools for medical diagnosis at the patient's bedside—the precise location at which most medical decision making occurs. Delays in access to essential medical information can result in an increased time taken for accurate diagnosis and commencement of appropriate medical management of patients. This chapter discusses the application of handheld devices into more powerful processing tools connected to a centralized hospital data repository that can support medical applications.

This chapter discusses the application of Mobile Web Services in the handling of emergency processes in the health sector. The proposed application implements a mobile system based on cellular phone networks in ambulances, where communication between a number of parties is critical in terms of time, efficiency, and errors. Furthermore, it equips doctors with mobile devices that enable them to get connected to the Internet and access the health record of a patient quickly. This chapter demonstrates the way in which the proposed Mobile Comprehensive Emergency System (MCES) application would work with both static and mobile servers. The implementation of this new system will enhance the current system communication and makes it more reliable, consistent, and quick, and would also free the human intervention otherwise needed to access information.

This chapter explores various advancements in mobile devices and related software applications that facilitate rapid diagnostics in healthcare. Furthermore, this chapter also provides an excellent discussion, based on the author's experience as well as study, on the incorporation and usage of mobile devices in healthcare administration. Finally, the integration and networking of mobile devices is presented as the next major and substantial level of development that would lead to comprehensive usage of mobility in healthcare.

Section III
Technical

Chapter X

Energy-Efficient Cache Invalidation in Wireless Mobile Environment / *R. C. Joshi, Manoj Misra, and Narottam Chand*... 132

This chapter discusses the caching at a mobile client as a technique that can reduce the number of uplink requests, lighten server load, shorten query latency, and increase data availability. A cache invalidation strategy presented by these authors can ensure that the data item cached into a mobile client has the same value as on the server of origin. Traditional cache invalidation strategy makes use of periodic broadcasting of invalidation reports (IRs) by the server. However, this IR approach suffers from long query latency, larger tuning time, and poor utilization of bandwidth. Updated invalidation report (UIR) is a method that replaces a small fraction of the recent updates as deemed necessary—thereby reducing the query latency. To further improve upon the IR- and UIR-based strategies, researchers present a synchronous stateful cache maintenance technique called Update Report (UR). The UR strategy outperforms the IR and UIR strategies by reducing the query latency, minimizing the disconnection overheads, optimizing the use of wireless channels, and conserving the client energy. This highly researched chapter is a 'must read' for readers researching and experimenting with caching techniques at the mobile client end.

Chapter XI

Review of Wireless Technologies and Generations / *Raghunadh K. Bhattar, K. R. Ramakrishnan, K. S. Dasgupta, and V. S. Palsule* 142

This is an excellent review of the basics of wireless communication and the corresponding wireless generations. While communications technology has advanced very rapidly during the last century, so has the needs and expectations of the people. The market has managed to balance the above scenario by providing the effective solutions as and when these became available, through a series of technological innovations. Furthermore, to encourage adoption and advancement in wireless communication technology, standardization of technologies and processes is also required. The era during which such technologies and standards are popular is generally termed as *Generations*. This chapter discusses the fundamentals of mobile technologies in relation to this mobile generation. Such a discussion of the communication technology generations not only provides an understanding of the past history of these technologies, but also creates the basis for understanding their future.

Chapter XII

Orthogonal Complex Quadrature Phase Shift Keying (OCQPSK) Spreading for 3G W-CDMA Systems / *Shailendra Mishra and Nipur Singh* 158

This chapter discusses the variety of digital modulation techniques that are currently being used in wireless communication systems, as well as various alternatives and advancements to these techniques. When using the 3G (third-generation) spread-spectrum systems, such as W-CDMA (3GPP) and cdma2000 (3GPP2), the handset can transmit multiple channels at different amplitude levels. However, modulation schemes such as OQPSK or GMSK do not prevent zero-crossings for multiple channels and are no longer suitable. There is a need for a modulation format or a spreading technique that can accommodate multiple channels at different power levels while producing signals with low peak-to-average power ratios. This is what the authors have proposed through their own OCQPSK (Orthogonal Complex Quadrature Phase Shift Keying) spreading technique for W-CDMA and cdma2000. Starting with the basic structure of the reverse link (uplink) for W-CDMA and cdma2000 with no scrambling, this chapter explains the transition through complex scrambling to OCQPSK. The chapter then describes the concept of complex scrambling and OCQPSK and how it works. Finally, this chapter describes how to measure modulation quality on the reverse link of 3G systems and a complete downlink physical layer model, showing various results of BER and BLER calculation and also various time scopes and power spectrums.

Chapter XIII

This chapter discusses the large-scale statistics of an Improved Cell Search Design (CSD) using cyclic codes and compares it with the 3GPP Cell Search Design using comma free codes (3GPP-comma free CSD) in terms of acquisition time for different probabilities of false alarm rates that would achieve faster synchronization at lower hardware complexity. The authors also propose design improvements in stage 2 of the 3GPP-comma free CSD by using a Fast Hadamard Transformer (FHT). Furthermore, masking functions are used in stage 3 of both the improved CSD and the 3GPP-comma free CSD to reduce the number of scrambling code generators required, as described in Chapter 13. This results in a reduction in the ROM size required to store the initial phases of the scrambling code and generators in stage 3, resulting in faster synchronization between the MS and the BS and improved system performance. The research results presented here indicate that for a channel whose signal-to-noise ratio is degraded with additive white Gaussian noise (AWGN), the Improved CSD achieves faster synchronization with the base station and has lower hardware utilization when compared with the 3GPP-comma free CSD scheme under the same design constraints.

Chapter XIV

This research-intensive chapter discusses the performance gains for communications systems resulting from turbo codes, turbo equalization, and decoding techniques. Turbo codes send digital data over channels that require equalization—that is, those that suffer from inter-symbol interference (ISI). Turbo equalizers have been shown to be successful in mitigating the effects of inter-symbol interference introduced by partial-response modems and by dispersive channels for code rates of $R>1/2$. The performance of iterative equalization and decoding (IED) using an M-BCJR equalizer is analyzed in this chapter. Furthermore, the bit error rate (BER), frame error rate simulations, and extrinsic information transfer (EXIT) charts are used to study and compare the performances of M-BCJR and BCJR equalizers on precoded and non-precoded channels. The authors predict the BER performance of IED using the M-BCJR equalizer from EXIT charts, and discuss in detail the discrepancy between the observed and predicted performances.

Chapter XV

The need for software standardization of mobile and wireless devices is crucial for successful component-based software engineering. Components in mobile and wireless devices require administration functionality that—despite existing standards, protocols, techniques, and tools—effective and efficient management may come up against odd component forms resulting from non methodical design. The availability of version 2 of the UML (Unified Modeling Language) and the prominence of executable modeling languages in the MDA/ MDE (model-driven architecture/model-driven engineering) open up opportunities for building manageable wireless software components as discussed in this chapter. This chapter also discusses a design method and library created on top of the Built-In Test (BIT) technology, and illustrates the concepts through a case study of home automation systems.

Chapter XVI

This chapter discusses the issues of mobile user data mining. Mobile user data mining is the collection of data collected from the activities generated from mobile users in order to analyze their behavior pattern to predict their future behaviors. The increasing adoption of mobile devices provides the ability for mobile user data mining to analyze data collected from mobile users. This can be used to determine the trends and patterns for decision-making purposes. It is applicable to marketing, banking, and retail industries. Finally, this chapter provides an insight to the underlying issues in mobile user data mining.

Mobile users desire customized bundles of services that need to be dynamically created from the service providers. However, services are unique; and since unused services do not generate revenues, they present a lost 'economic rent' for organizations that are not part of the network of service providers. As a result, they are not part of the customized bundle of services. The dynamic discovery of a bundle of individual services from such a network that meets the unique needs and constraints of the mobile user requires intelligent agent technology. Such agent technology would match personal needs of the user with the available services in a cost-efficient manner. This chapter provides a mechanism to create dynamic service bundles from ad-hoc user requirements using intelligent agents. The authors apply this technique to a mobile commerce environment and illustrate the composition of user-specific service "bundles" by intelligent agents that represent the interests of the m-commerce user. Such agent-based architectures provide users with customized solution "bundles" that reduce their cognitive burden, while improving the utilization of resources for organization that are part of the service provider network.

Chapter XVIII

In this information-centric age, an organization needs to access the most up-to-date and accurate information for fast decision making. Mobile access to the Internet provides convenient and portable access to a huge information space. However, loading and visualizing large documents on mobile devices is impossible due to their natural shortcomings such as screen space and computing power. In this chapter, the author introduces the fractal summarization model, based on fractal theory, for document summarization on mobile devices. This model generates a brief skeleton summary at the first stage, and the details of the summary on different levels of the document are generated on demand from users. Such interactive summarization reduces the computation load, which is ideal for wireless access. On the other hand, the hierarchical display in fractal summarization is more suitable for navigation of a large document, and it is ideal for small area displays. The automatic summarization, together with the three-tier architecture and the information visualization, are potential solutions to the existing problems in information delivery to mobile devices for mobile business.

The technological evolution of networks, together with the development of positioning systems, has contributed to the emergence of numerous location-based services. These services will be of major technical as well as economical interest in coming years. This aroused the interest of a great part of the scientific community which proposed to study these services with diverse requirements and constraints. One of the direct consequences in the database field is the appearance of new types of queries (mobile queries issued from mobile terminals and/or requesting information associated with moving objects such as vehicles). This chapter proposes a survey on mobile queries, with particular attention given to the location issue.

Section IV
Network

Wide area network (WAN) offers advantages such as providing myriad services available on globally diversified computers with reasonably simple process. The ability to dynamically create networks offers the processing powers of various processors at our command. With the advent of protocols like SOAP and Web Services, the consumption of services are more organized. In spite of various advances in communication techniques, the consumption of services through mobile gadgets is still only at the research level. The major impedances in implementing such systems on a mobile network are the latency factor, abrupt disconnection in service, lower bandwidth, and minimal processing power. The mobile agent's paradigm proves to be an effective solution to various issues raised. It has received serious attention in the last decade, and several systems based on this paradigm have been proposed and built. All such systems have been designed for a static network, where the service providers and the requestors are connected to the server on a permanent basis. This chapter presents a new framework of managing the mobile environment and the participating nodes with active intelligent migration. The functioning of the mobile agents in such a scenario is also presented.

This chapter develops the concept of load balancing that plays a key role in providing various advanced applications in the cellular mobile environment. Load balancing means the efficient distribution of channels among cells in accordance with their requirements to minimize call blocking. As the channels for these services are scarce, load balancing has emerged as a primary issue in today's scenario. Two different prominent schemes of load balancing are elaborated upon. This chapter is aimed at researchers and policymakers, making them aware of the different means of efficient load balancing, as well as underscoring the problem areas that need further vigorous research.

As mobile computing gains popularity, the need for ad hoc routing also continues to grow. In mobile ad hoc networks, the mobility of nodes and the error prone nature of the wireless medium pose many challenges, including frequent route changes and packet losses. Such problems increase the packet delays and decrease the throughput. To meet with the dynamic queuing behavior of ad hoc networks, to provide quality of service and hence improve performance, a scheduler can be used. This chapter presents a novel fuzzy-based priority scheduler for mobile ad hoc networks to determine the priority of the packets. The performance of this scheduler is studied using GloMoSim and evaluated in terms of quantitative metrics such as packet delivery ratio, average end-to-end delay, and throughput.

Next-generation wireless systems will provide users with a broad range of services, providing wireless technologies without any major interoperability issues. The recent growth in demand and deployment of WLAN/WPAN for short-range connections has been driven by the need to create ubiquitous networks, where one can be connected anywhere at any time making many services and applications a click away. These short-range access networks currently exist almost everywhere—at home, the workplace, hotels, hospitals, and so forth. Wireless local area networking standard (Wi-Fi) and the WPAN standard (Bluetooth and Zigbee) products utilize the unlicensed 2.4 GHz ISM band. Due to the dependence of these technologies on the same band, potential for interference exists. This chapter will focus on the characterization of these technologies, discussing differences and similarities, a wide range of applications and deployments, and the study of the potential interferences between such technologies when deployed within the same working space.

Mobile businesses are increasingly demanding high-speed facilities of multimedia services and Internet access "anywhere" and "anytime." Limited transmission resources (i.e., bandwidth) are the main obstacles to widespread use of mobility in business. Many mobile networks support advanced technologies, and mobile communications protocols have been developed to optimally utilize wireless resources. These policies support the heterogeneous access technologies for multimedia services in mobile networks. Many of these policies exploit the mobility information from the current and the neighboring cells to dynamically adjust the resource reservation, allocation, and call admission control policy to adapt quickly with the changing network traffics. Resource reservation is, however, necessary to support the migrating users from the neighboring cells. This chapter explains the key components of resource management mechanisms in mobile networks, including the fair distribution of resources among different users/clients involved in mobile business or use of wireless resources.

This chapter proposes a secure agent roaming scheme in the m-commerce agent framework. Intelligent agents are one solution to providing intelligence in m-commerce. However, merely having an agent that is intelligent is insufficient for m-commerce applications. There are certain tasks that are unrealistic for agents to perform locally, especially those that require huge amounts of information. Therefore, it is important to equip intelligent agents with roaming capability, as is discussed in this chapter.

This chapter introduces a secure online track-and-trace system for tackling counterfeiting. According to the Counterfeiting Intelligence Bureau (CIB), part of the International Chamber of Commerce, 7% of all world trade is in counterfeit goods, and the counterfeit market is worth $350 billion. Virtually every country in the world suffers from counterfeiting, which results in lost tax revenue, job losses, health and safety problems, and business losses. Furthermore, counterfeit goods do not only target famous brand names, but anything that will sell, such as bottled water. Counterfeiters are increasingly damaging businesses, and as such, businesses need to fight back against counterfeiting. Nowadays, there is an explosion of mobile wireless services accessible via mobile phones with a built-in camera. The mobile users can access the Internet at any time, from anywhere, using ubiquitous inexpensive computing. Mobile camera phones and other handheld devices are becoming indispensable. The aim of this chapter is to show how businesses can protect their products from counterfeiting by using a secure online track-and-trace system, which will allow their customers to authenticate the products in real time through a Web-enabled mobile camera phone. This will assist business and customers by confirming that the said product is genuine and not counterfeit, which can be accomplished at anytime and any location, and without any significant changes to the existing business operational systems.

This chapter reports on the development of a theory to increase the security of mobile business and its application to Australian information systems. To increase the growth of PKI, a theory called PORTABLEPKI is developed for the security of the wireless network. Furthermore, this chapter also discusses a framework for testing PORTABLEPKI and future research directions.

The emergence of wireless and mobile networks has evolved the domain of electronic commerce in to a new application and research area that we know as mobile commerce. However, applying mobile commerce to business applications is a challenging task since it involves a wide variety of disciplines and technologies. In order to make it easier to understand the application of mobile commerce, this chapter starts the discussion with the basics of mobile commerce from a technical perspective, followed by a discussion on Net-enabled mobile handheld devices such as smart phones and PDAs, and finally, mobile payment methods, including macro-payment and micro-payment methods.

Section VI
Strategy

Chapter XXIX

The convergence of the Internet and mobile networks has created new opportunities and applications. Considering mobile business only as an extension of the traditional Web can lead to missing out on unique and differentiable qualities for new value-added opportunities. Mobile banking is considered as potentially one of the most value-added and important mobile services available. The chapter examines the technological changes in mobile networks and the innovative attributes of a mobile Internet. It advances the theoretical framework of innovation in services to develop a customer-centric analysis of an m-banking value proposition. The chapter goes on to discuss critical factors in the diffusion of m-banking, and explores reasons of failure and further prospects of success.

Chapter XXX

This chapter discusses the opportunities and challenges of mobile commerce in emerging economies. It analyzes the profound impact of a mobile device on the way products and services are bought and sold in developing nations. The chapter argues that many mobile applications can have a much larger impact on emerging economies than those of the developed world. The chapter is aimed at creating an understanding of the unique social, technological, and economic drivers that can help entrepreneurs and solution providers to build and deploy compelling and revolutionary mobile commerce applications in these emerging markets.

Chapter XXXI

This chapter discusses the use of mobile, handheld computer devices that are connected wirelessly to a network for business and personal use across people, projects, tasks, or organizational units to infer a trend of general acceptance of m-business in the marketplace. The author describes the state of the mobile commerce industry from a worldwide perspective and the barriers to implementation of m-commerce, discusses the issues and challenges, and ends with conclusions and directions for future research.

Chapter XXXII

This chapter takes the reader through a step-by-step process of developing a mobile business initiative. Starting by describing the fixed as well as Mobile Internet environment, this chapter analyzes the characteristics of a wireless world and how to incorporate mobility in business. The discussion is supported by current examples of successful implementations around the world, made by big and unknown companies. Eventually a tool to design and deliver a wireless solution is provided with an eye on the business side, trying to make technology and business work together and speak the same language.

This chapter takes a look at mobile commerce riding on the wave of mobile computing applications. Mobile commerce, also known as m-commerce, is the new powerful paradigm for the digital economy. In view of that, this chapter examines issues relevant to the mobility of today's workforce. The meaning of mobility and its implications are explored in this chapter, along with the legal implications that arise in the pursuit of mobile commerce. Given its importance, this chapter also briefly delves into security issues related to m-commerce. Towards the end, a lighter side of m-commerce and mobile computing is provided, together with conclusions and future directions.

This chapter deals with the future of mobile technologies and applications in China. The effect of emerging technologies, especially mobile technologies, on the massive market of China cannot be ignored in the global context. This chapter gives the reader an insight into China's mobile telecommunication industry today. The authors first relate statistics about China's mobile business market, including user and device analysis that helps in providing an understanding of mobile business in China. This analysis is followed by a description of the major mobile technologies employed in China and a brief view of the Chinese market's status, followed by an insight into some newly rising industries which are potentially successful mobile sectors in China. Finally, a real-life example is examined—that of the M-Government Project in Gunagzhou, capital city of Guangdong Province.

Section VII
Application

Chapter XXXV

Applications with rich user interfaces and smart clients improve the user experience. As mobile enabling technologies advance in capability, affordability, and availability, users expect improved design of mobile devices that will leverage the advances and convergence in technology and the Internet to deliver richer applications and value-added m-services. They demand m-applications that facilitate communications, information retrieval, financial management, paying bills, trading, gambling, entertainment, and dating. The design and architecture of the next generation of mobile applications and browsers will be challenging, as developers must still consider the limitations of the small screen and input options, and the unreliable connectionless paradigm, and allow for backward compatibility with earlier protocols and formats. Mobile application developers must support various configurations and interface with a plethora of different mobile computing devices and platforms. Furthermore, designers must also address any environmental and/or health issues, and design a product that is socially acceptable and safe.

The establishment of the OntoQuery system in the m-commerce agent framework investigates new methodologies for efficient query formation for product databases. At the same time, it also forms new methodologies for effective information retrieval. The query formation approach implemented takes advantage of the tree pathway structure in ontology, as well as keywords, to form queries visually and efficiently. The proposed information retrieval system uses genetic algorithms and is computationally more effective than iterative methods such as relevance feedback. Synonyms are used to mutate earlier queries. Mutation is used together with query optimization techniques like query restructuring by logical terms and numerical constraints replacement. Also, the fitness function of the genetic algorithm is defined by three elements: number of documents retrieved, quality of documents, and correlation of queries. The number and quality of documents retrieved give the basic strength of a mutated query, while query correlation accounts for mutated query ambiguities.

Chapter XXXVII

One of the potential applications for an agent-based system has been in the area of m-commerce, and a lot of research has been done on making the system intelligent enough to personalize its services for the user. In most systems, user-supplied keywords are normally used to generate a profile of the user. In this chapter, the author proposes a design for an evolutionary ontology-based product-brokering agent for m-commerce applications. It uses an evaluation function to represent the user's preference instead of the usual keyword-based profile. By using genetic algorithms, the agent tries to track the user's preferences for a particular product by tuning some of the parameters inside this function. The author has developed a prototype in Java, and the results obtained from these experiments look promising.

Chapter XXXVIII

Web services (WSs) have become the industry standard tools for communication between applications running on different platforms and built using different programming languages. The benefits, including the simplicity of use, that Web Services provide to developers and users have ensured integration of Web Services architecture by almost all IT venders in their applications. As expected, with the proliferation of mobile phones, PDAs, and other wireless devices, the same requirements of making applications talk across platforms has become necessary on mobile devices. This has led to Mobile Web Services (MWSs), which are based on Web Services and related technologies like XML, SOAP, and WSDL, and which provide the best choice to be used in the architecture for integration of Web Services in mobile devices. This chapter discusses WS and MWS in the context of integration architecture, together with their advantages and disadvantages in usage. Since MWSs are deployed using wireless technologies and protocols, they are also presented and explained in this chapter.

Chapter XXXIX

Push technology is a kind of technology that automates the information delivery process without requiring users to request the information they need. Wireless has experienced explosive growth in recent years, and "push" is expected to be the predominant wireless service delivery paradigm of the future. For example, one would expect a large number and a wide variety of services such as alerts and messages as well as promotional contents and even e-mails to be regularly delivered to consumers' mobile devices such as phones or PDAS. As argued in this chapter, "pushing" information to a wireless device is a unique challenge because of the problems of intermittent communication links and resources constraint on wireless devices, as well as limited bandwidth. The authors in this chapter explore an efficient multicasting mechanism that "pushes" pre-specified information to groups of wireless devices with limited bandwidth and flaky connections. This chapter reports on the design and implementation of a prototype framework based on the concept of push technology to multicast information via wireless technology.

Section VIII
Method

Chapter XL

The advent of mobile technologies in recent times, coupled with the ever-increasing pressure for prices to drop, has opened up a whole new world of opportunities for business via the new medium for small and medium enterprises (SMEs). In particular, SMEs that have already embraced technology in many areas of their business find the move to embracing mobile technologies as the next logical step. This can be called m-transformation, and it consists of three ingredients: ICT infrastructure, new business process adoption, and a methodology to successfully lead to m-transformation and its many benefits for SMEs. The SME landscape presents some unique challenges however when it comes to attempting m-transformation. These challenges affect in turn all three aspects of m-transformation, causing the need for a methodology that is flexible and extensible in order to meet and surpass those challenges. In this chapter the author presents this methodology that has been used to successfully m-transform SMEs and shows that, although challenging, leading SMEs to a successful m-transformation is very possible, given sufficient background knowledge and a suitable, robust methodology to use.

Chapter XLI

In recent years, re-engineering of business processes is driven with a 360-degree view encompassing the customers, employees, suppliers, and partners. With the advancements in mobile technologies, mobile applications are swiftly making their way in enterprise business (processes). This chapter focuses on the application of mobile technologies in enterprise-wide business processes. The chapter particularly focuses on the use of mobile technologies to redesign or streamline business processes, including customer relationship management and supply chain management processes. The author has also succinctly highlighted how the "mobile layer" fit into the enterprise business architecture, and its subsequent potential.

Chapter XLII

Emerging mobile technologies have changed the way we conduct business. This is because communication, more than anything else, has become extremely significant in the context of today's business. Organizations are looking for communication technologies and corresponding strategies to reach and serve their customers. And mobile technologies provide the ability to communicate independent of time and location. Therefore, understanding mobile technologies and the process of transitioning the organization to a mobile organization are crucial to the success of adopting mobility in business. Such a process provides a robust basis for the organization's desire to reach a wide customer base. This chapter discusses the assessment of a business in the context of mobile technology, presents a brief history of mobile technology, and outlines an initial approach for transitioning an organization to a mobile organization.

Chapter XLIII

Mobile computing now encompasses the growing area of broadcast radio in data communication. This becomes an important criterion in providing good quality service with rapidly increasing mobile users. Policy-based approaches are widely used for security, quality of service (QoS), virtual private network (VPN), and so forth. In this chapter the authors examine the potential areas in mobile computing where policy-based approaches can be successfully implemented to enhance data communication.

Section IX
Customer

The growing popularity of the mobile phone and the diverse functionality of mobile services have forced mobile service providers to enter into a highly competitive business arena. In digital life today, mobile phone services are not restricted merely to communicating with people, but more and more value-added services have emerged to amalgamate disparate industries/businesses and open up greater market opportunities. These disparate industries/businesses may include recreational and travel services, mobile learning services, mobile banking services, and many others. Nevertheless the service providers must understand the consumer behavior in value-added services in order to enhance their product design. The key objectives of this research are to investigate and analyze the relationships between consumer behavior, consumer personality, and lifestyle in adopting mobile recreational services, and provide recommendations to the service providers for increasing competitiveness—in the context of Taiwan.

Chapter XLV

This chapter provides an introduction of using Mobile CRM to reach, acquire, convert, and retain consumers. Firstly, a definition of the term CRM is provided, and the author also gives an insight on extending CRM to the wireless world. Having presented the benefits of mobile data services and their benefits to businesses in terms of customer relations and marketing, however, businesses still faced the challenges of delivering the promise to consumers. More importantly, the adoption of mobile services is still low in business and consumer segments. The author identifies content appropriateness, usability issues, personalization, willingness to pay, security, and privacy as major challenges for businesses, and then recommends businesses to start segmenting their mobile consumers into Mobile Tweens, Mobile Yuppro, and Senior Mobile users; acknowledging that understanding the demographics, social, and behavioral issues of these three consumer groups is an initial step in Mobile CRM; before finally recommending the use of viral marketing as a mechanism to market mobile services. This is followed by matching relevant services to consumers to create a positive usability experience and always build a critical mass, but develop customers one at a time. The implementation of Mobile CRM will be fully addressed in the second part of the chapter.

This chapter addresses important factors for consideration when readying a mobile commerce business for global business, addressing both regional differentiation in demographics that influence classifications of customer segments, and differentiation in demographics within a region. Globally, not all customer segments have regular access to mobile commerce facilities, and even for those that do, other demographic factors can impede their potential as mobile-customers. When starting from an Anglo-centric perspective, it is vital to have awareness of global differences in culture, language, payment options, time zones, legal restrictions, infrastructures, product needs, and market growth that could either improve or inhibit mobile-customer uptake, and in the worst case, result in unexpected litigation.

Section X
Social

This chapter discusses the growing inappropriate use of mobile camera phones within our society. There are two areas of concern that are dealt with in this chapter. The first concern deals with individual privacy and the use of mobile camera phones as a tool of harassment. The second concern deals with organizations seeking to prevent industrial espionage and employee protection. This chapter outlines how these devices are being used to invade individuals' privacy, to harass individuals, and to infiltrate organizations. The author outlines strategies and recommendations that both government and manufacturers of mobile camera phones can implement to better protect individual privacy, and policies that organizations can implement to help protect them from industrial espionage.

This chapter explores how perceptions of the social context of an organization moderate the use of an innovative technology. This chapter proposes a research model that is strongly grounded in theory, and offers propositions that investigate adoption and diffusion of mobile computing devices for business-to-business (B2B) transactions. Mobile computing devices for B2B are treated as a technological innovation. The authors believe that such an extension of existing models by considering the social contextual factors is necessary and appropriate in light of the fact that various aspects of the social context have been generally cited to be important in the introduction of new technologies. In particular, a micro-level analysis of this phenomenon for the introduction of new technologies is not common. Since the technological innovation that is investigated is very much in its nascent stages, there may not as yet be a large body of users in a B2B context. Therefore, this provides a rich opportunity to conduct academic research.

With high optimism, the third-generation mobile communication technologies were launched and adopted by telecommunication giants in different parts of the globe. However, with an uncertain and turbulent social, economic, and political environment, and the downturn in the global economy, difficult conditions are pronounced for the initial promises of m-commerce technologies to be fully realized. The causes for this, determined so far, have been largely of a technical nature. This chapter shifts the focus of analysis from a pure technical approach to a socio-cultural one. The basic premise of this chapter is that cultural variations do play a very important part in shaping potential consumers' choices, beliefs, and attitudes about m-commerce services. The authors believe that to be an important way for the m-commerce industry to fulfill its potential.

This chapter discusses the mobile network as a new medium for marketing communications. It illustrates that the mobile medium, defined as two-way communications via mobile handsets, can be utilized in a company's promotion mix by initiating and maintaining relationships. Firstly, by using the mobile medium, companies can attract new customers by organizing SMS (Short Message Service)-based competitions and lotteries. Secondly, the mobile medium can be used as a relationship-building tool, as companies can send information and discount coupons to existing customers' mobile devices or collect marketing research data. The authors explore these scenarios by presenting and analyzing a mobile marketing case from Finland. The chapter concludes by pondering different future avenues for the mobile medium in promotion mix.

This chapter studies mobile business in its dynamic, historic, and evolving nature. The chapter offers discussions on the background of, need for, and concept of mobile businesses. Following the background review, the chapter discusses the current status of mobiles business and its model. In this part, some classification of mobile business is given, and most representative fields of mobile business are identified, followed by a discussion on the technical aspects of mobile business. Elements that make a business mobile, such as communication infrastructure and supporting networks, are also discussed.

Section XI
Case Study

Wireless technology is growing at a phenomenal rate. Of the many present challenges highlighted by the author, increased security is one of the main challenges for both developers and end users. This chapter presents this important security aspect of implementing a mobile solution in the context of Sydney International Airport. After tackling initial challenges and issues faced during the implementation of wireless technology, this chapter demonstrates how security issues and wireless application were implemented at this mobile-intense airport organization. The decision to deploy and manage the wireless spectrum throughout the airport campus meant that the wireless LAN had to share the medium with public users, tenants, and aircraft communications on the same bandwidth. Therefore, this case study also demonstrates an invaluable approach to protect unintended users from breach of existing security policies adopted by their corporate network. Authentication and data privacy challenges, as well as complete WLAN connectivity for tenants, public, and corporate usage is presented in this case study.

This chapter examines the adoption of radio frequency identification (RFID) technology in the commercial aviation industry, focusing on the role of RFID systems for improved baggage handling and security. Based upon secondary and trade literature, the chapter provides a timely overview of developments with regard to the implementation of the technology in commercial aviation. RFID technology holds distinct advantages over the currently used bar-code system for baggage handling. The chapter focuses on two major contributions that RFID promises commercial aviation: (1) improved customer service though better operational efficiency in baggage handling, and (2) increased airport and airline security. Particular attention is given to the initiative of Delta Airlines, an industry leader in the testing and development of RFID systems for improved operations in baggage handling. This chapter provides an avenue for academicians and business professionals to be aware of developments with RFID technology in this area.

Due to a significant cost advantage, mobile multicasting technology bears the potential to achieve extensive diffusion of mobile rich media applications. As weak performance of previous mobile data services suggests, past developments have focused on technology and missed customer preferences. Mobile multicasting represents a radical innovation. Currently, little research on consumer behavior exists regarding such services. The chapter addresses this gap by presenting results of qualitative and quantitative field trials conducted in three countries. It provides a continuous customer integration approach that applies established methods of market research to the creation of mobile services. Means-end chain analysis reveals consumers' cognitive reasoning and conjoint analysis drills down to the importance of service attributes. Desire for self-confidence and social integration are identified key motivators for consumption of mobile media. Services should aim for technological perfection, and deliver actual and entertaining content. Interestingly, consumers appreciate reduced but tailored contents, and price appears not to be a superseding criterion.

Due to a significant cost advantage, mobile multicasting technology bears the potential to achieve extensive diffusion of mobile rich media applications. As weak performance of previous mobile data services suggests, past developments have focused on technology and missed customer preferences. Mobile multicasting represents a radical innovation. Currently, little research on consumer behavior exists regarding such services. The chapter addresses this gap by presenting results of qualitative and quantitative field trials conducted in three countries. It provides a continuous customer integration approach that applies established methods of market research to the creation of mobile services. Means-end chain analysis reveals consumers' cognitive reasoning and conjoint analysis drills down to the importance of service attributes. Desire for self-confidence and social integration are identified key motivators for consumption of mobile media. Services should aim for technological perfection, and deliver actual and entertaining content. Interestingly, consumers appreciate reduced but tailored contents, and price appears not to be a superseding criterion.

Chapter LVI

This chapter discusses the impact of mobile technologies on service delivery processes in a banking environment. Advances in mobile technologies have opened up numerous possibilities for businesses to expand their reach beyond the traditional Internet-based connectivity, and at the same time have created unique challenges. Security concerns, as well as hurdles of delivering mobile services "anywhere and anytime" using current mobile devices, with their limitations of bandwidth, screen size, and battery life, are examples of such challenges. Banks are typically affected by these advances as a major part of their business deals with providing services that can benefit immensely by adoption of mobile technologies. As an example case study, this chapter investigates some business processes of a leading Australian bank in the context of application of mobile technologies.

This chapter focuses on Mobile GIS (MGIS), which uses wireless networks and small-screen mobile devices (such as PDAs and smartphones) to collect or deliver real-time, location-specific information and services. Such services can be divided into field and consumer (location-based services) GIS applications. The use of wireless networks and small-screen devices introduces a series of challenges not faced by desktop or wired Internet GIS applications. This chapter discusses the challenges faced by mobile GIS (e.g., small screen, bandwidth, positioning accuracy, interoperability, etc.) and the various means of overcoming these problems, including the rapid advances in relevant technologies. Despite the challenges, many efficient and effective Mobile GIS applications have been developed, offering a glimpse of the potential market.

Chapter LVIII

This chapter examines the potential of mobile technologies for the tourism industry. Mobile technologies have the capacity to address not only the pre- and post-tour requirements of the tourist, but also to support the tourist on the move. It is this phase of the tourist activity upon which mobile technologies can be expected to have the greatest impact. The development of applications for the mobile tourist will allow for the creation of a new range of personalized, location- and time-specific, value-added services that were not previously possible. Before such applications can be widely deployed, however, some fundamental technical and business challenges need to be addressed. Despite these challenges, mobile technologies have the potential to revolutionize the tourist experience, delivering context-specific services to tourists on the move as discussed in this chapter.

This chapter suggests creation of a model for the financial services sector of the international financial market through the components of telecommunication and information technologies. While telecommunication and information technologies have influenced activities related to business, convergence of these technologies is the crucial enabler that makes cross-border commerce in the present globalization scenario a reality. This chapter explains the use of knowledge-based financial systems and the process of incorporating mobile computing into these financial systems. The chapter discusses business process related to the financial services sector, creation of a knowledge-based financial system, and incorporating access to the system with devices that can be used in a wireless communication environment.

This chapter uses a case study approach to demonstrate how companies are adopting mobile technology in their business processes. The authors describe a company that has used mobile devices to distribute work orders to field staff, and allows them to input their travel and work times and material usage for processing by the company's ERP system. The chapter further examines the benefits obtained and the issues associated with the introduction of the system, and attempts to classify it according to an existing model.

This chapter makes an important contribution in discussing the use of mobility in providing local government services. In early 2002, a large local government agency, Penrith City Council (PCC), located on the western fringe of the Sydney metropolitan area, entered into a collaborative working relationship with the University of Western Sydney. Research and development work conducted under this arrangement led to some interesting experiences and resulting learning for the students, client, and academic staff. This chapter highlights the development projects involved in the evolution of the PCC Mobile Strategy and discusses several aspects of the learning experiences, including: release hype vs. the implementation realities of mobile technology, technological options for the introduction of mobility, user acceptance of new technologies, the management of client expectations, and local government standards and guidelines and their impact on development directions.

This chapter explores how a mobile tracking technology is able to further streamline the integrated supply chain. Previous technologies which have attempted to integrate suppliers, manufactures, distributors, and retailers have lacked the flexibility and efficiency necessary to justify the prohibiting costs. Radio frequency identification (RFID) technology, however, enables various organizations along the supply chain to share information regarding specific products and easily remotely manage internal inventory levels. These applications are only a sample of what RFID is able to accomplish for the integrated supply chain, and this chapter seeks to explore those applications.

This chapter reports on a system solution that has been developed by T-Systems' Solution and Service Center Ulm/Germany, within the service offering portfolio "Embedded Functions." Considering that an increasing number of goods will be "on the road" (on rails, on ship, in the air) for an appreciable percentage of the lifecycle, there is an urgent need to bridge the information gap between the automated systems at the factory sites and the storage control systems at the destination sites. This chapter discusses how the system solution has provided a synergy effect of connecting mobile communication solutions with auto ID services in the context of online surveillance during transportation, providing both downstream batch tracking, as well as upstream traceability.

Dedication

Keshav Raja

Foreword

As a conference chair of the recently completed International Conference on Mobile Business in Sydney, Australia, I had the unique opportunity of organizing, inviting, reviewing, and listening to a wide range of excellent researchers and practitioners in the area of mobile business. My understanding of mobility was further enhanced as I realised the phenomenal amount of research and industrial experimentation that is occurring in the area of mobile technologies and their application to business. Today, mobility encompasses themes such as devices, networking, architecture, design, applications, usability, security and privacy, entertainment, and mobile learning, to name but a few. These themes are embedded in this excellent book, edited by Dr. Unhelkar. This work, focused on the application of mobility in business, promises to become essential reading for all mobile researchers and practitioners. Therefore, with great pleasure and honour, I introduce this book to you.

While this handbook itself is research oriented, the contributing authors of the chapters within this book come from a cross-section of research *and* industrial expertise—a sensible and practical combination for a book that deals with *mobile business*. Through these internationally contributed chapters, the reader is exposed to in-depth discussions of the aforementioned mobility themes. The wide coverage of topics and the variety of contributors to this handbook make it a seminal addition to the body of literature and knowledge in the area of mobile business.

Mobile business, as the term indicates, deals with issues and challenges related to incorporation of mobile technologies in business processes. While the prior experiences gained from adopting electronic commerce can be helpful to business, mobile business provides some fundamentally unique issues and challenges of its own. For example, the customers in the mobile space are constantly on the move, requiring intensive research and experimentation in the area of location-based services to enable their tracking. Dropping of mobile connections is a common challenge in mobile transactions that also necessitates discussion of the security aspects of mobile applications. The sociological challenges of typing a text (SMS) reply on a small screen device or preventing an unauthorised photograph being taken in a swimming pool through a hand-held device open up a plethora of legal and ethical issues that are covered appropriately in the later chapters of this handbook.

Overall, I find this handbook an extremely comprehensive book dealing with the exciting and fast-moving domain of mobile business. Jonathan Withers, CTO of iBurst, reminded us (during his keynote address at the aforementioned conference) of the 10:30 Rule—in the next 10 years, technology will advance at a rate equivalent to the last 30 years! Researchers and industry practitioners will need to keep up with this challenge. However, this speed is what makes research in this area exhilarating. Cisco, the renowned networking company, further foresees fixed mobile convergence—where multiple applications will be seamlessly accessible from any access technology—as an area of research and application as we aim to provide a common set of services giving a consistent user experience regardless of the device.

My own philosophy of research in the mobile domain is that industry practitioners and academics *must* collaborate to understand, document, and provide practical solutions to the challenges of adopting mobility. To this end, I am particularly pleased to see the appropriate inclusion of numerous industry case studies in this book. Furthermore, each of the chapters is of excellent quality, containing up-to-date and relevant information. Each chapter has also been extensively referenced so that readers may follow the references to obtain an even deeper understanding of the issues concerning mobile business. Finally, the international aspect of this book is another great plus for a work of this kind. In today's shrinking world, it is fitting that we share our experiences globally—as has been achieved in this edited work.

I highly recommend this book to both researchers and practitioners in the industry as an invaluable desktop reference. This book will not only aid practitioners in what they are currently doing with mobile business, but will also open up numerous directions for further investigative research work.

Dr. Elaine Lawrence
University of Technology, Sydney, Australia
ICMB 2005 Conference General Co-Chair
August 2005, Sydney

Preface

Communication is the key!!

THE FOURTH WAVE

The search for extraterrestrial life is becoming more urgent—lest we humans end up with claustrophobia! Jet travel could not have shrunk physical distance at the speed with which the Internet has meshed the world of words, sounds, and pictures. Businesses of all types and sizes find a level playing field in the cyberspace and, barring a few troglodytes of no particular age group, the world is connected. Alvin Toffler was right when he asserted in his popular book, *Third Wave¹*, that not only are we all affected by change, but even the *rate* of change is *accelerating*. A quick look around you as you read chapters from this book—sitting, traveling, working, or sleeping—ratify Toffler's thoughts. And if, indeed, the rate of change is increasing, we are already *beyond* the third wave of information. This *Fourth Wave,* which is right upon us, is the wave of communication! Everyone is connected to everyone else *independent of time and location.* And mobile technologies provide the underlying basis for this age of communication, ensuring that businesses and people are connected directly and personally irrespective of where they are and what time of day or night it is. Riding on the back of the traditional Internet, mobile networks ensure that information that was available through a physical computing gadget at a fixed location is now available anytime and anywhere. This has obviously resulted in the tremendous popularity of mobile applications in the business domain. Furthermore, the *infrastructure* related to mobile technologies is also unique and, in some sense, distinctly different from the physical Internet. In simple business terms, setting of a transmission tower for mobile communications is relatively easier than installing physical lines to provide connectivity. Even the use of a satellite for communication purposes is becoming easier and cheaper for large organizations and governments than setting up of the costly fiber optics or similar physical wires and networks. Thus, because of their ease of usage from an end user's viewpoint, ease of setting up and usage of infrastructure, and rapidly dropping costs, mobile technologies are influencing global business like never before. Based on this phenomenal importance of mobile technologies, especially to mobile business, this book compiles contributions from a wide range of researchers and practitioners in their investigations and usage of mobile technologies in business.

PARADIGM SHIFT

Advances in mobile technology are reshaping the relationship between business and technology. There is a "paradigm shift"—it is now no longer just a matter of providing technological solutions for business problems. This is so because mobile technology is now becoming a *creative cause* for hitherto unknown new business models and business processes. This leads to complexities in adopting mobility in business, as both technology and business forces need to be balanced against each other. A similar balance is also aimed in the compilation of chapters in this book that bring together the technical, methodological, and social dimensions of mobile business to the fore. For example, while some chapters discuss the strategic needs of a business as a reason to use mobile technologies, other chapters demonstrate how mobile technology itself is considered as a driver for new business challenges as well as solutions.

The paradigm shift in terms of mobile technology adoption has occurred because, in practical terms, mobile technologies, including mobile applications, gadgets, networks, and content management systems, together work as a catalyst for deep structural change in how organizations accomplish goals. Although such a change requires significant effort in terms of BPM (Business Process Modeling), still the effort in applying mobile technologies to business processes and organizational structures is worthwhile as it allows organizations to gain greater reach and leverage new kinds of service delivery and interaction, culminating in significant productivity gains. In fact, it is not just organizations that stand to benefit with the advances in mobile technologies; the potential ease with which individuals can interact with organizations globally through mobile technologies also creates tremendous opportunities for easing the quality of life of people and society. Thus mobile technologies are becoming a creative cause for a paradigm shift in the business world that requires thinking, understanding, and case study experiences, as have been gelled together in this treatise.

MISSION

The aforementioned discussion highlights the crucial need to bring together the thinking and practical experiences of practitioners and consultants together with researchers and academics. This crucial need has been satisfied in this book, making it a significant contribution to the literature on usage and application of mobile technologies and networks in global business. This book provides significant *strategic* input into 'mobility', aiming to bring together thoughts and practices in technical, methodological, and social dimensions of this fascinating technology. Stated more succinctly, the mission of this book is:

To make a substantial contribution to the literature on 'mobility' encompassing excellence in research and innovation as well as demonstrated application of mobile technologies to mobile business.

CORE CONTENTS

For the sake of comprehensibility as well as enabling readers to focus on their area of interest, this book is divided into sections as follows:

- **Location:** Deals with location-based services that form the crux of the mobile revolution as applied to businesses.
- **Health:** Focuses on the phenomenal potential for application of mobility in the health sector.
- **Technical:** Focuses on the core mobile technologies.
- **Network:** Discusses the research and application of wireless and mobile networks.
- **Security:** Deals with the security aspect of mobile technologies.
- **Strategic:** Focuses on the strategic planning and management aspect of mobile technologies in business.
- **Application**: Discusses the incorporation of mobility in software applications.
- **Method:** Revolves around methodologies and processes related to mobile technologies including discussions on mobile transition processes.
- **Customer:** Focuses on the end user/customer aspects of mobility in business.

- **Social:** Handles the research as well as thoughts dealing the socio-cultural aspects of the influence of mobile technologies in society.
- **Case Study:** Discusses the practical application of mobility in practical business scenarios.

AUDIENCE

The following are the major categories of readers for this book:

- **Programmers and architects** of mobile-enabled software systems will find the discussions on technologies, networks, and security directly applicable to their work.
- **Business process modelers** and **information architects** will find the chapters dealing with incorporation of mobile technologies in business processes quite relevant.
- **Methodologists** and **change managers** will be interested in the chapters that describe the transition processes from existing to mobile businesses.
- **Sociologists** and **legal experts** will find the discussions on cross-border socio-cultural issues in applications of mobile technologies and the resultant globalization of businesses a fascinating read.
- **Strategic management** may find some of the earlier strategic discussions in this book quite relevant in setting the strategic directions of their organizations—especially because these chapters have been contributed by practicing senior managers.
- **Researchers** and **academics** will find numerous hooks in the research-based chapters of this book in terms of identifying areas of research, as well as following research methods when dealing with "mobility." Thus, the strong research focus of this book—especially the detailed and relevant references at the end of each contributed chapter, the research methodologies followed, and the discussions on research results (especially some excellent "action research"-based case studies)—make this book an ideal reference point for active researchers in this area.

ENDNOTE

[1]   Agricultural (first), Industrial (second), and Information (third) wave

Bhuvan Unhelkar
www.unhelkar.com

Critiques

Readers are invited to submit criticisms of this work. It will be an honor to receive genuine criticisms and comments on the chapters and their organization in this edited book. I am more than convinced that your criticisms will not only enrich the knowledge and understanding of the contributory authors and myself, but will also add to the general wealth of knowledge available to the ICT and mobile community. Therefore, I give you, readers and critics, a sincere *thank you* in advance.

Acknowledgments

First and foremost, I acknowledge all the wonderful contributing authors to this book. They come from a wide range of geo-cultural backgrounds, and they have enriched this book by making equally wide and varied contributions emanating from both intense research and practical industrial experience. This variation is reflected in the contents and presentation of the chapters. One thing in common, though—and which exudes through the individual chapters—is the dedication and hard work of each and every author in this book. Therefore, to these authors, I express my sincere gratitude and thanks.

Furthermore, I would specifically like to thank the following individuals (*my PhD students specifically for their help and work): Dinesh Arunatileka*, Bhargav Bhatt, Yogesh Deshpande, Anant Dhume, Samir El-Masri, Abbass Ghanbary*, Darrell Jackson, Vijay Khandelwal, Anand Kuppuswami*, Yi-Chen Lan, Girish Mamdapur, Javed Matin, Makis (Ioakim) Marmaridis, San Murugesan, Chris Payne, Anand Pradhan, Prabhat Pradhan, Mahesh Raisinghani, Prince Soundararajan, Ketan Vanjara, and Houman Younessi.

My thanks to my lovely family: wife Asha, daughter Sonki Priyadarashani, and son Keshav Raja, as well as my extended family, Chinar and Girish Mamdapur. This book is dedicated to my son, Keshav Raja, because of his disappointment and boredom with normal mobile phones at the ripe old age of 9+. All kids of around that age are going to grow up using mobile technologies in a way that is utterly unimaginable and hence I "pre-pay" my respect to them!

Bhuvan Unhelkar
www.unhelkar.com

Chapter XXXI
M–Business:
A Global Perspective

Mahesh S. Raisinghani
Texas Woman's University, USA

ABSTRACT

This chapter discusses the use of mobile, handheld computer devices that are connected wirelessly to a network for business and personal use across people, projects, tasks or organizational units to infer a trend of general acceptance of m-business in the market place. It describes the state of the mobile commerce industry from a worldwide perspective and the barriers to implementation of m-commerce, discusses the issues and challenges followed by the the conclusions and directions for future research.

INTRODUCTION

Only a few years ago, electronic commerce (e-commerce) visionaries were predicting the rapid acceptance of mobile commerce (m-commerce) as the evolutionary result of the e-commerce revolution. M-commerce is the buying and selling of goods and services through wireless handheld devices such as mobile phones and personal digital assistants (PDAs). The term m-commerce is used in this chapter to describe the adoption and use of mobile, handheld computer devices that are connected wirelessly to a network for business and personal use. Known as next-generation technology, m-commerce enables users to access the Internet without the need to find a place to plug in. According to these visionaries, most of the U.S. population should be paying bills and shopping with mobile phones or PDAs, while receiving updated flight information on the way to the airport.

In reality, a different situation took place. The wireless industry is lowering its expectations for revenue growth since consumers have not accepted m-commerce as widely as expected. There are signs that m-commerce is growing in popularity. Recently, ComScore Networks (2002) found that almost 10 million wired Internet users in the United States accessed the wireless Internet using mobile devices. Sprint has activated its 3G (third-generation) network in the U.S. market that it serves. Other nations, most notably China, are also moving forward with 3G initiatives. The Chinese are replacing their wireless infrastructures with packet Internet Protocol-based 3G systems in support of the 2008 Olympic Games (Lemon, 2001).

Senior managers will find this information useful for planning and adapting to this new horizon. Educators will also find the information useful as they re-focus academics away from legacy e-business to an m-commerce model in order to adapt themselves to new trends in the world economy. The timing of the shift toward a mobile environment model is important, as strategies must change in advance of the trend. This chapter is structured as follows: we first describe the state of the mobile commerce industry from a worldwide perspective and the barriers to implementation of m-commerce. This is followed by a discussion of issues and challenges. Finally we discuss the conclusions and directions for future research.

Mobile Commerce: State of the Industry

In 2004 we saw the convergence of Wi-Fi and VoIP to evolve a new technology that has been variously described as Voiceover Wi-Fi

(VoWiFi), Voiceover Wireless LAN (VoWLAN), Voiceover Wireless IP (VoWIP), Wireless Voiceover IP (WVoIP), Voiceover IP over Wireless LAN (VoIPoWLAN), Mobile VoIP, and Wi-Fi Telephony. According to a study by Frost and Sullivan Consultancy, about $25bn (E28bn) in trade will be generated through mobile payments in 2006, or about 15% of estimated online e-commerce consumer spending and mobile-accessed Internet and peer-to-peer payments will make up the bulk of payments, accounting for 39% and 34% of spending in 2006 respectively (M2 Communications, 2002).

The United States, to date tradition-bound by its extensive fixed-line network, is shifting towards wireless means of communication. There are signs that demonstrate an increasing interest by industry towards m-commerce models, but will m-commerce become a reality or will it be just another trend that soon will be outdated? Companies and consumers have fresh memories of past dot-com failures, and this might give one reason to doubt the imminent arrival of a significant level of m-commerce. Furthermore, the high prices of mobile services, together with slow access speeds, have not helped much to add to the luster of the mobile environment. M-commerce in the United States also faces other challenges such as lack of standards, lack of ubiquitous wireless network coverage, technical differences among wireless devices, and security, among others.

Yet despite these drawbacks, mobile wireless devices are a reality today, with expectations of up to 1.5 billion in 2005 (Van Impe, 2002). In 1990, there were five million wireless subscribers in the U.S. In 2000, the number improved to 90 million, and it is expected that by the end of the year 2005, this number will reach 140 million (Kalakota & Robinson, 2001, pp. 26-27). Some expectations go even further, for instance Nokia is predicting that cellular phone

Internet connections will outnumber PC Internet connections by 2004 (Smith, 2002). Leading companies such as Microsoft, Intel, Nokia, NTT DoCoMo, Sony, and Ericsson, among others, are preparing themselves to compete in this arena, taking the first steps that will ensure them their leadership in a future mobile environment.

Europe and most of the Eastern hemisphere have established standards for analog and digital cellular telephone networks. In the United States, however, the Federal Communications Commission (FCC) has set standards for only analog networks, preferring to let each carrier choose its own digital technology. The diversity of these standards (e.g., TDMA, CDMA, and GSM) from vendors means that a given digital handset may not work in all digital service areas, although multimode phones and roaming agreements have alleviated this problem to some extent. In addition, the Czech Republic and Europe in general have one major advantage over the U.S. when it comes to m-commerce: widespread SMS interconnection among mobile operators. Lack of cooperation among operators in the U.S., coupled with a large landmass with limited coverage, has helped Europe take the lead in the race for wireless superiority. SMS in the U.S. has also been hampered by negative pricing policies, since the users have to pay to both send and receive messages (M2 Communications, 2002).

In February 2003, Orange, Telefonica Mobiles, T-Mobile, and Vodafone signed an agreement to form a new Mobile Payment Services Association aimed at delivering an open, commonly branded solution for payments via mobile phones, designed to work across all operator networks. The aim is to make the new initiative available to the largest possible number of mobile phone users. The solution will work across country boundaries and will seek to complement existing industry solutions to be-

come the industry standard for m-commerce payments (M2 Communications, 2003). For customers, the aim is to provide the opportunity to purchase a wide range of digital and physical goods and services with their mobile phones using an easy, secure solution. Merchants and merchant acquirers will benefit from a standard set of interfaces through which they will gain access to a potentially huge international customer base. Software and solution vendors will benefit from published technical interfaces, enabling the development of compliant m-payment products and services. Operators will benefit from a standard way of integrating and efficiently managing their relationships with merchants, merchant acquirers, and content providers.

In early 2001, Alcatel, Ericsson, Nokia, and Siemens formed the Wireless World Research Forum to guide and promote research of wireless communications beyond 3G. With the deployment of third-generation systems already begun, long-term wireless research and development is focused on fourth-generation technologies, which promise to provide data rates of at least 10 Mbps and more likely 100Mbps; multi-standard capability and seamless interconnections between the multitude of home, workplace, and public wireless networks; quality of service enhancements; and efficient use of bandwidth by assigning frequencies and bandwidths dynamically to users based on services required. Since new-generation wireless phone technologies are introduced approximately once every 10 years, it probably will be 10 years before 4G is launched (PriceWaterhouseCoopers, 2002a).

Figure 1 summarizes the expected growth in m-commerce revenues over the period of 2001-2006.

IDC's forecast shows that total mobile data revenues are expected to be increasing by more than 31% per annum, whereby the CAGR

Figure 1. Western Europe mobile operator data services revenue 2001-2006

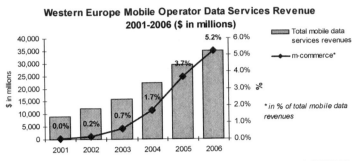

Source: IDC, "Mobile Data Platforms and Services in Western Europe Forecast and Analysis 2001-2006"

for revenues generated by m-commerce is estimated to be more than 265% per annum. As seen in Figure 1, despite rapid growth, income from m-commerce will remain a very small portion of total data revenues (highest value 5.2% in 2006). This implies that in the short run, m-commerce will not turn in profits to justify the investments in the new technologies that were initially believed to boost it. This applies to Europe and the United States. Asia, on the other hand, has developed more quickly with respect to m-commerce. Figure 2 illustrates the mobile operator data services revenue from 2001 to 2006 in Hong Kong and China.

The combined figures for Hong Kong and China show that the total mobile data services revenues are expected to increase for the period 2001-2006 by more than 70% per annum. The growth in m-commerce revenues in Hong Kong and China is currently 26.3% and is expected to outstrip 45% by 2006. This outstanding acceptance of m-commerce in Asia is only partially due to the currently used mobile technologies that require fewer investments for the upgrade to 3G. The major drivers behind this trend are the habits of the Asians who are keener on innovative technologies, and the variety of content providers that attracts an increasing number of mobile services users.

Interestingly in Asia, the investments were restricted to the lower priced 3G licenses. The service providers, therefore, are able to offer their services at a much lower cost. Especially in China and Hong Kong, where mobile tech-

Figure 2. Hong Kong and China mobile operator data services revenue 2001-2006

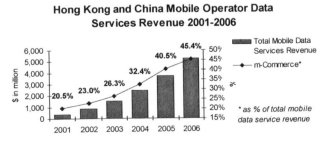

Source: IDC, "Acia/Pacific m-Commerce Forecast and Analysis: Opportunities Await"

Figure 3. Confidence in wireless e-commerce

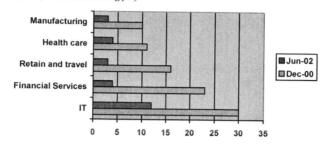

Does your company expect that wireless e-commerce will make a major contribution to its e-business revenue within the next 12 months?
(% of respondents answering yes)

Data: Information Week Research E-Business Agenda Study of 372 IT managers, June 2002. Additional data from E-Business Agenda Study, December 2000.

nology was introduced at a later stage and the 2G CDMA standard was adopted initially, the transition to 3G required little investment for the upgrade of the networks. This allows them to offer the services at a lower cost compared to the potential costs to the European consumers. The lower costs make it easier for consumers to try this new technology, become accustomed to and ultimately adopt it.

The mobile market penetration in the U.S. is around 41%, far less than countries like Finland with 75%, Hong Kong with 89%, or the United Kingdom with 74% (Magura, 2003). Besides, only 6% of users in the U.S. use their mobile phones to access the Internet; this is a much lower percentage compared with other countries like Japan with 72%, Germany with 16%, or the United Kingdom with 10% (Beal et al., 2001, p. 6). Business and private consumers in the U.S. use wireless devices mainly to place phone calls, read the news or the weather report, and trade stocks. A popular innovation in the U.S. is close to one million vehicles equipped with satellite tracking and communication devices, most of them with OnStar equipment from GM. OnStar had 800,000 subscribers in the U.S. and Canada in 2001, and ex-

pected four million vehicles equipped with this system by 2003 (Kalakota & Robinson, 2001, p. 6). Still, 61% of U.S. consumers think that wireless Internet access is too expensive, among other drawbacks including difficulty to read the screen (61%), slow access (37%), privacy concerns (16%), and unreliable service (12%) (Beal et al., 2001, p. 50).

Despite momentum in the consumer markets for m-commerce, many in the business community still remain pessimistic about its future for the enterprise. Consider Figure 3 which details the more pessimistic attitude prevailing in the U.S. marketplace for consumer applications.

Interestingly, there is an inverse relationship between pessimism and size of the company. Overall almost 75% of small companies did not expect m-commerce revenue, but only 58% of large companies felt the same way (Ewarl, 2002).

Barriers to Implementation

Among the key challenges in mobile and wireless information systems are issues related to internetworking and integration of different wireless technologies, reliable and secure communications, context and location-awareness in both applications and mobile devices, device and user interface issues, accessibility of wireless networks for continued business operation and support for group communication by allowing multicasting in the network, and application and/or middleware protocols (Varshney, 2003). For instance, the issue of PDA/cell phone and screen size, with respect to the amount of information that can be presented on such a small display, or infrastructure issues and problems of access, coverage, roaming, reliability, location management, and multicast communications are problems that are currently being addressed in the research community. Carriers will need to spend large sums to deploy sufficient networks, and the key question with the current state of the industry is: how much more money are telecommunications carriers likely to spend to ensure availability and performance of the carrier network and sufficient coverage for data transmission? Additionally, interoperability issues are of concern due to the lack of common standards (Baldwin, 2002). Many wireless networks are filled to capacity. For instance, AT&T struggles to keep the number of blocked calls under six million a day, not including the calls that are connected and then dropped, with spotty coverage increasing attrition among customers (Woolley, 2002).

Another concern regarding carriers is that there are three different and incompatible technologies deployed by U.S. wireless telecom companies that have resulted in three different paths to 3G. Sprint can offer 3G more rapidly and more cost effectively than AT&T because their wireless network is compatible, while AT&T would have to build a new network and was not able to launch 3G until 2003. Cingular is not planning a 3G roll-out and believes that demand will be weak, but other industry ana-

lysts believe Cingular lacks sufficient spectrum, and parents SBC Communications and BellSouth each use different types of wireless technologies, making a 3G solution much more expensive (Lanners, 2002).

Anecdotal evidence revealed that while there is a great deal of optimism for the eventual arrival of m-commerce, network infrastructure dominated as the most critical barrier to diffusion of m-commerce. Wareham and Levy (2002) provide additional insight in their study on mobile telecom diffusion and possible adopters of 3G computing devices. They noted that early adopters of wireless telephone devices correlated positively with executive-level positions and income levels. Of further significance to the authors was the study's conclusion that adoption of m-commerce applications would be triggered with the introduction of a 3G network into the marketplace.

ISSUES AND CHALLENGES

A familiar problem many companies face today is accessibility. Employees who work in areas where there is no cellular coverage cannot be reached. This adversely affects productivity. VoWLAN provides a solution to this problem. A company can setup an enterprise-wide Wi-Fi network used to deploy VOIP wirelessly. Successful deployment requires a comprehensive wireless coverage to enable all users in the organization to be connected at all times.

Unlike data traffic, voice traffic is very sensitive to connection and latency issues. A thorough site survey is required to plan a good deployment strategy. The latency factor requires the voice traffic to have uncontained connections. Thus, manufacturers specify and limit the number of connections at an access point to seven simultaneous connections, even though VoWLAN does not require a very high bandwidth. However, solutions are available with deployment strategies which can accommodate up to 30 connections. 'ON World' projects a market growth of 83% for VoWLAN handsets by 2007 and a Wi-Fi equipment market growth of 40% for the next five years. Newer technologies and standards in Wi-Fi and VOIP are emerging which will open doors to a whole array of applications for VoWLAN. With the development of standards like WiMAX and SIP and standardization of VOIP, cellular vendors are recognizing VoWLAN as a value-added feature, rather than a competitor, to their cellular phones.

Customers will use VoWLAN and cellular technologies interchangeably. When they work in Wi-Fi hotspots, they will take advantage of VoWLAN and make low-cost calls worldwide. Outside the Wi-Fi envelope, they can take advantage of the mobility and coverage of traditional cell phone technologies. Today, Wi-Fi is a dynamic and growing market, establishing its presence in both homes and enterprises. Wireless-enabled devices like laptops, handhelds, and even cellular phones are becoming increasingly popular, with more than 850 million Wi-Fi-enabled cell phones expected to be shipped by the year 2009.

VoWLAN is a new era in the evolution of personal mobile communication. Market potential and emerging technologies have driven many companies to plunge into the VoWiFi wave. Among the leading vendors are Cisco, Nortel Networks, SpectraLink, Symbol Technologies, Telesym, Texas Instruments, and Vocera. The huge market, converging technologies, emerging technologies, and value-added application are the driving forces for VoWLAN to be considered as the solution of the future.

CONCLUSION AND DIRECTIONS FOR FUTURE RESEARCH

Although m-commerce in the short term is facing some difficulties, companies have invested heavily into a range of technologies that could facilitate m-commerce. These technologies, especially the 3G, are able to offer bandwidth and speed that customers have become accustomed to through land-wired access. Unlike e-commerce, however, industry has not been able to adapt one common standard. These standards are necessary for the scalability and reach of services offered to the customer. Pricing, coverage, and functionality are the key drivers for consumers in the United States to adopt new wireless technologies and services. Without products and services that add value for the customer, users will not take to m-commerce. At this point, customers have not become acquainted with the technology and have not gained trust in the security options offered. To overcome these obstacles, intensive efforts by the service providers, content providers, and retailers are required to market m-business. In addition, a concerted effort is necessary in order to agree on a worldwide standard that will eventually trigger a wider variety of content offered. In the long term, customers are more likely to change their behavior and use services that offer an improvement to their lifestyle by saving time and/or adding convenience.

REFERENCES

Baldwin, H. (2002, September 20). The challenges of mobile development. *ZDNet*. Retrieved from http://techupdate.zdnet.com/techupdate/stories/main/0,14179,2881040-1,00.html

Basso, M. (2001, December 5). Will mobile portal's future be bright? *Gartner Dataquest.* Note AV-14-9502.

Beal, A., Beck, J. C., Keating, S. T., Lynch, P. D., Tu, L., Wade, M., & Wilson, J. (2001, June 4). *The future of wireless: Different than you think, bolder than you imagine.* Retrieved from http://www.accenture.com/xd/xd.asp?it= enWeb&xdn=_isc/iscresearchreport abstract_134.xml

Clarke, R. (1999, September 26). *A primer in diffusion of innovation theory.* Retrieved from http://www.anu.edu.au/people/Roger. Clarke/SOS/InnDiff.html

ComScore Networks. (2002, August 27). *Ten million Internet users go online via a cell phone or PDA.* Retrieved from http://www. comscore.com/news/cell_pda_082802.htm

Downing, C. E. (1999). Systems usage behavior as a proxy for user satisfaction: An empirical investigation. *Information and Management, 35*(4), 203-216.

Evans, N. (n.d.). *The m-business evolution: Business agility: Strategies for gaining competitive advantage through mobile business solutions.* Retrieved from http://www. developer.com/ws/other/article.php/1446771

Fichman, R. G. (2001). The role of aggregation in the measurement of IT-related organizational innovation. *Management Information Systems Quarterly, 23*(4), 4.

Harman, H. H. (1976). *Modern factor analysis.* Chicago: The University of Chicago Press.

Kalakota, R., & Robinson, M. (2001, October). *M-commerce: The race to mobility.* New York: McGraw-Hill.

Kort, T. (2002, January 21). Handheld market to grow 19% in 2002. *Gartner Dataquest.* Note HARD-WW-DP-0194.

Kwon, T., & Zmud, R. (1987). Unifying the fragmented models of information systems implementation. In R. J. Boland & R. A. Hirschheim (Eds.), *Critical issues in information systems research* (pp. 87-93). New York: John Wiley & Sons.

Lanners, J. (2002). Interview. *Sprint,* (October 1).

Lemon, S. (2001, July 17). Beijing's Olympic success spurs 3G vision. *IDG.* Retrieved from http://www.nwfusion.com/news/2001/ 0719olympic.html

Levin, R. I., & Rubing, D. S. (1994). *Statistics for management* (6th ed.). Englewood Cliffs, NJ: Prentice-Hall.

M2 Communications. (2002, May 31). Examining m-commerce in Central Europe. *Europemedia,* p. 1.

M2 Communications. (2003, February 25). New m-commerce association formed. *Europemedia,* p. 1.

Magura, B. (2003, Spring). What hooks m-commerce customers? *MIT Sloan Management Review, 44*(3), 9.

Mak, B. L., & Sockel, A. (2001). A confirmatory factor analysis of IS employee motivation and retention. *Information and Management, 38*(5), 265-276.

Mertz, C., & Serrell, M. D. (2002, October 15). Mobile application tools. *PC Magazine Online.* Retrieved from http://www.pcmag.com/article2/0,4149,545121,00.asp

Moore, G. C., & Benbasat, I. (1991). Development of an instrument to measure the perceptions of adopting an information technology innovation. *Information Systems Research, 2*(3), 192-222.

Nobel, C. (2001, March 19). IT is lukewarm to wireless consumer apps. *eWeek.* Retrieved from http://www.eweek.com/article2/0,3959, 104226,00.asp

Nunnally, J. C. (1978). *Psychometric theory.* New York: McGraw Hill.

PriceWaterhouseCoopers. (2002a). *Technology forecast: 2002-2004, 2,* 453.

PriceWaterhouseCoopers. (2002b). *Technology forecast: 2002-2004, 2,* 680.

Rai, A., & Howard, G. S. (1994). Propagating CASE usage for software development: An empirical investigation of key organizational correlates. *Omega, 22*(2), 133-147.

Rogers, E. M. (1983). *Diffusion of innovations.* New York: The Free Press.

Rogers, E. M. (1995). *Diffusion of innovations* (4th ed.). New York: The Free Press.

Sharma, S., & Rai, A. (2003). An assessment of the relationship between ISD leadership characteristics and IS innovation adoption in organizations. *Information and Management, 40*(5), 391-401.

Smith, D. (2002). *The pocket computer.* Retrieved from http://www.ed2go.com/news/ wireless.html

Tornatzky, L. G., & Klein, L. (1982). Innovation, characteristics and innovation-implementation: A meta-analysis of findings. *IEEE Transactions on Engineering Management, 29*(1), 28-45.

Uncapher, M. (2002, March). *Mobile commerce: WITSA/WIRG survey.* Retrieved from http://www.itaa.org/isec/pubs/e20023-04.pdf

Van Impe, M. (2002, June 19). *Nokia expects the number of mobile users to surge in the next three years.* Retrieved from http:// www.mobile.commerce.net/story.php?story_ id=1824

Varshney, U. (2003). Location management for mobile commerce applications in wireless Internet environment. *ACM Transactions on Internet Technology, 3*(3), 236-255. Retrieved from http://ftp.informatik.unitrier.de/~ley/db/ journals/toit/toit3.html#Varshney03

Wareham, J., & Levy, A. (2002). Who will be the adopters of 3G mobile computing devices? A profit estimation of mobile telecom diffusion. *Journal of Organizational Computing and Electronic Commerce, 2*(2), 162-174.

Woolley, S. (2002, May 27). Zeglis the Zealot. *Forbes,* p. 58.

Chapter XXXII
Mobile Strategy Roadmap

Francesco Falcone
Digital Business, Italy

Marco Garito
Digital Business, Italy

ABSTRACT

Convergence between business environment and technology solutions is a today challenge: is it possible to identify and adapt traditional business analysis tools into IT infrastructure and viceversa ? Mobile business has already dramatically changed our way of life: to get the most of it, it is necessary to understand how and when to take the next step in order to achieve the best possible results. After a short description of the current and available Mobile technologies, the chapter tackles the solutions that some companies have already implemented, enabling thus customers and stakeholder to efficiently and effectively cooperate: yes, because the bog news in mobile business is that such a technology makes business process and communications (either internal and external) visible. It is a true "one stop shop" where people are really always connected. It is now possible to create a framework where business environment and mobile solutions get together, enabling the development of a roadmap (or a mindset if you prefer so) that can be used to create new services. Enjoy the reading!

INTRODUCTION

This chapter outlines the convergences and the opportunities available in mobile business markets from both technology and business perspectives. An overview of the wireless options available to businesses is provided. This is followed by description as well as analysis of the mobile business in order to provide a possible framework enabling design, development, and eventual implementation of new business initiatives incorporating mobility.

THE MOBILE BUSINESS ENVIRONMENT

Mobile business is a broad definition that includes communication, transactions, and different valued-added services that are made available by using portable devices. Today, most of the attention is around consumer services, but business-to-business and business-to-employee segments are important too.

Another essential definition of "mobile commerce" is that it is referred to as "transactions with monetary value, conducted by mobile Internet." This definition covers business-to-business, business-to-consumer, and consumer-to-consumer transactions. Traditional voice calls are not included in the definition of mobile commerce, but the services using voice recognition systems to enable commercial transactions fall into the category. Mobile commerce is a subset of electronic commerce in terms of technical issues. However, the term "mobile e-commerce" is misleading because the business models and the value chain of mobility differ from electronic commerce.

Mobile commerce is not a truncated form of e-commerce, but an innovative way of conducting time-critical transactions regardless of location (May, 2001; Paavilainen, 2002).

Wirelessly Speaking

The wireless world is a complex environment consisting of different and competing technologies. Some of these technologies are as follows:

- **WiFi:** The first WLAN (wireless local area network) standard offering the capability to connect wirelessly to LAN; this technology developed rapidly with a wider offer of bandwidth (Gratton & Gratton, 2004).
  - **Strengths:** Expanding existing network without cables; expanding network where cables are difficult to install, rapidly evolving with users' need; largely used and close at hand; one of the fastest wireless technol-

ogy available; very flexible for home and small business
  - **Weakness:** Difficult set up and configuration
  - **Competitors:** HiperLAN
  - **Complements:** Bluetooth, WirelessUSB, and ZigBee
- **HiperLAN:** This technology is used in Europe and provides a different set of wireless communication specifications providing WLAN support. It is also compatible with 3G, enabling voice, and imaging communications (Gratton & Gratton, 2004).
  - **Strengths:** WLAN support provided, data rate of 54 Mbps, the 3G applications support
  - **Weakness:** Located only in Europe
  - **Competition:** WiFi
  - **Complements:** Bluetooth, WirelessUSB, and ZigBee
- **Bluetooth:** From the name of the Viking king Harald whose aim was to join together the Nordic European territories, a cable replacement technology to overtake the messy cables around laptops and desktops. Infrared was also engineered to tackle similar issues (Gratton & Gratton, 2004).
  - **Strengths:** Can be incorporated in many products and devices, low cost resulting in cheaper products, can make use of low power schemes, ease of use for consumer
  - **Weaknesses:** Small data rate available, uses the same frequency as other devices
  - **Competition:** WirelessUSB, some overlap with ZigBee
  - **Complements:** HiperLAN, WiFi
- **ZigBee:** Introduced to disseminate a large range of products and devoted to businesses willing to develop wireless prod-

ucts. It works with IEEE to set up a new standard (Gratton & Gratton, 2004).

- **Strengths:** Low power consumption and cost, affordable wireless solutions, fast wireless development, avoids coexistence issues
- **Weakness:** Uses the same frequency of other technologies and may overlap with Bluetooth
- **Competition:** WirelessUSB and some overlap with Bluetooth
- **Complements:** HiperLAN, WiFi
- **WirelessUSB:** A short-range wireless technology operating on the 2.4 GHz unlicensed spectrum; developed by Cypress Semiconductors to gap fill Bluetooth shortcomings (Gratton & Gratton, 2004).
  - **Strengths:** Low power consumption, affordable wireless solutions, allows fast wireless development
  - **Weaknesses:** There are restricted applications, uses same frequency of other technology, mostly unknown and not widely available
  - **Competition:** Bluetooth and overlap with ZigBee
  - **Complements:** WiFi, HiperLAN
- **Ultra Wide Band:** The FCC in the U.S. recently approved this technology, which is similar to Bluetooth, but can travel a distance up to 230 feet through obstacles by using minimal power. Currently two types of application exist: radar and voice/data communications.
  - **Strengths:** Low power consumption, can overtake obstacles, its radio waves travel further
  - **Weakness:** Early development stage and mostly unknown
  - **Competition:** HiperLAN
  - **Complements:** None yet (Gratton & Gratton, 2004)

- **WiMAX:** Represents the next evolution in broadband wireless technology and will be backed by Intel. This new technology, designed so that it does not require line of sight, should allow higher-speed downloads over much longer ranges than WiFi. In part this is because devices will support certain licensed spectrum bands, enabling them to transmit at higher power levels WiMAX should have clear advantages of speed and simplicity over 3G technologies for in-vehicle entertainment, flexible CCTV, and security systems; WiMAX devices could represent a user's second or third broadband connection

Does it make sense to have a converging strategy that combines fixed and wireless Internet business? The answer is yes, at least for DHL, the parcel delivery company: the fixed Internet tracking system took six months to be fully available, while the WAP version for the same service was made in seven days (Ahonen, 2002). What does it tell us? It is telling us that one of the success factors (with direct impact on bottom line and time constraints) is the reusability of technology, and it seems that converging technologies could deliver this effect.

Table 1 shows the advantages and disadvantages between a fixed Internet and Mobile Internet, combining the business side and behavioral side.

We personally recommend bearing in mind the above synopsis as it will be very helpful later. It is now time to take a closer look at what the current business scenario is in the mobile business: we are going to provide some examples encompassing at the same time general and common services and more specific industry-focused initiatives carried out in several companies.

Table 1. Advantages and disadvantages of fixed vs. Mobile Internet

ISSUES	FIXED	MOBILE
NUMBER OF USERS	Less	More
COST OF SERVICE	More because of infrastructure	Less
BILLING	Difficult to determine and many think it should be free	Easy: people are aware that making a call has a cost
MICRO-PAYMENT	Very difficult as commission fee for transactions, set up by credit card issuers, are or can be expensive	"Incorporated" in the mobile device and into the business model behind it

The Wireless World Today

We are going to take you into the world of m-commerce services with some simple descriptions of very promising applications. Let us start with *location-based services (LBSs):* these are an example of how service providers can make use of the inherent properties of mobile devices. Users are always carrying their devices with them and the mobile operator can localize the device, therefore providers can localize them with the help of the mobile operator (Ahonen, 2002; Eurescom, 2004).

Micro-Payment

Payment is an important issue when it comes to adoption and acceptance of services by customers. The customer can turn his mobile phone into a payment device and use it to pay for items and services at a real or virtual point of sale. A micro-payment example of converging technology is given by Coca-Cola vending machines: in Finland users can send an SMS message to the vending machine and pay through their phone bill; the same opportunity is available in the U.S., Poland, Australia, and Hong Kong (Ahonen, 2002, 2003). This is a confirmation that mobile business and micro-payments work well together, that people are available to use mobile phones to make small purchase of

goods and services. What about the business side? One of the strengths of Coca Cola is distribution (business side), while one of the strengths of a mobile phone is portability (consumer side): two sides of the same coin. The parallel lesson is that the same services or applications for either business or consumers can be easily and cost effectively reciprocated.

Gambling

Gambling and betting services, such as lotto, instant games, and sports betting, are very popular in the real world. "M-users" can place their bets using text-based technologies like SMS or WAP, or they can play games. Gambling is an excellent example of how an entertainment service could attract customers by offering a rich, though often mainly text-based contest, with a degree of user interactivity and a real-time user experience (Ahonen, 2002; Eurescom, 2004).

Intelligent Advertising

The basic idea of an intelligent advertising service is that customers receive advertisements (e.g., via SMS or MMS) from merchants on their mobile phones that are adapted to their personal preferences and location based (this is an example of how to combine two different services, thus providing a possibly unique experience for customers). The service requires a close collaboration between service providers (later on, some additional information about this point will be provided). The customer has to give permission (opt-in approach) in advance to receive ads from the mobile operator. The main advantage of the service is that merchants can be sure to reach the right person by knowing the user profile, so they are likely to be willing to pay for it; therefore, from a marketing point of view, it is possible to estimate how many prospects are in a given marketplace, the cost to get to them, and the ROI (return on investment) ratio (for more about the opt-in approach, see Roman & Hornstein, 2004).

It is time now to take a broader look at the wireless world as it is now.

Current Mobile Services Available in the Business Environment

New technology seems to suggest that mobile services will be the greatest opportunity for businesses to develop richer and more profitable relationships with individual customers by giving them what they actually want—once more a confirmation that wireless technology is an enabling tool. But how is business spelling this definition? Let us have a look at some examples.

Keebler Co., a subsidiary of The Kellogg Co., is the first consumer packaged goods company to use instant messaging to enhance its ties with customers of company brands: the service is called "RecipeBuddie" and allows customer to get recipes based on their mood or food preferences. The service is available for AOL and MSN platforms: the service starts as soon as the customers send an instant message to the screen name "RecipeBuddie" (Newell, 2004).

Another example comes from Land's End: the company launched a new service named Land's End Live, enabling customers to interact with service representatives; the service has been enriched by a new service called "Shop with a Friend," where two customers can exchange messages while shopping (Newell, 2004).

In the United Kingdom, Safeway, one of the largest retail companies, gives customers a Palm-powered device with small magnets on the back of them, so customers can put them on their fridge: when a product is finished, the customer checks off the product and brand, and a request is transmitted wirelessly to the nearest Safeway shop (Newell, 2004).

Now, just imagine a motor insurance that is calculated on how often, where, and when someone is driving: this is the approach of Norwich Union, the largest UK insurer and part of Aviva PLC. The service is called "Pay As You Drive": each customer's car has a black box to record location and time of each trip made by the car. The data is sent to Orange network, where the basic information is map-matched to a broad network database to enhance the data with the road types and numbers. Premiums are calculated accordingly, and customers can check the sum month by month, so they can change their driving patterns or behaviors. Norwich Union is thus able to adapt its value offer by matching customers' lifestyle by providing more than insurance products (check the company Web site for the latest deal).

Again in the U.S., Fidelity launched a wireless service based on the success of its Instant Broker initiative, called Fidelity Anywhere, which enables customers to manage their financial position; soon other services will be included (Newell, 2004; Harris & Dennis, 2002).

Abbey National offers e-banking services through the Genie mobile portal, which can be reached by WAP cellular phones by using any network (Newell, 2004; Harris & Dennis, 2002).

United Airlines in 1999 offered Palm-based service and now gives its customers the capability to be notified when flight status changes and a WAP-based service for last-minute updates and domestic booking for frequent flyers. Similar services are currently available in some hotel chains such as Holiday Inn, InterContinental, and Six Continents Products like i-SPOT Personal Item Locator by Digital Innovation and FINDIT, an electronic locator made by Ambitious Ideas, are already available. Similar scenarios are provided by RFIDs (radio frequency identification tags) with the smart management of supply chain, stock items, and procurement, now possible in real-time mode (Newell, 2004; Harris & Dennis, 2002).

So far we have seen mobile business as a part of the fixed Internet: can we imagine a use of wireless with other media, achieving a converging environment with a unique user experience? Of course we can.

A Set of Converging Technology Examples

In Germany, RTL-teletext, which offers a TV broadcast message board, claims to host up to 220,000 text messages a day. Almost 70% of the broadcasters in Europe have now launched their own SMS chat lines and enjoy similar success: SMS provides excellent indications of a show's popularity or potential, even though conversion rates (ratio viewer/SMS participant) vary widely by application and by the content of individual shows. McKinsey (Bughin, 2004) demonstrates that if more than 5% of a show's viewers interact with it, its audience is extremely engaged and more available in referral and word of mouth.

Viewers who use SMS-TV to vote for contestants on the hit Big Brother, for example, buy more show-related merchandise than do other viewers, and 70% of the teenagers who purchase Big Brother merchandise vote by text message, according to McKinsey research (Bughin, 2004). SMS interactivity can encourage ratings growth of 50 to 100% for niche cable and satellite channels. Advertising provides 20% of the revenues of the average thematic pay broadcaster, which can reasonably expect one out of every five shows to be interactive. A standard TV show is produced and packaged in an appropriate length: an SMS-TV (basically, sending an SMS to a given number displayed during a TV show) extends the lifecycle of the show. Let us see which are the key success factors.

- **SMS Must be Well Integrated and Synchronized with the Show's Content.**
- **A Show Host with "Push":** The host's suggestions should always be timely and subtle; viewers do not want to feel obligated.
- **Clear SMS Displays:** On-screen numbers and SMS displays are large and clear.
- **Sufficient Rewards:** Tangible rewards (right combination of online and off-line environment).
- **SMS Interactivity:** The broadcaster's production and technical departments should work together, under the same roof, and test the planned SMS interactivity (never underestimate people and process issues).

But how did the subject matter experts spell the words "mobile business"? NTT DoCoMo used the acronym "MAGIC," which stands for Mobile, Anytime, Globally, Integrated, and Customized. Ericsson developed the "0-1-2-3" approach: 0 written manuals, 1 simple button to the Internet, 2 seconds of delay waiting to

Table 2. Communication strategy (Luftman, 2002)

	STEP 1 AD HOC PROCESS	STEP 2 COMMITTED PROCESSES	STEP 3 ESTABLISHED PROCESS	STEP 4 IMPROVED PROCESS	STEP 5 OPTIMIZED PROCESS
UNDERSTANDING OF BUSINESS BY IT	IT management lacks understanding	Limited understanding by IT managers	Good understanding by IT management	Understanding encouraged among IT staff	Understanding required at all levels
UNDERSTANDING OF IT BY BUSINESS	Managers lack understanding	Limited understanding	Good understanding by managers	Understanding encouraged among staff	Understanding required of staff
ORGANIZATIONAL LEARNING	Casual conversations and meetings	Newsletters, reports, and group e-mail	Training, departmental meeting	Formal methods sponsored by senior management	Learning monitored effectiveness
STYLE AND EASE OF ACCESS	Business to IT only: formal	One way, informal	Two way, formal	Two way, sometimes informal	Two way, informal and flexible
LEVERAGING INTELLECTUAL ASSETS	Ad hoc	Some structured sharing emerging	Structured around processes	Formal sharing across the board	Formal sharing with partners
IT/BUSINESS LIAISON STAFF	None or use if needed	Primary IT/business link	Facilitate knowledge transfer	Facilitate relationship building	Build relationships with partners

Table 3. Metrics development (Luftman, 2002)

	STEP 1 AD HOC PROCESS	STEP 2 COMMITTED PROCESSES	STEP 3 ESTABLISHED PROCESS	STEP 4 IMPROVED PROCESS	STEP 5 OPTIMIZED PROCESS
IT METRICS	Technical only	Technical costs, metrics rarely reviewed	Review, act on technical ROI metrics	Also measure effectiveness	Also measure business opportunities, HR, partners
BUSINESS METRICS	IT investment rarely measured	Cost/unit rarely reviewed	Review, act on ROI, cost	Also measure customer value	Balanced scorecards, includes partners
LINK BETWEEN IT AND BUSINESS METRICS	Value of IT investments rarely measured	Business and IT metrics not linked	IT and business linked	Formally linked, reviewed, and acted upon	Balanced scorecards, includes scorecards
SERVICE-LEVEL AGREEMENTS	Use sporadically	With units for technology performance	With units, becoming enterprise wide	Enterprise wide	Includes partners
BENCHMARKING	Almost never	Sometimes informal	May benchmark formally, seldom act	Routinely benchmark, usually act	Routinely benchmark, act, and measure results
FORMALLY ASSESS IT INVESTMENTS	No assessment	Only when there is a problem	Routine occurrence	Routinely assess and act on findings	Routinely benchmark, act, and measure results
ONGOING IMPROVEMENT PRACTICES	None	Few, effectiveness not measured	Few, starting measuring effectiveness	Many, frequent assessments	Practices and measures well established

Table 4. Governance and policy (Luftman, 2002)

	STEP 1 AD HOC PROCESS	STEP 2 COMMITTED PROCESSES	STEP 3 ESTABLISHED PROCESS	STEP 4 IMPROVED PROCESS	STEP 5 OPTIMIZED PROCESS
FORMAL BUSINESS STRATEGY PLANNING	Undone or done when needed	At unit functional level, slight IT input	Some IT input and cross-functional planning	At unit enterprise, with IT	With IT and partners
FORMAL IT STRATEGY PLANNING	Undone or done when needed	At unit level, slight business input	Some business input and cross-functional planning	At unit enterprise, with business	With partners
ORGANIZATION STRUCTURE	Centralized or decentralized	Centralized or decentralized, some co-location	Centralized, decentralized, or federal	Federal	Federal
REPORTING RELATIONSHIP	CIO reports to CFO	CIO reports to CFO	CIO reports to COO	CIO reports to COO or CFO	CIO reports to CEO
HOW IT IS BUDGETED	Cost center, spending is unpredictable	Cost center by unit	Some projects considered as investments	IT = investment	Profit center
RATIONALE FOR IT SPENDING	Reduce cost	Productivity, efficiency	Also a process enabler	Process driver, strategy enabler	Competitive advantage, profit
SENIOR-LEVEL IT STEERING COMMITTEE	Does not exist	Meet informally as needed	Formal committee meet regularly	Proven to be effective	Also with external partners
PRIORITIZATION OF PROJECTS METHOD	Upon IT or business need	Determined by IT function	Determined by business function	Mutually determined	Partners' priorities included

access the service, and 3 keys to gain access to services and features (Ahonen, 2002).

What is a possible approach to develop a strategy for the next wave, without possibly reinventing the wheel?

MOBILE BUSINESS ROADMAP TO SUCCESS

In our opinion, it is necessary to have well clear in mind Porter's 5 Forces Analysis (Porter, 1980) and the SWOT (strengths, weaknesses, opportunities, and threats) scheme, and do the assessment twice: firstly at higher level (macro-economic level) and secondly at lower/specific industry level (micro-economic). This will help to determine the positioning and the market(ing) potential. This is nothing new: get big, get a niche, or get out (or do not enter at all). Is there a way to comprehend mobile business with its specific characteristics to create a tool enabling us to developing successful business initiatives? This is what the following considerations will try to do.

A preliminary issue needs to be solved before talking about mobile business: the alignment between Business and Technology and the following tables explain how to deal with the internal organization.

By matching rows and columns, it is possible for you to see what the situation looks like—what are the challenges and what is a foreseeable way out. Once you are done with the

Table 5. Partnership decision process (Luftman, 2002)

	STEP 1 AD HOC PROCESS	STEP 2 COMMITTED PROCESSES	STEP 3 ESTABLISHED PROCESS	STEP 4 IMPROVED PROCESS	STEP 5 OPTIMIZED PROCESS
BUSINESS PERCEPTION OF IT	Cost of doing process	Becoming an asset	Enabler of future activities	Driver of future activities	Partner with business of creating value
IT ROLE IN STRATEGIC BUSINESS PLANNING	Not involved	Enables business processes	Driver of business processes	Enabler or driver of business strategy	IT/business adapt quickly to change
SHARED RISKS AND REWARDS	IT takes the risks, no rewards	IT takes the most part of the risks, little reward	IT and business start sharing risks and rewards	Risks, reward always shared	Managers get incentives to take risks
MANAGING THE IT/BUSINESS RELATIONSHIP	Not managed	Managed upon need	Processes exist but not always followed	Processes exist and complied with	Ongoing improvement of processes
RELATIONSHIP/TRUST STYLE	Conflict and mistrust	Transactional relationship	IT as a valued service provider	Long-term relationship	Partner, trusted vendor of IT services
BUSINESS SPONSOR/CHAMPIONS	None	Other have a senior IT sponsor/champion	IT and business champion at unit level	Business sponsor/champion at corporate level	CEO is the business sponsor

Table 6. Technology implementation (Luftman, 2002)

	STEP 1 AD HOC PROCESS	STEP 2 COMMITTED PROCESSES	STEP 3 ESTABLISHED PROCESS	STEP 4 IMPROVED PROCESS	STEP 5 OPTIMIZED PROCESS
PRIMARY SYSTEMS	Office support	Transactional oriented	Business process enabler	Business process driver	Business strategy enabler/driver
STANDARDS	None or not enforced	Defined and enforced at functional level	Emerging coordination across the functions	Defined and enforced across functions	Also coordinated with partners
ARCHITECTURAL INTEGRATION	Not well integrated	Inside unit	Integrated across functions	Begins to be integrated with partners	Fully integrated
HOW IT INFRASTRUCTURE IS INTEGRATED	Utility, run at minimum cost	Start to be driven by business strategy	Driven by business strategy	Helps business to respond to change	Enables fast response to changing market

previous exercise, then you can take the next step.

Five Lenses to See Mobile

Mobile business can be analyzed into five different wedges: movement, moment, me, money, and machines. The first one, *movement*, is the most obvious, as movement or changing places is a natural characteristic of mobile phones and their networks. Mobility includes further concepts such as mobility, locality, global, home base, and positioning (Ahonen, 2002).

Table 7. Human resources (Luftman, 2002)

	STEP 1 AD HOC PROCESS	STEP 2 COMMITTED PROCESSES	STEP 3 ESTABLISHED PROCESS	STEP 4 IMPROVED PROCESS	STEP 5 OPTIMIZED PROCESS
INNOVATIVE AND ENTREPRENEURIAL ENVIRONMENT	Discouraged	Sometimes encouraged at unit level	Strongly encouraged at unit level	Also at corporate level	Also with partners
KEY IT HR DECISION MAKERS	Top business and IT management at corporate level	Same, with emerging functional influence	Top business and unit management, IT advises	Top business and IT management across firm	Top management across firm and partners
CHANGE READINESS	Tend to resist change	Change readiness programs emerging	Programs in place at functional level	Programs in place at corporate level	Proactive approach to anticipate the change
CAREER-CROSSOVER OPPORTUNITIES	Rare job transfers	Occasional transfer inside the unit	Regularly happens for unit management	Regularly happens at all unit levels	Also at corporate level
CROSS-FUNCTIONAL TRAINING AND JOB ROTATION	No opportunities	Decided by units	Formal program run by all units	Also across the enterprise	Also with partners
SOCIAL INTERACTION	Minimal IT/business interaction	Only business relationship	Trust and confidence start	Trust and confidence achieved	Fully achieved with partners and customers
ATTRACT AND RETAIN TOP TALENTS	No retention, poor recruiting	IT hiring focused on technical skills	Technology and business focus, retention programs	Formal programs for retention and recruiting	Effective programs for retention and hiring

The service can transfer with the user as long as the user moves; the changing user pattern highlights the need to have an easy-to-use device and be in contact with one's own personal relationships, get information and data when a need should arise, in one word: communicate. Therefore, the enabled device must be easy to use (the above mentioned 0-1-2-3 or Magic approach), portable (and current models of mobile phones bear this feature), and have the capability to perform simultaneous—convergent—purposes (not only phone calls, but also messages, pictures, data, or games, just to name a few). Moreover, mobility can assume concurrent faces depending on location either narrowly (time and map of a specific place) or globally (when someone wants to be informed about stock exchanges). The main task for developing and providing a smart service is to be user sensitive: understand where user is, and give useful information and connectivity. If a user happens to be in Sydney looking for an ATM, the system as a whole should provide consistent information about the closest ATM machine, not for the ATM machine located in his/her hometown in London. At the same time, the system should be able to provide timely information about the FTSE index, when requested; the roaming and the use of local infrastructure to underpin "that" specific need of information and connectivity can achieve this capability (Ahonen, 2002; Benni, Laartz, & Hijartan, 2003; Beck, 2001; Andersson, Talborn, & Werkert, 2002; Deprez, Steil, & Dahlstrom, 2004; Neimeyer, Pak, & Ramaswamy, 2003; Tsalagatidou, 2000).

The following dimension is *moment*—more simply, time. It is therefore possible to manipulate passing time, planning, scheduling, and postponing or coping with a sudden need or request, and even multitasking—that is, doing more than one thing at a time (talking on the phone while taking a picture of the monument in front of us to be sent to our friends at home, but two businessmen are passing by, talking about the next takeover of a company, so we have to catch up with our broker and tell him to buy more stocks of the same company).

The next dimension is *me*, a very personal area, but how much large it can be? Current mobile phones provide a quite remarkable amount of customization features (ring tones, colored screensavers, covers, and accessories to name a few): it is possible to think of more complex options, either serious or frivolous, and stretch the concept to an extension of our own personality. What is more, the content/service provided must be relevant for the user (example: do we really have to read through the whole horoscope or can we gain access to our own personal sign?). But this is not enough: the service as a whole should enable us to keep in touch with people we want to, give us the power to hold the keys (right now many mobile phones can ring different tunes according to the caller), or even allow many people to talk simultaneously. It easy to understand that *me* is the most powerful tool among the five (Ahonen, 2002; Benni et al., 2003; Beck, 2001; Andersson et al., 2002; Deprez et al., 2004; Neimeyer et al., 2003; Tsalagatidou, 2000).

Then we have *money*. This dimension is a further confirmation of the need/trend of converging technologies and converging multipurpose devices: a mobile phone that can easily replace wallet and coins, and enable mobile banking (many banks in several countries offer this service to their customers). On the business front, joining or offering this capability through sound alliances locally or globally can enrich the value chain. If this happens to be the chosen solution, it is more than ever necessary to adopt the tools described at the beginning of this section because, from a practical point of view, there are either bottom-line issues ("how to share the pie" revenues) and KPI (key performance indicator) matters to define (and therefore the value proposition of each side).

The last dimension is *machine*: we have already seen on the market many mobile devices that can perform different tasks.

But What about the Marketing?

The previous section offers us insights that could have an impact on design, development, and execution of the marketing strategy. From the five lenses of mobile business, it is possible to define segments of users and classify them according to their behaviors and patterns of use—whether or not they receive and make calls; if they use local network or 'roam' because they travel a lot; in which day and time of the week; do they just speak or send messages, pictures, and data. Or whether or not they are business users or private users, how much is their bill, and how many times do they recharge their pay-as-you-go/prepaid mobile phones?

Table 8 sums up the possible combinations of segments on which to develop a marketing strategy; it is easy to understand that joining these patterns and the calculation of profitability (net contribution marketing) is more reliable (Ahonen, 2002; Benni et al., 2003; Beck, 2001; Andersson et al., 2002; Deprez et al., 2004; Neimeyer et al., 2003; Tsalagatidou, 2000).

Therefore the analyses is based, once again, on real behaviors. Let us consider, for example, a direct marketing campaign where your company has to deal with the mobile "stuff". We can quickly design a top-down pyramid which includes all the cost/revenue situations, providing a sound metrics called "E:R" (expense to revenue) ratio for each media involved in this campaign (Roman & Hornstein, 2004).

Table 8. Marketing dimensions (for more details, see Ahonen, 2002)

OWNERSHIP	Personal phone, business phone, parents' phone, supplementary phone
CONTACTS	Few, many, random
LOCATION	National, overseas, movers, and shakers
TIME	Weekday, night, day, weekend
BEHAVIORS	Mostly caller, mostly receiver, sending data, images, files, music, call backer, avoider

The second consideration is that alliances and partnerships, as told before regarding the tools to use (Porter's 5 Forces and SWOT Analysis), enable a new definition of product/service in mobile business. Let us think, for instance, about Virgin, a global brand encompassing mobile services, travel, shops, music stores, to name a few fields filled by this company. What do we have at the end of this development process? A lifestyle proposition where customers and business cooperate to create a unique environment providing a unique experience: something which goes far beyond the so-called "+1 Factor" (a differentiating characteristic that makes customers choose one product instead of another) with a further positive impact on the bottom line, including brand equity (Ahonen, 2002; Benni et al., 2003; Beck, 2001; Andersson et al., 2002; Deprez et al., 2004; Neimeyer et al., 2003; Tsalagatidou, 2000).

The third consideration is that the overall value chain/value proposition needs to be properly reassessed, forcing all the parties to focus on what they do best: once again nothing new under the sun.

The last consideration is that, on the technical side, reusable technologies allow the roll out of quick and ongoing changes.

Current Trends

So, what is around the corner? Well, over the past few months, we have seen the number of mobile phone users overtake the number of residential lines in many countries in Europe. At the same time, telecommunication companies are experiencing an increasing demand for high-speed fixed Internet (DSL, ADSL, broadband) for households. Moreover, massive advertising and promotion campaigns have been deployed to convince people to shift from the old traditional line to the modern and eye-catching fast lines for quite affordable prices (value for money approach), and bundled with new services and content. Basically we can consider this phenomenon as a migration of technology, business opportunity, and money into a more sophisticated environment: mobile phones are becoming a substitute for the fixed phone; the existing fixed line is changing to become the entry gate for more powerful capability; we are also witnessing the convergence between TV and computer in hardware and software. What it is more difficult to estimate is the churn rate among mobile phone operators: many of them cover residential, business, high-speed, and wireless worlds. Moreover, the price of mobile phones is decreasing (maybe they will be a commodity and the fight will be for the services and the applications enabled and provided to the customers). Lately, many refer to "walled garden" to explain the impossibility for a customer to use the services of another mobile provider unless he or she does not become customer. The fix solution is to subscribe to many contracts, have more than one mobile or at least more SIM card, and switch from one to another depending on the need of the moment. Once again the key suc-

Diagram 1. Value chain for the wireless business

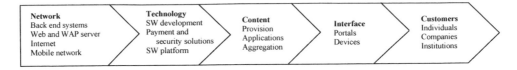

Table 9. Marketing dimensions revisited (Ahonen, 2002)

OWNERSHIP	Personal phone, business phone, parents' phone, supplementary phone
CONTACTS	Few contacts, many contacts, random
LOCATION	National, overseas, movers and shakers
TIME	Weekday, night, day, weekend
BEHAVIORS OF USERS	Mostly caller, mostly receiver, sending data, images, files, music, call backer, avoider

cess factor will be the quality of the service and its uniqueness, where customers will have strong bargaining power (Porter's 5 Forces are back again), and therefore the companies will need to find a strong value proposition to maintain their existing customers and attract new ones.

What Next?

Here we are at another crucial step: how to put the wheel on mobile business. Diagram 1 shows the value chain for the wireless business.

Now, let us try to swap the final step, "Customer," which consists of individuals, companies, and organizations, with Table 9.

The task is to develop a marketing-oriented strategy to match moody customers' demand. Why? Because happy customers return with frequent purchases; they are less available to change (the best way to keep a customer is to take him out of the market); they can be the most powerful marketing tool (word of mouth and referral have the highest impact among the different marketing and communication media).

Therefore there is an implicit high switching cost/barrier of entry in our favor—and finally, there is a tangible impact on the bottom line. It is now possible to have a clearer understanding of the possible combinations of marketplaces, but this is not enough: we have already covered the importance of establishing strategic alliances and partnerships. Now we have to select, rank, and classify developing compatible metrics and KPIs for each step of the other parts. The next diagram should help to fill the gaps and have a clearer understanding. Think about your business and try to properly allocate your partner and suppliers by taking into account your customers. A humble set of clues to success:

a. Do not think only about the technology issues, but consider the overall scenario (Plant, 1999), which is made by accounting (think about Dell: customers pay on the spot to buy a computer, but Dell pays suppliers 60 days or later, thus leveraging cash flow items), legal, distribution, and communication.

b. Try to implement what Kalakota and Robinson (1999) suggest by teaming up or partnering with your competitors to reduce costs or gain a kind of advantage. (Are you surprised? You should not be, as many other companies did the same such as Polaroid and the car makers including Covisint and others.)

c. If you happen to read a marketing book, once you get to the chapter dealing with the "4 Ps," remember that in our case, there are "7 Ps" (to product, place, price, and promotion, you need to include people, process, and physical presence).

d. To gain a competitive advantage, think about your services/products and, once you have defined all their "ingredients," do a displacement of one of them and combine those left accordingly (Kotler & Trias De Bes, 2003).

e. Keep the cannibal in the family (Piercy, 2002)—be ready to cannibalize yourself before someone else does.

f. Also be quick to reinvent your value chain. At the end of the game, you should have differentiated value propositions that can be further enhanced and improved.

A Thousand-Mile Trip Starts with a Single Step: But Where are We (Going) Now?

The potential for a mobile network is particularly high in the developing countries where it is difficult to set up a landline infrastructure for fixed Internet: for example, in the Philippines, SMS messaging has taken off rapidly.

Mobile devices are becoming mobile portals providing a wide range of customized services. The over-estimated 3G of mobile communication will—probably—deliver enriched experiences including sound and images and always be on (remember what we said about connectivity?).

On the business side, PDAs with diary, word processing, and e-mail features are becoming a killer communication tool; Mobliss, an American-based wireless marketing company has developed with Tribune Media, a wireless multiplayer game (by adapting Jumble, a scrambled word game) which can be an eye-catching advertising tool, as the words used to finish the puzzle can be redirected to advertisers' products or services, and the customers can also access the advertiser's call center (Harris & Dennis, 2002). Another example of advertising and promotion initiatives is given by PlanetHopper, a New York company which, after teaming up with General Cinemas theatres and a guide for bars and restaurants, is able to provide wireless coupons; now there are 20,000 users (Harris & Dennis, 2002).

If the sky seems to be pink, there are some clouds in the background: as we have already seen, there is the walled garden issue, the bandwidth, and the fact the mobile phones cannot compete with desktop computers in terms of display, computing power, and keyboard capabilities.

An interesting future development is SALT (Speech Application Language Tags), an initiative led by Microsoft, Intel, and Philips to realize a new platform enabling voice recognition-based services. It is still at an early stage but, on the business side, we can consider this project as a possible reengineering of the current call center. An English company, 365 Corporation, launched in 2001 the first comprehensive voice portal, Eckoh, which can be reached on the Web at http://www.08701101010.com or by phone (mobile or fixed). With Eckoh, users are able to e-mail, shop, listen to news, arrange conference calls, set up appointments, and more.

Some Tips About the Current Business Environment

An old Latin phrase says "Dum Romae consulitur, Saguntum expugnatur," (Livy.): translation "while in Rome the Senate is discussing, the enemies are conquering Sagunto" (a city in Spain). You may pose this sentence any way you like, but the meaning is quite straightforward: What is going on at the shop level NOW? What does the situation look like?

The Consumers' Behavior: Unpredictable Bills Make Them Move Away

A recent survey carried out by McKinsey consultants (Mc Kinsey Quarterly, 2004) tells us that "Mobile-telephony subscribers give many reasons for switching operators—prices, brands, friends' recommendations—but very few cite highly volatile monthly bills." But "subscribers whose monthly bills fluctuate substantially tend to churn much more than people with more consistent bills."

With an acid test on these movers and shakers' customers, operators could reduce the churn rate by "introducing pricing plans solely for subscribers with volatile bills": they could subscribe to an annual contract which includes a monthly predefined rate based on their previous pattern/phone bill, receiving timely alert as to any remarkable overdraft, thus enabling them to balance their position at the end of the contract.

This strategy should be combined with a more comprehensive program to "rise switching costs" from one operator to another (currently if you change operator, you can take the number but not, for example, the credit of your pre-paid mobile phone—another example of "walled gardens" approach).

Customers' Pain: Why Do They Complain?

The honeymoon between customers and telecommunication companies is over: during the past few years in Europe, the telecommunica-

tion sectors have been under the control of governments. The road to privatization and competition has been opened in the U.S. by a massive antitrust law case that gave birth to the so-called "baby bells"; Europeans followed slowly, and even though the technology gap between the Atlantic rims is almost filled, particularly in the last few years, consumers/subscribers are still facing a troubled situation with unclear terms and contracts, inaccurate bills, and unfair behaviors. Moreover, the dawn of new and converging technologies—where the same company is or can be provider of landlines, mobile, broadband, and interactive TV either as technology or content provider—makes the situation even more complicated.

But Why Do Customers Complain?

A complete survey was carried out late in 2004 by the Better Business Bureau (BBB) in the U.S. The majority of incidents involved billing problems, structured into three major categories:

1. setup and access, including difficulties retrieving statements online or getting detailed call logs;
2. errors, many involving calls made while phones (according to their owners) were not in use; and
3. failure of statements to reflect terms, such as credit or rate-plan changes, negotiated in discussions with customer service agents.

This has very little to do with technology, but with the questionable strategy carried out by many companies to outsource or employ contractors in customer-faced roles: in many cases these employees are not adequately trained, motivated (and paid of course) to do a good job and provide a good service.

Another set of complaints involves cases in which the customer believed that the carrier acted deceptively or otherwise misrepresented the terms of a contract. Such incidents are referred to as "miscommunication" to avoid the question of whether promises had really been made and broken or the customer misunderstood terms that were actually well explained—for instance, when a store's sales agent tells a customer that a certain house lies within a carrier's coverage area, without explaining that the quality of coverage varies within that area. This scenario is confirmed by the McKinsey survey (McKinsey Quarterly, 2004):

All of these issues are aggravated by the physical separation between the channel where the customer relationship originates (generally a store, and not always one under the carrier's direct control) and the channel (usually a call center) that provides customer service for the carrier. The different objectives of the managers who run these disparate operations can be tricky for the carrier to reconcile.

So, what to do? Customer service systems (called CRM, if you like) must provide agents with a customer's entire history of contacts from all channels. Agents should promptly refer calls to their managers when requested to do so, and managers must provide coaching for them in order to prevent rudeness. Most important, at the outset of every relationship, the carrier should communicate clearly to the customer how the contract period works. If it later changes, the carrier should explain the impact promptly.

Fixes require a mobile-telecom company to coordinate its efforts across departments: customer service, IT, marketing, and retailing (carefully select your partners at the shops level if you are involved in indirect sales). You will need to improve communication among these functions and to devise appropriate incentives for each of them (and, if any, replace the bad performer as soon as possible). And there must be clear consequences for stores that repeatedly fail to give new customers full explanations of the carrier's policies and pricing plans (once you select stores and more generally partner to complete your value chain, establish clear and well-understood KPIs in a carrot-and-stick way). Given the tremendous benefits of heading off intense customer dissatisfaction, companies should recognize the high cost to tackle these problems. In other words, preventing a problem is the best defense.

No-Frills and Low-Cost Operators

Denmark was among the first places to experience the trend. In 2000, Telmore bought unused mobile capacity from incumbent TDC Mobile. Telmore targeted college students with a fixed-price offer providing voice and Short Message Service (SMS) at rates that were initially more than 20% below those of the competition. Telmore minimized costs by using the Internet as a distribution channel, backed by a small call center, which resolved questions for a fee. This stripped-down approach eschewed expensive product offerings, mass advertising, and subsidized handsets. Despite the bare-bones offer, Telmore was consistently rated highest in customer satisfaction among Denmark's mobile providers. With an initial investment of just a few million euros, it captured almost 20% of the consumer part of the national mobile market (Beck, 2001; Braaf, Passmore, & Simpson, 2003; Deprez et al., 2004).

Germany is Europe's largest mobile phone market with 60 million subscribers, a fact that illustrates the broader interest in the no-frills option. A survey of thousands of German mobile users was conducted in order to estimate the importance they attach to the services and applications that wireless companies typically offer. The outcome shows that a third of the market has limited interest in advanced features or personal interaction with mobile operators: these customers want to make simple phone calls (and perhaps to use SMS) and would be happy to stick with handsets they already own or to pay for new ones. Since traditional mobile operators reduce up to 200 Euros a customer in handset subsidies, this finding is particularly important: no-frills customers do not want the large variety of products and services, such as access to news and weather bulletins, that are included by default in some standard contracts (Beck, 2001; Braaf et al., 2003; Deprez et al., 2004).

Attackers in several countries are experimenting with variations on the no-frills theme. Some of these players, such as Comviq (Sweden) and Telering (Austria), own their infrastructure, while others like CBB Mobil (Denmark) and Telenor's djuice (Sweden) buy capacity from other operators. Aggressive competitors have also made Hong Kong a price-sensitive market, and the U.S. is seeing some initial paths. In emerging markets like Asia and Eastern European countries, customers consider mobile phones as a substitute for fixed-line services, so the mobile market has matured rapidly, increasing the growth of price-sensitive segments; low personal incomes and the operators' need to pare costs to the absolute minimum have intensified the trend, and do not forget the digital divide issue for the developing countries and how mobile communication can fill the gap with developed countries (Beck, 2001; Braaf et al., 2003; Deprez et al., 2004).

Which are the drivers for this phenomenon? Several factors:

- household income,
- exposure to discounters, and
- Internet penetration.

In addition, advertising that focuses on price generally increases a market's sensitivity to it. The ability to use the same handset with different providers also makes markets more conducive to no-frills plans. Saturated markets with significant overcapacity tend to embrace them, since attackers can buy low-cost unused capacity from smaller players (Beck, 2001; Braaf et al., 2003; Deprez et al., 2004).

What happens next? As long as mobile phone markets mature and become saturated, the proportion of customers unwilling to pay for anything beyond basic services will keep growing. Incumbents must balance the potential threat of the attackers and decide whether and how to deal with the needs of these growing segments. The incumbents must improve the effectiveness and efficiency (the "do the right thing and do it right" motto) to stay cost competitive while facing declining prices. Many operators will invest in customer lifetime-management systems. Eventually larger operators will adopt more sophisticated market segmentation approaches (Beck, 2001; Braaf et al., 2003; Deprez et al., 2004).

CONCLUSION

Over the previous pages, we tried to outline the opportunities and the challenges of mobile business by describing and analyzing some of the most interesting initiatives, providing at the same time a possible set of tools to use to design and develop wireless solutions with a spot on the business aspects and challenges, perhaps the most challenging and intriguing part.

A recent survey (Kotler, 2004) highlights that anyone of us is hit by approximately 1,500 advertising messages everyday, and in this race the winner is the one able to provide fast, customized, and reliable solution. The DHL example, some pages ago, confirms that a mobile implementation can be a success: later in this book you will also find a collection of companies who successfully developed a mobile strategy—additional evidence that "wireless" does not mean "for the big guys only."

Now it is your turn: stop reading and try to do something.

REFERENCES

365 Corporation. (2004). Retrieved August 26, 2004, from http://www.08701101010.com

Ahonen, T. (2002). *M-profit. Making money from 3G services.* New York: John Wiley & Sons.

Ahonen, T. (2003). *Services for UMTS: Creating killer applications in 3G.* New York: John Wiley & Sons.

Andersson, T., Talborn, H., & Werkert, M. (2002). *Business models for mobile Internet.* Lund, Sweden: Institute of Economic Research, Lund University.

Bayne, K. (2002). *Marketing without wires.* New York: John Wiley & Sons.

Beck, H. (2001). Making money where it's scarce. *McKinsey Quarterly Special Edition on Emerging Markets.* Retrieved from http://www.mckinsey.com

Benni, E., Laartz, J., & Hijartan, K. (2003). The IT factor in mobile services. *McKinsey Quarterly*, (3). Retrieved from http://www/mckinsey.com

Better Business Bureau. (2004). *BBB advice: Don't ignore bogus billing during holiday rush.* Retrieved October 14, 2005, from http://www.frostillustrated.com/news/2004/1229/Consumer_News/014.html

Braaf, A., Passmore, W. J., & Simpson, M. (2003). Going the distance with telecom customers. *McKinsey Quarterly*, (4). Retrieved from http://www.mckinseyquarterly.com/article_abstract.aspx?ar=1356&L2=22

Bughin, J. R. (2004). Using mobile phones to boost TV ratings. *McKinsey Quarterly*, (4).

Deprez, F., Steil, O., & Dahlstrom, P. (2004). Meeting the no frills mobile challenge. *McKinsey Quarterly.* Retrieved October 15, 2004, from http://www.bangkokpost.com/mckinsey/McKinsey211004.html

Euresco. (2002). *Mobile electronic commerce.* Emporio—Project P1102—European Institute for Research and Strategic Studies in Telecommunication, Germany.

Eylert, B. (UMTS Forum Chairman). (2001, September 25). 3G chances and market opportunities. *Proceedings of the UMTS Forum,* Singapore GSM-GPRS in the Asia Pacific.

Gratton, S. J., & Gratton, D. (2004). *Marketing wireless products.* London: Elsevier Butterworth Heinemann.

Harris, L., & Dennis, C. (2002). *Marketing the e-business.* London: Routledge E-Business.

Hoque, F. (2002). *The alignment effect.* London: Financial Times Publisher.

IBM Institute for Business Value. (2004). Mobile portal strategy: When did business partnership become so critical to customer value? White Paper, IBM, USA. Retrieved August 26, 2004, from http://www-8.ibm.com/services/pdf/IBM_Consulting_Mobile_portal_strategy_Is_collaboration_the_key_to_customer_value.pdf

Kalakota, R., & Robinson, M. (1999). *E-business 2 roadmap to success.* Reading, MA: Addison-Wesley.

Kotler, P. (2001). *Marketing moves.* Englewood Cliffs, NJ: Prentice-Hall.

Kotler, P., & F. Trias De Bes. (2003). *Lateral marketing: New techniques for finding breakthrough ideas.* New York: John Wiley & Sons.

Luftman, J. (2002). Appendix F. Hoque (Ed.), *The alignment effect.* London: Financial Times Publisher.

May, P. (2002). *Mobile commerce.* London: Cambridge Press.

Neimeyer, A., Pak, M., & Ramaswamy, S. E. (2003). Smart tags for your supply chain. *McKinsey Quarterly, 4.*

Newell, F. (2004). *Why CRM doesn't work.* London: Hogan Page.

Paavilainen, J. (2002). *Mobile business strategy.* Reading, MA: Addison-Wesley.

Panis, S., Morphis, N., Felt, E., Reufenheuser, B., Boem, A., Nitz, J., & Saarlo, P. (2002). *Mobile commerce service scenarios and related business models.* Eurescom Project P 1102, Emporio–Project P 1102–European Institute for Research and Strategic Studies in Telecommunication, Germany.

Piercy, N. (2002). *Market-led strategy.* London: B&H.

Pippow, I., Eifert, D., & Strüker, J. (2002). *Economic implication of mobile commerce— An exploratory assessment of information seeking behavior.* White Paper, University of Freiburg, Germany.

Porter, M. (1980). *On competition.* Boston: Harvard Business Review Books.

Roman, E., & Hornstein, S. (2004). *Opt-in marketing.* New York: McGraw-Hill.

Siemens Application Marketing. (2002). *Mobile business—A task-based focus.* Retrieved from http://www.verista.com/mbusiness/mbusiness_wp_001.pdf

Tsalagatidou, A., & Veijalainen, J. (2000). *Mobile commerce emerging issues.* White Paper, Department of Computer Science and Information Systems/Information Technology Research Institute, University of Jyväskylä, Finland.

White, J. (2001). *Enabling e-business.* New York: John Wiley & Sons.

Chapter XXXIII
Relating Mobile Computing to Mobile Commerce

Nina Godbole
CQA, CISA, PMP, CSTE, ITIL (Foundation) Certified Professional Member—
Computer Society of India, India

ABSTRACT

In today's digital economy and the extended enterprise paradigm, mobility is on the rise. It is important to perceive mobility as an opportunity, rather than a threat. Although m-commerce is still at its infancy, it serves as an extension to e-commerce sites—it has been regarded as a value-added service. However, there are many issues and challenges while reaping full benefits of mobile computing solutions for m-commerce. This is because mobility and mobile computing is replete with many challenges on the business front, technical challenges as well as social challenges. This chapter undertakes discussion on understanding mobility and categories of mobile user types, understanding the meaning of m-commerce. Typical applications that support the m-commerce paradigm are illustrated through case studies. The chapter ends with a discussion on legal implications of mobile technology, and future directions for mobile commerce and mobile computing. The key message is that mobility is not just about connectivity—it is about function it provides and the way organizations work in today's digital economy.

INTRODUCING MOBILE COMMERCE

The Internet, especially combined with wireless technologies, has become more than just a communication media. Together, they form important business drivers for e-commerce (electronic commerce) and m-commerce (mobile commerce) and, as such, have become integral features of the global economy. The Internet, combined with wireless communications, enables flow of information among business players, increasing the speed and accuracy with which businesses exchange information and, simultaneously reducing the costs of transactions. For example, Internet economics facilitate a reduction in the number of middlemen involved. Along with the development of the Internet is the phenomenal growth of the mobile communications industry. In fact, the mobile communications industry is growing so rapidly that in 1999, there were more mobile phones sold than automobiles and PCs combined (Telecommunications Service Inquiry, 2003). During the Qualcomm BREW 2005 Conference (statistics on camera phone), a phenomenal rise in the use of SMS and picture messaging was reported. Kerner (2005) also states that by 2009, consumers worldwide will be buying and/or replacing their mobile phones at a rate of one billion per year. Thus, it should be anticipated that the trend in e-commerce, which has influenced the economy and lifestyle of today's culture, will extend to m-commerce enabled by the wireless technology and the wireless application protocol (WAP).

According to a Commonwealth Report (Telecommunications Service Inquiry, 2003), it is expected that by the year 2006, there will be 923 million Internet users, whereas 543 million will be mobile (wireless) users. Currently, there are almost 8.5 million mobile services in Australia after 13 years of operation, compared with approximately 10.64 million fixed lines after over 100 years of operation. The number of mobile phones sold in Europe has also grown rapidly. In the UK, there are approximately 30 million subscribers. In both Italy and Finland, 70% of the population owns a mobile phone. Consequently, this global trend has motivated businesses to adjust them with the change. Gartner raised its 2005 forecast (see Kerner, 2005) for mobile phone sales to 779 million units, an increase of 16% over 2004. As recently as May 2005, Gartner revised its 2005 mobile phone sales forecasts from 720 million to 750 million units, which is a 13% increase over 2004.

Thus, the mandate to mobilize business data is clear and, with global m-commerce reaching $200 billion by 2004 (Gartner, 2005), m-commerce is the new benchmark. Just as business

Figure 1.

organizations compete with one another to do business on the Internet, the evolution of technology continues through handhelds and smart phones. Examples of such handheld gadgets are shown in Figure 1.

The benefits of *pervasive information access* and the ability to do *business anywhere* are evident in organizations that have decided to embrace mobile commerce opportunities as part of their strategic choice. This poses certain critical issues for today's CEOs and CIOs: How will my company measure up against the m-commerce benchmark? Are we supplying business data to our workers via handheld devices? Are we supporting our critical business functions with mobile technology? Is our IT shop providing support to end users? These are some of the crucial questions explored in this chapter. This chapter undertakes discussion on understanding mobility and categories of mobile user types, understanding the meaning of m-commerce, and typical applications that support the m-commerce paradigm. Finally, the chapter ends with a discussion on legal implications of mobile technology, and future directions for mobile commerce and mobile computing.

THE NEW M-COMMERCE SCENARIO: THE MEANING OF MOBILITY

With the advances in the Internet, organizations make greater attempts to extend their "virtual reach." This results in greater mobility of the workforce. Thus, there is the need to deal with the requirements of employees as they move around (different locations in the building, between buildings, between companies, and while in transit traveling from one country to another). In the "Extended Enterprise" scenario of today, this provides immediate benefit in terms of workflow efficiencies. However, this also introduces problems with how we manage access and information sharing with suppliers and other external resources. "Mobility" is a means to change the way we do business. Thus, *mobility* is an attempt to extend corporate networks to connect to people who are increasingly on the move.

Mobility: The New Challenge

Over a period of time, "mobility" has affected us in a slow but sure way. One can argue that the workforce has always been mobile—commuting to and from work, completing sales transactions at a customer's site using handheld devices to access business information held on back-end databases at their parent organization, meeting suppliers and prospects, and making instantaneous changes to negotiating stances aided by access to customer history information accessed through mobile computing devices while working in the field. After all, organizations do have groups of geographically dispersed skills. However, today's workforce

does more than this in the paradigm of m-commerce aided by mobile computing.

Thus, today *mobility* impacts the workforce in a major way, as we have become more dependent on technology, the need to remain in contact, the need to be able to provide ever more up-to-date data and to be able to beat the competitors in the fast-moving marketplace.

Earlier in this chapter there was a discussion on the meaning of "mobility." Mobility can be considered as an opportunity for organizations to outpace the competitors, to add a differentiation by bringing a change to the style of working, to utilize the resources at our disposal in an optimum manner, without worrying about the underlying issues of technology. A June 2003 report commissioned by GRIC posed some interesting questions to organizations as part of a survey. A summary of the report is as follows:

- **Budget Spend:** The question posed in the survey was, "What is the percentage of overall mobile computing budget spent by on mobile applications and services?" A large number of respondents said that this was between 31% to 50%.
- **Priority Perceptions:** Another question posed in the survey was regarding "perceptions on mobile commerce and mobile computing. From the responses (in terms of priority for remote wireless access to various applications), it emerged that priorities in terms of strongest business case appear to be towards applications that cover *mobile office, sales force automation, automation of professional services, field service support, logistics,* and *transportation.*

When it came to the survey question on "*the relevance* of third-party mobile applications and services to organizations," *m-commerce* and *multimedia messaging* were found to be

Figure 2.

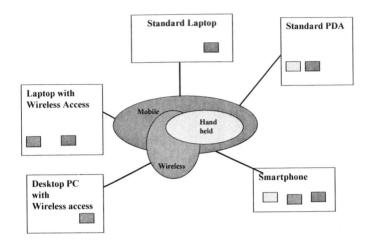

on top, while *collaboration* and *location-based computing* seemed to be rated much lower.

WHAT IS MOBILE COMMERCE?

Let us now try and understand mobile commerce in greater detail. M-commerce refers to the purchase of products and services using a mobile terminal. With the success of e-commerce for many types of purchases, m-commerce is considered the next logical wave after e-commerce supporting "business on the move."

Although it may appear that essentially m-commerce is e-commerce without the constraints of wired connection, many believe that m-commerce introduces time-location independence that never existed before—that is, the freedom to compute *anywhere, any time.* The "context sensitivity" feature is believed to be another added advantage to mobile commerce in the mobile computing paradigm. For example, according to Kolari et al. (2005), when mobile services and applications are made 'context-aware," they offer contextually relevant information to the users.

It is important to discuss *handheld devices* in the context of *wireless computing,* which is one of the main enablers of m-commerce. This helps us provide clear definitions for mobile computing and wireless computing. Figures 2 and 3 help us understand that.

Mobile commerce is best suited where the consumer is driven by a "sense of urgency"—that is, when they need to have their goods and services immediately for upcoming functions and events. Of course, logistic issues are the constraining factors in an m-commerce scenario, in the sense that having completed a purchase transaction using his mobile/wireless device, the consumer still has to wait for the physical delivery of the product. However, for certain kinds of products purchased through m-commerce, this limitation is overcome, for example, a movie ticket or information services, e-brochures, and so forth.

MOBILE COMMERCE ENABLERS

The previous section discussed what mobile computing is all about. This section addresses what promotes mobile commerce. Mallick (2003) writes of five major driving forces that act as enablers for the adoption of mobile and wireless solutions. In this section, we offer a brief overview of these factors:

- Wireless Networks
- Mobile Devices
- Software Infra-Structure
- Standardization
- Mobile Internet

Of the factors indicated above, the last one deserves a longer discussion. It is a "mega-enabler" for m-commerce in the mobile computing paradigm.

Wireless Networks

The initial marketing focus for *wireless carriers* was aimed at the consumer market. The emphasis was on inexpensive handsets and affordable calling plans for consumers. However, this conceived market did not generate as much revenue as expected, and since then, much of the focus of the wireless data services has moved to the corporate market. Currently, some significant improvements are in the offing for making wireless network improvements. They are summarized below. All these factors offer a great boost to mobile commerce.

Increased Bandwidth

Next-generation wireless networks will overcome the data capability limitations of voice-oriented wireless networks. These new generation wireless networks are designed for wireless data, and provide communication speeds between 56 Kbps and 384 Kbps. This speed is very adequate for the limited amount of data that is typically required from a wireless device.

Always-On Capability

This term refers to users' ability to access data at any time, without having to establish a connection to the wireless network for each session. This ability has been enabled by packet-switched networks. Users welcome this because, whereas for the *circuit-switched networks,* billing is on the basis of amount of time they are connected, *packet-switched networks* charge the users based on the amount of data transferred.

Lower Costs

This supports the above point and results from packet-switched networks allowing wireless operators to provide new offerings that are based on data usage instead of call times. Another factor is roaming charges. Many carriers have drastically reduced roaming charges for using other carriers' services.

Enhanced Services

New services are being offered that can add value to mobile solutions. Many carriers now allow users to download additional applications for their devices. In addition, some integrated services, such as *location-based services*, are providing companies with valuable features that can further contribute to the success of mobile solutions.

Inter-Operability between Carriers

Wireless operators are starting to work in a collaborative fashion to help promote benefits of mobility. The result is a new level of inter-operability for both data and voice communication. An example is the ability for users in North America to send text messages to users on

Figure 3.

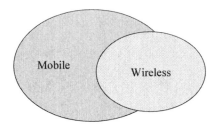

In most cases, "Wireless" is a <u>subset</u> of "Mobile", but in many Cases, an application can be mobile without being wireless

Figure 4.

other networks, a feature made available in 2002.

Mobile Devices

Incredible advances are being made for mobile devices. A few years ago, the choice was between a wireless phone and a simple PDA. Now there is a long list of options ranging from high-end PDAs with integrated wireless modems down to small phones with wireless Web browsing capabilities. Even the simplest of handheld devices provides enough computing power to run small applications, play games, and of course make voice calls.

A key driver for the growth of mobile solutions is the proliferation of devices in the enterprise. As more personal devices find their way into the enterprise, corporations are realizing the benefits that can be achieved with mobile solutions. The trend is for smaller devices and more processing power. Some such devices are shown in Figure 4.

Most of the devices depicted in Figure 4 work like "pocket PC," and they are very powerful. Users can count on them when it comes to portability, data security, and maximum operating time—crucial features for devices used for mobile commerce. Most of the devices shown above are featured by high-performance processors (with speeds ranging

up to 200 MHz), color display with front light, at least 64 MB of RAM, and 64 MB of nonvolatile Flash storage to protect data from resets or even complete power loss. Thus, today a device that fits in your hand (i.e., "handheld devices") has as much computing power as desktops did 10 years ago. The same device is also able to communicate over a wireless network and view office documents at the same time. This combination of size, power, and flexibility is a key enabling technology for enterprise mobile solutions. In today's mobile commerce paradigm, when employees are empowered by mobile solutions, a number of benefits emerge: increased employee productivity, faster response times to business changes, streamlined business processes, improved customer satisfaction, increased competitive advantage. A section on case studies later in this chapter illustrates this.

Software Infrastructure

To make a mobile computing solution successful, a strong software platform is required along with wireless networks and mobile devices. These platforms offer support for leading mobile computing models such as wireless Internet, smart client, and mobile messaging architectures. Software vendors have made significant advances in several key issues that enable mobile application development. For wireless Internet applications, the introduction of wireless application server frameworks, device emulators, WAP (wireless application protocol), and development tools has allowed many m-business solutions to be possible.

Standardization

As per industry reports (Quocirca, 2004), many organizations have cited the lack of standards as a major obstacle in developing mobile solutions. Organizations worry that that the technology they use will become obsolete before their mobile computing applications get deployed.

This is a valid concern because use of established standards provides an assurance that mobile computing applications will work now and in the future. However, as with any new industries, it takes time for standards to develop, especially in a continuously evolving scenario such as mobile computing for m-commerce. We are now at the point where standards are emerging at all levels of the mobile environment.

Mobile Internet

This is a phenomenon to reckon with and hence worth discussing in this section. There are many disputes and debates around this topic. For example, take iMode. It is an innovative solution that has revolutionized the Mobile Internet. It focuses on enhancing customers' lifestyles by providing users with an intuitive user interface for easy access to a wide variety of lifestyle Internet services over dedicated iMode-enabled handsets. These include multimedia content, push e-mail services, games,

mobile commerce, and banking. All this is integrated on an iMode gateway that will be developed especially to allow iMode users to have an optimized, end-to-end customer experience. Essentially, iMode, like WAP, is a way of providing information to mobile devices. iMode, however, is slightly different from WAP in that it uses cHTML—that is, compact HTML (http://www.webopedia.com/TERM/C/cHTML.html)—as a markup language, and uses more traditional Internet protocols to deliver it. The content is served using HTTP to a so-called iMode center, which in a way works like the WAP gateway. The iMode resource center (http://www.palowireless.com/imode/aboutchtml.asp) states that unlike WML used for WAP devices, c-HTML is a subset of HTML that leaves out coding for JPEG images, tables, image maps, multiple character fonts and styles, background colors or images, frames, and cascading style sheets. These things are excluded due to the low bandwidth and limited screen size of cell phones. Not only is cHTML simpler than WML for WAP phones, but developers need only make one version of the site for all iMode devices

Thus, iMode brings easy-to-use mobile data services to the user's fingertips in a transparent manner. For example, users need not be technology experts to use iMode services to make ticket reservations or explore mobile banking, stock trading, or restaurant bookings on the mobile phone. iMode was introduced in Japan in 1999; today it is the world's leading mobile Internet service. According to one source, (http://jgohil.typepad.com/minternet360/imode/), iMode currently offers 6,700 official sites worldwide with thousands of applications for the enjoyment of over 46 million mobile users worldwide. In Japan, in addition to the official sites, there are more than 80,000 independent iMode sites. iMode is a household name in Japan and the phenomenal success of iMode there provides a tested and proven ecosystem comprising close partnerships and strong associations with content providers, handset manufacturers, backend support partners, and global iMode alliance operators. The iMode platform enables content providers to develop or adapt their Internet content and applications easily for iMode devices.

The iMode service is provided on an open platform based on de facto Internet standards. To sum up, its easy-to-use functionality has attracted countless content and application providers, stimulating a vibrant market with thousands of sites to meet the needs of customers' diversified mobile lifestyles.

Although the iMode technology breakthrough may look promising, there are a number of problems. For example, consider the following from CapGemini (2001):

- In Europe, nearly 26% of employees in large companies have a GSM (Global System for Mobile Communication) handset for professional use, which is now higher than the rate regarding laptop equipment (16% of employees) and much more than equipment in personal digital assistants (PDAs) (4%).
- Voice usage is still dominant: more than two-thirds of companies have at least one WAP handset as a privilege for high-level managers and, on average, only one employee out of 24 has such a device. Furthermore, less than half of the companies equipped with WAP use it for data applications.
- WAP is used mainly for basic applications such as messaging, schedule consultation, or intranet access. Vertical applications such as maintenance, CRM, ERP, fleet management, and so on have been implemented only in a few companies.

- About one-third of the companies have already implemented Mobile Internet applications; 50% are said to be doing this in the near future.

The main reason for implementing Mobile Internet solutions is to facilitate exchange, notably through messaging. Indeed, many implemented applications are based on SMS solutions. But Mobile Internet is also used by some pioneer companies for advanced functions such as CRM, e-procurement, marketplaces, or supply chain. Even if only a few have implemented such applications, most companies are investigating Mobile Internet for the above mentioned functions in order to increase productivity and to acquire new clients.

The integration of Mobile Internet applications in existing information systems and middleware requires substantial efforts from companies and suppliers. As for the expectations, concerns of the users tend to be mainly around the types of devices, debit rates, tariffs, costs, and security. This is summarized below:

- In terms of devices, laptop computers are considered as the most suitable terminals, followed by cellular handsets and PDAs. Organizations may not want a unique device. It depends on the type of application and the work locations of the employees.
- Current data rate is not sufficient, but there is no immediate need for high speed; 60 kbit/s could be enough for most users, and consequently, GPRS (general packet radio service) could satisfy a large part of the market in the short term.
- For tariffing, basic voice services should be free of charge and companies would prefer a billing by volume of data or flat rate.
- In terms of investments, development costs are a major concern for most companies.

The cost of terminals is also seen as an obstacle for the smallest companies.

- Finally, security is a big challenge, as users want to be sure that all elements are protected (device, network, gateway, and server).

In the United States, the situation is quite different as the base of cellular users is significantly lower than in Europe (only 40 subscribers per 100 inhabitants), and the approach towards Wireless Internet is very pragmatic. Business users look for comfortable, easy-to-use, and relatively cheap solutions. The development of wireless packet network connections to PDAs is illustrative of that. As for operators, they prefer to upgrade their existing networks in the short term. Kerner (2005) from Gartner is also reporting that the PDA shipments are on the rise.

In Asia, there is a potential good outlook for Wireless Internet, with Japan and South Korea leading the market. In Japan, there are now more than 37 million Wireless Internet users (22 million users for NTT DoCoMo's iMode). However, in most cases there is no clear differentiation between residential and business users in terms of marketing approaches and offers. In the medium and long term, the Chinese market should create a strong dynamic as the cellular base is increasing rapidly (and should serve as a basis for advanced services, rather than fixed network).

Thus, to summarize, on the Mobile Internet, the key issues and challenges are (Quocirca, 2004):

- The success of Mobile Internet will be decided on three frontiers—technology, devices, and design quality of mobile commerce applications.
- Major players such as Ericsson are playing a key role in the mobile application

initiative as part of their strategy. Major stakeholders of the Mobile Internet community are: technology corporations, device/terminal manufacturers, software developers, mobile operators, content owners, ISPs and ASPs, start-ups, and venture capitalists.

M-COMMERCE APPLICATIONS IN MOBILE COMPUTING SCENARIO

Having discussed mobile commerce enablers, the power of mobile devices, and important phenomena such as the Mobile Internet and so forth, this section looks at the typical mobile computing solutions that drive the m-commerce market, including:

- **Digital Purchases:** These are products that can be downloaded and used immediately. Some of these m-commerce transactions are already happening (e.g., games and ring tones for cellular phones). The advances in mobile handheld devices make them great vehicles for game playing.
- **Mobile Banking:** Wireless devices provide two benefits to mobile banking. The first is providing access to personal bank accounts to view account history and execute transactions. This is an extension to Internet banking that is already very successful. The second benefit comes from use of mobile devices for online payments through e-cash.
- **Information Services on the E-Tap:** In today's dynamic world, "fixed-locations" are losing their meaning. Mobile users often feel out of touch with their daily routine as they travel around the globe in pursuit of business activities. Information services help address this need by providing information that the user is accus-

tomed to having. Typically, this information is stock quotes, weather information, entertainment, sports scores, and so forth. With mobile messaging technology, many forms of information can actually be pushed to the user in the form of an alert or notification.

- **Location-Based Services:** These offer the ability of the merchants to capture and provide services based on a user's current location and requirements are a new and powerful approach to selling services. Location-based services allow consumers to find the precise information they need at the exact time they want to use it. This will be an important enabler for m-commerce solutions, although privacy concerns will have to be addressed before location services enjoy wide acceptance in the user and business community.

- **Mobile Shopping:** Most forms of shopping are not going to be popular soon from the mobile devices. The reasons for people's discomfort using mobile devices for shopping is mainly due to the plethora of technical problems faced by mobile devices. For example, it is impractical to surf for items using constrained devices, while other methods of shopping are much more productive and socially enjoyable. Interestingly, there are some forms of purchases that lend themselves well to m-commerce. For example, purchasing travel and movie tickets for the same day. This form of m-commerce is found very convenient by users. Mobile devices can also be used for "comparison shopping" before making a purchase: a shopper in a retail store may want to first find the current price of a product from an Internet vendor to ensure they are getting a good price.

- **Mobile Advertising:** As mobile users start taking advantage of m-commerce solutions, mobile advertising is sure to follow. The mobile operators have access to several types of information that is attractive to advertisers. Once the advertisers find out where the users are located and what they use their mobile phones for, advertisers can send out personalized messages to these prospective customers. There are a number of threats for m-commerce performed on mobile handheld devices. Most notable among them is the risk of "customer backlash." This is the biggest obstacle to mobile advertising. If users start getting unsolicited messages (equivalent to "spam mails") and advertisements on their handheld/mobile devices, they are likely to switch service providers or keep their mobile devices off most of the times. Annoyed users may even go to the extent of stopping the usage of their handheld device. For this reason, in the near future, there could only be requested advertisements, such as the nearest gas station or restaurant.

Mobile User Types

Various sections, so far, discussed the meaning of mobility in the new m-commerce paradigm. The concept of mobility has a history. Mobility does not cover only telephony wireless standards, such as GSM and GPRS. Other technologies, too, have emerged. This includes the high-speed wireless networks of WiFi/802.11 and the Personal Area Network technology, Bluetooth. For the truly mobile user, these wireless technologies also need to be supported by "tethered" technologies including telephony and network usage, for example in hotels, in transit areas (e.g., airport lounges) and in customer/supplier offices.

In view of this background, mobility needs to be considered outside the constraining realms of the underlying technology. It is important to understand that mobility *is a way of working, not just a way of utilizing available technology*. Amongst others areas, mobility must cover those working from home, on the road, in hotels, and at airports, as well as those moving around within the organization boundaries. A mobile solution must provide not only the technology to provide access—it must also provide the functionality for the user to add value to the company.

Under this paradigm, "mobile solution" can be defined as:

The meeting of an individual's or a group of users' functional needs, enabling a company to gain added value through optimal utilization of each individual's time while they are not at a normal fixed work desk.

Within this definition, four main types of mobile users can be defined:

- **Tethered/Remote Worker:** This is considered to be an employee who generally remains at a single point of work, but is remote to the central company systems. This includes home-workers, tele-cottagers, and in some cases, branch workers.
- **Roaming User:** This is either an employee who works in an environment (e.g., warehousing, shop floor) or in multiple areas (e.g., meeting rooms).
- **Nomad:** This category covers employees requiring solutions in hotel rooms and other semi-tethered environments where modem use is still prevalent, along with the increasing use of multiple wireless technologies and devices.

- **Road Warrior:** This is the ultimate mobile user—spends little time in the office, but requires regular access to data and collaborative functionality while on the move, in transit, or in hotels. This type includes the sales and field forces.

For these new actors of *extended enterprise* in the world of mobility, the tools that are required to create an effective mobile infrastructure revolve around the need for communication and collaboration. Within this space, we need to consider the likes of e-mail, calendaring/scheduling, contact management, discussion forums, virtualized meetings, and so on. While the majority of these systems can be provided in-house, we should ask ourselves whether we should be providing them in this way. If we are looking to control costs and provide an optimal solution, then we need to consider opportunities for minimizing the skills required to manage these technologies, and removing our dependence on the need for application patches, upgrades, down-time, and so on. Given the emerging emphasis on "focusing on the core competency," today's organizations are looking for outsourcing opportunities to third parties for management of mobile computing/m-commerce solutions.

For these organizations, as the customer, it is imperative that they remain in control of their own strategy with focus on their core competencies. Organizations are getting to be demanding about what service level is guaranteed from the providers (to whom mobile computing solution management is outsourced). This way, organizations do not need to worry that do not have the technical skills to maintain the solution, as long as they know exactly how much this solution—which is backed by a service-level guarantee and has high levels of security—will cost them.

MARKET PERSPECTIVE FOR MOBILE COMPUTING AND M-COMMERCE

With ever-increasing sales of mobile computers, handheld PDAs, and now a new breed of Web appliances and smart telephones, mobile computing has been hailed as a hot new technology that will significantly change the way in which we conduct our work as well as non-work-related activities. Many vertical industries, such as financial services, public safety, healthcare, and utilities have adopted mobile applications since the early 1990s. Even horizontal applications, such as field-service dispatch and Internet e-mail access, have made significant gains recently. Since 1993-1994, early adopters of mobile computing—UPS, Federal Express, Sears, Xerox, IBM, and Merrill Lynch among them—have demonstrated the potential of the technology. Now mainstream businesses are ready to adopt this technology in a serious fashion.

This section provides a perspective on the journey of mobile computing progress. Major players in this journey have been the various committees for Internet governance and technical issues, the Internet service providers, the mobile handset manufactures, and so on.

According to one subject matter expert (Mallick, 2003), the history of mobile computing shows that solutions in the 1980s and early 1990s concentrated on pure access such as the use of modem racks and *remote access servers* (RASs). Although these solutions were inherently secure, requiring one-to-one connections between the user and the computer, the cost of sufficient telephone lines to meet anticipated peak demand, and the calling costs for users who needed to connect from abroad made the solution suitable only for the most important of tasks. There were bandwidth constraints too. This led to the rise of companies providing thin client solutions to existing applications.

Meanwhile, the Internet continued to evolve. With the advent of the Internet, solutions began to consider utilizing this network as a means of providing lower-cost access from around the world, utilizing *points of presence* (PoPs) provided by regional Internet service providers to access low-end hosted e-mail solutions for the exchange of information.

Large telecommunications companies and IT vendors such as BT (British Telecom) and IBM then enabled access from these PoPs to their networks, and enabled secure access back into corporates through the use of *virtual private networks* (VPNs). Again, the business models around these solutions were aimed at large organizations with leased line data solutions and depended on high user payments. Complexity was compounded with the advent of *mobile telephony* solutions driven in Europe by GSM. This technology provided low-speed, high-cost means of connecting through to corporate solutions while disconnected from any fixed network, and necessitated support for new mobile phone devices and connectivity between the phone and the laptop.

With the Internet providing a more accessible solution, *security issues* came to the fore. Driven by technologists, the usability of provided solutions tended to be marginal, and companies using the solution still needed to have plenty of bandwidth themselves to be able to support the needs of multiple users coming in through the firewall to access e-mail, calendaring, and scheduling, along with any other functionality they were trying to access.

Today, the market is demanding complete mobile solutions—at the right price, with the right levels of security, and with low management costs to the using company. These solutions must be flexible, so that the company can concentrate on its core competencies, without needing to employ expensive resources to track, implement, and maintain changes to the underlying transport technologies.

MOBILE COMPUTING CASE STUDIES

Having discussed the overall scenario on mobility and mobile computing, this section provides illustrations from some real-world reported examples (Longbottom, 2003) of how effective mobile solutions can be utilized to competitive advantage. These illustrations come from diverse scenarios such as Engineering Design, Logistics, Insurance, Sales, Field Service, and Home-Working.

Industrial Automotive Sector

The Business Scenario

A European automobile company has followed the trends within the market to the extent where very few components are now manufactured by the company itself. Component manufacture is now outsourced to the lowest cost provider, within the constraints of quality, and this has led to component manufacture being spread across the globe.

The Challenge

The lifecycle of a style of a car is about three to six months, and the company must be able to change the styling rapidly as the market dictates. Changes to the external styling of a car can force changes to underlying components, or to assemblies of these components. The company found that the need to keep returning to the component manufacturers for new designs, and then looking at how these impacted the assemblies, meant that windows of oppor-

tunity in the market were being missed. The company began to lose market share, yet it did not have the option of bringing manufacturing back onsite.

The Opportunity through Mobile Solution

The company decided on the use of a "virtualized solution," enabling the suppliers to be brought together using collaborative technologies for a Web-based meeting. Through the use of application sharing, component designs can now be compared and checked, ensuring that any inconsistencies can be rapidly dealt with. Also, specialized skills can be brought in to the meeting, which may change the design requirements.

The Benefits

The use of a *virtualized meeting capability* has enabled the required skills and attendees to share views and knowledge without the need or cost of travel, and the design can be finalized within a matter of hours, rather than days. The company involved has reduced the cost of a redesign from many millions of Euros by at least an order of magnitude, and can now cycle designs in days, rather than months, enabling them to hit the market windows more precisely.

Logistics

The Business Scenario

A pharmaceuticals logistics company had a specific, but intractable problem. Deliveries of high-value pharmaceuticals were being made through standard means—a driver turned up in the morning, took a sheaf of papers, loaded up a van, and drove around delivering the loads.

The Challenge

Everything was carried out on paper, which often got lost or was damaged in use. Several of the companies receiving the deliveries soon discovered that it was possible to report a non-delivery to the logistics company, who had little capability to identify whether this was the case or not. On the rare occasions where signed paperwork could be found, the complaining company could just say that they were at fault, and that they had just found the delivery. This fraud reached a level where the logistics company was heading for bankruptcy. A solution was required, and it was required rapidly.

The Opportunity through Mobile Solution

The company decided to move from a paper-based system to a mobile-enabled system. Within this solution, paper dockets were removed, with everything now being carried out electronically. The driver was now in touch with the central office at all times, and thus itineraries could be changed, extra information provided, or issues raised and dealt with at the point of delivery. The solution consists of a set of communication and collaboration components, integrated into the company's own ERP system and to a geo-location system that tracks the route of the vehicles.

The Benefits

The company involved believes that the solution has not only eliminated the fraud involved, but has improved the efficiency of their logistics, due to fewer non-deliveries and a higher degree of issues being dealt with at the point of delivery due to better information availability.

Insurance

The Business Scenario

A major insurance company carries out most of its business through intermediaries. These intermediaries spend a lot of time on the road meeting with prospects, who they are trying to get to sign up for life and health insurance. If a person is interested, they need to fill in a set of forms. The intermediary sends in these forms to the insurance company, who then provides a quote for the prospect.

The Challenge

This takes about three to four days, during which time the prospect has had time to re-think, and closure rates were low.

The Opportunity through Mobile Solution

The company decided to provide each intermediary with a mobile solution, consisting of a laptop with a suite of tools. Now, the intermediary can fill in the form for the prospect on the laptop, and can immediately send this back to the insurance company for an immediate quote. If the company has any issues, they can interact with the intermediary while they are in the prospect's home, ensuring that the issue is dealt with there and then.

The Benefits

With this solution, the company has persuaded intermediaries to concentrate on selling their product, as the closure rate is far higher than the manual systems still used by other companies, and has significantly increased its share of the market.

Sales Opportunity

The Business Scenario

A salesperson is en route to India for a major presentation to one of the company's biggest clients the next day. His departing flight is delayed and he finds himself sitting at the airport for several hours.

The Challenge and the Opportunity through Mobile Solution

During this downtime he inserts his Wi-Fi card and quickly locates the Wi-Fi hotspot in the airport lounge through the use of a suitable mobile client, such as GRIC's Mobile Office client. Once connected to the network, he downloads e-mail and discovers the client's requirements have changed substantially: to win the business he will have to change his offer considerably. He checks the hosted *Sales Force Automation System* (one of a range of personal and group productivity tools available to him wherever he is in the world) to verify some of the client's background information to make sure the new proposal will be appropriate. He accesses the ERP system back at the center via a VPN and personal firewall, and checks the warehouse for suitable parts and to block them from being sold in the meantime; he can request via e-mail that the logistics manager makes contact to show when delivery can be made.

He downloads the latest confidential pricelist from his company intranet. He then organizes a quick Web meeting with the technical department, the product manager, and the prospect to make sure all new information is understood and that the available parts will meet the prospect's requirements. After his Web conference has ended, he learns his plane will be delayed three hours. He sets up shop in a comfortable place and begins drafting a new presentation and proposal. After a few hours he has finished. He orders another drink, reconnects using his Wi-Fi and mobile access client, and sends the new presentation and proposal back to the product manager for input.

The Benefits

A quick call and they agree on terms. He shuts down his connection, boards the plane, and catches a few hours sleep. Once there, he races to his hotel, quickly freshens up, and connects using a local dial-up connection through the same mobile client. He has received an e-mail with final status on availability of parts and delivery time. When arriving at the meeting, the salesperson feels certain to have all final information and can proceed to give a targeted presentation that shows he knows his customer and does not need to walk away with a list of questions to check out "back at headquarters," thus ensuring that the deal is closed with a grateful customer.

Field Service

The Business Scenario

Field engineers generally receive their day's tasks prior to leaving a depot or home through dial-up data synchronization technologies.

The Challenge

Tasks can often not be carried out due to absence of customer, road and traffic problems, or lack of information needed to solve the issue.

Opportunity through Mobile Solution and the Benefits

Through a suitable mobile solution, the engineer can access mapping and traffic information, ensuring that they can actually access a field site. Should they not have the right information at hand, they can access information from a central source, whether this be technical documents or a knowledge base of other engineers' solutions. Should a specific person resource be required, that resource can be brought in through a Web meeting to share written information. Finally, should the engineer finish a job early or not be able to get to any job, the mobile solution can provide further tasks, enabling the company to minimize response times and maintain customer satisfaction.

Home-Working

The Business Scenario

A company has the need for further resources in dealing with customer requests.

The Challenge

There is a shortage of available space in the existing office for new resources to sit. The choices seem to be either to overcrowd the office or to rent further office space.

Opportunity through Mobile Solution and the Benefits

With the correct mobile solution, the correct resource can be found irrespective of geography. By providing access through to communication and collaborative functions, the new resource can work from home while maintaining contact with the rest of the existing team. Access can be provided to existing applications within the company, while tools such as instant messaging can be utilized to ensure that problems requiring input from other resources can be rapidly addressed. Also, with suitable reporting in place, central office can ensure that the home-worker is using the systems and is meeting targets.

Mobile B2E Outplaces M-Commerce

Mobile commerce has been the driver of much of the early wireless hype. M-commerce has also been the subject of debates and disputes, given that many consider it to be too early to assess and comment on benefits provided by m-commerce. Dholakia, Dholakia, Lehrer, and Kshetri (n.d.) provide some interesting discussion. According to them, mobile phones, mobile Internet access, and mobile commerce are growing much faster than their fixed counterparts. Several characteristics of mobile networks make them more attractive than fixed networks for less developed countries. Comparative analysis reveals the following:

- Global leaders in land-based telecommunications or Internet access are not necessarily the global leaders in mobile connectivity.
- When it comes to the design of m-commerce applications, multiple designs are likely to coexist and compete for an extended period of time.
- Sources and reasons for national leadership in the evolution of mobile commerce applications are likely to be significantly different from the national leadership patterns on the Internet, landline telecommunications, and computers.

A point to be noted is that this is different from most IT fields, where dominant designs have converged rapidly to one or two standards. When it comes to standards for mobile computing, a peculiar difficulty faced is that of "variety." Mobile phones alone have a variety of standards. To summarize the standards issue, listed below are the categories of standards for mobile phones:

- First Generation (1G)
- Second Generation (2G)
- Enhanced Second Generation (2.5G)
- Third Generation (3G)6
- Fourth Generation (4G)

The mobile computing scenario for m-commerce is also replete with socio-cultural issues. For example, as reported by Dholakia et al. (n.d.), in the Asian countries where mid-range income is becoming the new market target, the consumers from Middle Class strata, would be satisfied with smaller electronic devices and mobile handsets.

Notwithstanding this, proponents of m-commerce, though, have envisioned a world where consumers shopped happily via wireless Web devices including phones and PDAs. Consumers, particularly in the U.S., have been slow to shift buying patterns as expected. The most notable reasons for this seem to be limited screen size and slow performance.

An older study released in mid-2001 by management consulting firm A.T. Kearney quantifies the change in attitudes as the hype has died down. The Industry Standard summarized the study, publishing the following (Industry Standard 2001):

Researchers surveyed more than 1,600 mobile phone users throughout the U.S., Europe and Asia and found that only 12 percent said they intend to engage in m-commerce transactions. That's down from 32 percent just one year ago. And less than 1 percent has actually made any purchases with their phones in the past year.

The place that mobile technologies are gaining the most traction in the enterprise is with business-to-employee (B2E) applications. Companies are finding cost savings and productivity increases by providing their staff with mobile devices and mobile information access. In a recent article highlighting the results of a META Group study titled "Wireless Adoption, Trends, and Issues," CyberAtlas (2001) notes the following:

As might be expected, organizations with heavy use of pervasive devices by employees are more aggressive in implementing leading-edge wireless/mobile infrastructure components. Several studies have found that the first priority of implementation is for business-to-employee (B2E) applications, because these applications deliver the most immediate productivity return for organizations.

Issues and Challenges

Although mobility presents us with major opportunities, there are and will continue to be business and technical issues. We must deal with these issues, or at least be aware of them, so that we can maximize the effectiveness of the mobile computing solution. The issues for the mobility market are classified into two major parts: *business issues* and *technical issues*. These issues arise partly due to complexity of technologies, combined with the need to deal with multiple access points and concerns around corporate security and the impact of consumers purchasing handheld devices (Godbole, 2003). Higher-speed data telephony solutions, including GPRS, aided by the growth in Wi-Fi hotspot access points, will combine with the explosive growth in devices such as tablet PCs, smart phones, and hybrid devices to create a highly complex, fast-moving market. Here, the key to success will be in knowing how to create market differentiation through using the available technologies—rather than having the skills in-house on what the technologies are themselves.

Business issues of mobile commerce tend to rise due to a number of reasons. For example, organizations will not find it easy to dictate the kinds of devices their users will have. According to a white paper on the business case for device management of by SyncML (http://www.syncml.org), handheld devices used for mobile commerce can pose haunting and daunting issues. These devices also give rise to security concerns. Outside of specific areas where specialized devices from companies are required, device choice is rapidly becoming predominantly a personal lifestyle choice. Business organizations will need to recognize this and frame appropriate security policies that advise on minimum capabilities that a device should have for the company to provide any level of support for it. This allows usage of mobile computing devices with minimal computing features that are called for within organizational context for mobile commerce. As the Web has become ubiquitous, even small companies have increased their global reach. Whereas it was expected that the technology solutions that emerged along with the Web (e.g., video/teleconferencing, instant messaging) would minimize the need for travel, the amount of international travel continues to increase. Although the face-to-face experience does provide value, one person cannot always have all the answer, and it is here where collaborative and communications technologies can provide the missing pieces of information.

Market forces tend to play a part in a sluggish market, where the cost of setting up full facilities for staff can be prohibitive. The use of home-based workers enabled through a set of virtualized functions can provide fast response to market requirements at minimum costs. Similarly, resources can be optimized through the use of part- or shared-time workers working from home in the same manner. Overall, to enable the uptake and adoption of remote and mobile tools, there is a need for a single, easy-to-use client on the user's device enabling access not only to central networks, but also for responsive, hosted collaborative functionality. Given this scenario in the mobile computing market, within business, profitable market opportunities are created through one of three ways:

1. eliminating organization costs from fixed network solutions (i.e., the conventional wired network-based access),
2. controlling variable costs within an organization, or
3. creating higher margins through the more effective sale of existing or new goods

Of these, the need to control variable costs is the one with the least risk. However, attempting to control costs, without the changing of business processes and business models, could prove dangerous. Truly controlling variable costs needs a new look at how business is conducted, why it is done that way, and how it can be done differently in today's mobile computing paradigm to facilitate conducting business in the future.

With mobility, the issues must be grasped and resolved in the earliest possible stage before it gets too late to control. Mobility of workforce provides opportunities for organizations in the digital economy to create new business processes, re-engineer some of the non-value-adding business processes with the aim to create higher profit margins, and develop new commerce channels within the extended enterprise while controlling implementation costs, as well as costs for support and maintenance. We also have the opportunity to create a flexible mobile infrastructure, and the foundation to mobile computing operators for gaining market share from their competitors.

The key to using mobility as a differentiator involves more than just looking inside the firewall; it involves extending thinking to company users outside the firewall. By looking at how mobility can influence and facilitate working differently—not just automating what we already do—we can look at how we can maximize the utilization of our expensive human resources. We can ensure that decisions are made on the right information, provided at the right place, at the right time. In this game, the players include suppliers and customers in our *collaborative work paradigm* in an open, yet secure manner. Organizations need to consider utilization of virtual teams consisting of disparate expertise to resolve issues irrespective of location.

Thus, mobility and m-commerce need to leverage this in the market for profitable differentiation. To make the most of mobility, we need to understand the differences between effectiveness and efficiency. Efficiency is just being able to do more with the same resources—even if what we are doing is wrong. Effectiveness is being able to increase the value provided to the company by doing more and more things correctly. Effectiveness is often governed by access to real-time or near-time information between workers, and those with the highest requirement for this access tend to be the most mobile. For example, task-driven groups such as field engineers can plan their routes more effectively when they have access to the latest information. Sales forces can ensure that customer expectations are accurately set through good knowledge of inventories and supplier lead times. Managers can make the correct strategic decisions while on the move. The entire organization can become more responsive, more effective, and therefore more profitable through collective organizational intelli-

gence enabled by the correct utilization of mobility across the whole organization.

As part of this effectiveness, we also need to look at how we use resources for knowledge sharing, utilization, and knowledge management—often we have in-house skills that we only occasionally utilize. If we can provide a flexible collaborative infrastructure enabling resources to exchange views and information irrespective of geography, we can optimize the use of these resources through making their skills more globally available, or outsource certain areas of skill so paying only when these skills are utilized.

When it comes to the implementation of a mobile infrastructure as an opportunity to outsource the solution, it is unlikely that 100% outsourcing of all existing solutions may happen at the same time. Therefore, it is important that any mobile solution provide the tools required for connecting back into organizations' central networks, so that existing applications can be accessed by users through appropriate means.

Consider, for example, secure VPN connection, driven through a non-technical, user-friendly front end. Again, although providing access through to existing applications can provide short-term benefit, one needs also to look at how to be able to evolve the solution through the use of emerging technologies to be more effective in the mobile context. The most important of these emerging technologies will be *Web services*, where applications will be able to make their functional components available to other applications through a series of standardized calls. For outsourced solutions, the emergence of Web services provides a means of enabling information to be interchanged between the managed solution and the existing internal applications. There are many vendors in the market (e.g., Attachmate, IBM) looking at how to take existing "legacy" applications and enable them for Web services.

Although older mobility solutions used *direct-dial solutions* with relatively high levels of security, the very nature of today's mobile needs predicates the use of the public Internet. Therefore, security will be at the top of everyone's concerns. For mobile solutions, the use of VPNs provides a secure means of "tunneling" through the public Internet. However, additional security can be obtained through the use of additional encryption, and external attacks to the client can be prevented through the use of personal firewall technologies. As the solution will also be enabling users to connect back into your own environment, you should consider the possibility of attempts to hijack the connection, and look for mobile solutions that guard against Trojans, Piggy-backs, Worms, and so forth.

Outsourcing is the current vogue. However, many organizations worry about "letting go" of their solution. Often, their concerns are around the inherent dangers with letting an outsourcing company take control of hardware, application, and data. There are several factors involved in this. Firstly, it is likely that any outsourcer will have better physical security than can be provided in-house by most companies. Secondly, the outsourcing company is likely to positively vet its employees—again, an area where few organizations have the capability to do other than cursory checks. Thirdly, the skills that the outsourcer employs will be dedicated skills, ensuring higher availability and better current knowledge of technology than can be afforded within the company itself.

It must also be appreciated that "technology," "data," and "function" should not be utilized interchangeably. Technology is essentially a set of plumbing which can easily be outsourced. Data can reside anywhere and can be accessed by the technology. Function is what the technology and the data support, and is what the company and the end user must have.

Mobility: Opportunities and Issues to Organizations

As per Quocirca reports (2004a, 2004b), the main barriers to entry into m-commerce are *security*, *performance*, and *skills scarcity* (i.e., the need for specific skills to manage a rapidly changing technology environment); and in most mobile computing-related surveys, the main concerns of people are how to open up their own systems to provide this access, without compromising on security, performance, and system availability.

Thus, the inclination seems to be towards use of third-party, managed, hosted solutions.

ALLUDING TO LEGAL CONSEQUENCES OF MOBILITY IN BUSINESS

This turns out to be a serious issue in today's mobility and mobile computing paradigm. Legal aspects of mobile agents, given their increasing deployment in mobile computing solutions for m-commerce, become entities for attention. "Agents" are a special form of (or improvement on) components. They are goal directed and usually self-activating in the sense that they possess their own thread of execution and can initiate actions without intervention from the outside (incoming communication, user interface, etc.). An important distinction between agents and components (on which they are usually based for implementation) is that agents can decide whether to fulfill requests made to them, while components just execute commands.

Mobile agents can travel from one host to another, taking their code, data, and state with them. Because of this mobility, a number of legal problems can arise. Stationary agents, which always remain on the server on which they were created, are less of an issue: they remain completely under the control of their owner if locally created. There are many agencies to which a finger can be pointed in case of disputes arising out of malfunctioning of these mobile agents, which could result in loss to the customer. Potential entities for this blame are the network provider, programmer of the agent's code, owner of the mobile agent, and owner of the server. Sonntag (n.d.) provides interesting discussion on the topic of usage of agents.

Certain open issues come up:

- What are the legal consequences of mobile agents traveling through different countries? In contrast to ordinary Internet traffic, they are not just transported, but probably executed on servers in other countries. If they possess some information, which is illegal in this country but lawful in the countries of its origin and destination, could the agent be stopped? What actions are required with this information to constitute an offense, or is the mere existence sufficient?
- The technical side is even more challenging. It is often difficult to identify the server, which is responsible for the destruction or a certain change in an agent. Even if extensive logs are used, following them is extremely expensive and difficult, especially if they are located in other countries (often a court order is required for disclosure).
- In some cases electronic signatures and logs would allow automatic arbitration with the evidence collected. However, what should be the result: a fine, an e-mail to the parties? How would it be enforced?
- Similar to users on the Web, the nationality of an agent is easily defined: its owner's nationality. But how can this be detected and/or verified? One possibility is digital certificates, but the nationality is usually not verified and included in them. This is especially important for blocking or disallowing some actions for agents of a certain nationality, for instance, to enforce taxes or not to sell certain goods to some countries. A solution could be using attribute certificates certifying the nationality, similar to the authorization for signing for legal persons.

CONCLUSION

In essence, m-commerce is the convergence of wireless communication technology and Internet-based e-commerce. It can be expected that m-commerce will be more attractive to customers (or online users). M-commerce has two distinctive advantages over brick-and-mortar business as well as over e-commerce, namely flexibility and ubiquity. Through wired networks, consumers with mobile handsets can conduct business transactions without being fixed at a computer terminal. And as the number of mobile phone users is much more compared to the number of Internet users, businesses can expect to reach a larger market. For the information commerce sector, m-commerce also offers personalization and real-time speed. While the Internet provides a huge amount of information on virtually any topic, mobile devices can disseminate information.

FUTURE DIRECTIONS

With m-commerce and mobile computing applications, users are able to customize and personalize their mobile devices to retrieve information specific to their needs. The speed in which m-commerce transactions can take place is also worth noting. Consumers typically receive almost instant response for information-related services. Compared to wireless voice or the Internet alone, the wireless Web offers to its users greater convenience. For example, its users do not have to waste their time looking up phone numbers whenever they want to place business orders. Database access for company listings, product availability, and order information can be readily available on the wireless phone. Searching for competitive prices when shopping can be more conveniently done through the Web.

However, m-commerce is still at its infancy. The most popular applications so far have been for entertainment and information-related services, such as the instant messaging service. Although wireless billing is a promising practical application, the security issue must be tackled carefully for this application to be viable. Early implementations of m-commerce have been extensions of e-commerce sites. Technologies such as WAP and Wireless Mark-up Language (WML) enable just that. For businesses, this would translate into better communication with their customers, unlimited by trading hours and geographical distance.

Although m-commerce merely serves as an extension to e-commerce sites, it has been regarded as a value-added service: offering specific information, entertainment, or a transaction over public or private mobile telecommunications networks. According to Worldwide Market Analysis and Strategic Outlook 2003-2009 (2004), there are certain industry estimates suggesting that between 2002 and 2005, 600 million Internet-enabled mobile phones will be sold. As for revenue, it turns out that the European m-commerce market was $24 billion by 2003, while end users' spending on mobile e-commerce services were expected to reach more than $200 billion in 2005. These figures show that m-commerce has many positive value-added services to offer in a huge market. This

indicates that with the high likelihood of m-commerce reaching more than 600 million by the end of 2006, businesses cannot afford not to pay attention to this new medium.

In the future, when m-commerce security is more robust, the use of mobile devices as digital wallets will become more popular. This would enable consumers to shop and pay using mobile devices much like a credit card, as the credit information is stored in those mobile devices. Future and potential applications of m-commerce will involve "information-rich" products and applications with multimedia capability.

A few more concluding remarks need to be in place to end this chapter. Mobility is on the rise—it is important to see it as an opportunity, not a threat. Many organizations may look at mobility as purely an opportunity to extend their existing systems—those who take the opportunity to change business processes and look to mobility as a means to an end will be the winners. Also mobility is not just about connectivity—it is about function. Further, it is important to understand that MOBILITY can be highly technical, so organizations must concentrate on making it easy for the user. If the solution is seen as being of little complexity to the user and is couched in terms that make business sense to them, they will use it. There are many issues and challenges: mobility and mobile computing is replete with its own business and technical challenges.

Today belongs to "business-driven IT." That means it is important to keep in mind that IT is not a core competency. Thus, organizations are looking for outsourced solutions to provide a more flexible system that should be more responsive, available, and manageable than an in-house solution. Also, in-house is constraining. Mobile technologies are changing—and putting in place an in-house solution will require continuous change in the coming years. This will adversely impact organizations' companies capabilities to compete and will incur heavy costs. This will further enhance the trend towards outsourced solution because that can provide a more flexible environment.

REFERENCES

CapGemini. (2001). *Internet mobile for business: So far from expectations.* Retrieved April 18, 2005, from http://www.capgemini.com/news/2001/0618mobile.shtml

Cellular. (2005). *'Third generation' mobile technology 3-G key* and *WAP glossary.* Retrieved May 25, 2005, from http://www.cellular.co.za/technologies/3g/3g.htm and October 7, 2005, from http://www.refreq.com/WAPTech/wap_glossary.htm

CyberAtlas. (2001, September). Business-critical applications driving wireless initiatives. *CyberAtlas.*

Dholakia, N., Dholakia, R. R., Lehrer, M., & Kshetri, N. (n.d.). *Patterns, opportunities, and challenges in the emerging global m-commerce landscape.* College of Business Administration, University of Rhode Island, Kingston, USA.

Godbole, N. (2003). Mobile computing: Security issues in hand-held devices. *Proceedings of the NASONES 2003 National Seminar on Networking and E-Security.*

Godbole, N., & Unhelkar, B. (2003, December). Enhancing quality of mobile applications through modeling. *Proceedings of the Computer Society of India Convention,* IIT-Delhi, India.

Hansmann, U., Merk, L., Nicklous, M., & Stober, T. (2003). *Principles of mobile computing.* New Delhi, India: Springer.

Industry Standard. (2001, May). Analysts change their tune on m-commerce. *Industry Standard.*

Intel. (n.d.). *Intel Technology Journal, archival of past journals.* Retrieved from http://intel.com/technology/itj/q22000/articles/art_6.htm

Jgohil.Typepad. (2005, January 18). *A case on NTT DoCoMo.* Retrieved from http://jgohil.typepad.com/minternet360/imode/

Kerner, S. M. (2005). *PDA market up or down?* Retrieved May 23, 2005, from http://www.internetnews.com/stats/article

Kerner, S. M. (2005, July 22). *Cell phones rising.* Retrieved May 19, 2005, from http://www.internetnews.com/stats/article.php/3522076

Kolari, J., Laakko, T., Hitunen, T., Ikonen, V., Kulju, M., Suihkonen, R., Hu, S., Virtaneen, T., & Tytti. (2005). *Context aware services for mobile users: Technology and user experiences.* Retrieved May 20, 2005, from http://www.vitt.fi/inf/pdf

Longbottom, C. (2003). *Mobility as an opportunity.* Report commissioned by GRIC.

Ly, F., & Sugianto, S. S. (n.d.). Business models in the digital economy. Retrieved April 15, 2005, from http://aisel.isworld.org/subject_by_publication.asp?Subject_ID=17

Mallick, M. (2003). *Mobile and wireless design essentials.* New Delhi, India: Wiley-Dreamtech India.

Mobile Applications Initiative. (n.d.). Retrieved from http://www.mobileapplications initiative.com/

MobileInfo. (2005). *One stop Web site for mobile computing and wireless information to access hot topics in the mobile computing industry.* Retrieved May 25, 2005, from http://www.mobileinfo.com/3G/3G_Wireless.htm

Quocirca. (2004a, Summer). Enterprise wireless update. *Quocirca Insight Report.*

Quocirca. (2004b, November). Optimizing the mobile workforce—A business value perspective. *Quocirca Insight Report,* 6.

Quocirca. (2005). *Mobile devices and users.* Retrieved April 18, 2005, from www.quocirca.com

Reiter's. (2005, June 6). *Reiter's camera phone report.* Retrieved April 18, 2005, from http://www.wirelessmoment.com/statistics_camera_phones/

Sonntag, M. (n.d.). *Legal aspects of mobile agents with special consideration of the proposed Austrian e-commerce law* (pp. 2, 3). Institute for Information Processing and Microprocessor Technology (FIM).

SyncML Initiative. (n.d.). *Business case for device management.* White Paper. Retrieved May 25 and October 7, 2005, from http://www.syncml.org

Telecommunications Service Inquiry. (2003). *Data and mobile services. 2000—Connecting Australia.* Commonwealth Department of Communications, Information Technology and the Arts. Retrieved May 20, 2004, from http://www.teleinquiry.gov.au

Worldwide Market Analysis and Strategic Outlook 2003-2009. (2004, August). Future mobile computing: Device trends and wireless solutions. *Worldwide Market Analysis and Strategic Outlook 2003-2009.*

Chapter XXXIV
The Future of Mobile Technologies and Applications in China

Xiao Chen
Nanjing University of Chinese Medicine, China

Wei Liu
Nanjing University of Chinese Medicine, China

ABSTRACT

This chapter deals with the future of mobile technologies and applications in China. The effect of emerging technologies, especially mobile technologies, on the massive market of China cannot be ignored in the global context. This chapter gives the reader an insight into China's mobile telecommunication industry today. The authors firstly relate statistics about China's mobile business market including user and device analysis that helps in providing an understanding of mobile business in China. This analysis is followed by a description of the major mobile technologies employed in China and a brief view of the Chinese market's status, followed by an insight into some newly rising industries which are potentially successful mobile sectors in China. Finally, a real life example is examined—that of M-Government Project in Gunagzhou, capital city of Guangdong Province.

INTRODUCTION

After the arrival of the Internet in China in 1987, there were huge changes to the way the Internet developed further in China. Earlier, without the popularity of the Internet, both people and businesses used dial-up modems to log onto the Internet. Later, Digital Subscriber Line (DSL) technology came into common people's lives. A few years later, it was the application of broadband, which made the connectivity to the Internet very popular. Finally, in 2004, wireless technology further extended the application of the Internet in people's lives as well as in businesses. These various methods cover almost all markets of Internet access; they are both beneficial for Internet service providers (ISP) and Internet operators. They bring in plenty of profits for the two major Internet operators—China Telecom, and China Netcom—and the two make the Internet market more ubiquitous. However, traditional access methods like dial-up and DSL technology need stable network facilities for the support, therefore it will not be convenient for the users to use the Internet without distance limit, and it might be a limit for business use and ensure the business processing in anywhere customers needs. For the high demand of moveable and convenient business, a new-generation connecting method called "Mobile Internet" (Wireless Internet) has been developed; its the main technologies are wireless application protocol (WAP), General Packet Radio Service (GPRS), Code Division Multiple Access (CDMA), Centrino, and Chinese Wireless Authentication and Privacy Infrastructure (WAPI). WAP was the first trial of Mobile Internet in China, but it seemed to be a failure; Centrino and WAPI are the two latest trends of Mobile Internet in China; both have advantages and disadvantages. Which will be chosen by the future, and which will be dominant in China's Mobile Internet industry? Let us take a look. In this chapter, we discuss the importance of mobile technologies and applications in this modern era with relevance to mobility in China.

MOBILE STATISTICS IN CHINA

Business Statistics

The information industry has been the one of the major drivers to impact economic development in China. And within the information industry, the communication and network markets have continued to maintain high-speed progress. Up until the end of 2003, the number of telecommunication users in China exceeded 400 million, and the number of Internet users

Figure 1. Trend of increment of China's common phone users and mobile phone users from 2001 to 2004, and forecast for the number in 2005 (CLI4 Report, 2003)

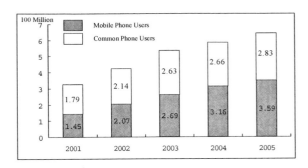

reached 79.5 million. These figures establish the foundation of the development of mobile technologies and applications in China (OECR.com Report, 2004).

In China's domestic telecommunications market, mobile communication equipment and services are widely accepted and adopted not only by individuals but also organizations. At present, mainstream mobile communication equipment includes mobile phones and PHSs (Personal Handphone Systems). As the largest mobile phone market, China had 269 million users in 2003, which increased 30% from 2002. Figure 1 shows the trend of increment of China's common phone users and mobile phone users from 2001 to 2004, and it makes forecast for the number in 2005 (CLI4 Report, 2003).

Along with the upgrade of technologies, the number of PHS users also goes up dramatically from 2002. In September 2002, the number of PHS users exceeded 10 million, and at the end of 2002, the number increased to 11.13 million. At present, this number continues to increase at high speed (CLI4 Report, 2003). From Table 1, we find the number of PHS users in 2002 was over 11 million, and in 2003 the number boosted to 24 million, double that of 2002. Also we find the numbers of mobile phone users and other

Table 1. Development trend of different kinds of China telecommunication users (CLI4 Report, 2003)

USER	UNIT	YEAR 2002	YEAR 2003
Common Phone Users	100 million	2.14	2.63
PHS Users	10,000	1,113	2,400
Message Users	10,000	14	45
Broadband Network Users	10,000	357	580
Defense Data Service Users	10,000	46	45
FR and ATM Users	1	9.9	12.6
Internet Users	10,000	4,970	9,200
Mobile Phone Users	100 million	2.07	2.69

Table 2. Capability of China Mobile and China Unicom from 2003 to 2005 (Broadcom Consulting, 2004)

	YEAR 2003	YEAR 2004	YEAR 2005
China Mobile GSM (10,000)	21100.0	23600.0	24000.0
China Unicom GSM (10,000)	8000.0	9000.0	9000.0
China Unicom CDMA (10,000)	3900.0	5000.0	6500.0

kinds of communication users increased by geometric series. Typically, we can see that Internet users in China increased so much that the number reached 92 million in 2003.

After the split and re-unification of the department of China's information industry at the beginning of 2002, two organizations were formed to compete for the mobile communication operation market. These are China Mobile and China Unicom. China Mobile owns the largest Global System for Mobile communication (GSM) and General Packet Radio Service network in China, which covers all the provinces and cities in China. Nevertheless, China Unicom established more than 300 subsidiary companies in 31 provinces in China, and its GSM network also spans most of China, excepting some parts of Tibet and certain villages. Meanwhile, China Unicom also finished its CDMA network, thus China Unicom owns two different networks which both cover the whole China. After the first phase of the project, the capability of the new operation network

reached 15.81 million people, and there are 188 large switches, 9,000 direct stations, and 14,458 base stations. Nowadays, CDMA covers 92% of all of China's 332 cities and is affiliated with more than 1,800 villages and towns.

Similarly, China Mobile invested RMB 300 million for the 3G communication network GPRS in 16 China provinces whose capability was 414,000 users in 2001. Up until the end of 2002, the total capability of GPRS was three million, covering 241 major cities in China. The income of GPRS users comes to more than RMB 80 million.

In the network market, according to relevant data statistic, the global sale income of Wireless Local Area Network (WLAN) was $7 billion in 2003, and it will increase to $44 billion in 2008, which means a 44% hybrid increment ratio per year. Although the start-up of China's market is a little late, the average increment ratio is up to 51% per year. In 2003, the total sale income of WLAN in China was $58.06 million. Furthermore, as a result of

Table 3. Advertisement investment for WLAN in China from 2001 to 2003 (CNIIC, 2003)

YEAR	2001	2002	2003
Cost (RMB 10,000)	283.8	468.2	1547.8
Increment Ratio	---	64.98%	230.59%

coming to an agreement about a WAPI standard with the U.S., the scale of China's wireless broadband network market is two times as large as it was at the end of 2003.

The WLAN operation in China also drives the market of networking access equipment sale. Thus, because of earlier participation, foreign providers could provide a wireless access point, wireless router, wireless bridge, and other system equipments. For instance, Airconnect of 3Com, Aironet series of Cisco, Intel PRO/Wireless 2011LAN, and Intel PRO/Wireless 5000LAN could meet the needs of individuals and different sizes of organizations. Also, domestic manufacturers such as Huawei, ZTE, and DTT start to focus on the WLAN market. They finish their own products in succession to supply to customers; these products include an inner wireless network interface card (WNIC), a wireless client adapter, a solution scheme, and so on.

Also, manufacturers increase their advertisement investment in order to build up visualization and occupy the WLAN equipment market. The advertisement costs in China from 2001 to 2003 for WLAN products are shown in Table 3.

In addition, the new technology standard also becomes the force to make the WLAN market develop fast in China. Since IEEE802.11 is executed in due form, WLAN technologies grow fast and become mature gradually. Figure 2 lists several standards that are accepted and used in the WLAN market (TechTarget, 2005).

At present, theses standard and technologies are widely used in all manufacturers' products. These factors improve the development of China's WLAN market at a high speed.

MOBILE USER ANALYSIS IN CHINA

Generally, the mobile user in China could simply be divided into two parts. One part is the mobile communication equipment user, which means mobile phone and PHS consumers. There is no doubt that the users of mobile phones and PHSs will be continuously increasing. The call service will keep developing quickly. However, we want to introduce and analyze the mobile users who manage data transmission and Internet surfing.

High Development Speed of Data Transmission Operations

In China's mobile communication market, the data transmission operations contain Short Message Service (SMS) based on GSM and Multimedia Message Service (MMS) based on GPRS. Table 4 shows the income of China's telecommunication industry from 2000 to 2002.

Figure 2. IEEE wireless speed model (CNNIC, 2003)

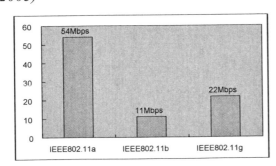

Table 4. Major telecommunication operations income from 2000 to 2002 (CNIIC, 2003)

OPERATION TERM (100 MILLION)	YEAR 2000	YEAR 2001	YEAR 2002
Telecommunication Total Incomes	3006.5	3535.2	4115.82
Long-Distance Communication Income	618.3	528.7	563.8
Ratio	20.56%	14.96%	13.7%
Data Communication Income	72	131	181.1
Ratio	2.39%	3.71%	4.4%
Local Communication Income	928	1027.7	1212.3
Ratio	30.87%	29.15%	29.5%
Mobile Communication Income	1253.4	1617	1954.4
Ratio	41.69%	45.74%	47.5%
Other Operation Income	70.8	181.1	163.02
Ratio	2.35%	5.12%	3.9%

Figure 3. Rate of Internet users and mobile phone users in several countries (CNIIC, 2003)

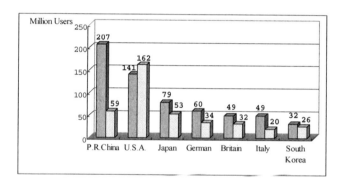

In 2002, the income of data communication was RMB 18.11 billion, an increase of 38.1% from 2001. Moreover, the percentage of data communication in the total income of China's telecommunication industry rises quickly. Though the proportion is still not very large, the potential increment and development is imponderable.

Improvement of Combination with Internet and Mobile Phone

In 2003, the rate of mobile phone users and Internet users was 3.5:1 in China; no other market achieves this rate (see Figure 3).

Only according to quantity, interest, and development trend is it possible for China to achieve advantage in the field of the combination of mobile phone and Internet.

The manner than young people use China's Internet may indicate the trend of manner for all people. In China, 70% of Internet users are under 30 years old.

The combination of entertainment value, fashionableness, and low price of the mobile gadgets provides an excellent opportunity to cater to the younger demographics of China. Web sites usually provide mobile phone users with the services listed in Table 5.

However, to these young people, the price of the combination of mobile phone and Internet

Figure 4. Age analysis of China's Internet users (CLI4, 2003)

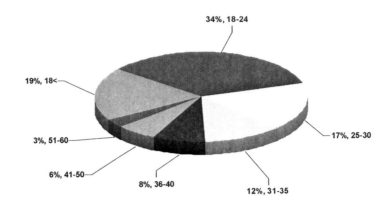

Table 5. Major operation terms that Web sites provide (CNNIC, 2004)

PRODUCT	DESCRIPTION
SMS	Elementary words short message
MMS	Photos/graphic short message
WAP	Online scanning (include searching)
IVR	Obtain voice message and entertainment

is acceptable nowadays. For example, the phone ring and game download, appointment, and short message meet the needs of young people and have large data communication traffic, but the price is not so high. For the major Web sites in China, the income flow related to short message and game starts to increase (CNNIC, 2004).

Also important is wireless network users, including wireless networking access equipment users and WLAN users.

Enterprise Users Increase Sharply and Individual Users Keep Steady Increment

The mobility of WLAN equipment could bring users convenience and higher access speed. Along with the price decline of WLAN equipment such as Access Point and NIC, the number of China's WLAN users will mushroom.

Moreover, as the terminal unit of WLAN equipment, most notebook computers begin to install WLAN NIC. From Figure 6 we can clearly see sales information of notebook computers during the first three quarters in 2003 and 2004.

It is not difficult for us to find that enterprise users provide the major purchasing power, which is about two-thirds of total sales num-

Figure 5. Internet major Web site income category, fourth quarter 2003 (CNNIC, 2004)

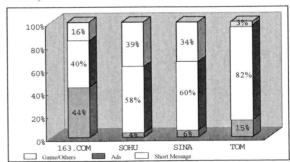

Figure 6. Notebook sale amount of first three quarters in 2003 and 2004 in China (CNNIC, 2004)

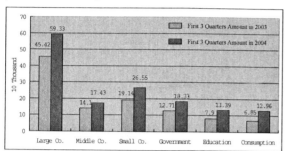

Figure 7. PDA and notebook sales from 2000 to 2003 (Blogchina, 2004)

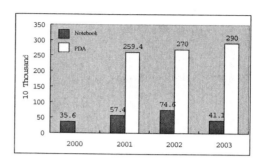

bers. As more and more districts are covered by the WLAN network, there is no doubt that in the increment of WLAN users, enterprise users will occupy a sizeable percentage. At the end of 2002, there were 25,000 WLAN users: 15,000 enterprise users and 10,000 individual users. As predicted, the scale for enterprise application will exceed the application of the public operation market in 2005.

Industry Users Are the Power for WLAN Market Development

At the present time, WLAN application focuses on public service, campus network, and family and individual fields, but as application use broadens, the focus of the market will gradually transfer to industry application. According to the experience of other countries, the aim for WLAN industry application is high-level business managers. However, the final goal of the WLAN market is to make it popular;

therefore, industry users are the largest group of potential WLAN users.

Potential User Market Is Quite Huge

As the terminal units of WLAN equipment progress, the sale of PDAs and notebook computers is increasing constantly. As one professional comments, the notebook computer is the most suitable terminal unit for the WLAN application. Meanwhile, sales numbers of notebook computers indicate that the notebook is one of the products whose sales are increasing at high speed (see Figure 7).

Up to the second quarter of 2003, the sales number of PDAs was 2.9 million; thus, the number of users who use PDAs to access the Internet is only 8% of all Internet users. Also, in the same timeframe, about 17.6% of notebook users become WLAN users, while among PDA users who use PDAs to connect to the Internet, there are 30% WLAN users.

Table 6. Movement of different WLAN application ratio markets in China from 2002 to 2005

	Public Service Market	Industry Application Market	Family Application Market
2002	42.2%	24.6%	7.6%
2005	12.4%	39.2%	18.9%

USAGE OF MOBILE TECHNOLOGIES IN CHINA

WAP (Wireless Application Protocol), Maybe Too Good for China

Dating back to 1999, the first sales of wireless data services were provided by former China Tele (China Tele and China Mobile Tele today), who shared connected mobile network and paging with Chinanet, the largest Internet infrastructure for business use in China. This service, however, was not balance-developed and could not be served in some provinces in faraway China. Therefore, the customers of this service were a minority, and few people knew about it. With the high-speed growing economy, Internet technology, and mobile telephony in China, a new application was introduced to mainland China—wireless application protocol (WAP). WAP made more and more people accustomed to the new concept of accessing the Internet wirelessly via mobile phones. In 1999, the hotspot for the Internet was WAP. Both operators and equipment suppliers such as Ericsson, Nokia, and Motorola strived to create and provide the trial service of WAP. In March 2000, China Mobile launched its trial service of WAP in six major cities in China, including Beijing, Shanghai, Shenzhen, Tianjin, Guangzhou, and Hang Zhou; meanwhile, China Unicom launched the same service in Guangdong and Shanghai, the two fastest developing provinces in China (Xiangdong, 2000). WAP was finally abandoned by the market due to low speed of transmission, lack of sufficient content and fewer competitors, and high cover charges. The two major operators quickly found that only new, effective technology could satisfy the needs of consumers.

GPRS and CDMA: The Right Choice and Rising Star in China

Eventually, China Mobile and Unicom both found that WAP was not their final product and the correct choice for the m-Internet market. They knew that WAP was the "DOS" of the wireless technology world, and the "Windows" operating platform had not yet been established. GPRS technology is one of two developments chosen by China Mobile after the unsatisfied WAP services in China. The main benefits of GPRS are that it reserves radio resources only when there is data to send, and it reduces reliance on traditional circuit-switched network elements. The increased functionality of GPRS will decrease the incremental cost to provide data services, an occurrence that will, in turn, increase the penetration of data services among consumer and business users. In addition, GPRS will allow improved quality of data services as measured in terms of reliability, response time, and features supported. The only way to succeed in m-Internet is to find the right model for it. Both of these attempt some development.

The other mobile operator, Unicom, chose a totally different service called CDMA, which is a large, successful business in South Korea. With the slogan of "Green Telephone, Less Radiation," and with the marketing strategy of "less pay, more calls," in the years that followed—especially in the year 2001—millions of mobile phone users were attached to Unicom as the newest innovation of the m-Internet industry. Another reason Unicom chose CDMA was that Unicom was the sore licensed operator who could provide CDMA service in mainland China. In its long-term blueprint, Unicom deployed its CDMA network in more than 200 Chinese cities, with network capacity of up to

494

about 15 million users within one year. In 2001, China Unicom updated its IS-95 CDMA to CDMA2001x, s CDMA2001x services, which was the latest CDMA technology worldwide at that time (Xiangdong, 2000).

To this point, we have discussed Mobile Internet service via mobile phone. However, many people like to surf the Internet via computers, especially laptop computers; yet, they prefer to get rid of the limit of wired connection and become truly free. The following technology will focus on the wireless technologies for computer use.

The Newcomer from the U.S.—Intel Centrino Technology: Great Waver in the Mobile Computer Market

In 2002, Intel, the largest microprocessor manufacturer, introduced a new wireless technology called Centrino, which integrates the wireless connection function into the Pentium CPU. This technology is based on IEEE standard 802.11b, and it is proven that Centrino can make the real wireless Internet available without help from a CID card. Also, general performance of PCs seemed to be upgraded, therefore customers gained a more comfortable and multifunctional laptop computer. When first introduced into the China market, Centrino was quickly adopted by many major laptop computer manufacturers like IBM, HP, TOSHIBA, and some other domestic computer manufacturers, like Tsinghua, Founder, and Lenovo. The promotion stage was also done very well, so that Centrino is the representation of wireless technology, ensuring the availability of real wireless technology. Three years into the use of Centrino, users have calmed down and are ready to verify whether Centrino is really "precious" or "rubbish" for the wireless world. Some people say Centrino still needs much support, like a connection base, access portal,

and so on. Furthermore, because the infrastructure construction in China is ongoing, users cannot access the Internet at any place they want, so it cannot be said that the use of Centrino can make users free of distance. In this case, the road ahead for Centrino is still so long that timeframes cannot be determined.

Security Issue: China's Own WAPI

WAPI is aimed at device manufacturers around the world hoping to produce mobile-ready goods for sale in China, and is a competitor to the Wi-Fi security standard. According to a notice issued by the Standardization Administration of China, which manages standards in various industries, as of December 1, 2001, Chinese government agencies prohibited the import, manufacture, and sale of Wi-Fi gear that did not use China's new security specification (Nan, 2001).

Presently, all Wi-Fi gear makers are to start using the Wired Authentication and Privacy Infrastructure specification, support for which is not included in current or upcoming security specifications, such as Wi-Fi Protected Access or 802.11i, developed and enforced by the Institute of Electrical and Electronics Engineers and the Wi-Fi Alliance, effectively making it incompatible with the standards developed by technology industry groups. WAPI is to be used with Wi-Fi standards in the 2.4GHz radio band, according to the standards, adding confusion to the market by presenting another obstacle for manufacturers looking to sell products in China, as WAPI adds another security specification that needs to be considered as companies start to install Wi-Fi networks.

Security experts say that, by prohibiting gear that does not use WAPI, the Chinese government is throwing an obstacle in the way of manufacturers looking to enter the Chinese market. "It would be unfortunate if we are not able to resolve this so that the China standard and others in the world can't coexist," said Dennis Eaton, chairman of the Wi-Fi Alliance. "Wi-Fi vendors may have to use special requirements for products that sell into China." According to Eaton, it is clear that WAPI does not use Advanced Encryption Standard (AES), a key encryption component for wireless networks.

RISING BUSINESS IN CHINA: EDUCATION AND MOBILE

Together with the high-speed development of the SMS market in China, a new concept of mobile education (m-education)—which combines today's hi-tech communication technologies and new patterns of education—has come out in this continuous changing time. Here we are going to analyze this interesting topic in the range of market possibility analysis, fundamental frame of the m-education, and common applications of the m-education in China. First, let us look at the market analysis.

Market Analysis for M-Education in China

As discussed in pervious sections, the volume of transactions processed in the terms of SMS was huge in the year 2003 in China; it is a definite to say that people in China grow accustomed to using SMS as one of their common communication channels to gather useful information they need. However, if we go through various types of SMS business in China, we may find that the majority of those business transactions were entertainment-based and living-information businesses, like entertainment news, weather forecasts, and so on. There are limited services that could provide people with useful information about education and learn-

ing. In this way, there must be a blank for SPs (service providers) to explore this market (Xiangliang, 2002).

Today in China, two main objectives are to improve education quality and continue to expand the volume of education. How high-tech can help China to achieve those two goals is the hot topic in China today. After analyzing four fundamentals of education—teacher, student, teaching content, and teaching media and facilities—people realize that the thing high-tech can do is change teaching media that Chinese people have used for thousands of years—that of "Chalk and Backboard." In China, correspondence courses have proved the possibility of remote education, and upcoming broadcasting and Internet education is becoming more and more popular in China. In this case, it seems that education in China is becoming more versatile, and education in China is getting beyond the limits of time and geographic location. However, within the competitive environment, the need for lifetime education is also increasing in China: people who are eager to study but have no fixed time need a more moveable platform to gain knowledge and education information. Furthermore, for instance, students who are on vacation, but who need to keep up with school, also need an effective platform to get such information. In the meantime, with a decrease in costs, a high popularization of mobile devices (e.g., cell phones) among students in China has occurred. Furthermore, the capability of networking and computing technologies improved so much that the bandwidth for communication and data processing is enough for a large number of people to query for various information about education in various formats, like SMS and MMS. In such a background, with high integration of mobile technologies and education, m-education in China seems to be blooming; the market will be very interesting and profitable for people to explore.

Basic Formats of Mobile Education in China

There are many different formats of m-education in China; some are categorized by content, some by technology used in mobile learning, and others by objects. Here we focus on technology-based mobile education formats (Huang, 2002).

SMS-Based Mobile Education

In this model, SMS—this well-developed and widely used format—could not only enable the data connection (encoded characters) among users, but also the data communication between user and Internet server. By using this feature, people could enable communication via wireless networks and the Internet; therefore, the interaction between students and the school will be activated as follows:

a. Officers (administrative side) could send teaching notes and announcements to teachers.
b. Teachers could send teaching and study notes and announcements to students.
c. Teachers or officers could read questions which are sent by their students or clients.
d. Teachers or officers could answer those questions which are sent by their students or clients.
e. Both teachers/officers and students could query for their own purposes to gather particular information.
f. Students could use SMS to check marks and credit points they have achieved.

In a word, via SMS, the information flow among students, teachers, and officers will be direct and unblocked, and people who use SMS could achieve the objective of "study any-where, and at any time." However, we still need to solve some problems. At first, the server for the connection needs to be enlarged. Then, workable and effective SMS-oriented software should be developed. After that, a customized interface should also be taken into account.

Internet-Access-Based Mobile Education

In this model, people could use their multimedia mobile devices to gather educational information from the Internet. Despite the SMS-based model, users could acquire multimedia information by using Web explorers like those installed on their devices (e.g., Internet Explorer and Netscape). With technologies such as WAP, GPRS, CDMA, and even 3G, the speed for wireless connection is become faster and faster, and either the capacity, or capability, or the file format will become larger, more advanced, and compatible. As with SMS-based mobile education, the issues of high-level service fees and software needed to support the service seem to be a long way off for the explorer.

Local Server-Based Mobile

In this model, a mobile workstation is used as its server to provide information sharing among different terminals by using IR or Bluetooth technology. In this case, with a limited number of mobile devices available with IR or Bluetooth features, a WLAN with a limited range could be set up so the data flow among those terminals could be OK, and data and information sharing will be available (Guangzuo, 2004).

Without doubt, the future of mobile learning will flourish; the market of mobile learning will be a big cake waiting for someone who is hungry.

Figure 8. Main fundamental frame of m-education (based on Guangzuo, 2004)

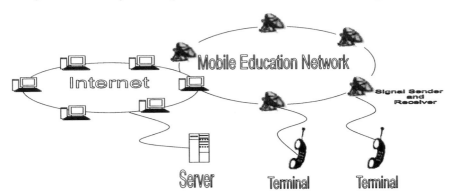

Cases for Mobile Education in China: Peking University's Mobile Virtual Campus (MVC) Project

As the leading university in China, Peking University takes an early step to implement the latest wireless technology as an operating wireless system on their campus; it is called "Mobile Virtual Campus Project." This project was based on the mobile telecommunication and computing technology, and it adopted a distributed infrastructure. As Figure 9 shows, the MVC model consists of five components: Data & Information Center, Mass Information Pro-

cessing Module, General Agency, Networking System, and End Users (Guangzuo, 2002). The Data & Information Center is the main storage device (logical and physical) of MVC, and it is linked to the university's main severs, adding in some open module especially for mobile usability. The Mass Information Processing Module processes all the services required. General Agency is a layer that contains several networking and data communication protocols that enable the users to use MVC. And the users—WLAN users, and PDA and mobile phone users—could use different Network Access Methods to visit and get useful information from MVC. We can conduct MVC mainly based on a top-down model—information was sent from the top, and the end users received the information from their agencies.

Peking University launched this project by implementing 250 wireless routers which covered main buildings like students dorms. Students and teachers could access the intranet and Internet by using their accounts, so that mobile education could become possible. Users such as students could access to their own sites to submit assignments and other academic documents, and teachers could use wireless networking to access the Web to view students'

Figure 9. The MVC infrastructure model (Guangzuo, 2002)

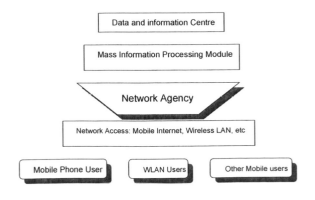

work, and view some teaching materials. From the statistics provided by the Peking University Students Union, there are more than 5,000 MVC users using the mobile networking system, and they have been satisfied by such an effective wireless-based system.

Moreover, the SMS-based platform for mobile phone and PDA communication—called Large-scale Distributed Parallel Mobile (LDPM) platform—is still under construction. It is based on the Mass Information Processing Module, and is especially for the implementation of mobile telecommunication. This system supports various kinds of software developing language like C++ and JAVA, so it is possible to develop various numbers of specific mobile systems on it. Several project research teams have finished the development of SMS-based mobile education systems, including Mobile Question-Answering System, Mobile Office Automation System, Mobile BBS System, Mobile News System, and Mobile Teaching Material Administrative System (Guangzuo, 2002). By the year 2002, the Mobile Office Automation System and Mobile Teaching Material Administrative System were, in their BETA-released version, being used at Peking University. Although some of their functions need to be improved, it is still believed that in the near future, teachers and administrators could edit short messages to send information to their target, and students could use their own mobile phones to get information from the MVC. At such time that MVC is well implemented, both teachers and students at Peking University will enjoy their digital life better.

GOVERNMENT SECTOR AND MOBILE

Whenever talking about electronic commerce or electronic business in China, many people are very interested in government, and they treat government as an independent but great market which has a lot of potential demands on the information service. In recent years, a new conception called *mobile government* came into being. As we know, the Chinese government has come a long way to achieve an electronic-technology-based working style, and officers have grown accustomed to monitoring and operating their businesses through the computer and the Internet. However, because of the immovability of LANs, which have been deployed in almost every office nationwide, officers and governors are not able to get wanted information if they left their office and without prepared notes. This has caused much inconvenience for the officers and has lowered work efficiency. Those drawbacks make the appearance of mobile government possible.

How can mobile government benefit users? Take a look at a successful case. Guangzhou, capital city of the Guangdong province in China, has launched an information project mainly concerned with how to build a mobile-technology-based platform to run services for the citizens. If the project were successful, people in Guangzhou could use their own mobile terminals to get services from the local government. China Mobile (Guangzhou Branch), leading and largest mobile telecommunication operator in China, was assigned to develop the project. After analyzing the real case, they changed four high-rated government services into a mobile model (Bing, 2004).

1. **Fanyu Project:** Fanyu government used SMS to automatically announce applicants who applied for the permits whether they passed the evaluation and audit process. This service shortened the waiting time for each individual applicant.

2. **Taxation Payment:** In order to get taxation records, a taxpayer could edit SMS

and send it to National Revenue. The system automatically feeds back the enquiry content and provides all related information.

3. **Passport and Visa Enquiry Service:** China Mobile and the Guangzhou Police Department opened service for people to enquire about their passports and Visa records; also, people could apply for a passport by sending SMS to the Immigrant and Emigrant Department of the Guangzhou Police Department.

4. **Custom Declaration Service:** After installed specialized facilities on every ship, custom officers could get geographic, audit, and video data, like location about every ship. This service can also process the custom declaration service by sending particular forms to the server. This service may avoid smuggle affairs and facilitate the processes of Customs.

As we can draw from the above, mobile government could:

- Facilitate processing speed and save a lot of time.
- Save government operating costs.
- Improve citizens' satisfaction.
- Produce a better feeling about local government.
- Improve the decision-making correction level for the government officer.

Besides two newly mobile business models described above, there are still many blank areas awaiting exploration, including mobile business in the health sector, catering sector, and so on. It is easy to foresee that a few years in the future, mobile business could infiltrate into every industry sector in China.

CONCLUSION AND FUTURE DIRECTION

We plan to extend this literature survey and experiment with how China can benefit from use and application of mobile technologies. As we can see from this chapter, all statistics about mobile business shows that figures are improving, capacity is improving, and the environment for the development of mobile business is still improving in China. Thus, it is easy to generate from this chapter that China has been adopting every leading mobile technology all over the world, such as GPRS, CDMA, 3G, and Bluetooth, and not only adopting the advanced technologies; China is trying to put those mobile technologies into practice to produce productivity in many industry sectors, such as Entertainment, Education, Government Service, and so on. It is clear that mobile technologies benefit people throughout the world, including the people of China; we believe that the future of mobile business in China will only get better and better. We will continue to create more technologies and wait for the day when it blossoms out.

REFERENCES

Beal, A., Beck, J. C., Keating, S. T., Lynch, P. D., Tu, L., Wade, M., & Wilson, J. (2001, June 4). *The future of wireless: Different than you think, bolder than you imagine.* Retrieved from http://www.accenture.com/xd/xd.asp?it=enWeb&xd=_isc/iscresearchreportabstract_134.xml

Bing, W. (2004). Mobile government: The trend of e-government. *Nanfang Daily Newspaper,* (November 17), A-13.

Blogchina. (2004). Retrieved October 5, 2004, from http://www.blogchina.com/

Chen, L., & Nath, R. (2004). A framework for mobile business applications. *International Journal of Mobile Communications, 2*(4), 368-381.

China Mobile. (n.d.). Retrieved October 5, 2004, from http://www.chinamobile.com/

Choices. (n.d.). Retrieved October 15, 2004, from http://choices.cs.uiuc.edu/MobilSec/

Christensen, G. E., & Methlie, L. B. (2003). Value creation in e-business: Exploring the impacts of Internet-enabled business conduct. *Proceedings of the 16th Bled E-Commerce Conference: E-Transformation,* Bled, Slovenia.

CLI4 Report. (2004). Retrieved October 5, 2004, from http://www.cli4.net/

CNNIC. (2003). *13th statistical survey on the Internet development in China.* Retrieved from http://www.cnnic.com.cn/download/manual/en-reports/13.pdf

CNNIC. (2004). *Report.* Retrieved October 5, 2004, from http://www.cnnic.com

Department of Informatization Promotion with the MII. (2003). *E-business development in China* (in Chinese). DIP Information Center.

Feiwen, H. (2001, September 29). Good news about the PHS local wireless service by China Telecom (in Chinese). *Communication Information News.*

Ghosh, A., & Swaminatha, T. (2001). Software security and privacy risk in mobile e-commerce. *Communications of the ACM, 44*(2).

Guangzuo, C. (2004). *Mobile education: A new way for modern education technique.*

Chongqing, China: Southeast Normal University Publishing House.

Guangzuo, C., Doyong, S., Zhangliang, Zhangbo, & Zhangbaoli. (2002a). Simulation design for wireless mobile home robot based on LDPM. *Proceedings of the Korea-China Workshop of FIRA2002,* Korea.

Guangzuo, C., Ypngsheng, H., Zhangliang, Dongyong, S., & Hu, C. (2002b). A case study on home robot based on Mobile Internet. *Proceedings of FIRA2002,* Korea.

Huang, C. (1999, May). An analysis of CDMA 3G wireless communications standards. *Proceedings of IEEE Vehicular Technology Conference* (pp. 342-345), Houston, TX.

Lange, D. B., & Serven, O. M. (1999). Good reasons for mobile agents. *Communications of the ACM.*

MII. (n.d.). Retrieved November 20, 2004, from http://www.mii.gov.cn/

Mobile Commerce World. (n.d.). Retrieved February 5, 2005, from http://www.mobilecommerceworld.com/

Pahlavan, K., & Krishnamurthy, P. (2002). *Principles of wireless networks.* Englewood Cliffs, NJ: Prentice Hall.

Rao, A. S., & Georgeff, M. P. (1995). BDI agents: From theory to practice. *Proceedings of the 1st International Conference on Multi-Agent Systems* (ICMAS-95). San Francisco: ACM Press.

SETC. (n.d.). Retrieved February 12, 2005, from http://www.setc.gov.cn/

Shu, W. (2003). Wireless LAN and its application in internal and international education. *Modern Educational Technology, 13*(5), 18-21.

Stats. (n.d.). Retrieved February 12, 2005, from http://www.stats.gov.cn/

TechTarget. (2005). Retrieved from www.techtarget.com

Walke, B. H. (2002). *Mobile radio networks: Networking, protocols and traffic performance* (2nd ed.). New York: John Wiley & Sons.

Xiangdong, W. (2001). *Report on mobile communication & Mobile Internet in China.* Retrieved from http://www.telecomvisions.com/articles/pdf/china_mobile_internet.pdf/

Xiangliang, H. (2002). *Mobile education, a great vision* (in Chinese). Retrieved February 17, 2002, from http://www.ccw.com

Yuanrong, Y. (2002, November). An adaptability e-learning system based on Web. *Proceedings of the 8th Joint International Computer Conference.* Zhejiang, China: Zhejiang University Press.

Zmijewska, A., Lawrence, E., & Steele, R. (2004). Classifying m-payments—A user-centric model. *Proceedings of the 3rd International Conference on Mobile Business* (ICMB), New York.

Section VII

Application

Chapter XXXV
Developing Smart Client Mobile Applications

Jason Gan
University of Technology, Australia

ABSTRACT

This chapter examines the convergence of mobile technologies based on smart client architecture. To improve the usability and accessibility of mobile applications and services, the smart client architecture extends the capabilities of the mobile computing platform with support for mutimodal interfaces, smart client database and synchronization, presence awareness, location awareness and identity management. Its broad impact on business communication and productivity is highlighted as a tangible benefit.

INTRODUCTION

In the highly competitive mobile market, a key differentiator is provided by improving the user experience of the mobile application. To improve the user experience, common usability and accessibility problems in mobile applications can be mitigated by multimodal interfaces and smart client architecture. For instance, the provision of multimodal interfaces for browser-based applications can help to overcome the limitations of small viewing areas and input options, whereas smart clients based on rich application interfaces can be utilized to push processing load onto the mobile device. Furthermore, smart clients that enable presence, context sensitivity, location awareness, and real-time collaboration promise a new paradigm for mobile communications, delivering far richer, dynamic user experiences.

SMART CLIENT

As mobile enabling technologies advance in capability, affordability, and availability, users expect improved design of mobile devices that will leverage the advances and convergence in technology and the Internet to deliver richer applications and value-added mobile services (a.k.a., m-services). A key enabling technology for delivering on the promise of mobile applications with high levels of functionality, performance, flexibility, and integration is the *smart client*. This is a type of application model that bridges the gap between the thick and thin client models, providing the responsiveness and adaptability of a thick client model with the manageability of a thin client.

Dave Hill, from the Microsoft .NET Enterprise Architecture team, defines five characteristics of a smart client application (2004):

1. **Utilizes Local Resources:** Smart clients exploit local resources such as hardware for storage, processing, or data capture to deliver a richer user experience.
2. **Connected:** Smart clients are ready to connect and exchange data with various systems across the enterprise.
3. **Off-Line Capable:** Off-line capability using local caching and processing enable operation during periods of disconnection or intermittent network connectivity. Smart clients can send data in the background, resulting in greater responsiveness in the user interface.
4. **Intelligent Install and Update:** The smart client interface allows the remote update of the smartphone software to repair bugs, change characteristics, or incorporate new features.
5. **Client Device Flexibility:** Smart client applications support multiple versions that target specific device type and functionality.

The smart client architecture supports multimodality, data integration, Bluetooth interoperability, presence awareness, location awareness, and identity management. Each of these features extends serviceable functionality to the mobile application, from voice-activated commands to authentication and non-repudiation services. Moreover, the integration of serviceable functionality promises to deliver rich user experiences. For example, the voice-activated smart wireless device will automatically connect, authenticate, and show the identity and location of the receiver.

The convergence of presence, location, and identity management is an emergent technology integrating the services that support a secure mobile network and an online community environment with applications that facilitate information retrieval, communications, dating, gambling, financial management, trading, paying bills, games, and entertainment. As the mobile market is highly competitive and dynamic, and driven by the mass market demand for high-performance m-applications, the impact of technology convergence highlights the need for common industry standards.

INDUSTRY STANDARDS FOR SMART CLIENT MOBILE APPLICATIONS

As the specifications for smart client mobile application interfaces are complex, it is necessary for wireless developers to adhere to industry standards. The Mobile Industry Processor Interface (MIPI) Alliance is a non-profit organization that spearheaded the initiative of industry specifications for smartphones and application-rich mobile devices.

Wireless and embedded software developers have a choice between Microsoft .NET and Sun Java development frameworks and runtime environments for designing and delivering next-generation mobile applications.

The Microsoft .NET Compact Framework is a subset of the developer software for PCs and servers that allows powerful .NET applications to run on handheld computers, and specifically supports: Pocket PC, embedded solutions running on Windows CE .NET for smart mobile devices, and Microsoft Smartphone 2002. The .NET Compact Framework can be extended to support additional mobile device interfaces. The inclusion of SQL Server CE 2.0 provides developers with a powerful, local relational database for creating dynamic, client-side mobile applications with database replication that enables remote devices to edit data in parallel. SQL Server 2000 supports two methods of replicating information from a back-end master to a remote client database: Remote Data Access (RDA) and merge replication (Thews, 2003).

The Sun Java platform includes the Micro Edition specification for building rich and smart client applications for mobile and embedded devices.

SMART CLIENT AND MULTIMODALITY

Smart client mobile applications equipped with multimodality and speech capabilities represent the next generation in portable office communications and wearable computer technology. Multimodality is defined as the optional presentation of the same information content in more than one sensory mode (European Communications Standards Institute, 2003). The multimodal interface on a multiple context-aware device enables interaction through different communication channels (a.k.a., modalities) to overcome the inherent input/output limitations of mobile devices. For example, a user

might use voice-only commands rather than the standard keypad input, while audio output can deliver additional information that cannot fit on the screen. The multimodal information is processed by the interpreter component of a multimodal host server.

Multimodal integration is the combination of different modalities to form a flexible user interface. There are seven types of modalities: visual, auditory, tactile, olfactory, gustatory, vestibular, and proprioception. Of these, visual and auditory are the most commonly used in m-applications. However, users who are blind and/or mute stand to benefit from the availability of tactile feedback and gesture modalities that address their physical disabilities.

SMART CLIENT DATABASE AND SYNCHRONIZATION

Reliable, secure, and immediate access to enterprise data irrespective of time and place is paramount in a time-critical business application, and it is especially critical in completing an online session—for example, when you have to close a bid on time. The problem with wireless technology is that data access and connectivity are intermittently connected and unreliable due to dropped connections, coverage issues, low bandwidth, and high latency. Smart client off-line access to data using a localized mobile database and synchronization with a central database (a store-and-forward mechanism) can help reduce and eliminate the performance bottlenecks related to slow and unreliable networks, and thus improve the user experience.

Besides improving the user experience, mobile applications can also affect the work environment. In particular, the collaborations of disparate teams stand to benefit from presence awareness.

PRESENCE AWARENESS

A study on disparate software project teams discussed the potential impact on productivity and the bottom line from leveraging presence and collaborative technologies in multi-site environments (Herbsleb, Mockus, Finholt, & Grinter, 2001, p. 9).

Presence is defined as a collection of real-time data describing the ability and willingness of a user to communicate across specific media and devices (Schneyderman, 2004). Presence awareness, a vital property of instant messaging applications, allows users to know when other users in a community are online and willing to exchange messages, what devices can be used for communications, and the real-time status of these devices. This can result in time and cost savings and improved productivity in many enterprise environments. For example, customer service environments stand to benefit from decreased operational costs and the ability to connect knowledge experts in real time across geographical locations and time zones. The adoption of presence technology into the workplace enables disparate workers to connect and communicate more efficiently by overcoming the lack of context and absence of informal communication (Herbsleb, Atkins, Boyer, Handel, & Finholt, 2002). A presence-based publish-and-subscribe channel can be deployed for discovery of available managers in workflow collaborations and for push content delivery.

It is important for developers to address the security and privacy concerns regarding presence technology. Building security and privacy controls into the presence-enabled device allows such features as access control, visibility, message blocking, and message encryption. Privacy policies restrict communication channels and prevent unsolicited conversations or uninvited listening. Biometric scanners built into the presence-equipped device help to prevent such exploits as identity masquerading and spoofing. In addition, each receiving device would be configured to transmit its location and identity information to the mobile network.

LOCATION AWARENESS AND PERVASIVE COMPUTING

Location awareness in smart mobile applications is provided by mobile positioning technology such as the Global Positioning System (GPS). The GPS is a satellite navigation system that allows a mobile device with a GPS receiver to pinpoint a location on Earth by measuring the distances from a number of satellites simultaneously. Integration with GPS technology provides a mobile device with a pervasive-computing interface to location-based services (LBSs) such as emergency assistance and personal navigation. Assisted GPS (AGPS) describes a mobile positioning system that consists of the integrated GPS receiver and network resources such as an assistance server and reference network. The assistance server communicates with the GPS receiver via the cellular link and accesses data from the reference network, resulting in boosted performance of the receiver.

IDENTITY MANAGEMENT

Smart client mobile applications may require that access privileges to specific resources are granted only to users who have the right to use those resources. In an e-business system, access to certain resources is restricted to those who can supply the proper identity credentials by a process called authentication. Identity management is the federation of trusted endpoints that are able to authenticate each other's presence at these endpoints with the intention of having a secure transaction (Smith, 2002). The banking and finance industry is a strong driver for this technology to reduce online fraud. Moreover, businesses and customers stand to benefit from expedited billing and payments. As identity and trust form the basis of success-ful e- and m-business, there is a strong demand from business and government sectors for effi-cient mobile integration of identity management services that will provide the end user with a more satisfactory online experience, enriched with higher levels of personalization, security, and control over identity information.

CONCLUSION AND FUTURE DIRECTION

The smart client provides an application model for developing richer, more responsive, and more usable mobile applications. Key areas of technological development based on the smart client are multimodality, presence awareness, location awareness, and identity management. These technologies describe a future to be defined by voice-activated systems, real-time collaboration, location-based services for in-formation or emergencies, and mobile integra-tion of identity services. Developing smart mobile applications that leverage the advances and convergence in technology is a step closer to realizing the full potential of a wireless Internet.

REFERENCES

Antonopoulos, A. M. (2004, May 10). Location and presence take identity management to the next level. *Network World Data Center News-letter*. Retrieved December 28, 2004, from http://www.nwfusion.com/newsletters/datacenter/2004/1004datacenter1.html

European Communications Standards Institute. (2003). Human factors (HF); multimodal inter-action, communication and navigation guide-lines. *ETSI EG 202 191, 1.1.1*(August).

Herbsleb, J. D., Atkins, D. L., Boyer, D. G., Handel, M., & Finholt, T. A. (2002). Introduc-ing instant messaging and chat in the work-place. *Proceedings of the SIGCHI Confer-ence on Human Factors in Computing Sys-tems: Changing our World, Changing Our-selves*. Minneapolis, MN. New York: ACM Press.

Herbsleb, J. D., Mockus, A., Finholt, T. A., & Grinter, R. E. (2001, May 12-19). An empirical study of global software development. *Pro-ceedings of the 23rd International Confer-ence on Software Engineering* (ICSE'01), Toronto, Canada. Retrieved from http://www-2.cs.cmu.edu/~jdh/collaboratory/research_papers/ICSE_01_final(2).pdf

Hill, D. (2004). *What is a smart client any-way?* Retrieved December 12, 2004, from http://weblogs.asp.net/dphill/articles/66300.aspx

Schneyderman, A. (2004). *Presence in mobile VoIP networks*. Retrieved from http://www.tmcnet.com/voip/0904/featureshneyderman.htm

Smith, R. (2002). *Identity management—give me liberty or give me passport?* Retrieved December 28, 2004, from http://www.giac.org/practical/GSEC/Robert_Smith_GSEC.pdf

Thews, D. (2003, October). Create mobile database apps. *Visual Studio Magazine*. Re-trieved January 23, 2005, from http://www.ftponline.com/vsm/2003_10/magazine/features/thews/

Chapter XXXVI
Ontology–Based Information Retrieval Under a Mobile Business Environment

Sheng-Uei Guan
Brunel University, UK

ABSTRACT

The proposed OntoQuery system in the m-commerce agent framework investigates new methodologies for efficient query formation for product databases. It also forms new methodologies for effective information retrieval. The query formation approach implemented takes advantage of the tree pathway structure in ontology, as well as keywords, to form queries visually and efficiently. The proposed information retrieval system uses genetic algorithms, and is computationally more effective than iterative methods such as relevance feedback. Synonyms are used to mutate earlier queries. Mutation is used together with query optimization techniques like query restructuring by logical terms and numerical constraints replacement. The fitness function of the genetic algorithm is defined by three elements: (1) number of documents retrieved, (2) quality of documents, and (3) correlation of queries. The number and quality of documents retrieved give the basic strength of a mutated query, while query correlation accounts for mutated query ambiguities.

INTRODUCTION

Mobile computing will be the next buzzword of the twenty-first century. Presently, consumers demand personalized wireless computing services while they are mobile. This infantile paradigm of mobile computing is opening up new markets. Corporate power users who are at the cutting edge of technology are always armed with an arsenal of mobile equipment.

CURRENT SITUATION AND MOTIVATION OF RESEARCH

According to Reuters and NUA Internet surveys in 1997 (Wieerhold, Stefan, Sergey, Prasenjit, Yuhui, Sichun et al., 2000), about 1.81% people worldwide surf the Internet for information daily. In the same year, according to surveys done by the Forrester Research (Wieerhold et al., 2000) and the Yankee Group, there was a significant increase of online retail sales, from $600 million in 1996 to more than $2 billion in 1997. This value reached $282 billion in 2000 and is still increasing. With the exponentially growing number of Internet users over these few years, the International Data Corporation (IDC) expected an increase to $4.3 trillion by 2005. Thus, as can be seen, trading online has become increasingly important to the commercial world. It is inevitable that e-commerce will be the next strategy that companies will adopt.

At the same time, with the introduction of new technologies such as WAP, HSCSD, GPRS, UMTS, and Bluetooth, together with new and personalized applications, it is believed that the e-commerce arena will sooner or later merge its applications with handheld devices to create more opportunities for the birth of mobile commerce. In fact, research from the IDC expected the mobile portal to reach 55 million users by 2005.

However, according to the IDC, there is a 26% drop in the sales of handheld devices in the first quarter of 2002. One of the reasons why the potential of mobile commerce is largely unrealized to date is because a single killer application that can attract wireless users to use wireless services still does not exist. According to a recent survey by Gartner, Inc., besides the importance of coverage of wireless network and pricing issues, the wireless Internet and data services is the next crucial factor that attracts users to use wireless service. As such, there is a need to improve data services over the wireless network. One of these services is the information retrieval service.

Most electronic product information retrieval systems are still not efficient enough to cater to the increasing needs of customers. Typically, as product information in the Web soared eminently, reusing and sharing of product information has become extremely important. This is especially true in the m-commerce arena where the bandwidth of mobile devices is low and large data would not be possible. Thus, the discovery of new information retrieval techniques is inevitable. Also, observations and studies have shown that the average user often selects inappropriate information retrieval resources and uses them inefficiently and ineffectively. People seem to be content to retrieve any information on their topics, regardless of quality. Few people currently recognize the need for improving their information retrieval skills. Hence, there is a need to simplify the way people form queries to retrieve information.

OBJECTIVES AND RESEARCH CONTRIBUTION

The main objective of this chapter is to improve information retrieval services for the m-commerce arena. After considering the flaws in current information retrieval systems, this chapter proposes a methodology for efficient query formation for product databases in m-commerce. In addition, this chapter proposes a methodology for effective information retrieval systems, which includes the evaluation of retrieved documents to enhance the quality of results that are obtained from product searches.

This chapter discusses the usage of ontology to create an efficient environment for m-commerce users to form queries. The establishment of a method that combines keyword searches while using ontology to perform query formation tasks further allows a more flexible m-commerce environment for users. Also, with the use of genetic algorithm, it is hoped that query effectiveness can be achieved, at the same time saving computational time. In this chapter, new ways of defining the fitness function of the genetic algorithm are also explored.

DEFINITION OF ONTOLOGY

In philosophy, ontology is a theory about the nature of existence, of what types of things exist; ontology as a discipline studies such theories. In the world of artificial intelligence, ontology (Fensel, 2000; Braga, Werner, & Mattosso, 2000; Hendler, 2001) is defined as a design of a conceptualization to be reused across multiple applications. A conceptualization is a set of concepts, relations, objects, and constraints that define a semantic model of some domain of interest. In other words, ontology is like the structure that describes or encodes the conceptualization in any relational aspect.

QUERY FORMATIONS

Query Formation Using Keywords

The traditional way of searching is to use keywords as a guide to form and process queries. Query formation using keywords is just fine if the user knows exactly what he or she is looking for. However, it may be difficult or impossible to use keyword-based search when the vocabulary of the subject field is totally unfamiliar to the user. Furthermore, using the same keywords in a different order or at a different time might change the number as well as the order of sites listed. As can be seen, there is usually some form of uncertainty when forming a query using keywords. In addition,

because a word can contain different meanings, ambiguities will arise due to missing context. Moreover, no single engine searches the entire Internet, so it is often necessary to search several engines. Essentially, the element that is lacking in keyword-based queries is the user's sense of linkage between multiple keywords and context. Also, if too many keywords are used, the constraints may become too tight to retrieve anything.

Query Formation Based on Agent Ontology in Bioinformatics

A relatively new approach is to form a query using ontology where ontology serves as a context and structuring mechanism for keywords. Functional bioinformatics is an emerging sub-field of bioinformatics that is concerned with ontology and algorithms for computing with biological functions (Karp, 2000). It uses regulatory pathways to define ontology structures. In graph theory, a regulatory pathway is usually represented as an oriented graph with vertices corresponding to substances and edges corresponding to interactions (Rzhetsky, Tomohiro, Sergey, Shawn, Michael, Sabrina, et al. et al., n.d.). This is similar to what we find in tree ontology—the nodes represent the substances and the links between nodes represent the interactions.

Functional bioinformatics focus on large-scale computational problems such as problems involving complete metabolic networks and genetic networks (Karp, 2000). This allows information retrieval algorithms to evaluate queries easily (McGuinness, 1998). Also, functional bioinformatics is concerned with representing, visualizing, and computing functional descriptions of individual genes and gene products (Karp, 2000). Having such an archive with organized ontological structures, ontology can be used to design knowledge portals for manual browsing (McGuinness, 1998). Thus, any user can easily generate a query by just a few mouse clicks on the displayed ontology and then submit it to the database to get required information. Many Web catalogs, such as Yahoo, employ ontology in order to organize their contents (McGuinness, 1998). This type of query formulation would not be possible without semantic representation of the ontology. Ontology often plays the role of controlled indexing languages. Moreover they may be used in assisting the query formulation process, or for expanding queries with synonyms, hyponyms, and related terms in order to improve recall (McGuinness, 1998). Ontology can grow and shrink as necessary based on the context. In different contexts, parts of an ontology can be hidden or another made visible so that a new view of the same information space can be generated to suit a certain group of audience. A variety of smaller ontologies can be created by experts from various fields independently of each other, and then later can be merged with little effort to create a bigger ontology that can serve a different or larger audience.

INFORMATION RETRIEVAL

Relevance Feedback

Relevance feedback (Salton, 1989) is still the main technique for query modification. This technique has been investigated for more than 20 years in various information retrieval models, such as the probabilistic model (Robertson & Sparck Jones, 1976; Robertson & Walker, 1997; Haines & Croft, 1993) and vector space model (Salton, 1970; Boughanem, Chrisment, & Tamine, 1999; Salton, 1989). Relevance feedback is based on randomly changing the set of query terms as well as the weights associated with these terms according to the document retrieved and judged during the initial search. Conventionally, retrieved documents

are not evaluated. This means that a user might end up getting junk instead of the things he wants. Also, when there are too many documents available, the user has to filter them manually. These problems are fundamental to the motivation of this research.

Recently, many scientists were gathering a genetic approach to solve the information retrieval problem. However, there is still a great deal of room for research on genetic algorithms as there are many ways the fitness function can be defined.

Genetic Algorithm

A lot of research has been done on how genetic algorithm can be used in information retrieval. One popular approach is the restructuring of queries. Restructuring queries is necessary to improve the efficiency and effectiveness of the queries formed. This boils down to the fundamental concepts in relevance feedback. Hence, genetic algorithm actually extends the concepts of relevance feedback. The difference is that genetic algorithm uses more than one query and compares the fitness among these queries. The fittest query will survive in the end. This is much more effective than when only one query is used where hill-climbing search is used (Boughanem et al., 1999). Here, mutation and crossover take place during restructuring.

Yang and Korfhage (1994) use genetic algorithms to search the space of possible queries to generate better queries based on relevance feedback for the vector space model, and they obtained remarkable results. Kraft (Salton, 1989) explored to extend the concepts to deal with Boolean queries. Another approach uses evolutionary algorithms to evolve agents instead of queries (Kouichi, Toshihiro, & Hiroshi, 1999). These agents contain search parameters that will search all Web databases and are tuned up by genetic algorithms.

In this chapter, the focus is to extend the concepts of using genetic algorithms in query restructuring. The difference lies in the type of restructures. Basically, three types of restructures—using synonyms, logical operators, and numerical constraints—are used in the search process to enhance query effectiveness. Each type of restructure is rendered under different situations depending on the suitability in that situation.

The Fitness Function

Salton and McGill (1983) consider that the two ultimate measures of query fitness are namely precision and recall. Precision is the percentage of documents retrieved that are relevant, while recall measures the percentage of the relevant documents retrieved (Kraft, Petry, Buckles, & Sadasivan, 1994; Salton & McGill, 1983). These two tend to be inversely proportional so that one is traded for one another in most situations. They are complementary and competitive (Kraft et al., 1994). A more general query retrieves more documents, and recall increases because of more relevant documents retrieved. As the query becomes less general, fewer documents are retrieved and thus precision increases. Therefore, usually these two measures are not used directly. In Kraft et al. (1994), two arbitrary constant weights were used to balance the fitness function.

Another measure is the average search length (Losee, 1991). The average search length is the average number of documents or text fragments examined in moving down a ranked list of documents until arriving at the average position of a relevant document (Losee, 1988, 2000). Evaluating the performance of a filtering or retrieval process with the average search length provides a single number measure of performance. Also, it is capable of being predicted analytically (Losee, 2000) and is easily

understood by a system's end users. This is unlike most of the other retrieval and filtering measures based on precision and recall that are used to evaluate retrieval systems.

Another fitness function is the average maximum parse length (Losee, 2000). The average maximum parse length is the average (over a set of sentences) of the largest number of terms in a parse for each sentence. There are also measures that combine both the average search length and the average maximum parse length. The motivation for it is that having different genes producing the same average search length value is very common, and adding the average maximum parse length breaks the ties that exist when using average search length alone as the fitness function (Losee, 2000). When two genes with identical average search lengths are compared, the gene with the larger average maximum parse length is selected. This computational method is useful when the average search length is not a very good measure of the quality of parsing performance due to the limited improvements obtained with disambiguation. In Losee (2000), a useful fitness measure was obtained by weighting the two values such that that one-hundredth of the average maximum parse length was added to the negation of the average search length. The greater the value for this combined function, the fitter the gene.

In review to the present ways of defining the fitness function for a given genetic algorithm, the relevance of the documents retrieved has been greatly emphasized. Typically, present methods only dealt with the relevance of the document retrieved. This is reasonable but inefficient, because it is rather difficult to indicate the relevance of a document when the number of documents could be very large. In fact, relevancy can also exist at the query stage. This chapter measures the relevance of queries instead of documents retrieved. When this happens, the efficiency will improve significantly as the number of queries will be much smaller than the number of documents retrieved. Of course, it might be true that a relevant query might not retrieve a relevant document. However, in the context of this chapter, where the retrieved documents are products from product databases, it is reasonable to say that almost all the documents retrieved are relevant to the query formed. This is because most product databases have rather fixed formats, unlike documents retrieved from the Internet where the variations of formats can be infinite. In addition, when a query is formed from a product ontology, it is obvious that the linkage between the terms has rid off much of the irrelevancy of the documents retrieved. Thus, the fitness functions used in the present methods are not used in this chapter.

THE PROPOSED APPROACHES

Combining Keyword Queries with Ontology

Both keyword- and ontology-based approaches have their advantages and disadvantages. Ontology provides the structure, context, and visual aid, while keyword provides a direct search mechanism. Thus, by combining keyword-based queries with ontology, it is possible to achieve a better and more effective query formation. Basically, most would think of a parallel combination that means the keyword approach can be accessed as a backup method to the ontology approach. For example, when a user cannot find a term in the ontology, he may still be able to use keywords to search whatever is not covered by the ontology.

The parallel solution can be considered in terms of its coverage, but still lacks efficiency in some ways. For example, as the ontology gets larger, finding a term becomes a chore and slows down everything. To tackle this problem,

a serial combination will be needed. This is the part where most research contents can be found, as this method—although similar—does differ from the most current research.

On the whole, before the ontology terms are accessed to form the queries, there will be a keyword search to find the required ontology term. For example, "ps2" can be hidden in the node "mouse" when presented in the ontology. The user will not be able to know where "ps2" can be found intuitively without eyeballing the ontology. With the help of keyword search, the term "ps2" can be found easily.

Query Restructuring

In forming queries, there can be a great chance that the vocabulary used by the user to describe a query does not exactly match the vocabulary used by a query system. This will result in getting insufficient information. For example, a person looking for televisions might key in "television" for his search, but the product ontology might describe televisions in a node called "tv" or "tv set." Intuitively, "television" is equal to "tv" and "tv set." Therefore, restructuring dealing with domain ontology relationships might be useful. These relationships involve semantic links such as hyponyms and synonyms (Braga et al., 2000). Here, using synonyms is an adequate option to restructure queries.

Also, one major problem about information retrieval is that either too little or too much information is retrieved when making a query. When too little information is retrieved, it is logical to relax the constraints of the query. Thus the use of synonym or hyponym might be necessary. However, this approach has a major disadvantage. By relaxing the constraints of a query using synonym or hyponym to increase the number of documents retrieved, one could actually deface the meaning of the original query such that it could drift away from the user's intention. This concern can be alleviated by having user feedback along the process; there is still much to research on the use of the synonym or hyponym. Another new approach that we have considered in this research is to relax constraints step by step. This option can better eliminate the chances of constructing far-fetched queries from the use of genetic algorithm. For example, step by step means to relax numerical constraints in steps of 5% or 10%. For instance, price<2000 becomes price<2200. In addition, instead of relaxing constraints, this approach can also tighten constraints. The only limitation that this option has is that it is only applicable to numerical values. When the constraints are words, this method might not work.

Of course, one can mutate wordy constraints by removing or truncating a few letters or even words from them. But, the same problem about ambiguities will set in.

The Fitness Function

As mentioned earlier, since the relevancy focus has been shifted to the query stage, and product documents are mostly relevant when the query is relevant, it is not necessary to include measurements like precision or recall, which measure relevance at the document-retrieved stage. In this chapter, a totally new design of the fitness function is schemed.

Instead of favoring a query by giving it credits for relevancy, it is a good idea to work from the irrelevancy of the query by giving it demerits since most documents are relevant. Here, the irrelevancy of the query will set in after the mutation of the queries is applied when using genetic algorithm. This irrelevancy will form one of the elements in the fitness function.

Also, addressing the issues of finding too many or too few documents, it is functional to include a weight element that measures the

number of product documents that a query can obtain in its fitness. Another weight element is the quality of the product documents that a query can retrieve. This element should be included simply because it is rather easy to measure the quality of product documents than to measure other documents from the Internet because of some numerical values. In making queries on products, it is common to encounter numerical constraints where the values are measurable. Of course, this element is limited such that it can only measure the fitness of queries with numerical constraints. The details of the complete design of this fitness function are discussed later in this chapter.

PROTOTYPE DESIGN AND IMPLEMENTATION

Overall Architecture

Basically, the architecture of the research problem lies largely on the design of the agents. It is important to know exactly what the agents are that will be needed before drilling into the problem. Figure 1 shows a system architecture of the relationships among the modular agents.

Each mobile user can activate the Main Agent of the OntoQuery system. On the whole, three major application agents—the Synonyms Editor Agent, the Ontology Editor Agent and the Query Formation Agent—are interfaced to the Main Agent. The Synonyms Editor Agent merely interacts with the Synonyms Agent by editing the semantics of the synonyms. The Ontology Editor Agent interfaces with the Product Ontology Agent to allow the user to edit his own product ontology. At the same time it saves the new ontology terms with the help of the Synonyms Agent to prepare the synonyms for editing. The Query Formation Agent uses the Product Ontology Agent to aid the user to form queries. It also calls for the Information

Figure 1. System architecture among agents

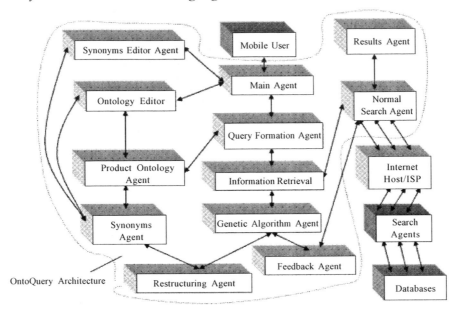

Retrieval Agent, which is interfaced directly to the Normal Search Agent and Genetic Algorithm Agent. The Genetic Algorithm Agent employs the Restructuring Agent to restructure query population according to the feedback and requirements received from the Feedback Agent. Meanwhile, the Restructuring Agent does query restructuring with the help of the Synonyms Agent. The Feedback Agent also links up the Normal Search Agent, which will submit queries to retrieve information wirelessly from the databases in the Internet through the Internet service provider. In addition, there will be Search Agents residing at the host or the Internet service provider, which provides the information retrieval service its connected databases. Here, many mobile users are connected to the Internet via the OntoQuery architecture, while the Internet connects to many databases. The Normal Search Agent will then display the retrieval results via the Results Agent.

Query Formation Using Ontology

Query formation is a rather straightforward application; there will be multiple requirements just like present query systems in the industry. The challenges here lie in the way queries are created and the format in which they will be

Figure 2. Illustration of using ontology to form queries

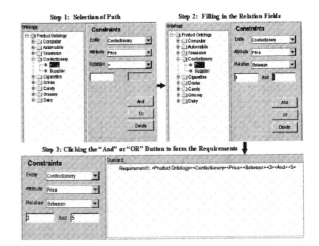

Figure 3. Illustration of the sequence of events for finding ontology terms

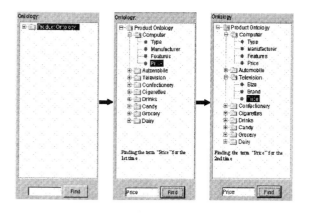

Figure 3. Illustration of the sequence of events for finding ontology terms

sent to the Search Agents. Here, query formation will be done with the aid of tree ontology. Multiple requirements help form a complete query. An illustration of the query formation process is shown in Figure 2.

As can be seen from this illustration, using ontology to form a query has become a simple, three-step procedure. This helps the user save several steps by forming a query using the ontology path that is selected. Thus, it can be claimed that forming queries using ontology is more efficient than using keywords.

Combining Keywords and Ontology

The design of parallel combination is rather straightforward. An ontology does not cover everything. Thus, besides having ontology for the user to click on when forming a query, there should be some fields present for the user to fill in. When these fields are being filled in, they can either replace the use of ontology either partially or completely.

For a serial combination, keywords are used to look for ontology terms in the ontology. This is necessary because when the ontology is too large, search for an ontology term by manual clicking becomes difficult. Thus, there would be a field that allows the user to highlight the terms in the ontology itself. This process is similar to how we search for words in Microsoft Word documents, except that now the search is to transverse in an ontology tree structure, as shown in Figure 3. From this illustration, it can be seen that using keywords to search ontology terms in the ontology creates an efficient environment and context for the user. It increases the chances for getting results from a tedious search of ontology terms, especially when the product ontology gets bigger.

Information Retrieval

Using the query formed by the query formation application, an application searches the databases to retrieve information. Intuitively, this application would first do a normal search before allowing the user to proceed with a genetic algorithm search. This is because a

Figure 4. Flowchart of the searching process system architecture among agents

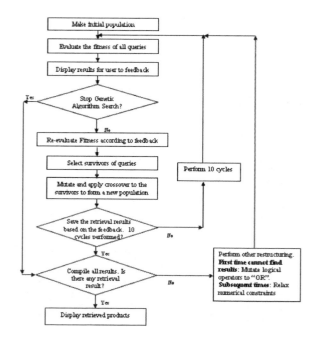

genetic algorithm search would definitely take a much longer time than a normal search because of its expensive iterations. The retrieval results are presented to the user and if he is not satisfied, he can then choose to proceed with a genetic algorithm search.

Genetic Algorithm

If the user requests the use of a genetic algorithm, the system will request some input from the user to perform genetic algorithm computation. The system then creates a population of queries from the original query. Figure 4 shows the flowchart of the sequence of events.

Basically, genetic algorithm will mutate the queries according to the synonyms of the terms in the ontology. This is necessary in order to retrieve all the relevant documents that are related to the query. Thus, there will be a lookup table of synonyms. This creates semantics to relate the synonyms with the ontology terms. Each term is a gene, and each query forms a chromosome.

The Fitness Function

The main concern of using genetic algorithm is the design of the fitness function. In this application, three major elements are used to define the fitness function. They are the fitness of the number of documents retrieved (f_d), the fitness of the average quality of the query results (f_q), and the overall correlation for the query (f_r). The fitness of each chromosome is calculated as follows:

$$Fitness = |f_r.(f_d + f_q)|$$

$| . |$ indicates that the fitness function is normalized to form a population distribution function.

The calculation of the value of f_d is not as straightforward. Since, the drawback is getting either too few or too many documents, the user should be able to specify the ideal number of documents (i). Typically, if the user does not know the value of i, the default value will be 20. Using the value of i as the mean value, two 'band pass filters'-like functions, namely the triangular and Gaussian functions, are used to create a more flexible mapping from number of documents retrieved (d) to f_d. The triangular function gives a constant drop in "gain" (decrease in fitness) from the 'center frequency' (the mean). This function is good when the user wants to give an equal amount of demerits for every document that is away from his expected or ideal number. By specifying the value of the gradient for the right side of the triangle, the user can determine how many demerits to give for every document that exceeds his expected number. When the value of this gradient is 0, the triangular function will look like a "high pass filter". The Gaussian function is a more robust or high-ordered "band pass" such that its "bandwidth" could be specified by the value of the standard deviation. Thus, this function is useful when the user wants to give heavy demerits to queries that do not fall near his expected or ideal number. Also, when the standard deviation is large, the function will look like a 'low pass filter'.

Only requirements that are specified by numerical constraints will have the value f_q. For example, documents retrieved according to *price<2000* have numerical results and would retrieve results under the columns of price while documents retrieved according to *brand=Sony* do not *have* numerical results under the columns of brand. Here, it is required that the numerical values are summed up and averaged. Then, they are normalized. In addition, there is another consideration. The signs "<" and ">" formed during the query indicate the direction towards which the quality is favored. For example, when encountered with a "<" sign, it is logical to give higher fitness to the chromosomes with smaller values in the numerical results. Here, a reciprocal function (1/x) is used before normalization takes place.

The next interesting portion that contributes to the fitness function is the correlation of the synonyms (f_r) with the ontology terms. A value from 0 to 1 is assigned to each relation between the ontology terms and their synonyms. For example, the discrete correlation between the terms "Television" and "TV" can be 0.9, while the discrete correlation between the terms "Price" and "SP" can be 0.7. When a requirement is <Television><Price><<><2000>, the value or f_r will be the product of all the discrete correlations. In this case, the value of f_r is 0.63. Also, the user should be able to edit the correlation values to his preference. In addition, when there are many requirements in the query, these requirements will be linked with an "OR" or "AND" term. In this case, multiplying these correlation values together would not be rational. This is because, when too many terms are different, the value of f_r would become very small. In order to alleviate this problem, using maximum and minimum approaches would be more suitable than the product approach. For example, when there are two requirements A and B with correlations f_{r1} and f_{r2} respectively, a query of A "OR" B would give an overall correlation of $\max(f_{r1}, f_{r2})$. Conversely, a query of A "AND" B would return an overall correlation of $\min(f_{r1}, f_{r2})$.

In a multi-objective requirement, the long term will be split as two or more terms whereby the paths of the tree become longer. For example, <Alarm Clock><Selling Price><<5> is single objective while <Clock><Alarm><Price> <Selling><<5> is multi-objective. In fact, the maximum and minimum approaches should be combined with the product approach if the requirement is multi-objective. This is because if we multiply the correlation of <Clock><Alarm> and <Timepiece><Ring> because both terms are different, the problem of small r-value will surface again. However, in this chapter, the requirements are single objective. Thus, this combination of approach is not employed.

Mutation and Crossover

The concept of mutation is to replace some terms with synonyms when parsing the results. By doing so, there will be a lesser chance that some information is missed because the query will be more comprehensive. Basically the mutants are the terms that are included in each query. These terms are mutated randomly according to the synonyms so that new populations will be formed. Crossover will only be interchanging the different genes between two different chromosomes. A one-point crossover will be performed. This is also done randomly.

Feedback and Selection of Survival

The survivors are selected according to their overall fitness in the roulette-wheel selection

Figure 5. Screenshot of a feedback frame

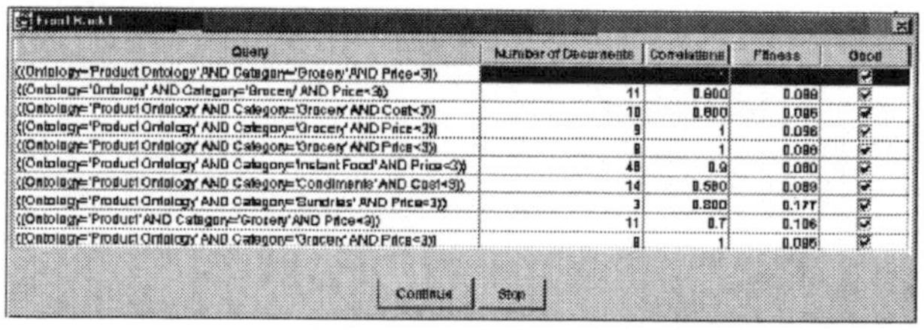

Figure 6. Illustration of mutating numerical constraints

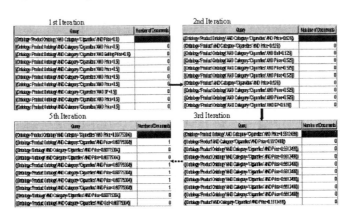

manner. However, before this is done, the system will prompt for feedback from the user. The feedback will show the user some quality of each query. From this quality metric, the user may choose to kill queries that do not meet his requirements. Figure 5 shows a screenshot of a feedback presented to the user.

Those selected query results from user feedback will have their fitness re-evaluated and then go through a selection of survival. In this way, feedback will more effectively serve the user's needs, such that it will never display results that are not selected by the user. In addition, if the user is satisfied with the results, he can choose to end the genetic algorithm by clicking on the "Stop" button. In this way, he can look at the retrieved results immediately. The final result will show the product items that were retrieved according to those queries that evolved in each round and survived the feedback selection.

DISCUSSION OF USING FEEDBACK

There is a trade-off between accurate feedback and computational time. As our feedback displays the fitness results of the user, it gains accuracy but loses efficiency in terms of computational time. When computational time is more important than accuracy, instead of displaying the fitness results after evolution, it might be better if the system just displays the mutated queries before evolution. Thus, the feedback could be provided earlier to the stage when the queries are just mutated. Also, if the combination of the two types of feedback is to be adopted, early feedback can kill some queries first and the survivors will go through the second feedback.

Figure 7. An OntoQuery design using scrollbars for handheld devices

Other Restructures

Besides replacing query terms using synonyms, logical operators and numerical constraints can be restructured. This is necessary when the system cannot retrieve any document with its synonyms. However, it is still necessary to inform the user and consult his permission to carry on with this restructuring.

The system will mutate "AND" operators that link each requirement to "OR" operators. Logically, no documents will be retrieved if two far-fetched requirements are submitted as a query. Thus, mutating a logical "AND" term to an "OR" term can be a good choice to relax query constraints and yet it does not affect the originality of the query by much. In addition, the system can mutate numerical constraints. The system applies step-by-step relaxation to numerical constraints in a query until some documents are found. An illustration is shown in Figure 6.

This illustration is performed at steps of only 5%. As can be seen, because of the compounding effect, one item could be found during the fifth iteration of the process. Obviously, this method is still an effective method in terms of query restructuring because it is able to retrieve something which may be useful for the user.

Application of OntoQuery in M-Commerce

The screenshots used for illustrations are implemented using Java. One major consideration here is that mobile devices tend to have a much smaller screen. Therefore, in order to realize the full OntoQuery architecture, there is a need to scale down some of the display size. Another possible solution is using scrollbars to view the screen, as shown in Figure 7.

PROTOTYPE TESTING AND EVALUATION

Effectiveness of the Genetic Algorithm

It is believed that the effectiveness of the genetic algorithm chosen is mainly determined by its supremacy in query effectiveness amplification. This is because its evolution power allows more retrieval results. The system was tested with a product list database. The effectiveness was measured by testing a series of queries with and without using genetic algorithm. For example, a query "<Product Ontology><Grocery><Price><<> <3>" can only retrieve nine items when a normal search was performed, but can retrieve 98 items when genetic algorithm was performed. Table 1 shows the other results obtained by other queries.

By comparing the results shown in Table 1, it is obvious that using genetic algorithm does in fact retrieve more items than using a normal search. Thus, it is logical to say that using genetic algorithms will provide a more flexible and effective solution to information retrieval systems.

Table 1. Results showing effectiveness of GA

Query Formed	Without GA	With GA
<Product Ontology><Drinks><Price><<><2>	25	95
<Product Ontology><Diary><Price><<><3>	3	92
<Product Ontology><Candy><Price><<><3>	13	178
<Product Ontology><Confectionery><Price><<><3>	20	248
<Product Ontology><Confectionery><Supplier> <Contains><Ho>	1	52

Figure 8. Trends for the correlations between mutated queries and original query

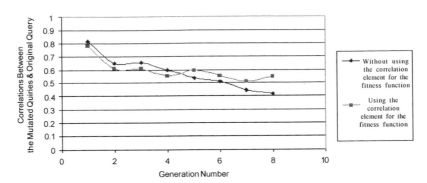

Effects of the Fitness Function

The fitness function in a genetic algorithm determines how well it can optimize a query. The OntoQuery system tested out various fitness functions to improve the power of the genetic algorithm. The usage of triangle or Gaussian functions to evaluate the fitness for the number of documents retrieved suggested some ways to counter the "too many or too few retrieved documents" dilemma in typical search engines. The fitness for quality allows the assessment of queries based on what they can retrieve. This gives a good weight to prevent the fitness for the number of documents retrieved to dominate the fitness function.

It is thought that the introduction of using correlation prevents the original query from mutating into irrelevant queries. Implementing a choice to select whether to include the correlation fitness in the fitness function in OntoQuery creates a chance to test this claim. Figure 8 shows the graphs of the trends. From the trendlines, it can be seen that a converging trend was achieved such that the queries will still be quite relevant when correlation fitness is used. A more diverged or decreasing trend for the mutated queries was obtained when correlation fitness was not included. This proves that the use of correlation can prevent the original query from mutating into irrelevant queries.

Efficiency of the Genetic Algorithm

Although using genetic algorithm allows a more flexible and effective platform in retrieving information, there is no doubt that it trades off efficiency due to its expensive iterations. Thus, the only study that can be made here is about its improvement over relevance feedback. In relevance feedback, query expansion is achieved by modifying a query. Similarly, genetic algorithm extends the relevance feedback techniques with an addition rule, the survival of the fittest.

In this research, it is thought that several factors affect the efficiency of genetic algorithm. These anticipated factors include the population size used, the number of generations, and the length of the query. Here, the efficiency of the system is measured as follows:

$$\varepsilon \text{ (Efficiency)} \approx \frac{E}{t}$$

$$\approx \frac{D}{I}$$

where E denotes the effectiveness of the system

t denotes the time taken for the system

D denotes the number of relevant

Figure 9. Graph of efficiency vs. population size

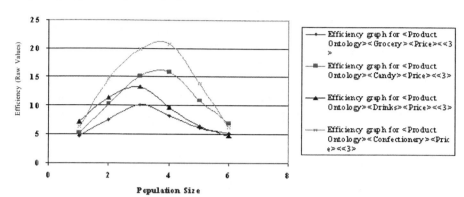

Figure 10. Graph of efficiency vs. number of generations

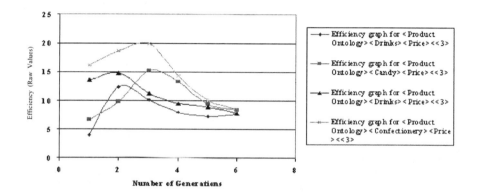

documents retrieved

I denotes the number of iterations

Efficiency is formulated as above because it is believed that the number of documents retrieved is linearly proportional to the effectiveness of the system. Also, the number of iterations is directly related to the time taken to retrieve the results.

Effect of the Population Size

This test involved the comparison of the results obtained by varying population size from values 1 to 6. A few queries were used to find out the optimal population size by comparing the efficiency obtained when each population size was used. The number of generations is set to 3.

Figure 9 shows the results obtained by averaging three samples.

From Figure 9, it can be seen that the efficiency initially increases with the population size, but eventually decreases for all the queries used in this test. Every query has an optimal population size. However, these optimal values are not the same. In addition, the optimal efficiencies peaked at different values. A study was made on these findings, and it was found that the optimum value of the population size depends much on the uniformity of the database. For example, in the category "Confectionery" or "Candy", the number of items that are contained under the category of each synonym in the database is more uniform than in the category of "Drinks" or "Grocery". Also, the optimal efficiency value is higher when the

Figure 11. Effects of longer queries

Effect of the Number of Generations

This test is similar to the test on the effect of population size except that here, the number of generations was varied from values 1 to 6. The population size of this test is set constant at a value of 3. Figure 10 shows the graph of the results obtained by averaging three samples. Again, with resemblance to that of the population size test, efficiency initially increases with the number of generations, but eventually decreases. The difference between the two tests lies in the optimal values. It is observed that the optimal numbers of generations are smaller than the optimal population sizes. Indeed, by comparing Figures 9 and 10, it can be observed that the efficiency for a larger number of generations is better than the efficiency for greater population sizes. The explanation for this is that if queries go through more generations of competition, the queries become fitter after each round and hence gives better results.

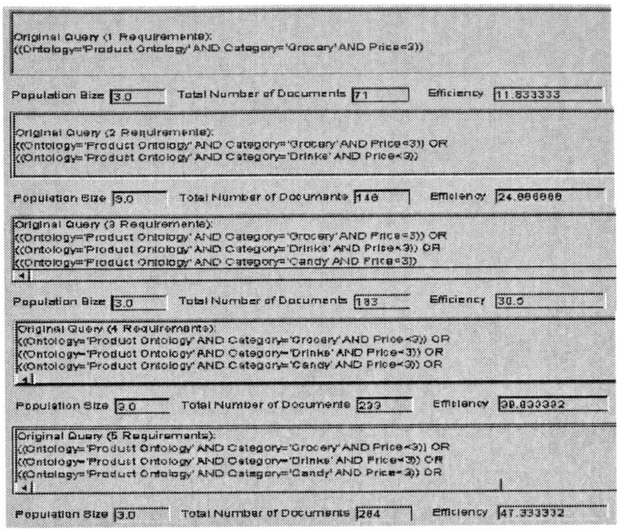

database is bigger. This is reasonable, as more results would be returned.

Effects of Longer Queries

Logically, a longer query would unleash more power in genetic algorithms. This is because there will be more synonyms qualified for query modification. In this test, queries with two or more requirements are tested. The population size is set to 3 and the number of generations is set to 2. The result is compared with the result with only 1 requirement. The screenshot of the results is shown in Figure 11. As can be seen from Figure 11, efficiency increases as the number of requirements increases. Thus, it is reasonable to say that our genetic algorithm is more suitable for longer queries to be efficient.

CONCLUSION AND FUTURE IMPROVEMENT

In summary, this research work investigated the OntoQuery system within an m-commerce agent framework against current query formation and information retrieval systems.

The prototype implementation results showed that querying formation using an ontology approach is efficient as it provides a friendly environment to the user. In addition, by combining the keyword and ontology approaches, a more efficient and effective way of forming queries could be achieved. Thus, the objective to propose efficient query formation for product databases is successful.

It was found that genetic algorithm is able to optimize queries effectively. Also, using genetic approaches, the OntoQuery system proposed and tested out various fitness functions for searching product databases. The use of correlation prevents the original query from diverging. In addition, restructuring of the logical terms and numerical constraints in queries served as an effective way of constraint relaxation for mutated queries to respond to the situation of no documents retrieved. Moreover, adding feedback to the system helps it to cater to the needs of the user more closely. It also helps to maintain a converging query trend.

It was also found that the efficiency of our genetic algorithm initially increases, reaches a maximum value, and then decreases as population size increases. This is the same for the number of generations. Thus, there is an optimal value of population size and number of generations when genetic algorithm is applied in query optimization. Also, by having longer queries, the real power of genetic algorithm will be unleashed.

Learning of Synonyms

With the usage of synonyms, it is rather tedious and difficult for a user to add synonyms to his ontology terms. Thus, there is a need to automate the learning of synonyms from other agents. The synonyms table of each agent may start off as one with only a few synonyms. Since the synonyms table is an ontology, learning of synonyms can be achieved by ontology exchange. During the interaction among agents, the agents will exchange their synonyms for a

particular query or even during product ontology exchange. For example, Agent1 requests for "Television" and in its synonyms table, "Television" only has 1 synonym "TV." Agent2 received queries from Agent1 and in its synonyms table, "Television" can be found as either the ontology term or as one of the synonyms. It has the term "TV Set" as a synonym. Now, Agent2 recognizes that Agent1 does not have "TV Set" in its table. Thus, the term "TV Set" is added to the synonyms table of Agent1. In this way, the synonyms table can be expanded easily as more interaction is made.

Training the Genetic Algorithm

As mentioned, there is an optimal value of population size and number of generations when genetic algorithm is applied in query optimization. These optimum values would probably depend on the type and size of database. However, in most situations, it is quite impossible to predict the type of database that the system will be accessing. So it is suggested that the genetic algorithm undergo some training such that fine tuning of these parameters can be achieved when they revisit the same database.

Nevertheless, some users might wish to trade off some efficiency for that extra bit of optimization. Thus, the system should also allow some tolerance for the user to decide.

Improvement in Feedback

The feedback allows more interaction with the user. There are still many ways in which the environment can be made more friendly. For example, during the feedback to the user, if the user finds that too many documents would be retrieved, it will be better if the system can allow the user to add in more requirements to his original query. At the same time, the user can delete some requirements from his original query if too few documents are retrieved.

REFERENCES

Boughanem, M., Chrisment, C., & Tamine, L. (1999). Genetic approach to query space exploration. *Information Retrieval, 1*(3), 175-192.

Braga, R. M. M., Werner, C. M. L., & Mattosso, M. (2000, September 6-8). Using ontologies for domain information retrieval. *Proceedings of the 11ᵗʰ International IEEE Conference on Database and Expert Systems Applications,* Greenwich, London, UK (pp. 836-840).

Fensel, D. (2000, November). The Semantic Web and its language. Trends & controversies. *IEEE Intelligent Systems, 15*(6), 67-73.

Hendler, J. (2001). Agents and the Semantic Web. *IEEE Intelligent Systems, 16*(2), 30-37.

Karp, P. D. (2000). An ontology for biological function based on molecular interactions. *Bioinformatics, 16*(3), 269-285.

Kouichi, A. B. E., Toshihiro, T., & Hiroshi, N. (1999, September 21-24). An efficient information retrieval method in WWW using genetic algorithm. *Proceedings of the International Workshop on Parallel Processing,* Wakamatsu, Japan (pp. 522-527).

Kraft, D. H., Petry, F. E., Buckles, B. P., & Sadasivan, T. (1994, June 27-July 2). The use of genetic programming to build queries for information retrieval. *Proceedings of the 1ˢᵗ IEEE Conference on Computational Intelligence,* Florida (pp. 468-473).

Losee, R. M. (1991). An analytic measure predicting information retrieval system performance. *Information Processing and Management, 27*(1), 1-13.

Losee, R. M. (1998). Parameter estimation for probabilistic document retrieval models. *Journal of the American Society for Information Science, 39*(1), 1-16.

Losee, R. M. (2000, July 2). Learning syntactic rules and tags with genetic algorithms for information retrieval and filtering: An empirical basis for grammatical rules. *Information Processing & Management, 32*(2), 185-197.

McGuinness, D. L. (2001). Ontological issues for knowledge-enhanced search. *Proceedings of FOIS'98,* Italy (pp. 302-316). Washington, DC: IOS Press.

Robertson, S. E., & Sparck-Jones, K. (1976). Relevance weighting of search terms. *Journal of the American Society for Information Science, 27*(3), 129-146.

Rzhetsky, A., Tomohiro, K., Sergey, K., Shawn, M. G., Michael, K., Sabrina, H. K et al. (n.d.). *A knowledge model for analysis and simulation of regulatory networks.* Retrieved from http://www.columbia.edu/cu/cie/GeneWays_fact_sheet.pdf

Salton, G. (1989). *The transformation, analysis and retrieval of information by computer.* Reading, MA: Addison-Wesley.

Salton, G., & McGill, M. (1983). *Introduction to modern information retrieval.* New York: McGraw-Hill.

Wieerhold, G., Stefan, D., Sergey, M., Prasenjit, M., Yuhui, J., Sichun, X., et al. (2000). *OntoAgents—A project in the DARPA DAML PROGRAM.* Retrieved from http://www-db.stanford.edu/OntoAgents

Yang, J. J., & Korfhage, R. R. (1994). Query modification using genetic algorithms in vector space models. *International Journal of Expert Systems in Research Applications, 7*(2), 165-191.

Chapter XXXVII
Intelligent Product Brokering Services

Sheng-Uei Guan
Brunel University, UK

ABSTRACT

Agent-based system has great potential in the area of m-commerce and a lot of research has been done on making the system intelligent enough to personalize its service for users. In most systems, user-supplied keywords are normally used to generate a profile for each user. In this chapter, a design for an evolutionary ontology-based product-brokering agent for m-commerce applications has been proposed. It uses an evaluation function to represent the user's preference instead of the usual keyword-based profile. By using genetic algorithms, the agent tries to track the user's preferences for a particular product by tuning some of the parameters inside this function. A Java-based prototype has been implemented and the results obtained from our experiments look promising.

INTRODUCTION

In this age of information technology, there has been an increasing demand for more and more sophisticated software that are capable of integrating and processing information from diverse sources. Traditional software technologies have failed to keep pace with these increasing demands, and alternative solutions are been considered. Agent-based systems (Nwana & Ndumu, 1996; Aylett, Brazier, Jennings, Luck, Preist, & Nwana, 1998) have been proposed as a potential solution, and much research has been done on this relatively new technology.

One of the potential applications for agent technology is in the area of m-commerce. According to a study done by Frost and Sullivan[1], it has been projected that electronic commerce conducted via mobile devices such as cellular phones and PDAs (personal digital assistants) will become a whopping $25 billion market worldwide by 2006. Some of the driving factors behind the m-commerce "revolution" have been attributed to the compactness and high penetration rate of these mobile devices. This, along with the relatively low cost of entry for most service providers, has made m-commerce the buzzword of the next century.

CURRENT SITUATION AND MOTIVATION OF RESEARCH

However, despite all the hype and promises about m-commerce, several main issues (Nwana & Ndumu, 1997; Morris & Dickinson, 2001) will have to be resolved before agent technology can be fully adopted into any m-commerce systems. Clumsy user interfaces, cumbersome application, low speeds, flaky connections, and expensive services have soured many who have tried m-commerce. In fact, a usability study done in London by the Nielsen Norman Group[2] has found that about 70% of the participants have said that they would not want to use a WAP- (wireless application protocol) enabled phone again within a year, after they tried it for a week. Security and privacy concerns have also dampened enthusiasm for m-commerce. One of the concerns has been the fact that mobile devices such as PDAs are very easy to lose. They are also an easy prey for thieves, and unauthorized personnel can have easy access to the valid user ID and passwords stored in these devices to make fraudulent transactions.

Taking all these concerns into account, it seems like good old e-commerce will remain as the preferred choice for online transactions for many years to come. Customers will only use wireless mobile device to access the Internet if they have a good reason to do so. Therefore, in order to entice customers to participate in m-commerce, the developers will have to offer something that is unique and which no self-respecting consumer can live without. One of the potential "killer" applications for m-commerce could be an intelligent program that is able to search and retrieve a personalized set of products from the Internet for its user.

Currently, when a user wants to search for a particular product on the Internet, what he will normally do is to use popular search engines such as Altavista[3] or Yahoo![4], and enter keywords that describe the product. These search engines will process these keywords and churn out a large number of links for the user to visit. On the other hand, if the user already knows of some URLs that might have the product information, he will visit these Web sites and hopefully get the information that he is looking for.

Although these are the more common methods of searching for information on the Internet, it need not necessarily be the best or the most efficient ones. Neither the search engine nor the Web site knows the preference of the user and hence might provide information that is totally irrelevant to the user. For example, if the user wants to search for information about "software agents," the search engine could return links to "insurance agents" instead. A significant amount of time could be wasted on such irrelevant information which could have been better spent on other, more important tasks.

In an agent-based m-commerce, agents act on behalf of their users by carrying out delegated tasks automatically. Currently, there is no single agent that can perform all the tasks meted out by the user. Like humans, specialized agents are required that are able to work in a specific type of environment. A product brokering agent seems to be a potential solution for this scenario. The agent will search for the products in the background with minimal user intervention, thereby allowing the user to concentrate on other tasks. It could be programmed with the user's preferences in mind and filter out irrelevant products automatically. The agent could also detect shifts in the user's interest and, through some evolution mechanism, adjust itself accordingly to suit the user.

OVERVIEW OF THE CHAPTER

This chapter describes the design of an intelligent ontology-based product brokering agent capable of providing a personalized service for its user. It does this through *user profiling* (Soltysiak & Crabtree, 1998). Such agents are able to learn the preferences of the user over time and recommend products that might interest the user. The agent achieves this either by user feedback or through its own observation. This technique has been used quite successfully for specific types of agent tasks, typically those that are information intensive and often involve the World Wide Web (WWW).

We first highlight some of the related works that have been done by other researchers. A proposed design for an evolutionary product brokering agent will then be presented to the reader. A prototype of the product brokering agent has been implemented using Java, and the system is put through a series of tests. Results obtained from these tests are discussed later in this chapter. Although the results are encouraging, some limitations of the system will also be highlighted. Potential applications for a product brokering agent in m-commerce will then be discussed. Finally, the last section presents some concluding remarks along with a discussion on the possible extensions to the current work.

BACKGROUND

Personalized product brokering agents require a profile of the user in order to function effectively. The agent would also have to be responsive to changes in the user's interests, and be able to search and extract relevant information from outside sources. The rest of this section will highlight some of the works done by other researchers which are closely related to a product brokering agent.

At MIT Media Labs, Sheth and Maes (1993) have come up with a system that is able to filter and retrieve a personalized set of USENET articles for a particular user. This is done by creating and evolving a population of information filtering agents using genetic algorithms (Zhu & Guan, 2001). A screenshot of their implemented system is as shown in Figure 1.

Some keywords will be provided by the user representing the user's interests. Weights are also assigned to each keyword, and the agents will use them to search and retrieve articles from the relevant newsgroups. After reading the articles, the user can either give positive or

Figure 1. Screenshot of the information filtering system

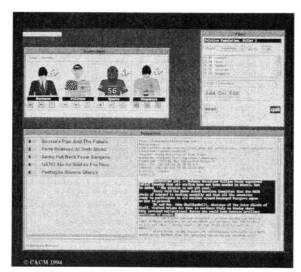

negative feedback to the agents via a simple GUI. Positive feedback increases the fitness of the appropriate agent(s) and also the weights of the relevant keywords (vice versa for negative feedback). In the background, the system periodically creates new generations of agents from the fitter species while eliminating the weaker ones. Initial results obtained from their experiments have been encouraging and showed that the agents are capable of tracking its user's interests and recommend mostly relevant articles.

While the researchers at MIT require the user to input their preferences into the system before a profile can be created, Soltysiak and Crabtree (1998) believed that the user's profile can be generated automatically by monitoring the user's Web and e-mail habits, thereby reducing the need for user-supplied keywords.

Their approach is to extract high information-bearing words, which occurs frequently in the documents that are opened by the user. This is achieved by using ProSum[5], which is a text summarizer that can generate a set of keywords to describe the document and also determines the information value of each keyword. A clustering algorithm is then employed to help identify the user's interests, and some heuristics are used to ensure that the program could perform as much of the classification of interest clusters as possible, thereby minimizing the amount of user input required in the profile generation process.

However, they have not been completely successful in their own experiments. The researchers admitted that it would be very difficult for the system to classify all of the user's interest without the user's help. Nevertheless, they believed that their program has taken a step in the right direction by learning user's interest with minimal human intervention.

A new product brokering agent usually does not have sufficient information to recommend any products to the user. Hence, it has to get product information from somewhere else. A good source of information will be the Internet. In order to do that, a method suggested by Pant and Menczer (1998) is to implement a population of Web crawlers called *InfoSpiders* that searches the WWW on behalf of the user. It will gather information on the Internet based on the user's query and indexes them accordingly. It behaves much like a personalized search engine, but is designed to evolve and retrieve only relevant Web pages for its user.

These agents initially rely on traditional search engines to obtain a starting set of URLs, which are relevant to the user's query. The agents will then visit these Web sites and decode their contents before deciding where to go next. The decoding process includes parsing the Web page, and by looking at a small set of words around each hyperlink, a score is given based on their relevance to the user. The link with the highest score is then selected and the agent visits the Web site.

No further details have been provided on how they extract or analyze the contents of the Web pages, but it has been mentioned that they use neural networks, HTML, and XML parsing tools that are commonly used by other Web crawlers. The agent stops after they had visited a pre-determined number of Web pages or when it could no longer find any relevant Web pages. The user can also terminate the search anytime he wishes while the program is still running.

DESIGN OF PRODUCT BROKERING AGENT

A product brokering agent can be used to search for all kinds of products. In our application, the agent will be used to search for some computer products, namely *CPU, Mainboard,* and *Memory.* It is possible to extend the application to search for other products. All the

530

codes are written in Java, as it is object oriented in nature and is compatible across multiple operating systems.

Similar to the information filtering agents done by Sheth and Maes (1993), an initial population of product brokering agents will be created and evolved using some form of genetic algorithms. However, in this design, the *profile* of the user is not based on any keywords supplied by the user. In fact, no keywords are required to be entered by the user. Instead, each agent will have an evaluation function that will be used to calculate the *value* of each product. Products that have a higher *value* will have a higher chance of being recommended by the agent. This evaluation function has some tunable parameters, which characterizes the user's preferences for a particular category of products. Initially, these tunable parameters will be randomly generated based on some heuristics, but they will evolve over time to match the user's preferences.

In this design, some assumptions have been made about the system. One of the most important assumptions is that the user of the system is a rational person and will select a product rationally. Another important assumption is that the value, which a user places on a product, can be calculated mathematically. The product values that we are focusing on will be those that can be calculated by using some tangible attributes (e.g., *price*) of the product. The agent will not be able to calculate the intangible value (e.g., *branding*) that a user has placed on the product. If these assumptions are not met, the agent will not be able to track the user's preferences successfully.

Ontology

Before the product brokering agent is able to explore the Internet and retrieve product information for the user, the agent needs to have some prior knowledge such as the URL of some relevant Web sites, keywords, or quantifiable attributes that can be used to describe the product. It could be very tedious if the user has to enter such information into the agents when he wants to search for a particular product. Imagine the amount of data he will have to enter if he wants to search for several different products.

The *product ontology* has been implemented in a tree-like structure, with the leaf nodes representing the products and the parent node representing the product category. Each leaf node actually contains a Java class called *productInfo*, which has some prior information about the product. Different products will have a different *productInfo* class. New products can be added as a leaf node to the parent node easily. When the leaf node is selected, it will pass the product information to the product brokering agents. Currently, selection of the leaf node will pass the URL of the product's Web site and its attributes to the agents automatically.

Product Brokering Agent

After describing how the agents are going to obtain their product knowledge, the next stage is to define the agent itself. The agent will basically be a Java programming thread that will be running continuously in a while-loop until some terminating conditions have been met. A unique agent name will be given to each agent

Figure 2. Screenshot of product ontology

so that we can identify and differentiate the agents from one another.

Agent's Fitness Function

To calculate the fitness of the agent, the proposed fitness function has been defined by using the following equation:

$$Fitness = \frac{\sum_{n=window\_size} points\ earned\ in\ the\ recent\ n\ generations}{n}$$

(1)

This fitness function basically sums up and calculates the average amount of points earned by the agent in the current and the previous *n-1* generations, where *n* is the window size of the agent's short-term memory. For example, assume an agent has a *window_size* of 3 and the following diagram shows the amount of points earned in each generation:

In this example, the total amount of points earned by the agent in the recent three generations is 9.5. Therefore the fitness of the agent is about $\frac{9.5}{3} = 3.1667$. The amount of points earned before that will not be considered. The rationale for this is that more emphasis should be placed on the agent's current performance instead of its past performances. As the fitness of an agent would be used to determine which agent to evolve, we do not want its past performances, which might be irrelevant now, to influence the evolution process.

An agent's fitness will always be a positive value, and a new agent would start off with some default fitness. The fitness of an agent can also be a good indicator about the agent's performance. Therefore, in order to keep track of the agent's performances, each agent will have a list called *fitness_history,* which is used to store the fitness of an agent for each generation. Hence, after an agent has been awarded some points, it will calculate its new fitness using Equation 1 and insert the value into the *fitness_history* list. Details on how an agent earns its points will be discussed in the next few sections.

Agent's Lifecycle

Once an agent has been created, it can be in any of the four different states, namely *Dormant, Active, Evolve,* and *Death.* During agent creation, the agent will register itself to a database and its default state will be the *Dormant* state. It would also de-register itself before it is removed from the system. This is to allow the user to keep track of all the agents running in the system. The database also allows the agents to store any product information that they have found on the Internet. Figure 3 shows how the four different states are implemented.

Dormant—The agent is not doing any task at the moment. It is waiting for the user to give it instructions. Note that this will be the default state of the agent once it has been created. At this state, the user can modify the agent's parameters before starting the agent.

Figure 3. Agent's lifecycle

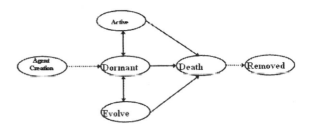

Active—The agent has received some instructions and is currently performing some tasks for the user. The types of tasks performed by the agents will be discussed further on.

Evolve—The agent has received some user feedback regarding its performance. It is now analyzing this feedback and making the appropriate adjustments.

Death—The agent has been killed and it can no longer perform any tasks for its user. Note that it is still present in the system and will only be removed completely when instructed by the user. This allows the user to recycle any information that he might find suitable.

Agent's Task

Once the user has passed some instructions to the agent, it will switch to the *Active* state and activate the appropriate tasks. As an agent might need to perform different types of tasks simultaneously, these tasks are implemented as independent and self-contained programs, which are separated from the agent. Therefore, instead of implementing several agents from scratch to perform different tasks, we only need to implement a basic agent and a few task programs. What the basic agent needs to do is to call the appropriate task program, pass some information to it, and the task programs will handle the rest.

For our application, the agent's task has been designed specially to parse information from Hardwarezone.com[6], a Web site hosted in Singapore that displays up-to-date information of various computer products in table form. The task program allows the agent to establish a connection to the Web site and download the HTML document onto a local computer. The program then parses the document and extracts the relevant information for the agent by looking for specific tags within the HTML docu-

Figure 4. Screenshot of an agent's database

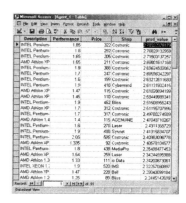

ment. In our application, the program will be able to extract information such as the description of the product, its price, and the name of the shop that is selling this product.

Agent's Knowledge

After an agent has retrieved some product information, it needs a place to store this piece of information. As mentioned, when an agent is created, it will register itself to a database. A Microsoft Access database is used in this application. Within the database, a table will be created for each agent to store all the information that it has retrieved. In addition, the agents will also store this data on a *global* database. The *global* database will contain all the products that have been retrieved by the agents in the system. Figure 4 shows a screenshot of an agent's database.

Product Recommendation

Before recommending a product to the user, the agent should first be able to evaluate which product would best fit the user's requirements. A proposed method is to use some quantifiable attributes such as *cost, performance,* and so forth to evaluate the products. An example of an evaluating function could be the following equation:

$$product\_value = perf\_weight*performance - cost\_weight*cost \qquad (2)$$

The attributes used in Equation 2 try to model the two types of factors that can influence a user's choice. The first attribute (*performance*) represents the performance of the product, while the second attribute (*cost*) represents the cost of the product. It has been assumed that the better the product, the higher will be its *performance,* and a better product usually results in a higher *cost.* From Equation 2, it can be seen that a product with a higher *performance* and/or a lower *cost* will result in a higher *product_value.*

The two weights *perf_weight* and *cost_weight* represent the weights that the user could give to each attribute. These two parameters are actually used to represent the user's preferences and are incorporated inside the agent. If *perf_weight* has a higher value, it means that the user places more emphasis on the performance of the product. Likewise, if the user has a higher value for *cost_weight*, it means that the user is more concerned about the cost of the product. Note that, for different products, a different set of attributes and weights could be defined for Equation 2, and all these could be defined inside the *product ontology.*

When an agent is created, these two weights will be initialized based on some heuristics and would be used to calculate the value of each and every product found in the agent's database. The agent will then rank the products according to their values and select the top three products to be presented to the user. The value of *perf_weight* and *cost_weight* will be allowed to change when the agent undergoes evolution.

Agent's GUI

It will be useful if the user is able to observe what is happening inside an agent when required. To facilitate this, each agent will have a simple GUI that shows information such as the name of the agent, its current status, products recommended, and so forth. It would also allow the user to change some of the parameters inside the agents.

The agent's GUI is implemented as shown in Figure 5. The GUI allows the user to see what is the top product inside the agent's database and also some of its internal parameters. The user can kill the agent from this GUI by clicking the *kill* button or update some of the agent's parameters by using the *update* button. When the user clicks the *history* button, the agent's history will be shown to the user (Figure 6). The *database* button will pop up a GUI (similar to Figure 4) which shows all the products inside an agent's database.

Figure 5. Agent's GUI

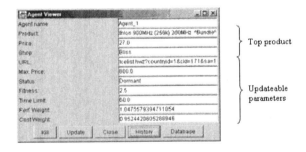

Figure 6. Agent's history

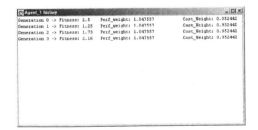

Figure 7. Screenshot of the monitoring tool

Figure 8. Screenshot of an agent's database after sorting

Monitoring Tools

A monitoring tool will be provided for the user which allows him to observe and control the behavior of his agents while they are searching for products on the Internet. This tool will be the main interface between the user and his agents.

The user can choose from a list of products provided in the *product ontology* and enter some parameters (e.g., number of agents, etc.) before starting the search. Once all the parameters have been entered into the system, the appropriate number of agents will be created to search for the product on the Internet. While the program is running, the user can start, stop or provide feedback to the agents anytime by using this tool. A text message will also be provided to allow the user to track the progress of the agents in the system. A screenshot of the implemented system is as shown in Figure 7.

User Feedback

During user feedback, each agent in the system will select the top three products in its database and add them into a *recommended* list. A sorting function will be implemented to allow the user to sort the list according to his preferences. If the user cannot find any product that he fancies in this list, he can look at the *global* list, which contains all the products that have

been retrieved by the agents in the system. When the user sees a product that he likes, he can select the product by clicking on it and all the agents in the system will be informed about the user's selection.

The agents will take note of the product that the user has selected and search for that product inside its own database. At this stage, each agent would have already assigned a product value to each and every product in its database. To determine the amount of points to award to an agent, it will be asked to rank the products in an ascending order according to this value. Hence, products with a higher value will be located at the bottom of the table. The agent will then determine the position of the user-selected product and take note of its row number. The formula to calculate the exact amount of points to give to an agent is as follows:

$$points\ awarded = \frac{row\ number\ of\ user\ selected\ product}{total\ number\ of\ products} \times maximum\ points$$

(3)

As an example, assume Figure 8 shows the agent's product list after it has been sorted in ascending order.

In this example, there are a total of 13 products found inside the agent's database, and the product with the highest product value is located at row 13. This will be the top product inside the agent's database. However during

Figure 9. Evolution process

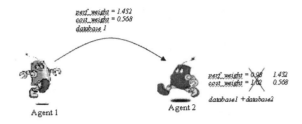

Agent 1   Agent 2

feedback, the user might have actually chosen the product at row number 7 instead. Therefore, assuming that 5 is the maximum amount of points awardable, the amount of points that the agent earns in this case will be:

$$points\ awarded = \frac{7}{13} \times 5 = 2.692$$

Using this example, it can be seen that if the rankings of the two selected products are far apart, the agent will actually receive less points. Also, if the user-selected product is not inside the agent's database, the agent will not receive any points at all!

Evolution Process

The fitness of an agent will be used to decide which agent will undergo the evolution process. In conventional genetic algorithm, the agent with a higher fitness will have a higher chance of survival as compared to an agent with a lower fitness. However, in this application, there will be a slight variation in the algorithm. Instead of killing the weaker agents, they will simply copy over all the parameters of the fitter agents. The weaker agent will also copy over the database of the fitter agent and merge the two databases together to form a larger database. Let *Agent 1* be the fitter agent and *Agent 2* be the weaker one. Figure 9 shows what happens between the agents during this evolution process.

However, the parameters inherited by the new agent in this evolution process might not necessarily be the most optimal. Therefore, the new agent will try to adjust its newly acquired *perf_weight* and *cost_weight* to better reflect the user's requirements. First, it will use the newly acquired parameters to re-evaluate all the products found inside its new database. Then the agent will select the best product based on these new parameters. If it is the same as the user-selected product, no further changes will be required, but some small and random mutations in the parameters will be allowed.

However, that will not usually be the case. In this case, the agent will compare the *performance* and *cost* attributes of the products that are selected by the user and the agent. Let *p1* and *p2* denote the *performance* of the products selected by the user and agent respectively. Also let *c1* and *c2* denote the *cost* of the products selected by the user and agent. Four possible scenarios will have to be considered:

1. p1 > p2 and c1 > c2
    The user has selected a product that has a much better performance but more expensive than what the agent has suggested. The agent can deduce that the user places more emphasis on the performance rather than the cost of the product. Therefore, it will increase its *perf_weight* and reduce its *cost_weight*.

2. p1 < p2 and c1 < c2
    The user has selected a product that is of a lower performance but cheaper than what the agent has suggested. The agent can deduce that the user places more emphasis on the cost rather than the performance of the product. Therefore, it will reduce its *perf_weight* and increase its *cost_weight*.

Figure 10. Recommended list of products

No.	Description	Price	Shop
1	AMD Athlon 1.0GHz (2...	165.0	IMS Systems
2	AMD Athlon 1.2GHz (2...	89.0	Bliss
3	AMD Athlon MP 1600+...	385.0	IMS
4	AMD Athlon MP 1900+...	490.0	Superpet
5	AMD Athlon XP 1600+ (...	123.0	io Data
6	AMD Athlon XP 1800+ (...	180.0	io Data
7	AMD Athlon XP 1900+ (...	295.0	io Data
8	AMD Athlon XP 2000+ (...	475.0	Sysnet
9	AMD Athlon XP 2000+ (...	309.0	Video-Pro.com
10	AMD Duron 800MHz (8...	88.0	Laser
11	INTEL 667A Celeron-S...	100.0	BEAM
12	INTEL 950A Celeron-S...	65.0	Video-Pro.com
13	INTEL Pentium-4 1.8A...	379.0	MediaPro
14	INTEL PIII 750E (SSE)	165.0	HardwarePlace
15	INTEL XEON 1.7GHz (...	520.0	IMS

Select   Delete   Global List   Agent's list   Close   Sort   ASC

3. p1 < p2 and c1 > c2

   The user has selected a product that is of a lower performance and more expensive than what the agent has suggested. The agent will be confused over such a selection and will prompt the user if it should carry on the evaluation. If the user still wants the agent to carry on, it will either reduce its *perf_weight* or *cost_weight*. This might happen when the user has placed some form of intangible attributes/values on the product which are not present inside the agent's evaluation function.

4. p1 > p2 and c1 < c2

   The user has selected a product that is of a higher performance and cheaper than what the agent has suggested. This scenario will not arise during evolution. Looking back at Equation 2, a product with a higher performance and/or a cheaper product will result in a higher value being assigned to that product. Since using *p1* and *c1* will definitely result in a higher product value as compared to *p2* and *c2*, this scenario will not happen.

The evolution process as described in this section is actually quite similar to the use of the reproduction and crossover operator to clone the fitter agents and then use the mutation operator to mutate some of the parameters within the agents.

SYSTEM EVALUATION

To evaluate the performance of the implemented system, some simple experiments have been conducted to see if the agents are able to track the user's preference. All the agents in these experiments are running inside a single computer. The computer used in this experiment is a Pentium 4 with 384MB of memory and operating under the Windows ME environment with JDK1.3.0. The system connects to the Internet via a 56.6kbps modem.

Product Recommendation

In this experiment, a group of 20 randomly generated product brokering agents are instructed to search for one of the products on the Internet. The product chosen for this experiment is the *CPU*. For this part of the experiment, the user wants to get the best *CPU* possible and he does not mind the price. After instructing the agents to search for the product, the system is allowed to run on its own for about 10 minutes so that the agents can retrieve sufficient products before the user gives feedback. After 10 minutes, the user clicks on the *result* button and the recommended list is as shown in Figure 10.

From the recommended list, the user selects the current best product at row 13, which happens to be Pentium 4 1.8GHz as shown in Figure 11.

Figure 11. User selection

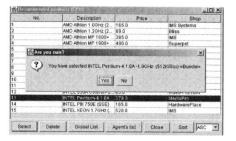

Figure 12. Recommended list after feedbacks

Figure 13. New products recommended by the agents

While the feedback is been made, the system continues to search for products in the background. After making a few similar selections, the agents evolved and re-evaluated their list. The new recommended list is now as shown in Figure 12. The list now only shows the best *CPU* that has been retrieved by all the agents.

When the user is satisfied with what the system has learned, he allows the system to carry on searching the Internet for new products on its own. After some time has passed, the agents have found an even better performing *CPU* and it is reflected in the agent's recommended list.

Two other scenarios have also been tested on the system. One of them is to search for the cheapest *CPU* available, while the other is to find a right mix of *performance* and *cost* for the user. The steps used in these scenarios are similar to those used in the first part of the experiment, and the results obtained are encouraging.

Tracking User's Preferences

In this experiment, the objective is to test if the system is able to detect a change in the user's preferences. If it is able to do so, we need to determine how fast the system will be able to respond to these changes. This could be observed by looking at the average fitness of all the agents in the system. The average fitness of all the agents should remain high if the system

is able to track and respond to the changes effectively.

An initial population of 20 randomly generated agents is created, and the response of the system is observed by changing the number of agents to evolve in the population. For each test case, the same set of test data are used and will be described in the next few sections.

Gradual Changes in User's Preferences

In this part of the experiment, the user starts by selecting the best *CPU* available. After a few selections, the user will gradually choose cheaper and cheaper *CPUs*. The experiment stops after all agents begin to recommend the cheapest CPU available. Figure 14 shows the

Figure 14. Tracking gradual changes in user's preferences

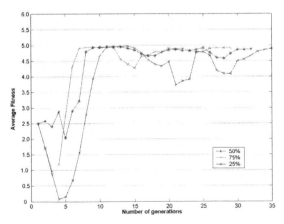

Figure 15. Tracking abrupt changes in user's preferences

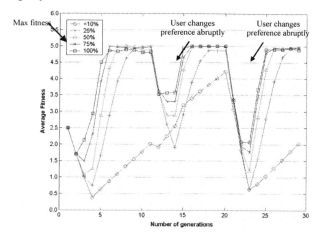

average fitness of the agents, when the user gradually changes his preferences.

The results obtained from this experiment have shown that the system is capable of tracking gradual changes in the user's preferences. Although some "dips" are observed during the experiment, the average fitness of the agents in the system remains high while the user changes his selection. These dips could happen because some of the agents might not have the products that the user has selected in their database. Therefore, these agents do not receive any points and could "pull down" the average fitness quite significantly.

Abrupt Changes in User's Preferences

In this part of the experiment, the user makes two abrupt changes in his preferences. Initially, the user starts by selecting the best *CPU* available. After a while, he abruptly changes his preferences by selecting the cheapest *CPU* and then reverts back to his original selection. Figure 15 shows the average fitness of the agents when the user changes his preferences abruptly.

Interestingly, the results obtained suggested that as we increase the proportion of agents to evolve, the response of the system would be much better. The best result is obtained when 100% (black line) of the population is allowed to evolve. It can be seen that the system in this scenario attains the maximum average fitness in a shorter time as compared to others. Also, when the user makes an abrupt change in his selection, the average fitness of the system does not drop as much as compared to the rest, and it recovers much faster. The system could not track the user's preference when less than 10% (red line) of the population is allowed to evolve.

Although allowing a larger proportion of agents to evolve will result in a faster response, it might cause the system to converge to a solution prematurely. This has been observed when 100% of the population is allowed to evolve. The type of products recommended by the agents is not as diverse as compared to when only 50% of the population is allowed to evolve.

Hence, there is a tradeoff between them. From the experiment, it has been observed that allowing 50% of the population to evolve will be

Figure 16. Diversity of products recommended

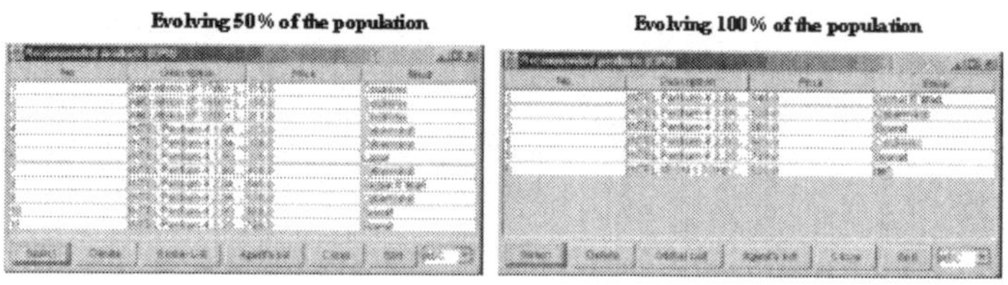

a reasonable compromise between the response time and the diversity of products recommended.

Limitations of the System

Although preliminary test results have shown that the product brokering agents are quite successful in tracking the user's preferences after a few evolutions, it has been noticed that sometimes it does not follow the user's selection. A possible explanation for this could be due to the mismatch of the evaluation function that is used by the agent to calculate the perceived value of a particular product. Currently, the function only takes into account the performance and cost of a given product. However in reality, there might be other product attributes, which could also affect the user's decision. Some of them could also be intangibles, which might be difficult to represent in the function.

M-Commerce Applications

The proposed design of a product brokering agent has been implemented using Java in a desktop computer. However, mobile devices such as phones and PDAs tend to have a smaller screen, slower processors, as well as limited memory. Hence, this will pose some serious constraints when we want to transfer the software into these mobile devices. There is also a serious lack of standardization, as these mobile devices use different OS platforms, which makes it difficult for the developer to create a single program that can run on all devices.

After taking all these into consideration, a possible solution is to use software that is compatible across multiple operating platforms. A good candidate is Java, which has been used to implement the system as mentioned in this chapter. Java is a platform-independent software technology, allowing the same code to be run on any system. This greatly reduces the time and cost of software development, and is particularly attractive for Internet or local network applications, because instructions can be sent over the Net without prior knowledge of the characteristics of the target device. However, the disadvantage of Java as compared to other programming languages, such as C, is that it is less efficient and has slower program execution. Faster processors and more memory are needed to compensate for this. This results in higher cost, and, for wireless applications, shorter battery life. However, this disadvantage has been slowly reduced by the introduction of more efficient JIT (just-in-time) compilers. Currently, the developers of Java have also introduced some highly optimized and micro versions of the Java software to cater specifically for small devices such as cellular phones and PDAs.

As for hardware, Philips Research[7] has developed a co-processor that improves the execution of Java embedded software by a factor up to 10. This obviates the need for powerful processors and/or more memory that this programming language often requires, while maintaining the advantage of enabling fast and economic product development and easy integration with the Internet. The invention supports the use of Java and related languages in a rapidly growing and developing market, with applications ranging from smart cards to mobile phones and set-top boxes.

Application of Product Brokering Agent in M-Commerce

A PDA is an ideal device for m-commerce applications. It tends to have a larger screen and a more powerful processor as compared to a cellular phone, but is less bulky than a laptop. Making an existing application viewable in any wireless device—a process known as

Figure 17. Screenshot of a PDA with the implemented GUI

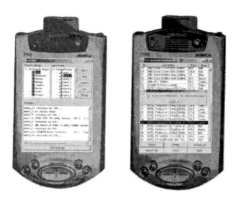

the largest viewable screens in the market and also an integrated Bluetooth for wireless links to Bluetooth-enabled cellular phones. This device also supports the Java Virtual Machine, which will allow our software to be integrated into the PDA easily. The specifications of the PDA are shown in Table 1. For more detailed information, the reader is encouraged to visit the manufacturer's Web site[8]. A comparison of the iPAQ Pocket PC with other PDAs on the market can be found at PCWorld.com[9].

CONCLUSION AND DISCUSSION

This chapter has demonstrated that by using genetic algorithm and an evaluation function, it is possible to design and implement an intelligent product brokering agent for m-commerce applications.

A simple prototype of the system has been implemented using Java, and the preliminary results obtained from the experiments have been encouraging. However, there are some limitations in the current prototype that might hamper the system's performance. More re-

transcoding—is among one of the biggest challenges of m-commerce. In order to fit the screen of the PDA, the GUI implemented in this chapter will have to be scaled down to the appropriate size. A possible solution to fit all these into the PDA screen is to use scrollbars that allow the user to scroll the GUI. A possible screenshot of a PDA with the GUI is shown in Figure 17.

The PDA selected for our application is the Compaq iPAQ Pocket PC H3870. It has one of

Table 1. Specifications for Compaq's iPAQ Pocket PC H3870

iPAQ Pocket PC H3870

Operating System	Microsoft Pocket PC 2002
Processor	206 MHz Intel StrongARM 32-bit RISC processor
Display Type	Color reflective thin film transistor (TFT) LCD, 64K colors
Resolution/Viewable Image Size	240 x 320/ 2.26 x 3.02 inches
Pixel Pitch	0.24 mm
RAM	64MB
ROM	32MB
Input Method	Handwriting recognition, soft keyboard, voice record, inking
Wireless Connectivity	Bluetooth™, Infrared port (115 Kbps)
Dimensions	5.3" x 3.3" x .62"
Weight	6.7 oz. including battery

search will have to be done before a truly robust system can be implemented for m-commerce.

One of the possible improvements to the current work will be to allow the agents to be distributed in a network instead of being hosted entirely by the same computer. As the host computer might not have sufficient resources (processing power, bandwidth, etc.) to support all the agents in the system, it will be advantageous if some agents can be hosted by another computer. For m-commerce applications, this would mean the agents could now be hosted by a commercial Internet service provider (ISP).

An m-commerce user would not want to spend huge sums of money on maintaining a wireless connection to the ISP or a phone company. Likewise, it is unrealistic for mobile devices such as cellular phones and PDAs to be always "online." Currently, some ISPs do provide some form of storage spaces for their subscribers to store files inside their servers. In extension to this, an ISP could now also offer—with a reasonable fee—to host the agents that have been created by/for their subscribers. These agents could perform their tasks inside these servers and report back to its user when he re-establishes another connection with the ISP.

However, allowing agents to be distributed over the network will raise some issues, which the developer should look into before the system could be implemented. Since the agents are distributed, some form of communication protocol and *ACL* (Agent Communication Language) will have to be designed and incorporated into the system. One way is to upgrade the monitoring tool that has been implemented in this chapter so that it can communicate with the remote agents using some *ACL* via a socket connection. Currently, there exist some high-level agent languages such as *KQML* (Knowledge Query and Manipulation Language) (Finin, Fritzson, McKay, & McEntire, 1994) and *FIPA*

ACL, which has been developed for inter-agent communication.

Security issues also arise when agents are hosted by other computers. There is now a need for us to distinguish between agents that are sent by different users. In m-commerce, security is of paramount importance. Sensitive and private information of the user will have to be safeguarded from other hostile entities. This is especially important in the case of mobile agents. As they travel from host to host, we have to prevent them from been intercepted and its contents "core-dumped" by hostile hosts. Malicious agents could also masquerade as the original agent and trick an unsuspecting user into giving up his personal information.

REFERENCES

Aylett, R., Brazier, F., Jennings, N., Luck, M., Preist, C., & Nwana, H. S. (1998). Agent systems and applications. *The Knowledge Engineering Review, 13*(3), 303-308.

Finin, T., Fritzson, R., McKay, D., & McEntire, R. (1994). KQML as an agent communication language. *Proceedings of the 3rd International Conference on Information and Knowledge Management.*

FIPA Specifications. (n.d.). Retrieved from http://www.fipa.org/repository/fipa2000.html

Maes, P. (n.d.). *Agents that reduce work and information overload.* Retrieved from http://pattie.www.media.mit.edu/people/pattie/CACM-94/CACM-94.p1.html

Morris, S., & Dickinson, P. (2001). *Perfect m-commerce.* London: Random House Business Books.

Nwana, H. S., & Ndumu, D. T. (1996). An introduction to agent technology. *BT Technology Journal, 14*(4), 55-67.

Nwana, H. S., & Ndumu, D. T. (1997). Research and development challenges for agent-based systems. *IEEE Proceedings on Software Engineering, 144*(1), 2-10.

Pant, G., & Menczer, F. (2001). Evolve your own intelligent Web crawlers. *Autonomous Agents and Multi-Agent Systems, 5*(2), 221-229. University of Iowa.

Sheth, B., & Maes, P. (1993). *Evolving agents for personalized information filtering.* Boston: MIT Media Lab.

Soltysiak, S., & Crabtree, B. (1998). Automatic learning of user profiles—Towards the personalization of agent services. *BT Technology Journal, 16*(3), 110-117.

Soltysiak, S., & Crabtree, B. (1998, March). *Identifying and tracking changing interests.* Retrieved from http://www.btexact.com/projects/agents.htm

Soltysiak, S., & Crabtree, B. (1998). Knowing me, knowing you: Practical issues in the personalization of agent technology. *Proceedings of PAAM'98,* London (pp. 467-484).

Zhu, F. M., & Guan, S. U. (2001, October 7-10). Evolving software agents in e-commerce with GP operators and knowledge exchange. *Proceedings of the 2001 IEEE Systems, Man and Cybernetics Conference,* Tucson, AZ (pp. 3297-3302).

ENDNOTES

1. http://www.infoworld.com/articles/hn/xml/02/03/22/020322hnmcommerce.xml
2. http://www.nngroup.com/reports/wap/
3. http://www.altavista.com
4. http://www.yahoo.com
5. Profile-based text summarization
6. http://www.hardwarezone.com
7. http://www.philips.com.sg/news.shtml #5January
8. http://www.compaq.com/products/handhelds/pocketpc/H3870.html
9. http://www.pcworld.com/features/article/0,aid,82004,pg,5,00.asp

Chapter XXXVIII
Understanding Mobile Web Services (MWS) and Their Role in Integrating Mobile Devices

Samir El-Masri
University of Western Sydney, Australia

ABSTRACT

Web services (WS) have become the industry standard tools for communication between applications running on different platforms, and built using different programming languages. The benefits, including the simplicity of use, that Web services provide to developers and users have ensured integration of Web services architecture by almost all IT vendors in their applications. As expected, with the proliferation of mobile phones, PDAs and other wireless devices, the same requirements of making applications talk across platforms has become necessary on mobile devices. This has lead to the mobile Web services (MWS), which are based on the Web services and related technologies like XML, SOAP and WSDL, and which provide the best choice to be used in the architecture for integration of Web services in mobile devices. This chapter discusses WS and MWS in the context of integration architecture, together with their advantages and disadvantages in usage. Since MWS is deployed using wireless technologies and protocols, they are also presented and explained in this chapter.

INTRODUCTION

Web services (WSs) represent the next major chapter of online computing that has enabled seamless integration of application services across the Internet. WSs are the cornerstone towards building a global distributed information system, in which many individual applications will take part. The centre for that global system will obviously be the WS. As there is no place today for a stand-alone computer, there will be no place for stand-alone applications in the future. Therefore, building a powerful application whose capability is not limited to local resources will unavoidably require interacting with other partner applications through WS. The Internet has been the revolution in networking that links computers and people in a manner that changes the way we live and work and do business forever. It is believed that WSs will be changing the way we develop software and build applications, in a way that one application will depend on and use many other applications online (El-Masri & Unhelkar, 2005).

The strengths of WSs come from the fact that WSs use XML and related technologies connecting business applications based on various computers and locations with various languages and platforms.

The increase of applications in the mobile world (mobile phones, personal digital assistants (PDAs), etc.) makes it necessary for those applications to communicate with other applications residing on computers on the Internet. WS is a successful architecture for building software applications on the classic network. Recently, Java readied itself for wireless Web services (Yuan & Long, 2002). Microsoft, the leading company in building computer applications, and Vodafone, the leading group in the MS world (Microsoft & Vodafone, 2005a, 2005b) agreed to work together to build standards to facilitate the integration of mobile applications with other applications using a new architecture called mobile Web services (MWSs). Many papers have been published recently in this area (e.g., El-Masri, 2005; El-Masri & Unhelkar, 2005; El-Masri & Suleiman, 2005). The proposed MWSs are to be the base of the communications between the Internet network and wireless devices such as mobile phones, PDAs, and so forth. The integration between wireless device applications with other applications would be a very important step towards global enterprise systems. Similar to WS, MWS is also based on the industry standard language XML and related technology such as SOAP, WSDL, and UDDI. These technologies will be presented with more details in the next sections. Practically, Microsoft .NET framework makes it a simple task to build

a mobile Web application consuming Web services (Arora & Kishore, 2002). Java and IBM also have their own environment for the same purpose.

EXTENSIBLE MARKUP LANGUAGE (XML)

Understanding XML forms the starting point of this discussion, leading into Web services. This is because XML is at the core of WS. XML is a simplified version of SGML (Standard Generalised Markup Language), on which HTML (HyperText Markup Language) is based (Quin, 2004; Ray, 2001). HTML has its own defined tags, which cannot be modified. On the contrary, XML allows users to choose their own tags and elements depending on their need. HTML is a data and presentation language viewed by humans via browsers. XML is a data carrier document and is independent of any presentation using a related technology like XSLT (Extensible Style Sheet Transformation), which in turn is an XML document. The XML document can be viewed with different formats via Internet browsers, PDAs, or mobile phones, or it can be transferred to another XML document. Because of those features, XML represents for IT vendors a brilliant future as a common language and a medium to exchange data independently of the languages and the platforms used by applications. XML is dramatically and rapidly changing technology, and many believe that XML is the next revolution in technology. Below is an example of a simple XML document of a health record of a patient:

```
<?xml version=”1.0"?>
<Patient_ Health_Record>
  <Patient_Name>      Peter      Lee</
Patient_Name>
```

```
    <Patient_Mobile>0405060708</
Patient_Mobile>
    <Patient_Adress>20 Pit Street Sydney
NSW 2000</Patient_Adress>
    <Patient_Local_Doctor>
        <Doctor_Name> Michelle Fouler</
Doctor_Name>
        <Doctor_Mobile>        0415161718</
Doctor_Mobile>
    </Patient_Local_Doctor>
    </Patient_ Health_Record>
```

In the example above, there is an element called <Patient_Health_Record> that contains four sub-elements: <Patient_Name>, <Patient_Mobile>, <Patient_Mobile>, and <Patient_Local_Doctor>. The last sub-element <Patient_Local_Doctor> in turn contains two sub-elements: <Doctor_Name> and <Doctor_Mobile>. Each element or sub-element contains data such as <Patient_Name> Peter Lee</Patient_Name> where the data are *Peter Lee*. As shown above, each XML document has its own tags that could be introduced to others by XML Schema (Priscilla, 2002).

WEB SERVICES

The most successful use of XML comes from its role in WS. Earlier technologies like DCOM, CORBA/IIOP, and RMI have been used in integrating applications. These technologies could make two or more applications based on different languages and platforms communicate with one another. The problem is that these technologies are complex and expensive to implement, and they have to be implemented specifically for the applications in question. On the contrary, the WSs that are based on SOAP messaging are simple, easy to deploy, and can be used by any application. WS and SOAP use the ubiquitous Internet protocol HTTP as a transport protocol, which makes it even more popular. When an application uses Web services provided on the network, it does not actually integrate into the application computer; it instead invokes or calls from their locations on the network. Once services are called, they will be processed on their own machines, calculate the results, and pass them on to the client application over the Internet. Therefore, in the Web services environment, when we run an application, many machines will be working together to achieve the request¾that is, the application machine and the services' providers machines (Chatterjee & Webber, 2004)

Web services are resources available on the Internet, while Data Link Library (DLL) is made up of local resources, as shown in Figure 1.

In Figure 1(a), the application uses some DLL and .exe local files, while in (b) the client application uses and consumes some services located on remote machines. The application and service providers' machines connect together via the Internet.

Having mentioned the advantages of WSs, it is fair also to mention their disadvantages. As the Internet is directly involved in WS calls and return results, the performance of the application will depend primarily on the speed of the network. Therefore a slowing network could put the application down. The performance of the application will also depend heavily on the performance of the providers' machines. The consistency of the services provider is also questioned, as sometimes services are not available. Security could also be a concern when hackers may intercept the data on the network. Those disadvantages present real challenges for the future of WS. A lot of research is going to overcome those challenges and make WS a mature and practical technology.

Figure 1. The difference between using integrated and local resources (DLL) (a) and invoking remote Web services over the Internet (b)

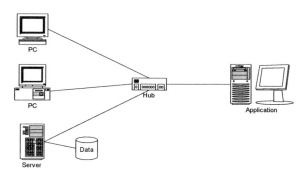

(a) Monolithic application using integrated and local services

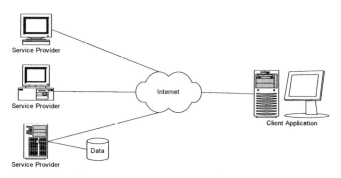

(b) Client application invoking remote Web services over the Internet

UNIVERSAL DISCOVERY AND DESCRIPTION AND INTEGRATION (UDDI)

WSs are usually registered and published through UDDI, which is in turn an XML-based document. UDDI is similar to the well-known yellow pages of the telephone directory. In other words, UDDI represents the yellow pages for WS, where most WS providers register their services. Developers who build applications based on WS resources search UDDI to find suitable and useful capabilities or services represented by functions (methods or behaviours) and classes. UDDI could also provide specifications about the WSs, which can be useful for potential client applications.

WEB SERVICES DESCRIPTION LANGUAGE (WSDL)

WSDL, which could be integrated into or separated from UDDI, provides complete details about WSs in terms of how to use, invoke, and connect to potential client applications.

WSDL describes all the services programmatically available for use, as well as the ports, messages, and protocols. In addition it should contain the XML Schema explaining the structure and the data type of the documents. It specifies in detail the parameters you should send in the message heading for the WSs and the expected response message if any. In fact WSDL describes the structure of the message that should be sent and received by the client application. WSDL is an XML-based document, which could be generated automatically by some tools like Microsoft .NET framework. Although service providers could communicate with the client applications directly, there is a strong need to register with UDDI which expose them to more customers and make them easier to be found (see Figure 2).

As shown in Figure 2, the application could be a normal Internet client application or a wireless/mobile application. The service provider can also be a mobile provider as well as normal services. Figure 2 shows that an application developer starts the process by searching the UDDI for required services needed in their application. The found methods representing the services must have been registered in advance with UDDI. The second step starts when UDDI provides specifications (WSDL) to the application about how to use and call the services.

SIMPLE OBJECT ACCESS PROTOCOL (SOAP)

Like UDDI and WSDL, SOAP is an XML document that contains an envelope as the root

element representing the message, and which contains an optional header element and a body element. The header is often used for security and encryptions. SOAP is used as a protocol of communications mainly between WS providers and client applications that consume WS. SOAP is carried by HTTP and is firewall friendly (El-Masri & Unhelkar, 2005).

SOAP-RPC

A SOAP message becomes more attractive when it is associated with RPC (remote procedure call), especially to those who suffer from the complexity of CORBA and DCOM. A client application can send a SOAP message containing a call and parameters for an operation (method) that belongs to some service providers. The WS centre or operator passes on that call to the operation or method of a class belonging to the WS provider. The operation returns a response carried back to the client by the WS centre (El-Masri & Unhelkar, 2005).

Figure 2. Discovering WS, developing application through WSDL, and registering by application developer

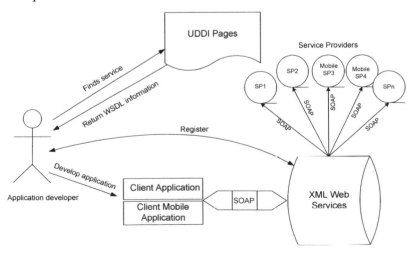

Figure 3. Mobile Web services—client and provider may be mobile devices

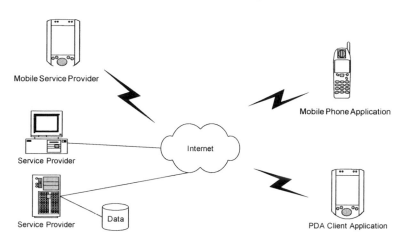

MOBILE WEB SERVICES

Let us start with an attempt to understand the meaning of the terms *mobile* and *wireless* devices. Mobile is wireless, but wireless is not necessarily mobile. Wireless devices do not require wires and cables to be connected to a network, and can be static and not moving such as a PC in a Wireless LAN (WLAN), wireless keyboard, and wireless mouse. As for mobile devices, in addition to being wireless, they can move freely in a wider environment and still be connected to the network such as a mobile phone. This chapter deals with the mobile term as it covers the other wireless one.

The scenario where a mobile device such as a mobile phone or PDA can wirelessly access the Internet network, and use and consume Web services, will claim the mobile Web services use. It is also possible that the WS provider would be hosted by mobile devices (El-Masri & Suleiman, 2005) (see Figure 3).

This case exists when a mobile phone can provide its location to some applications, for example parents like to locate their children when they go to school. Another example of a

Web service provided by a mobile phone could be details and information about the owner of the phone for employers so that an employer using a client application can look for the right employees by accessing some services provided on mobile devices (El-Masri & Suleiman, 2005). This will obviously require in both cases the authorisation of the user. It can be concluded that mobile Web services are used whenever there are mobile devices using or providing Web services to consumer applications (Chatterjee & Webber, 2004).

MOBILE WEB SERVICES CHALLENGES

Many constraints make the implementation of WSs in a mobile environment very challenging. The challenge comes from the fact that mobile devices have smaller power and capacities as follows (Chatterjee & Webber, 2004):

- Small power limited to a few hours
- Small memory capacity
- Small processors not big enough to run larger applications

- Small screen size, especially in mobile phones, which requires developing specific Web sites with suitable size
- Small keypad that makes it harder to enter data
- Small hard disk
- The speed of the data communication between the device and the network, and that varies

Having mentioned the current limitations of mobile phones, it is believed that people will use more PDAs than traditional mobile phones, as the former can have a wider screen, a more powerful processor, and larger memory and storage. As for the power and battery, a lot of improvement has been made.

PROXY-BASED MOBILE WEB SERVICES

The most popular MWS is a Proxy-based system where the mobile device connects to the Internet through a proxy server. This system is specifically useful when the mobile device is a mobile phone that has a limited processor capacity, limited memory and battery (see Figure 4).

Most of the processing of the business logic of the mobile application will be performed on the proxy server that transfers the results to the

mobile device that is mainly equipped with a User Interface to display output on its screen. The other important advantage a Proxy server provides in MWS is, instead of connecting the client application residing on the mobile device to many service providers and consuming most of the mobile processor and the bandwidth, the proxy will communicate with service providers, do some processing, and send back only the final result to the mobile device. An example of a client application on a mobile device that needs more than one service provider is a travel planning application, where airlines, hotels, and rental car services are needed at the same time.

MOBILE DEVICE PROTOCOLS

In the following sections, some mobile device protocols will be explained and discussed to show the infrastructure on which MWSs could be implemented:

TCP/IP

The transmission control protocol/Internet protocol was originally used to transfer data over the Internet network. This platform-independent protocol breaks data into smaller packets, transfers them over the network, and reassembles them back to reproduce the original

Figure 4. Proxy-based mobile Web services

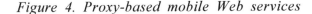
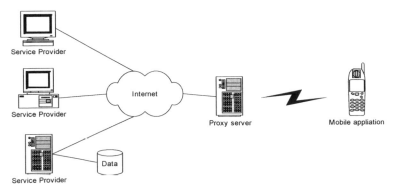

data at the destination computer. TCP/IP is the transfer protocol used to access the Internet and MWSs from PDAs or other devices that support TCP/IP-based network communication (Chatterjee & Webber, 2004).

WAP (GPRS)

Wireless application protocol (WAP) or general packet radio service (GPRS) is used as a transfer protocol for new mobile phones and other WAP-enabled mobile devices to access the Internet and MWSs. WAP has been introduced specifically to access the Internet from mobile phones with limited battery, memory, and processor capacity. In this context, the mobile phone will be used mainly as a user interface (thin client) to access the Internet through WAP Gateway software that deals with encoding and decoding Web sites using XML and Wireless Markup Language (WML) to present data to mobile phones with a text format (Chatterjee & Webber, 2004).

WIRELESS NETWORKING EVOLUTION

MWS operations may be carried out by two different infrastructures. The first is the Wi-Fi wireless environment, which is supported by the increasing number of hot spots implemented by telecommunications companies such as Telestra in large cities to provide access to the Internet for mobile users. Wi-Fi technology offers direct access to the Internet by covering cities with wireless access points connected to the network. The more Wi-Fi hot spots, the better signal and coverage we get.

The other technology is based on the mobile phone network, which is based on cells and satellites. By improving the bandwidth of com-

munication on mobile phones and make it reasonable, the Internet and all its services can be used on that network. In the following sections new technologies will be presented to improve the bandwidth of the cellular network.

GSM

The most popular mobile network is Global System for Mobile (GSM) communications. Within the GSM networking system, each mobile phone is allocated with a single slot of 9.6 Kbps. In new GSMs, this slot has been updated to 14.4 Kbps. It is obvious that this speed is very slow for accessing the Internet from mobile devices no matter what protocols (TCP/IP or WAP) will be used (Mobile Info, 2001).

3G

There is also a new high-speed implementation of GSM called High-Speed Circuit-Switched Data (HSCSD), which can multiply the previous speed (14.4 Kbps) by 4 to reach a transfer rate up to 57.6 Kbps. This is the 3G or *third-generation* emerging technology of mobile networking with a rate a bit higher than dial-up Internet access. 3G technology is increasing, but it is not yet fully implemented (Mobile Info, 2001).

4G

The growing need to access the Internet from mobile devices pushes networking developers to develop faster mobile networking with an aim to reach a broadband speed over a mobile network. A mobile broadband speed can be reached by *fourth generation* (4G). 4G is under test and will be implemented by different technologies such as FLASH-OFDM for Nextel's iDEN, EDGE for GSM, and EV-DO for CDMA (Mobile Info, 2001).

CONCLUSION

In this chapter, the principle of Web services was addressed, along with related technologies such as XML, SOAP, WSDL, and UDDI, and the great benefits they have brought to the development and integration of software applications over the Internet. The extension of Web services into mobile Web services, where mobile devices can provide services to other applications and can consume services from other computers on the Internet network, was also demonstrated. To effectively access the Internet and Web services from mobile applications built on mobile devices, there was a need for a larger bandwidth on the mobile wireless network. It was shown that larger bandwidth is already available using 3G and 4G technology. The new Wi-Fi technology has also been explained as an extension to the Internet network. The main challenges to the mobile Web services would be the consistency of the services, their availability, and the coverage of the mobile network as well as the security issues, as the wireless environment is even more vulnerable.

REFERENCES

Arora, G., & Kishore, S. (2002). *XML Web services—professional projects.* Cincinnati, OH: Premier Press.

Chatterjee, S., & Webber, J. (2004). *Developing enterprise Web services.* Englewood Cliffs, NJ: Prentice-Hall.

El-Masri, S. (2005). Mobile comprehensive emergency system. In *Proceedings of the 2nd International Conference on Innovations in Information Technology,* Dubai, UAE.

El-Masri, S., & Suleiman, B. (2005). Providing Web services on mobile devices. In *Proceedings of the 2nd International Conference on Innovations in Information Technology,* Dubai, UAE.

El-Masri, S., & Unhelkar, B. (2005). Modelling XML and Web services messages with UML. In *Proceedings of the Information Resources Management Association 16th International Conference,* San Diego, CA.

Microsoft & Vodafone. (2005a). *Convergence of PC and mobile applications and services.* Retrieved February 10, 2005, from http://www.microsoft.com/serviceproviders/mobilewebservices/mws_whitepaper.asp

Microsoft & Vodafone. (2005b). *Mobile Web services technical roadmap.* Retrieved February 10, 2005, from http://www.microsoft.com/serviceproviders/mobilewebservices/mws_tech_roadmap.asp

Mobile Info. (2001). *4G—beyond 2.5G and 3G wireless networks.* Retrieved August 8, 2005, from http://www.mobileinfo.com/3G/4GVision&Technologies.htm

Priscilla, W. (2002). *Definitive XML Schema.* Englewood Cliffs, NJ: Prentice-Hall.

Quin, L. (2004). *Extensible Markup Language (XML).* Retrieved July 10, 2004, from http://www.w3.org/XML/

Ray, E. T. (2001). *Learning XML.* Sebastopol, CA: O'Reilly & Associates.

Yuan, M. J., & Long, J. (2002). *Java readies itself for wireless Web services.* Retrieved January 20, 2005, from http://www.javaworld.com/javaworld/jw-06-2002/jw-0621-wireless.html

Chapter XXXIX
Push–Multicasting to Wireless Devices Using Publish/Subscribe Model

Jon Tong-Seng Quah
Nanyang Technological University, Singapore

Chye-Huang Leow
Singapore Polytechnic, Singapore

ABSTRACT

Push technology is a kind of technology that automates the information delivery process without requiring users to request the information that they need. Wireless has experienced explosive growth in recent years; "push" will be the predominant wireless service delivery paradigm of the future. We can expect a large number and a wide variety of services, alerts and messages, such as promotional content, to be delivered to consumer's phones or PDAs. To push information to wireless device becomes a challenge because of the problem of intermittent communication links and resource constraint on wireless devices as well as limited bandwidth. This chapter describes an efficient multicasting mechanism that "pushes" pre-specified information to groups of wireless devices. The mechanism is able to operate with limited bandwidth and also overcome the connectivity problem. A framework has been designed that implements the concept of push technology to multicast the sales information via wireless technology. The design of a message-oriented system for wireless information is described and that is followed by the implementation details that are compliance to Java Messege Service (JMS).

INTRODUCTION

Today, the vast scale and scope of current online information sources on the Internet makes it difficult to find and process relevant information. Finding a specific piece of information on the Internet requires time-consuming search. Hence, automation to push pre-specified information to the user seems like the next logical step to solve these problems.

We are moving towards third-generation wireless technology where multimedia applications are supported in wireless handheld devices and hand phones. It is believed that push will be the predominant delivery methodology in wireless device services. This is due to the problem of servers being unable to push data to clients who are disconnected. However this is not an issue in GPRS wireless networks. In GPRS wireless networks, the users will be always connected to the Internet. Hence, we can expect a large number and a variety of services, alerts, and messages, such as promotional content, to be delivered to consumers' mobile phones or PDAs in real time. In addition, there are some constraints in wireless technology, such as the small memory capacity in the devices, limited bandwidth, and the high cost of information searching on the wireless network. Push information to wireless handheld devices will save a great deal of time and money compared to surfing the Internet via WAP technology. Thus, this brings forth the idea, to create a "wireless push" channel to push information to wireless devices in real time.

Furthermore, from a software point of view, in order to cope with the limited bandwidth problem, this research will study an efficient multicasting mechanism to push sales information to a group of members over the wireless network. A simple application and framework about the Internet selling process has been designed, whereby the sales information like product catalog will be multicast to the interested mobile users in real time. The Publish/ Subscribe messaging model has been used as a message delivery mechanism in this framework because of the capability of the messaging system to multicast information to a specific group of recipients. The following three criteria must be satisfied in order for a piece of information to be considered suitable for delivery using push mechanism:

- The kind of information desired must be known ahead of time like stock quotes, new headline, and so forth.
- Searching for such information must be an inefficient use of the user's time.
- The user must want this information regularly.

DATA DELIVERY MECHANISM

The paper "Data in Your Face: Push Technology in Perspective" (Franklin & Zdonik, 1998) presents some ideas on data dissemination in order to provide a framework for thinking about push technology. The authors have outlined several options for data delivery, and the comparison of their characteristics is illustrated below:

- Client Pull vs. Server Push—Pull based is a request-response operation, which is client-initiated transfer of information from server to client, whereas in push-based operation, the server initiates the transfer.
- Aperiodic vs. Periodic—Aperiodic is an event-driven operation, where the transfer of information is triggered by an event. In periodic delivery, the transfer information is performed according to a pre-arranged schedule.
- Unicast vs. 1-to-N—With unicast data communication, the data to be transferred is sent from one source to one destination, while multicast and broadcast are 1-to-N data communication. In multicast data delivery mechanism, the transfer data is sent to specific subsets of clients, whereas in broadcast data delivery mechanism, the transfer data is sent to an unidentified set of clients that can listen to it.

Periodic push has been used for data dissemination in many systems, for example an Internet mailing list that regularly sends out mail to users. Aperiodic push, which is based on Publish/Subscribe protocol, is becoming a popular way to disseminate information to end users.

Benefits of Push Technology

The paper "'Push' Technologies: Reborn for Business" (Farber, 2000) summarized the benefit of push technology into one sentence:

It's mirroring what a really good human assistant would do if all they had to do was sit around and watch out for you.

Push technology has advantages for both end users and content providers. For end users, it significantly improves the response time of accessing Web content. Clients do not have to waste cycles and network traffic to poll servers.

For content providers, they are allowed to target their audiences in a more direct way, which results in a cleaner business model. Servers can use more processor time for data production rather than to process numerous client requests and send much data over the network. Furthermore, servers can better manage the amount of data transferred over the network as it delivered useful and interesting information to clients.

Constraint and Challenges in Push Technology

"Data in Your Face: Push Technology in Perspective" (Franklin & Zdonik, 1998) has identified some issues regarding push technology. Current Web casting applies pull technology instead of push due to limitation of HTTP protocol. HTTP protocol is a kind of pull-based protocol. Broadcasting and multicasting are not widely used as it potentially causes bandwidth problems if multi-clients request the same data through unicast data communication. In addition, the article "What's All Wrong with Today's Push Technology?" (Berst, 1997) has pointed

out that push technology should be built on top of multicast data communication. It is not advisable to implement push technology on top of unicasting data communication due to inefficiency that will result as the user based grows. Furthermore, the article "The Push Technology Rage…So What's Next" (Gerwig, 1997) also pointed out that the next step of push technology is a multicasting delivery mechanism that will enable true push technology to millions of users across the Internet. The Internet still uses the basic one-to-one ratio of one request for information sent to one computer at a time. Therefore Internet service providers (ISPs) will have to update their networks to handle multicasting, which will enable content providers to "broadcast" data to a large number of users rather than sending content using a one-to-one model. Furthermore, this article highlighted a need for a standard message in pushing technology as a common way to tell users what is in a content channel and what the users need to view it. XML as a standard of W3C has been briefly discussed in this article.

The article "When Push Comes to Shove: Push Technology Is All the Rage—What Does This Mean for Java?" (Blundon, 1997) discussed "Web casting" or "push" technology and how it can dramatically increase the productivity of the Internet or intranet in data communication. In short, the main challenges in push technology are:

- Multicast or broadcast information to those users who are interested to prevent clogging the network traffic.
- Standardize the push message to all channels, especially to push information to different platforms or devices, like various manufacturers' wireless phone, Palm OS, IPAQ, Visor, and so forth.
- Security will be an issue because the user allows information or software components to be downloaded to his local devices.

Figure 1. Publish-and-Subscribe messaging

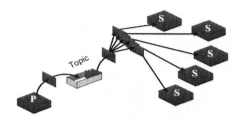

PUBLISH-AND-SUBSCRIBE MESSAGING—FRAMEWORK FOR PUSH SELLING IN WIRELESS NETWORK

The Pub/Sub messaging model is where there is only one sender and one or more receivers. Rather than queuing for sending and receiving messages, there is a topic, which is equivalent to an individual newsgroup. Interested clients subscribe to receive all messages sent to those topics, and they can publish to those topics if they wish to. The number of publishers and subscribers grows and shrinks over time. Figure 1 illustrates the process of Pub/Sub messaging where "P" is publisher and "S" is subscriber. This model is suitable for the push model application, as consumers are delivered messages without having to request them.

Wireless JMS

JMS provides asynchronous and message-based transportation. Using message queues hosted on both the client and the server side, JMS applications can be operated in disconnected mode, and data synchronization occurs transparently and immediately without user intervention. In addition, it provides an ideal abstraction layer for developing mobile applications and also increases the scalability of the system. Many mobile devices can simultaneously send messages to a server. When messages arrive at the server, they are added to an inbound queue

and can be dealt with when resources are available, or can be forwarded to other servers for load sharing. In practice, JMS allows wireless services to operate more responsively, to recover from sporadic network outages easily, and to allow mobile applications to be operated off-line. Finally, JMS allows messaging to be implemented elegantly atop Bluetooth (Salonidis, Bhagwat, Tassiulas, & LaMaire, 2005), Wireless LAN (Wu, Long, & Cheng, 2002), GPRS (Vergados, Gizelis, & Vergados, 2004), UMTS (Yang & Lin, 2005), and Mobitex (Wei & Jost, 2003).

Mobility Issues

In this section we will look at the mobility issues of developing mobile applications. Mobile applications present many challenges for software developers and require them to deal with the issues not present in wired line systems. The following summarizes the issues:

- **Intermittent Communication Links:** Mobile devices often lose network coverage. To deal with these intermittent communication links, the mobility application should provide disconnected operation and guarantee important data reaching mobile devices.
- **Resource Constraint:** Mobile applications must be optimized aggressively for small ROM and RAM footprints, as well as for low usage of CPU cycles and battery power.
- **Multiple Bearers:** Currently there are many different bearers in wireless networks like SMS, GPRS, Infrared, Bluetooth, and HTTP. Developing one mobile application and running it in different bearers becomes the challenge of developers.
- **Multiple Platforms:** There are various platforms in the mobile communicator

Figure 2. Architecture design of Push Selling—Multicast Message to Wireless Devices

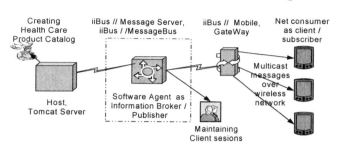

device market such as Palm OS, Pocket PC, Symbian, and Windows CE. It is advantageous to develop applications in such a way that they can be run on various platforms, without requiring substantial modifications.

Building mobile applications using a middleware with JMS implementation can solve most of these issues and problems. iBus//Mobile JMS middleware has been selected to prove the concept in this chapter. The following sections will elaborate on the implementation details of the application model proposed here.

Architecture Design

A simple application and framework has been designed as shown in the Figure 2. This push-based application implements the Internet selling process whereby the product catalog created by the seller will be multicast to mobile users. The proposed model mainly comprises three types of nodes:

1. The *host* provides the base data that is to be disseminated. In our example system, we assume the core business of the system is to sell health care products and services. The system will disseminate the products' information in the form of a catalog. The system provides a GUI for

creating the product catalog. This GUI is a JSP (Java Server Page) interface running on top of a Tomcat servlet engine. This GUI enables the user to create the product catalog, which will then be sent by the *host* to an agent known as an "information broker."

2. The *information broker* plays a vital role in acquiring information about the product catalog, and adds on additional value or data to that information and then pushes this information to clients. The information broker is running on top of iBus//MessageServer and iBus//MessageBus. In the event that this broker is being triggered, it will create a topic and all users who are subscribing to this topic should be able to receive the messages. This broker behaves like a "publisher" to multicast the product catalog to those active subscribers. The data delivery mechanism is based on the Publish/Subscribe model, which will deliver the information to a specified set of interested clients. In addition, the clients' sessions and configuration data will be kept and maintained in this node (i.e., the information broker server).

3. *Clients,* which are net consumers, will receive such selling information without requesting it via iBus//Mobile. iBus//Mobile provides a gateway for transferring

Figure 3. Publish/Subscribe Messaging in 1-N push technology

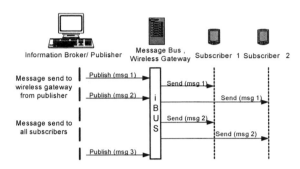

the messages to wireless devices. From the JMS provider's point of view, the gateway is a regular JMS client. From the mobile device's point of view, the gateway is a communications hub and message format translator. A simple MIDLET application has been developed using Light-Weight JMS Client Library. This library is provided by iBus/Mobile. The MIDLET will be loaded into a client's PDA and runs as a client's agent to receive the incoming push information. The MIDLET is built on top of a PALM operating system.

Message Delivery Mechanism— Publish/Subscribe Model

The message delivery mechanism implemented in this system is based on the Publish/Subscribe model, illustrated in Figure 3. The transfer data is pushed to specific subsets of clients. The following describes the message flows:

- Once a seller has created a product catalog, the system will trigger a Servlet to send this product catalog in the form of message to the information broker.
- The broker will create a topic named "push_info." In order to receive this mes-

sage, the mobile users need to subscribe to this topic.

- The broker will start publishing its first message, and iBus middleware will start sending the message to all subscribers. The middleware ensures that all subscribers receive the message of this topic "push_info."
- The broker is allowed to publish a second message, although the first message has not completely been sent to all subscribers.
- As time passes, new topics may be created by the broker. Users can log on using their user identification registered with the broker to browse through the Web pages displaying available topics and select/unselect topics according to their interests.

Result

This section presents the results of the implementation as a proof of concept. Figure 4 shows a user interface for sellers to create their product catalog. The GUI is developed in JSP and runs in Tomcat Servlet Engine. The page consists of several input textboxes for users to key in catalog information like product code,

Figure 4. User interface for creating product catalog

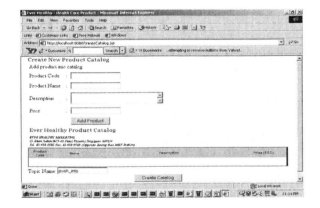

product name, product description, and price. The seller is allowed to create multiple items per catalog by clicking on the "Add Product" button. The seller publishes the catalog under a specific topic by keying in the topic name at the bottom of the page. Once "Create Catalog" button has been clicked, a product catalog will be created and the page will trigger a Servlet known as "triggerPushAgent." The Servlet writes the catalog message to a file, named "productcatalog.txt," and then it will instantiate the Information Broker. The broker will read the message from that file and multicast the message to all subscribers. The subscriber needs to enter the topic name to which he wishes to subscribe. His device will then receive all messages under this topic.

Once the subscriber has logged on, the application will be in listening mode and wait to receive incoming messages.

Figure 5 shows that the two subscribers, User001 and User002, have received the product catalog.

CONCLUSION AND FUTURE DIRECTION

This push-based selling system enables mobile users to receive product information in real time. The crucial element of the system is that the sender assigns messages to topics and not to a particular remote object, and then the receivers couple the messages based on the topics. The system provides a company as content provider to simultaneously push its selling information to multiple handheld devices such as PDAs. Therefore, from the perspective of the sender, it is more effective and efficient in transmitting large amounts of data to a group of receivers in real time. No additional user interfaces or efforts are required in selecting or customizing the group of receivers. In other words, subscribers received the selec-

Figure 5. Clients' application—receiving and displaying product catalog

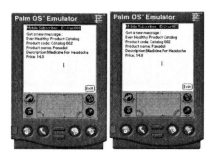

tively streamed information based on their indicated interests—as and when contents are added to topics at the broker, once connectivity can be established while they are on the move.

Furthermore, from the sellers' perspective, it allows targeting of their audiences in a more direct way, which results in a cleaner business model. The server can better manage the amount of data transferred over the network as it delivers information on topics that clients have subscribed to. This significantly enhances scalability and uses the available network bandwidth with maximum efficiency.

From mobile users' point of view, they do not need to spend a lot of time and money in order to get the information they want. The information is automatically pushed into their wireless handheld devices. We are currently working with two groups of early adopters of the technology. The first group consists of investors who want real-time feeding of stocks information into their devices for decision making. The second group consists of medical practitioners who need to get the latest condition updates of their inpatients and use the information to decide whether urgent attention is needed by those patients under their care. Both groups of users have given their thumbs-up for the trial system.

We are working on ways to better the current implementation. Research on IP multicast for mobile hosts, especially in a wireless environment like Mobile IP, is recommended to enhance the system and to realize true multicasting, which is not just from a software perspective, but also from a network infrastructure point of view. A multicast agent scheme for mobility support of IP multicast in the Internet is possible. The approach is to use three-layer architecture for multicasting to mobile hosts, and introduce multicast agents that serve as access points to the multicast backbone by mobile hosts and are located close to the current locations of mobile group members. At the Internet level, multicast agents are simply multicast routers participating in multicast routing.

Besides the multicast delivery mechanism, it is useful to also study the broadcast delivery mechanism. Broadcast is considered a type of push-based data delivery mechanism. The basic idea is, the server periodically broadcasts the sales information, while client agents monitor the broadcasted messages and only receive messages that they require. The advantage of this approach is that other clients who are also monitoring the broadcast do not directly affect the performance of any other client receiving data from the broadcast.

Furthermore, we are also researching the use of mobile agents in wireless handheld devices because they will overcome wireless network limitations such as low bandwidth and intermittence disconnection problems. Agents can act as buyers and sellers, and communicate and negotiate with each other autonomously (Schmid & Quah, 2004; Quah & Seet, 2004; Quah, Leow, & Soh, 2004).

REFERENCES

Berst, J. (1997). *What's all wrong with today's push technology?* Retrieved from http://www5.zdnet.com/anchordesk/story/

Blundon, W. (1999). *When push comes to shove: Push technology is all the rage—what does this mean for Java?* Retrieved from http://www.javaworld.com/javaworld/jw-04-1997/jw-04-blundon.html

Farber, D. (2000). *Push technologies: Reborn for business.* Vineyard Haven, MA: Vineyardsoft Corporation.

Franklin, M., & Zdonik, S. (1998). Data in your face: Push technology in perspective. *Proceedings of the ACM SIGMOD International Conference on Management of Data,* Seattle, WA (pp. 516-519).

Gerwig, K. (1997). The push technology rage...so what's next? *ACM Digital Library, The Craft of Network Computing, 1*(2) 13-17.

Quah, J. T. S., Leow, W. C. H., & Soh, Y. K. (2004, June 21-24). Mobile agent based e-learning system. In *Proceedings of the 2004 International Conference on Information and Knowledge Engineering,* Las Vegas, NV (pp. 256-265).

Quah, J. T. S., & Seet, V. L. H. (2004, June 10-13). Improving usability of WAP portal through adaptation. In *Proceedings of the 7th International Conference on E-Commerce Research,* Dallas, TX (pp. 407-420).

Salonidis, T., Bhagwat, P., Tassiulas, L., & LaMaire, R. (2005). Distributed topology construction of Bluetooth wireless personal area networks. *IEEE Journal on Selected Areas in Communications, 23*(3), 633-643.

Schmid, A., & Quah, T.-S. (2004). Synergetic integration of aglets and e-speak in e-commerce. *Informatica—An International Journal of Computing and Informatics, 27*(4), 391-398.

Vergados, D. D., Gizelis, C., & Vergados, D. J. (2004). The 3G wireless technology in tactical communication networks. *IEEE Vehicular Technology Conference, 60*(7), 4883-4887.

Wei, S., & Jost, A. G. (2003). Virtual socket architecture for Internet access using Mobitex. In *Proceedings of the IASTED International Conference on Wireless and Optical Communications* (Vol. 3, pp. 549-554).

Wu, H. T., Long, K., & Cheng, S. D. (2002, June). Performance of reliable transport protocol over IEEE 802.11 wireless LAN: Analysis and enhancement. In *Proceedings of IEEE INFOCOM 2002* (pp. 599-607).

Yang, S. R., & Lin, Y. B. (2005). Modeling UMTS discontinuous reception mechanism. *IEEE Transactions on Wireless Communications, 4*(1), 312-319.

Section VIII

Method

Chapter XL
A Methodology for M–Transformation of Small and Medium Enterprises (SMEs) and its Application in Practice Using CBEADS©

Ioakim (Makis) Marmaridis
University of Western Sydney, Australia

ABSTRACT

Organization worldwide come to realize that in the ever changing business world, survival and success is closely linked to adopting information and communication technologies (ICTs). Along with the technology however, organizations have to also adjust their processes to take full advantage of the potential ICTs have to offer. This process, of technology adoption, linked with process adjustment and re-engineering is called e-transformation. For organizations that have successfully e-transformed, it is now necessary to become more agile through the adoption of mobile technologies. This adoption leads to the need m-transformation which is the next logical step from e-transformation. In this chapter we define m-transformation and present a methodology that SMEs can adopt in order to m-transform. The methodology takes into account the special characteristics SMEs have and allows them to leverage their strengths towards a smoother m-transformation process. Furthermore, we show how m-transformation can be practically applied, and in doing so we introduce our technology platform called CBEADS. Finally, we present some of the lessons learned and demonstrate how SMEs may progress through the adoption of mobile technologies into their operation into gaining increased competitiveness and a global reach.

INTRODUCTION

Mobile technologies are a key influence in any attempt at globalization of business (Unhelkar, 2004). Therefore, what was once understood under the banner of "e-transformations" (or electronic transformations of organizations) now needs to be understood within the context of mobility. This leads to the idea of mobile-transformation or "m-transformation." M-transformation is the process of transitioning an existing organization into the mobile business world. The earlier e-transformations capitalized on the connectivity accorded by the ubiquitous Internet (Arunatileka & Ginige, 2003; Ginige, 2002). M-transformation is the next logical step for these organizations.

However, the many diverse benefits as well as the challenges of m-transformation are not well known, especially in the SME sector. As a result, there is uncertainty and trepidation. Furthermore, the mobile technology itself is in the process of development and is not fully matured. Based on our spot interviews with some of the SMEs that we had earlier e-transformed, we discovered that the aforementioned reasons were contributing to their reluctance to take up m-transformation. This has led to our interests and desire to investigate further this challenge and discover a set of simple yet consistent requirements leading to a methodology that can be applied in m-transformation. The scope of our study is mainly focused right now on SMEs, given their unique characteristics as well as our knowledge base of working with them in the past (Arunatileka & Ginige, 2003; Ginige et al., 2001; Marmaridis, Ginige, & Ginige, 2004a; Marmaridis, Ginige, Ginige, & Arunatilaka, 2004b; Kazanis, 2003).

This chapter starts with our working definition of m-transformation. It then explores some of the benefits that are due as a result of m-transformation. This is followed by a discussion on how SMEs perceive m-transformation—this section is meant to help us uncover and investigate potential disconnects between our understanding of mobile benefits and that of SME users. Next, we almost exclusively put the focus on the unique characteristics SMEs present and how they affect the uptake of m-transformation in the context of currently available mobile technologies and corresponding

business process mainstream thinking. With a solid understanding of the unique SME characteristics, we then proceed to describe the requirements for each aspect of m-transformation, namely ICT infrastructure, business process adoption, and m-transformation methodology. Finally, we describe the methodology in detail and provide insight from our experiences in applying this methodology to a number of SMEs wishing to m-transform their business.

BACKGROUND TO THIS WORK

Before diving further into the details of SME m-transformation as they are expressed via requirements for IT systems and technologies, it would be wise to offer some background information about the projects that we have so far undertaken and some of the technology used in those, namely CBEADS©. A number of years we identified the need for SME organizations in the greater Western Sydney region of NSW, Australia, to increase their uptake of technology in order to become competitive in the global economy and survive. A vehicle for us achieving this was through the process of e-transformation, where SMEs are encouraged and assisted in their uptake of computers and other information and communication technologies (ICTs) to better their business overall.

In order for us to facilitate the uptake of ICTs, we have also developed a framework known as CBEADS©, which stands for Component-Based E-Application Development (and deployment) Shell and provides a low overhead, low-cost infrastructure for e-business applications development and deployment. Along with the technology we also developed a comprehensive methodology for e-transformation, a concise map of which is our e-transformation roadmap that offers step-by-step guidance to organizations embracing e-transformation.

Having successfully applied the methodology and the CBEADS© technology framework to e-transform a number of SMEs, we are now seeing the need to assist those same organizations move into the next step of m-transformation. The overall approach to doing this is similar to that of e-transformation where we start with a methodology to m-transformation backed by ICTs and tools to achieve those goals faster. We include in this chapter our work in the area of deriving the methodology for m-transformation so that it can meet the unique requirements for SME organizations wishing to undertake an m-transformation journey. A lot of the findings and ideas presented here we strongly believe apply universally to SMEs elsewhere in Australia and the rest of the world. With the hope that the information we provide here will be useful to you, the reader, let us proceed with painting a picture of what SME m-transformation looks like.

DEFINING AND UNDERSTANDING M-TRANSFORMATION

The area of m-transformation is actively researched (Unhelkar, 2004), while on the other hand there are many m-enabled software products that try to capitalize on the high growth and demand for this area in businesses. Consequently, the process and scope of m-transformation remains undefined and not properly understood. For the purposes of this chapter, we shall use our working definition of m-transformation which is consistent with our view of it as the next logical extension to e-transformation.

We define m-transformation as "the evolution of business practices via the adoption of suitable processes and technologies that enable mobility and pervasiveness." In accordance with this definition, we see the adoption of m-transformation as having two sets of prerequisites: those of business process adoption and those of adoption of suitable mobile technologies and products. Without proper understanding of both these aspects of mobile transformation, embarking on an m-transformation trip can result in a reduced set of benefits flowing back into the organization or, at worst, failure of the transformation process.

Despite its challenges, though, it is essential that m-transformation is seriously considered by organizations. This is because the advantages of transitioning to a mobile business are also not unfounded. Some of the main benefits that we see stemming from a successful m-transformation are as follows:

- Better information flow between systems, since decisions can be taken quicker by making the necessary information available to the key decision makers faster.
- Improved customer projected image, since the adoption of mobile business processes enables the business to serve the customer in a personalized manner—adding to the projection of a forward-thinking and dynamic image of the company.
- Rapid and dynamic customization of products and services being offered by the organization—depending on the location and density of potential customers at a particular time and place.
- Increased internal efficiency in managing human resources. This primarily results from the fact that in an m-organization, information, and people are not being deskbound anymore.
- Improvement in employee morale coming from flexibility in the workplace—especially with the possibility of telework (Ranjbar & Unhelkar, 2003).

The aforementioned benefits can be summarized and grouped into three major areas

Figure 1. Major benefits from m-transformation of SMEs

Global M-Enabled SME

positively influencing m-transformation, as shown in Figure 1 (Marmaridis & Unhelkar, 2005).

With these major benefits—both tangible and intangible—that m-transformation promises, it comes as no surprise that there are a large number of SMEs—especially in the Western Sydney region that the authors have worked with—wishing to m-transform themselves. This may well extend the reach of these SMEs to global markets, as envisaged in Figure 1. However, as against a typical medium to large organization (like a bank, an insurance company, or an airline), SMEs have some special characteristics and present special needs that can be an impediment to their entry into the mobile business arena.

The section that follows describes in more detail what those characteristics of the SMEs are and how they could be overcome. However, let us start with putting m-transformation in the context of the SMEs.

M-TRANSFORMATION AND THE SME

Irrespective of our academic definitions of m-transformation and our view of its current status and future directions, the actual understanding of what it means to the SMEs has a major influence on its success or failure. The end recipients of our work can have a different view and interpretation of mobile transformation than the ones we formulate. From our experience, this has certainly been the case with SMEs and their view of m-transformation in general. Therefore, further discussions and explanations beyond formal definitions are required in this study.

SME Interpretation of M-Transformation

Whereas we, the researchers, saw opportunities for growth and a host of benefits stemming from m-transformation, some of the SMEs we worked with saw disturbing changes to their business and their ways of doing things associated with high costs of implementation. They could certainly appreciate that having the abilities to remotely access information could be, at times, invaluable. However, SMEs are not ICT experts, nor do they wish to understand the intricacies of mobile technologies. They want something that is simple and easy to use, and just works. Talking to SMEs that are fixated in this mindset about moving into m-transformation was a challenge. It was obvious that we needed to understand the real cause that established this mindset in the first place and then the reasons that were fuelling it. The answers to this came into two parts. Firstly, SMEs have a unique "results-based" view of doing business. The SME characteristics are unique compared to larger enterprises, particularly due to significant lacunae in strategic planning. Secondly, their understanding of ICT technologies that could be suitable for mobility is very fragmented and incomplete; what they seemed to be missing the most was the reassurance that those technologies once implemented will work with each other and be consistently and easily accessible.

Table 1. SME characteristics affecting m-transformation

SME Aspect	SME Characteristic
Staff	Small headcount, affected by seasonal conditions. For example, departure of an IT staff member or power user may mean the entire IT support has vanished. This can create challenges if mobile transformation depends on a particular staff member.
	Limited time and critical availability of business staff. Training users can prove a significant challenge.
	Lack of ICT knowledge and lack of time and interest in gaining that knowledge.
Clients	Each client is highly valued. Client primarily dictates the business. Views of clients on mobile business processes critical in attempting m-transformation.
ICT Systems	Very small budget available covering mostly ongoing running and maintenance with no provisions for new developments.
	Lack of in-house ICT expertise and minimal development in-house.
	Ad-hoc systems establishment and lack of any strategic planning for ICT. Mostly off-the-shelf, point solutions used, making mobile systems integration a challenge.
Decision Making	Decisions are quick, usually taken by the owner/proprietor, and the conditions for decision making are very volatile. Individual personalities may prop in the m-transformation decision-making process.
	Organisational knowledge is tacit and centralized in few key staff. Concern for "loss of turf" or area of influence due to m-transformation is significant.

SPECIAL CHARACTERISTICS OF SMES

Let us consider the special characteristics of SMEs that influence their m-transformations. We were able to glean these characteristics through closely working with numerous SMEs in the corresponding e-transformation processes mentioned at the start of this chapter. Table 1 illustrates those in relation to staff, clients, IT systems, and decision-making needs.

Let us now examine how each of these impact on the SME operations and could therefore influence m-transformation.

Small Staff Count Affected by Seasonal Conditions

SMEs come in all sizes, from one-man operations to over 200 staff. The typical SME size however is between 5 and 100 people. Two distinct characteristics exist in relation to the SME staff, the fluctuation in numbers and casual nature of work, as well as the over-reliance on particular staff. SME staff numbers greatly fluctuate in response to seasonal factors and varying demand for products and services. Therefore most staff tend to be employed on casual or short-term contract basis. There are only a few people within each company that have a sense of continuation with it.

Due to the few permanent staff, there tends to also be over-reliance on some of those. For instance, the departure of an IT staff member or a computer power user may significantly drain the IT expertise of a particular company. This can lead to challenges in the mobile transformation, as it may be heavily dependant on particular staff members within the organization.

Small to Non-Existent ICT Budget

The typical budget for an SME of average size is $3,000 AUD or less per annum, based on our experience. This amount, which for a large enterprise might equate to just three days of onsite work for an external consultant, will allow an SME to refresh some of their aging hardware for instance. This budget is typically aimed at ongoing running and maintenance of the ICT infrastructure, as opposed to development or acquisition of new technologies for m-transformation.

Lack of Permanent In-House ICT Expertise

In-house ICT expertise is very scarce in SMEs. Organizations of 100 people and under typically have no in-house IT staff due to cost implications. Instead, they rely on external IT contractors to provide, install, and maintain equipment for them as required.

In our experience, reliance on IT contractors and lack on in-house IT staff can significantly hinder the ICT infrastructure planning for m-transformation. IT contractors are only called in when there is a problem requiring their attention, and they tend to operate in "fire fighting" mode. As a result, there is little done in planning of technologies for mobility and other areas.

Staff Have Limited Time and Critical Availability

Because of the small headcount in SMEs, roles tend to be broad, with no backup people to fulfil those should the primary person become unavailable. Also most staff are occupied with carrying out important, day-to-day business functions. These two facts pose significant challenges to the uptake of m-transformation since staff training can prove a challenge. Unlike a large corporation, most SMEs cannot afford to have some of their business staff taken out of their productive capacity for days at a time. A company might significantly suffer,

for instance, if their accountant was being trained in a new IT system for two days in a row. A lot of the accounting work would either lag behind or not get done at all.

ICT Systems Establishment is Done Ad-Hoc

There is a significant lacunae in strategic planning in SMEs, especially in the area of ICT systems to aid m-transformation. Most ICT systems in use by SMEs tend to be off-the-shelf, point solutions with very poor if any integration with each other.

The rapidly changing business environment requires SMEs to select and implement ICT solutions fast and with minimum cost. Coupled with their lack of expertise to forward plan those solutions, they tend to rely on commodity hardware and package-based software for particular functions.

As a result, in a typical SME one can find a host of assorted applications, with no standards in regards to the hardware used, no standard operating environment (SOE), or any thoughts of interoperability provided between applications or data stores.

In such an environment, mobile systems' integration can prove to be a challenge, especially given the resources SMEs have available towards such integration work.

Organizational Knowledge is Centralized and Tacit

The organizational knowledge in an SME tends to be tacit and centralized in a few key staff. It is not uncommon in our experience that only a single person in the company (typically the owner) has full knowledge about the company's production, sales, financial, and client information. In larger SMEs this key knowledge might reside with two to three staff in total.

With the knowledge being so centralized, there are obvious and significant concerns for "loss of turf" or area of influence due to m-transformation. The few key staff members that hold a lot of influence in the organization may view m-transformation as taking away from their influence and power. It is therefore possible for some to oppose the move for m-transformation, significantly hindering its progress and chances of success.

Very Limited Diffusion of ICT Knowledge among Staff

To most SME staff, ICT is just another tool that can be used to enhance their productivity, allowing them to perform their work tasks faster and easier. As most of the staff, however, is under time pressure to perform their day-to-day tasks, they are left with very little time for gaining additional ICT knowledge. Most SME staff have in fact very limited knowledge of IT and, given the time pressures already on them, a very limited or diminishing desire to increase their know-how in that area. Most people are content with what they already know as long as it is enough for them to perform their work tasks.

Considering the limited ICT knowledge and desire of staff to expand it, there are concerns with regards to the uptake of new technologies and rate of adoption of m-transformation. Very few staff will see added value in the new mobile technologies at the beginning. For most, these systems will initially be yet another thing they must learn about, adding onto their existing workload.

Decision Making is Quick and Reactive

The fact that today's business environment is ever changing is well accepted. Even more so is the fact that fearsome competition demands fast-paced action and decisions to be handed down as quickly as possible (Ginige, 2004). This is of course all true with any company, but the need to deliver faster, be more agile, and respond to signals from customers as soon as possible is most pressing for SMEs. Those increased pressures lead SMEs into a mode of decision making that is very quick, based on a limited set of facts and appearing very reactive.

Unlike larger organizations, where decision making is a formal and rather rational process, in an SME, most decisions are made by the owner/proprietor who tends to be involved hands-on with the company operations.

In our experience, decision making in the context of SMEs is influenced by the personality and character traits of individuals. Individual personalities, for instance, may prop in the m-transformation decision-making process.

Each Client is Very Highly Valued

SMEs place a lot of value on their clients and are very sensitive to their wants and needs. Although clients are very important for organizations of all sizes, this is particularly true for SMEs. They lack global reach and generally have to compete with each other to maintain and grow their client base from a limited pool of prospects.

Therefore, the client primarily dictates the business, and their voice is heard. Views of clients on mobile business processes are critical in attempting m-transformation. M-transformation aims at providing the SME with greater efficiency while enhancing their clients' experience in dealing with them. It is therefore very important for SMEs to closely monitor the needs of their clients and implement the right mix of technologies in their m-transformation to effectively fulfil those needs. They cannot afford to leave their clients outside m-transformation decisions, as the risk of dissatisfying them is imminent and great.

As per the working definition that we have adopted, m-transformation has two sets of prerequisites: those of business process adoption and those of adoption of suitable mobile technologies and products. To coordinate the introduction and mix of these two prerequisites throughout the process of m-transformation, there is also a need for an overarching methodology that can act as a roadmap and guide to the process. The unique characteristics of SMEs that were discussed previously will certainly impact on the business process adoption, the technology mix and adoption, as well as on the methodology itself. The next section discusses how these characteristics of SMEs affect the entire m-transformation process and its constituent parts while describing the requirements those parts have to meet in order to be effective towards the m-transformation process of an SME organization.

ICT SYSTEMS REQUIREMENTS FOR SME M-TRANSFORMATION

Stemming from our critical analysis and close work with SMEs, we have derived a set of requirements that ICT systems for m-transformation should adhere to in order to be used effectively by SMEs. Meeting these requirements means that they can better fit with the SME way of thinking and approach towards m-transformation, hence being effective to address some of the particular characteristics of SMEs that were discussed previously. These requirements for ICT systems to assist in SME m-transformation are summarized in Figure 2.

As Figure 2 shows, there are six different requirements that ought to be met.

Figure 2. M-transformation IT systems requirements for SMEs

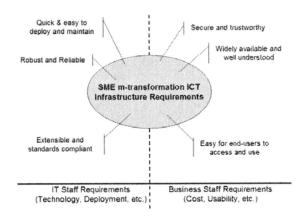

Solutions Quick and Easy to Deploy and Maintain

Solutions for m-transformation suitable for SMEs must be relatively straightforward and quick to deploy while requiring very little maintenance, if any at all. Whilst it is fine for a large company to choose to write custom software or standardize its IT system on particular languages and databases seeking peak performance, these things matter little to SME users. A lot of researchers and many large organizations see it quite fit and natural to build solutions for mobility based on XML (Naedele, 2003), Web services (Staab et al., 2003; Dustdar, Gall, & Schmidt, 2004), and large Java frameworks (Kawashima & Ma, 2004). While this is quite fitting for larger organizations, in the context of an SME, these approaches are mostly unusable due to their complexity and cost.

Secure and Trustworthy

Whatever solutions are adopted to fulfil the technical part of the m-transformation must be secure and trustworthy. Each ICT system an

SME adopts will typically handle a significant part of their business needs, hence security is paramount. This includes security at both levels, internal and external. Every system should be capable of preventing external unauthorized access to itself and its data. It should also however have safeguards in place to limit the access of its current trusted users to the designated data and application areas only. Any attempts a user makes to access information beyond their privileges should be logged. Subsequently, these exception logs should be readily available to designated staff who can take action accordingly. In addition to this, ICT solutions must also have built-in mechanisms of enforcing integrity of their data and prevent unauthorized changes to access logs, security logs, and other such system information. This feature will make them trustworthy and provide a lot more credibility to their output information and reports.

Robust and Reliable

ICT systems to assist in the m-transformation process are bound to be heavily used and relied upon. As such, in order for users to gain confidence in the system which will in turn guarantee its continuing use, it has to be reliable. There is little that can be done to draw users into using a system in the long run if they perceive it as unstable or erratic. Also for users to lose confidence in a system is much easier and faster than for those to gain the same amount of confidence. Beyond appearing reliable, ICT systems must also be robust. By this we mean that they should cater for incorrect use and be prepared to deal with end users that are not sufficiently trained in using the system. Instead of cryptic error messages or crashes, or worse still giving out the wrong information to users, the system should never trust user input, provide informative error messages, and in general

expect the unexpected from its users as much as this is practically possible. A robust and reliable system that covers only some basic functions has far more chances of success in an SME organization compared to a system that is feature packed with a lot of pleasing user interfaces, but which fails often or demonstrates an erratic and inconsistent behavior.

Extensibility and Standards Compliance Are Paramount

IT systems developed or purchased must be extensible, able to cope with change, and standards compliant. This is particularly important for SMEs since they do not have the know-how or funds necessary to develop or keep maintaining proprietary solutions. Whatever systems they adopt towards their journey into m-transformation should be based on widely accepted standards, so that the chances of future interoperability with other systems are much higher. Also because of the changing business environment, requirements for mobile access to applications and data are also bound to change often. It is therefore crucial that the systems an SME uses, no matter how simple they are, can be extended to accommodate for those changes over time.

Widely Available and Well Understood

Stemming from the fact that in-house IT expertise is scarce in the SME space and the cost of engaging external experts can be prohibitive for some SMEs as well, it is important that IT systems selected to assist with m-transformation are widely available elsewhere and well understood. It stands to reason that the more commonly used a piece of technology is, the more know-how will be readily available for it in the market, hence at an affordable price. On the other hand, less known or used technologies getting expert support might prove to be quite a costly exercise. An example of technologies that we feel meet these parameters are the Linux vs. the netBSD operating systems. On one hand, Linux is widely used and gaining in momentum every single day, with a very significant following of IT experts backing it up. This means that finding a consultant that understands Linux and can provide maintenance or other services around it is not very difficult at all, and because of the number of people available, the hourly rates for such consultants are very reasonable. On the other hand, netBSD is another free and open source operating system that in many respects provides superior performance than Linux out of the box. It has a significant number of followers as well which, however, are much less compared to those for Linux. For an SME to get by netBSD expertise would be a hard task, especially when they are in a hurry or in the event of a system problem. Accordingly, even if they find an expert in netBSD, it is expected that the cost for such a person would be higher than the equivalent one for a Linux system.

ICT Solutions that are Easy to Use

Whatever solutions are adopted to fulfil the technical part of the m-transformation have to be easy to use. The more common the technologies used are, the easier it is convincing an SME to embrace them since the perceived risk stemming from their usage is reduced and their effectiveness largely proven. Also the training burden will be less, and for very common technologies and systems, it is possible that staff might already be well versed in their use from past roles they have had. Such an example is the Microsoft Office suite of software, for instance, where most sales and administrative staff would typically have enough knowledge

of how to use it to be productive with it straight away. For less common systems, we found that integrating additional functions onto existing applications that staff are using on a daily basis makes for a very quick way to have this functionality adopted by end users. For instance, in most of the SMEs we work with, a simple e-mail to the SMS gateway system would boost productivity significantly, allowing for faster, reasonably priced communication that most staff could use with virtually no training at all. On the other hand, offering SME staff a state-of-the-art PDA to use over GPRS to access their company's systems might prove an error-prone, complex undertaking with a lot of associated costs which could also face resistance from end users since they have to get significant training in it before they can be productive using it.

Technology aside, business process adoption is the second prerequisite ingredient in a successful m-transformation, and it is equally affected by the unique SME characteristics identified previously. The section that follows discusses the requirements for business process adoption during m-transformation.

BUSINESS PROCESS ADOPTION REQUIREMENTS FOR SME M-TRANSFORMATION

M-transformation is typically a long process during which an organization undergoes quite a few large and several smaller changes. These changes include, of course, ICT systems and infrastructure, but also they relate to business process changes and the way staff go about performing their daily tasks.

As such, it is important to establish the requirements to be met for the smooth and successful adoption of business processes by SME organizations undergoing m-transformation. These requirements are discussed below.

Small and Incremental Changes

It is generally true that the big-bang approach works for only very few things when it comes to dealing with SME organizations. Because of their size and capabilities, SMEs tend to be very much influenced by change. Hence, adopting new business processes that allow for mobility has to be done as a series of small and incremental changes, as opposed to the big-bang approach. Business processes in SMEs tend to be long standing, and they change only when they absolutely have to. In our experience working extensively with SMEs, it is very rare that they will pursue best practice or that they will knowingly change business processes of theirs that "work" for better, more efficient ones. For most companies their current processes are the result of years of evolution and trial and error; expecting therefore that they will abandon those overnight for mobile-enabled processes is fairly unrealistic. When on the other hand changes are introduced slowly and are small enough to be easily understood by staff, while at the same time compliment the existing process and not upturn it, they are far more likely to succeed.

Adoption Must Minimize Risk by Changing Non-Core Processes First

Another requirement for adopting business process changes during m-transformation is that these are done to non-core business processes first. As the case is with every type of change, the exact results of the change are unknown until it is done. In some cases the effect can be predicted, but when it comes to business processes, it is best that the risk is minimized. This can be done through the introduction of mobile capabilities to non-core processes first. Support processes, as they are called, can provide

a great test bed for putting the technology through its paces, as well as allowing staff to get accustomed to change in the way things are done, while realizing the benefits of mobile technologies. As staff become accustomed to the new technologies and the abilities of performing tasks faster and smarter with increased flexibility, they are more likely to welcome changes to some of the more critical, core business processes. Also by the time those get to be adopted to a mobile-enabled environment, the return on investment for m-transformation typically has already started materializing as well, proving a further incentive for continuing with m-transformation.

Change Introduction Must Be Gradual and Reversible

Another important requirement for the process of adopting new business processes during m-transformation is the ability to be gradual and the changes reversible. SMEs tend to place a lot of value on their existing processes and be attached to them. Therefore, gradually introducing the changes allows staff to settle into the new routine of doing things, as well as over time detach themselves from the old habits. Additionally, as yet another means of reducing risk, both real and perceived, changes should be reversible. This means that business process changes ought to be well defined and measurable, so that if in the course of monitoring the change it turns out that it is underperforming or has bad side effects, it can be reversed, leaving things to their previous working state. This is nearly commonplace in larger organizations; however, in the SME space it is very often overlooked. It is frequent in our experience for the management of the SME to take drastic, nearly ad-hoc decisions under pressure in response to an adopted business process that is under performing. In those scenarios, instead of reversing the process to its predecessor, they often try to "patch" it, with even worse results sometimes.

Return on Investment Must be Continuous and Apparent

The discussion of the SME characteristics made clear that they like a large, dedicated budget to put towards their efforts of m-transforming. Hence, every dollar spent towards that goal has to offer and also be seen to offer value as soon as possible. To that end, adopting existing business processes to work in a mobility-enabled environment is no different. It is essential therefore that new or improved business processes offer continuous and apparent return on investment (ROI). It is hard to keep the SME management interesting in perusing the m-transformation goals if they see no obvious value stemming from the changes their business processes are undergoing along the way. This is particularly true with the adoption of business processes, given that they are intangible—unlike PCs or networking gear, for instance, which people can readily see and touch as soon as they are acquired and deployed.

Change Should Gradually be Applied throughout the Organization

While trying to minimize risk, we identified the requirements of business process adoption for mobility to target non-core processes first. Another requirement now is added, that of ensuring that changes to business processes will cascade over time to reach as many staff members across the organizations as possible. The reason for this requirement is to ensure that there is enough penetration and visibility of the changes that are happening towards m-transformation throughout the staff population.

Without this taken into account, business process adoption and therefore m-transformation may be perceived as something smaller than it really is, affecting only a handful of staff—especially when it is first applied to secondary business processes that are typically inside a single department or between a handful of individuals within the company. On the contrary, the changes have to eventually cascade down and flow onto as much of the staff as possible. This way the benefits of m-transformation are more readily realized, while at the same time everyone gets to see the true impact that m-transformation is making across the organization, hence gaining more buy-in and support at all levels.

METHODOLOGY REQUIREMENTS FOR M-TRANSFORMING SMES

So far we have discussed the requirements for ICT systems and the adoption and change to business processes as part of m-transformation as they are dictated by the SME characteristics. There is however one more component that must be looked at very carefully when dealing with SME m-transformation. This is the methodology. The methodology for SME m-transformation is very important because it is what will "glue" together the technology on the one hand and the new or improved business processes on the other to produce the desired set of outcomes in the end—namely, increase the efficiency of the organization, offer better management and utilization of resources, and allow for more responsive customer service. As the case has been with both technology and business process adoption, the SME characteristics also impact the methodology itself. Therefore, it needs to also meet particular sets of requirements in order for it to be well suited to the task within the SME space. Some of the most important requirements are shown as follows.

Staged, Well-Communicated Process

Taking an SME from its current state to being fully m-transformed will certainly take time. The process duration can vary depending on the size and awareness of the SME, its level of technical competence, and many other factors. What is certain however is that m-transformation is not going to happen overnight for anybody. To that end, the methodology must adopt an overall staged process that leads to full m-transformation. For instance, there is no need to try and implement the entire infrastructure that has to be put in place all at once. In fact doing so is a sure way to fail quickly. Instead a gradual ease into the new technologies would be the best path to their effective adoption and use. The same holds true of course for the business process adoption as we have already discussed. The methodology must also allow for communicating the process in advance with the stakeholders in the organization. M-transformation requires quite a few resources and continuous commitment to it in order to succeed, hence people ought to be kept informed about what is going to happen next before they are expected to embrace change and promote it.

Benefits of M-Transformation Must be Shown Quickly and Continuously

Very much like the case is for ICT systems and business process adoption, the methodology for m-transformation as a whole should ensure that the benefits of the efforts can be easily seen and that they start to come into existence as early into the process as possible. If results are not shown soon, people are bound to lose faith

and confidence in the process, as well as lose sight of the end goals of m-transformation that will only materialize after a relatively long time and through persistent investment.

Methodology Must be Balanced Between ICT and Business Processes Adoption

In order for the methodology to be suitable, it has to offer a balanced approach between technology adoption and business process re-engineering or adoption. The two prerequisites of m-transformation must be kept in sync, therefore the methodology has to have provision for monitoring these two very important items and track their progress over time. The actual ratio or "mix" between the two will certainly always be different, as it depends on a lot of parameters like the current levels of technology adoption within the SME and their natural capability to respond to changes well or not. By applying a methodology that does not make provision for measuring and tracking the amount of each ingredient used, it is nearly impossible to keep track of the current mix, let alone optimize it to better suit the particular SME undergoing m-transformation.

Plan for the Worse and Expect Resistance to Change

Humans have a natural tendency to resist change and in many situations will go out of their way to ensure that the status quo is maintained. This is also something that might happen during the journey of m-transformation inside an SME. Some people will be receptive to change, while others will be less so and typically a minority will actively oppose any kind of change. The reasons behind the resistance to change are many and diverse, and we shall not cover them here. Suffice to say that the m-transformation methodology has to be built taking the worse case scenario in mind, as well as the fact that people shall resist changes necessary for the full implementation of m-transformation. Therefore the methodology has to be flexible to deal with such situations and provide multiple levels of fallback in order to effectively manage those accordingly.

Methodology Evolution Must be Based on Feedback Received

Just like m-transformation is a dynamic process that over time has to adjust to the changes of the environment until it helps the organization reach its longer-term goals, so should the methodology for m-transformation be as well. More specifically, the methodology must have built-in feedback loops where input from the SME itself should be actively solicited and used to adjust the future steps or timing of those as prescribed by the methodology for m-transformation. If the methodology was set in stone and was completely isolated from the environment, it would have very few chances of actually being successful. The requirement for gathering, analyzing, and incorporating feedback into the next steps of the methodology along the entire m-transformation journey guarantees that

changes in the environment will be taken into account, hence increasing the chances of success for the entire process of m-transformation.

The section that follows describes the actual steps in the methodology for m-transforming SME organizations.

OUTLINING THE METHODOLOGY FOR M-TRANSFORMING SME ORGANIZATIONS

After discussing the requirements of the methodology for SME m-transformation, we will now move on to outlining the actual steps involved in the methodology itself. As Figure 3 shows, there is a total of eight steps involved and a prescribed sequence in which those steps need to be taken. You will also note that some of these steps are repeated more than once throughout an m-transformation project.

With Figure 3 in mind, setting the overall framework for when and where each of these steps is taken, let us now look into each of the steps in more details.

Identify the stakeholders. The first step in the process it to identify the stakeholders in the SME organization. In other words, identify and establish good working relationships with the

Figure 3. Considerations in selecting ICT for SME m-transformation

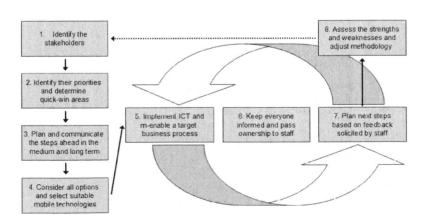

Figure 4. Considerations in selecting ICT for SME m-transformation

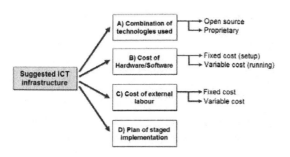

key staff in the company that can provide you with information and support for the m-transformation process. Some of these people of course would include the managing director and owner(s) of the company. In addition to those, other key staff might include the financial controller or chief accountant of the company or perhaps the production manager if the SME is involved in the manufacturing sector.

Identify their priorities and determine the areas where mobility could offer a quick win or significant competitive advantage. Via interviews with the key stakeholders identified previously, one should be able to identify a few important pieces of information, including the priorities these people have (and collectively the organization has) in regards to areas in which they believe the introduction of mobile-enabled technologies will be more appropriate. Also through understanding more about the organization, one could determine what particular processes, departments, or other areas of the organizations could realistically offer a good initial test bed for trialling the changes and technology associated with the m-transformation effort. These areas or processes would ideally be able to provide either a quick-win situation of a significant competitive advantage once they become mobile enabled.

Put together a medium- to long-term plan of the steps involved in m-transforming this

particular SME and make it available to all staff. This is an extremely important step that should be noted accordingly. It puts the foundation of establishing channels of open communication and information dissemination throughout the SME, allowing as many people as possible to be informed and also get excited about the upcoming e-transformation. Ideally these steps should establish m-transformation as the process through which the company will realize some major benefits as opposed to a black box process that involves a lot of time and money investment whilst offering unclear or undefined results.

Consider all options when selecting suitable mobile technologies. The selection of technologies capable of fulfilling the ICT requirements for m-transformation is an important step. During this, one should consider all options within the SME budget and decide which combination of technologies will provide the best value, while meeting the requirements set out previously with regards to ICT systems. In addition to those requirements, Figure 4 also offers a comprehensive overview of additional things that would need to be considered before arriving at a final decision for the technology mix of choice.

Implement the technology infrastructure selected in an incremental manner while using these added facilities to enhance aspects of the business processes previously identified as quick-win situations. This is one of the key steps that will repeat over and over for each targeted business process during the entire course of m-transformation for the organization. In this step the theory and planning done so far is put to the test. The ICT framework is established or enhanced, and the added functionality is used to improve existing business processes or replace old ones with new, more efficient processes. It is also during this step that resistance will be felt from some of the staff, since now they would realize that m-transformation is actually affecting how they do or they used to do things, as opposed to being just an initiative on paper with no actual implications to their roles and duties. Because this step will be repeated for as many business processes as will become mobile enabled, ongoing management buy-in is crucial along with wide acceptance of the changes and progress by the majority of staff.

Ensure everyone is kept informed and takes ownership of the process as appropriate. Key to the ongoing success of the effort towards m-transformation is keeping everyone informed, making progress visible, and also passing the ownership of certain subtasks involved in m-transformation to staff of the organization directly so that they can relate to those and push them further ahead. Although we already mentioned this numerous times before, we cannot stress enough the importance of establishing contact and a communications channel to the staff of the SME organization through which they can find out about the progress of the m-transformation process as a whole. Without this, people will quickly lose sight and faith in m-transformation which, in turn, will have a detrimental effect to it.

Solicit feedback and use it in planning the next steps towards the goal of m-transformation. This step is also one that repeats over and over in the methodology, and it is important to do so in order to ensure that the next few steps planned will be effective and carry the entire process forward. There are numerous ways of getting the feedback; what is more important to understand at this stage, however, is that there is a need to collect feedback from staff at all levels and incorporate it into the future planning of activities and tasks in the m-transformation journey.

Assess the strengths and weaknesses of the project. At the completion of the process, carry out an assessment of what the strengths and weaknesses of it were so that they can be provided as feedback in the overall model of SME m-transformation and improve the methodology over time, as well making it even more suitable for the task at hand when engaging with a new SME wishing to undergo m-transformation. This step will help ensure that the methodology steps are kept in sync with the actual needs and set of parameters that affect SMEs in their journeys to m-transformation. That way the methodology can better serve the cause and also grow and evolve over time, increasing in its completeness that will hopefully lead to it leading to better results for subsequent m-transformation projects.

Armed with the actual methodology steps, the next section will help you put them in context by offering some insights we gained from applying this particular methodology to a few different SMEs that we are closely working with.

PUTTING THE METHODOLOGY INTO ACTION—EXAMPLES FOR SUCCESSFULLY M-TRANSFORMING SMES

Through the years we have worked with many SME organizations, first assisting them with their e-transformation journey and later with their m-transformation. Through these engagements we have learned quite a few valuable lessons. In this section we give some brief insights to the things we have learned in the hope that you will not have to learn to avoid those the hard way, but you will rather know of some of the pitfalls ahead of time. Here are some of the lessons learned, along with a brief outline of each one.

Do not rely on management buy-in alone to promote m-transformation. Even though you have heard us talk about management buy-in for m-transformation in numerous places throughout this chapter, one of the things that we found out in practice was that on its own it is not enough. Management buy-in is essential in every way for m-transformation to start and keep going. If however there is no buy-in of e-transformation from the wider staff base, it can be seen as yet another management initiative "pushed down peoples' throats" and lead to increased resistance against it from staff members. This is why the methodology puts so much emphasis on gaining the management buy-in up front, but also open and maintain an open channel of communication with all staff, not only the senior management of the organization.

Give ownership of the e-transformation process to people, and help them progress it further. We have already spoken of the need to pass the ownership of individual business process adoption on to staff members, and allow them to provide feedback and contribute to their future shape. What we also found to be very important is the need to, as a whole, push the ownership of e-transformation on to staff within the company. In cases where we acted just as a mediator between technology and business users, with the latter carrying the ownership of m-transformation, the results have been very good. On other hand, acting as a gatekeeper and protecting the overall process while pushing the ownership of individual processes only to people can have the wrong effect. It sometimes sends the message to staff that they are not trusted to handle the larger process of m-transformation and that they are instead micro-managed throughout the entire journey.

Keep a balanced position against technology and business users throughout. Depending on your background, whether technical or more business oriented, you may have a natural tendency to favor either the technology or the business process end of things in m-transformation. While this is understandable, you should refrain from expressing this while engaged in an actual m-transformation. If IT or business staff perceive you as being partial to the other one respectively, your credibility might be damaged in their eyes. This will of course lead to making the remaining m-transformation journey more difficult. Therefore, even though the mix of technology vs. business process adoption is not always equally divided, your attitude towards approaching each should be that of a person well informed, but open minded, without preconceived ideas about either aspect.

Select technologies based on realistic value, not current trends or hype. Selecting the ICT framework necessary to power the efforts for m-transformation—and in particular for an SME with limited knowledge of IT and very small budget for such purchases—can be a real challenge. From our experience it is best to look at solutions that are within the SMEs' budget and offer the best value for the money. Even though the IT industry puts a lot of emphasis on current trends and "hot" technologies from time to time, you should not let hype drive your decisions. Remember that these decisions can make or break the chances of success of m-transformation for the particular organization.

Look for components-based technologies that can expand and scale. Selecting technologies suitable for m-transformation can be a very hard task in itself. The sheer breadth and depth of solutions currently available both as commercial software and at the state of research ideas and prototypes is very substantial. In jointly working with SMEs, we looked at several available solutions—commercial and others—ranging from simple gateways between e-mail and SMS or MMS over the common mobile phone networks, all the way to new environments for mobile collaboration over spe-

Figure 5. CBEADS© high-level architecture

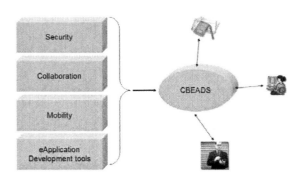

cific domains (Dustdar & Gall, 2003; Litiu & Zeitoun, 2004; Looney & Valacich, 2004) and mobile disconnected business processes (Sairamesh, Goh, Stanoi, Li, & Padmanabhan, 2002). One thing that we found missing was a common framework that could bring together the disparate set of technologies that are needed in order to provide enough IT facilities to better business processes and in turn reach m-transformation.

As an answer to this gap, and to fill it while assisting several SME organizations in achieving their m-transformation goals, we have developed and successfully used a home-grown technology framework known as CBEADS©. The name stands for Component-Based E-Application Development (and deployment) Shell. As its name suggests, CBEADS© is component based and can be extended easily to accommodate changes in the requirements from one engagement to the next. Being fully Web enabled allows for quick and easy deployment, while all the user interfaces are put together via HTML controls that end-users are familiar with through their general use of the World Wide Web. CBEADS© is multi-platform and so far is supported on both MS Windows™ and Linux operating systems.

Through CBEADS©, the SMEs we have worked with covered their needs for a central framework for deploying their e-business mobile-enabled applications, as well as an environment to develop and extend those. With user management and centralized authentication and access control services also offered by CBEADS, the overheads associated with creating and deploying new applications to select groups of staff in the organization are largely alleviated.

Figure 5 shows the four major subsystems that make up the CBEADS© framework and how it provides a consistent, device-independent view of the applications and data it has access over.

When selecting technologies for m-transformation, it is very important to ensure that they meet some common characteristics. As is the case with CBEADS©, you should also strive to establish a common platform for managing users and access control to system applications. Managing users and groups in one place greatly reduces overhead and makes the process of deploying applications significantly faster. Also a World Wide Web-accessible system should be preferred over a client-server one, again due to deployment being easier and not dependant on the operating system of the client machines, which in SMEs could vary significantly. Finally, having an integrated development and deployment environment can prove helpful when small changes are needed to existing applications or extensions are required for applications to become mobile enabled.

We hope that these additional guidelines stemming from our experiences in working with SMEs undergoing m-transformation will prove useful and assist others as well to carry out the process with great success.

CONCLUSION AND FUTURE DIRECTIONS

The next logical step from an organization's e-transformation is that of m-transformation, which proves to be particularly difficult to succeed in especially in the case of SME organizations. SMEs have unique characteristics that set them apart from larger companies. These characteristics impact upon all aspects of m-transformation, including the requirements for a suitable mix of ICT solutions, the adoption of business processes that can take advantage of the new technologies, and the methodology of m-transformation as a whole.

We have presented the individual requirements for each of these aspects of m-transformation along with reasons for each to succeed in the SME space. Stemming from these we looked at the exact steps involved in the methodology for m-transformation applicable to SME organizations in particular. Finally, from our experience in helping SMEs m-transform, we have offered a set of guidelines on how to select technologies that could assist in the process of m-transforming an SME, as well as how to deal with management buy-in and more.

In view of all the benefits—both tangible and intangible—that m-transformation promises to deliver to its embracers, it should come as no surprise that it is an already "hot" topic that will no doubt continue to gain in popularity for quite some time. Hopefully, this chapter has armed you with the necessary understanding of the unique terrain of the SME landscape in relation to m-transformation. The list of requirements we presented may prove useful to SMEs themselves, software vendors, and systems integrators who are involved with assisting companies along their m-transformation journey.

REFERENCES

Arunatileka, S., & Ginige, A. (2003, June 3-6). The seven e's in e-transformaon—A strategic e-transformation model. In *Proceedings of the IADIS International Conference,* Lisbon, Portugal.

Dustdar, S., & Gall, H. (2003, February 5-7). Architectural concerns in distributed and mobile collaborative systems. In *Proceedings of the 11ᵗʰ Euromicro Conference on Parallel, Distributed and Network-Based Processing,* Genova, Italy (pp. 475-483).

Dustdar, S., Gall, H., & Schmidt, R. (2004). Web services for groupware in distributed and mobile collaboration. In *Proceedings of the 12ᵗʰ Euromicro Conference on Parallel, Distributed and Network-Based Processing,* A Caruna, Spain (pp. 241).

Ginige, A. (2002). New paradigm for developing evolutionary software to support e-business. In S. K. Chang (Ed.), *Handbook of software engineering and knowledge engineering* (Vol. 2, pp. 711-725). Hackensack, NJ: World Scientific.

Ginige, A. (2004, March 26-27). Collaborating to win—Creating an effective virtual organisation. In *Proceedings of the International Workshop on Business and Information,* Taipei, Taiwan.

Ginige, A., Murugesan, S., Khandelwal, V., Pollard, T., Costadopouls, N., & Kazanis, P. (2001). *Information technology in Western Sydney: Status and potential.* Western Sydney: University of Western Sydney.

Kawashima, T., & Ma, J. (2004, March 23-24). Tomscop—A synchronous p2p collaboration platform over JXTA. In *Proceedings of the*

24th International Conference on Distributed Computing Systems, Tokyo, Japan (pp. 85-90).

Kazanis, P. (2003). *Methodologies and tools for e-transforming small to medium size enterprises.* Unpublished PhD, University of Western Sydney, Australia.

Litiu, R., & Zeitoun, A. (2004, January 5-8). Infrastructure support for mobile collaboration. *Proceedings of the 37th Annual Hawaii International Conference on System Sciences,* Big Island, HI.

Looney, C. A., & Valacich, J. S. (2004, January 5-8). Mobile technologies and collaboration. *Proceedings of the 37th Annual Hawaii International Conference on System Sciences.*

Marmaridis, I., Ginige, J. A., & Ginige, A. (2004a). Web-based architecture for dynamic e-collaborative work. In *Proceedings of the International Conference on Software Engineering and Knowledge Engineering,* Banff, Canada (p. 445)..

Marmaridis, I., Ginige, J. A., Ginige, A., & Arunatilaka, S. (2004b, May 23-26). Architecture for evolving and maintainable Web information systems. In *Proceedings of IRMA04,* New Orleans, LA.

Marmaridis, I., & Unhelkar, B. (2005). Challenges in mobile transformations: A requirements modelling perspective for small and medium enterprises. In *Proceedings of the 4th International Conference on Mobile Business* (ICMB 2005), Sydney, Australia (pp. 16-22).

Naedele, M. (2003). Standards for XML and Web Services security. *Computer, 36*(4), 96-98.

Ranjbar, M., & Unhelkar, B. (2003, December 16-18). Globalisation and its impact on telecommuting: An Australian perspective. In *Proceedings of the International Business Information Management Conference (IBIM03),* Cairo, Egypt. Retrieved from www.ibima.org

Sairamesh, J., Goh, S., Stanoi, I., Li, C. S., & Padmanabhan, S. (2002, September 28). Self-managing, disconnected processes and mechanisms for mobile e-business. *Proceedings of the 2nd International Workshop on Mobile Commerce,* Atlanta, GA (pp. 82-89).

Staab, S., van der Aalst, W., Benjamins, V. R., Sheth, A., Miller, J. A., Bussler, C., et al. (2003). Web Services: Been there, done that? *Intelligent Systems, IEEE [see also IEEE Expert], 18*(1), 72-85.

Unhelkar, B. (2004, December 15-18). Globalization with mobility. In *Proceedings of the 12th International Conference on Advanced Computing and Communications* (ADCOM 2004), Ahmedabad, India.

Chapter XLI
Business Process Mobility

Harpreet Alag
Agilisys Limited, UK

ABSTRACT

This chapter introduces the concept of business process mobility. Mobility in this case refers to the ability of a human resource to work from multiple locations and in non-office environments; business process mobility involves enabling that resource to carry out specific aspects of a business process while mobile. It attempts to explain where and how mobile enabling processes and systems can benefit. The chapter argues the need for redesigning business processes to support mobility instead of simply adding mobile systems. It further attempts to explore the approach for analyzing and redesigning processes to support mobility. The author also hopes to provide an understanding of mobile systems and their role in enterprise mobility. The chapter touches upon the essentials of mobility strategy and concludes by discussing key contents for a business case for mobile enabling business processes.

INTRODUCTION

In the last decade, advancements in technology, particularly the Internet and electronic commerce, have changed the way people work and live. Technology has helped businesses around the globe to bring about drastic improvements in the how they conduct business, and helped them produce new products and services faster than ever. Some of these products and services have changed the face of industry completely, for instance Internet banking.

Mobile technologies have been at the forefront of all technological developments in recent years. The growing number and use of mobile phones and personal digital assistants (PDAs) is a good indicator of overall potential of mobile applications. Globally, the number of mobile phone users reached 1.5 billion in June 2004 (IT Facts, 2005). Mobile phone sales in

2004 increased by 30% to 674 million units compared to the sales in 2003, while the sales growth in 2003 compared to 2002 was only 20.5%. The increasing sales of mobile phones also include replacements or upgrades. A wide range of mobile devices and applications are available in the market, including laptops, handheld computers, tablet PCs, PDAs, and mobile phones with PDA capabilities. Some of the leading names in PDAs are PalmOne, HP, RIM, and Dell. A number of devices having wireless capability is rapidly increasing. In 2004, 44% of PDA devices offered integrated wireless network support (IT Facts, 2005). Some of the popular mobile applications are personal information management applications such as contacts, calendars, and more recently wireless e-mail. Short Messaging Service, popularly known as SMS or "texting", is still the most widely used mobile application. Convergence of Internet and wireless technologies—also known as "wireless Internet"—is considered the fastest growth area in technology industry. The possibilities offered by this convergence are virtually unlimited. Large-scale use of mobile technologies is expected to have a large impact on business and consumers in the coming years.

This chapter focuses on the use of mobile applications in core business processes across an enterprise. It is an attempt to understand how mobile applications can be used to provide mobility to business processes and extend the philosophy of mobility to build a mobile enterprise.

The following topics will be discussed in detail:

- Evolution of mobile business applications
- The concept of mobile business process
- Motivation for mobile enabling business processes
- Enterprise mobility and enterprise mobile systems
- Redesigning for mobility
- Enterprise mobility strategy
- Business case for mobile enabling

EVOLUTION OF MOBILE BUSINESS APPLICATIONS

The first generation of mobile applications worked in a disconnected or off-line mode: they were synchronised with a desktop computer by physically connecting using a USB or serial port connection. Applications available on compact mobile devices ranged from personal information management (PIM), basic word processing, and spreadsheets, while laptop computers offered almost the same capabilities as desktop computers, but without mobile connectivity. The first generation of applications helped to some extent in improving individual productivity by providing users with the ability to do basic tasks even when away from the workplace. This only worked well with personal information such as contacts and calendars, as data integrity was not at risk.

With the next level of evolution in mobile communications, it became possible to build and implement mobile applications that operate in near real time. The second generation of applications worked in an online, but not always-on mode, and could transfer data in near real time. These applications were basic in functionality due to limitations of mobile connectivity, hardware capability, and data transfers, and relied primarily on wireless application protocol (WAP) and Short Messaging Service. Such applications included checking account balances via mobile phones, receiving alerts/notifications, and checking order status. These applications mostly obtained data from the Internet or a specific server and presented it in a suitable format on mobile devices by means of device-specific user interfaces. Such applications mainly worked by requesting and retrieving information from enterprise systems and relaying it on mobile devices using specific communications standards/protocols. The use of mobile applications was largely restricted to non-core activities for productivity gains and not core business processes. Calendar, contact, and time management are some of the popular mobile-enabled non-core activities. The reason for this is that core processes in a business can be very complex in terms of number of activities, decision making, and number of people involved. Introducing a mobile element in the IT systems of such processes was often not feasible due to one or more of the following reasons:

- Limited capabilities of technology
- High cost of mobile enabling to address complex requirements of core processes, implying ROI cannot be achieved
- High risk or disruption in implementation

With further developments in mobile infrastructure and proliferation of mobile technology types and providers, the third generation of "always-on" mobile applications is now possible. These applications are always connected in near real time to a network to send and receive data. Even complex business tasks requiring "always-on" connectivity and data transfer can be supported by these applications. Mobile enabling complex core processes of a business have now become feasible and economically practical with "always-on" connectivity and wireless Internet. There are multiple connectivity options that when used in combination can provide an effective mobile solution for most requirements. These include General Packet Radio Service (GPRS), Code-Division Multiple Access (CDMA), WAP, Bluetooth, and Wireless LAN (Wi-Fi). GPRS, a part of the GSM (Global System for Mobile

Communications) family, is fast evolving as the leading standard for wireless data communications around the world. According to Forrester Research, in March 2005, 78% of the total 1.5 billion mobile telecom users worldwide were connected using a GSM network (IT Facts, 2005). Wireless e-mail is by far the most popular mobile application across the world. RIM's Blackberry™ is one of the leading wireless e-mail applications; it works on automatic push technology. According to Radicatti Group, around 3.2 million business users were expected to have used wireless e-mails in 2004. Gartner reported 30 million users were expected to use Wi-Fi hotspot worldwide by the end of 2004, up from 9.3 millions users in 2003 (IT Facts, 2005).

These new-generation mobile applications are increasingly finding their place in core business activities beyond improving personal productivity. Across industries, many business functions are exploring mobile applications to increase business efficiency. Customer relationship management (CRM), supply chain management (SCM), and (physical) asset management are some of the prime business areas that can benefit from mobile applications. In addition to core horizontal processes, mobile applications are also finding success in industry-specific vertical processes such as pharmaceutical sales and insurance. Globally, businesses are exploring mobile technologies to support their mobile workers and strengthen the concept of mobile offices. Growing capabilities and robustness of technology are expanding the potential role of mobile applications in core business. With the growing use of mobile applications, the term 'm-commerce' has already found cognizance. Mobile commerce is the process of channelling electronic commercial transactions through mobile devices.

Some common applications in use are listed below:

- "Office" workers checking e-mails and viewing documents on Blackberry™
- Handheld computers with bar code scanning software used to check deliveries in and out of warehouses
- Maintenance engineers downloading their day's job list and uploading the job completion status

MOBILE BUSINESS PROCESS

The early use of mobile applications in automating business processes was primarily technology driven. In order to realise potential benefits of mobile technology, the design of mobile applications was driven by possibilities offered by technology, namely availability of mobile devices and mobile connectivity. Process managers changed their processes to utilise technology options available, rather than ensuring that technology worked within their business requirements and achieved all the business objectives. However, with the latest technological developments, processes can be mobile enabled using a business-driven approach, where mobile technology is just an enabling tool.

The term "business process" has been defined by a number of authors including popular definitions by Michael Hammer and Thomas Davenport. Davenport's definition is perhaps simplest, according to which a business process is "a specific ordering of work activities across time and place, with a beginning, an end, and clearly identified inputs and outputs" (Davenport, 1993). Business processes can be complex and large scale or smaller processes made up of few activities. A good example of a high-level process is procure-to-pay, which covers end-to-end process, beginning with requisitioning and ending with the payment for goods received. Completing an expense form is a common example of a low-level activity applicable to all businesses. Processes can be decomposed to smaller sub-processes and viewed at lower levels of detail. The lowest level sub-process is an activity with a well-defined input and output. Gruhn and Kohler (2003) proposed the term *mobile business process,* according to which mobility is given for a business process, when at least for one of the process activities there is externally determined *uncertainty of location* (Valiente & van der Heijden, 2002) and the process needs cooperation with external resources for its execution. I will discuss this in more detail in one of the later sections of this chapter. To simplify the above definition: a mobile business process is one that consists of one or more activities being performed at an uncertain location and requiring the worker to be mobile. Such a process can be supported by mobile systems to increase process efficiency. For processes that are supported by mobile systems, the term *mobile-enabled business process* is more appropriate, to differentiate from a *mobile business process.*

MOTIVATION FOR MOBILE ENABLING PROCESSES

This section explores the motivations behind mobile enabling business processes. The key motivation for mobile enabling business processes is the need to serve customers faster and reduce costs. Globalisation and intense competition demand that businesses respond faster to changing market and customer needs, by reducing time-to-market for products and services and serving their customers in a near instantaneous fashion. Businesses are continuously striving to become more efficient and effective. The ever-increasing pressure is only forcing businesses to be more innovative. Businesses considering mobility of processes will have to build a business case for having their processes mobile enabled, justifying how it will help their business. The essence of a mobile-enabled process is that a worker should be able to conduct essential business regardless of their location. The extent to which an enterprise addresses mobility of processes depends on the following factors:

- Need for *mobile enabling processes,* in other words the need to support mobile information systems
- Technological feasibility of the mobile information system
- Financial justification based on cost-benefit analysis model
- Operational feasibility—human factors and so forth

As far as the motivation for mobility is concerned, it can be argued that there are two basic drivers for addressing process mobility.

1. **Process Efficiency:** A set of factors that demand cost reduction, improved customer service, and response time. An example of mobile enabling a process for efficiency is sales staff being able to create online quotations and orders at the customer site using their mobile devices.
2. **Increased Personal Productivity of Employees:** Time and travel management are some of the common processes for mobile enabling and achieving significant productivity improvements.

In some areas of business, the benefits of enterprise systems cannot be fully realised, as a large number of mobile workers cannot access wire-bound systems. As a result, many organisations cannot realise the expected return on investments in expensive enterprise systems. In addition to that, most business processes and supporting systems are designed around office-based employees and are not friendly to mobile workers. A good example is an expense claim form, which is typically made available on company intranets and cannot be accessed by sales staff when they are away from the office. Mobility of processes supported by systems can help achieve additional return on investments in enterprise systems such as ERP, CRM, and SCM. Business managers do not introduce new technologies into the business processes to become technology leaders; they are interested in the added business value and how technology contributes to it; in how they can achieve the goals of streamlining their processes. From the point of view of mobility, managers are interested in what value mobile enabling can add to the business processes.

MOBILE ENTERPRISE: PROCESSES BEYOND THE WORKPLACE

Businesses that aim to support mobile workers and enhance process effectiveness will need to consider extending their process and systems beyond the workplace. In order to achieve this, they will have to change their processes and systems in line with the objectives of process mobility. An enterprise that can transform its processes to make its mobile workers and processes more effective can be considered a *mobile enterprise.*

Smart organisations will aim to leverage process mobility for strategic advantages. Such businesses derive tactical and strategic value from mobile enabling processes. In order to gain maximum benefits from mobility, organisations should have a mobility strategy defined—aligned to its business strategy—and it should complement the IT strategy.

MOBILE ENTERPRISE SYTEMS: SYSTEMS FOR MOBILE PROCESSES

Today, the use of mobile systems is ubiquitous in our daily lives, for instance the courier delivering a parcel and recording delivery details on a handheld device, and the recipient signing on the handheld device using a stylus. The receipt details are instantly sent to the central systems, thus providing real-time status of deliveries for their billing and customer service departments. Restaurants are also benefiting from mobile systems by providing handheld computers on which serving staff input orders and view the status of their orders. Kitchen staff can see the orders instantly on the kitchen computer and update this status when an order is ready. This saves time moving between the dining area and the kitchen, and thus reduces the overall time taken to serve food to the customers. Restaurants and bars using Chip 'N' Pin technology for credit card payments have chosen to use mobile devices to save customers needing to walk to the counter to key in their PIN. This is an example of a mobile application requirement arising out of regulatory changes.

Mobile systems can be beneficial across a number of processes in most business areas. Large corporations are embracing mobile systems in almost all major areas of business such as sales, procurement, warehouse management, and so on. A number of mobile solutions available in the market are truly enterprise encompassing in nature as they fully integrate with the existing enterprise systems and bridge across major processes of the enterprise. Com-

panies like SAP, Oracle, and Siebel offer mobile solutions that build on their existing enterprise systems offerings, primarily ERP, CRM, and SCM applications. These solutions address mobility in areas such as sales, field service, procurement, supply chain, and asset management. Mobile applications can be used to redesign or improve processes at a specific activity level (e.g., e-mail), or can be used to mobile enable large end-to-end processes that cut across functions (e.g., procure-to-pay processes). Similarly, mobile applications can be utilised for most processes in sales, supply chain, asset management, and plant maintenance. Systems that mobile enable core processes and key activities across multiple functions in an organisation can be seen as *mobile enterprise systems.* Mobile enterprise systems can either be enterprise systems extended to support process mobility or separate mobile systems integrated with existing enterprise systems. There are two key aspects to a *mobile enterprise system.* First, the mobile system should support one or more core business processes, and second, it works on the existing enterprise data.

As discussed earlier, the drivers for process mobility are location uncertainty and user mobility. As most sales and service staffs are highly mobile with activities carried out at external (uncertain) locations, these areas make a strong case for mobile systems. With mobile systems, the sales and field service staff can access business information wherever they like and capture data wherever it is generated. For example, sales personnel can view order status from a customer site and create new orders online using their mobile devices; service engineers can input job completion details on handheld computers that update centralised databases in real time. With the effective use of mobile systems, sales teams can spend more time with customers and prospects. Table 1

Table 1. Mobile enabling Sales and Service processes

CUSTOMER RELATIONSHIP MANAGEMENT (CRM)		
Sales:	**Order Management:**	**Account Management:**
• Product stock availability check • Up-to-date product pricing information • Delivery schedule • View all customer information: Contacts, needs, decision makers, interaction history, etc.	• Order entry • Order status • Order delivery • Payment collection	• View contact information • View past interaction • View planned events, meetings, etc. • Negotiate and amend contracts in real time, printing these off, and e-mailing back to head office once agreed
Opportunity Management:	**Activity & Task Management:**	**Field Service:**
• View existing opportunities • Capture new leads and opportunities	• View planned sales activities and tasks • Create plan for new activities and tasks • Capture customer signatures on a handheld device to confirm agreements	• Create service orders • Capture assignment completion details • Record customer complaints • Manage service tasks and activities

Table 2. Mobile enabling Asset & Material management

ASSET & MATERIALS MANAGEMENT	
Work Order Management:	**Materials Management:**
• Work order creation • Work order tracking • Capture actual completion of tasks and activities • Capture labour and material consumption • Issue material against work order • Raise requisition for material	• Check material stock availability • Issue and return material • Physical stock count and balance adjustments
Equipment Management:	
• View and update equipment repair history • Record technical measurements of equipment • Replace equipment at functional locations • Manage asset hierarchy • Install or uninstall equipment at functional locations	

lists some illustrative activities under sales and customer services that could be mobile enabled to improve process efficiencies.

Other high-potential areas are asset management, plant maintenance, and materials management. Mobility of workers in these processes is usually within a limited area, but requires movement around that area—for example, capturing technical measurements of equipment around the plant and updating equipment repair history on a handheld device. Mobile technologies, when combined with other technologies such as bar code and more recently RFID (Radio Frequency Identification), can offer more appealing solutions and bring about substantial efficiency improvements. For instance, radio frequency (RF) tags can be used to store maintenance and service data pertaining to equipment. With the help of mobile devices, users can instantly view the equipment maintenance information and repair history stored on the RF tag attached on the equipment. Table 2 lists some of the activities that require

Figure 1. Mobile enterprise systems

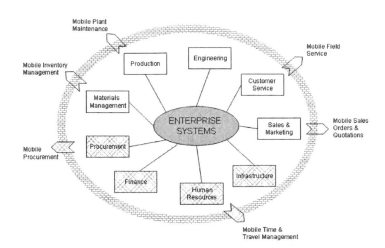

local mobility and can obtain productivity gains from mobile enabling.

Supply chain management and procurement management processes can also benefit from mobile enabling. Mobile enabling employee-oriented processes such as filling and approving time sheets, travel expense forms, and leave requests can increase employee efficiency by making effective use of unproductive time such as travel by train, taxi, or air, as well as waiting periods.

With the growing use of mobile systems, mobile technologies are finding their place in enterprise systems' architecture and technology strategy. Organisations considering mobile systems for their core processes are viewing mobility as a strategic and not just technological element. From a systems perspective, mobile systems can be seen as a virtual mobile layer around the enterprise architecture. Similarly, the business process architecture can be understood to have a mobility layer that represents the mobility of business processes and supporting systems. Business organisations striving to be competitive will have to address mobility requirements and capture the opportunities arising from mobile enabling business

processes. Such organisations will require not only a mobile layer in their technology stack, but also a corporate-level strategy for mobility. Figure 1 shows an architectural view of the enterprise systems of an engineering business. The figure shows typical business areas using mobile systems as an extension of the enterprise systems. The mobile systems in the diagram are shown as a virtual layer around the enterprise systems.

REDESIGNING FOR MOBILITY

As mentioned earlier, mobile applications are now available for almost all key areas of business that require process mobility. With the proliferation of technology, the applications designed for mobile devices are nearing their desktop cousins in terms of functionality and performance. In order to leverage the capabilities of mobile systems, the design of business processes needs to be assessed from a mobility perspective and, if required, to be redesigned to maximise the advantage of mobile enabling. Mobile applications should not just be utilised to extend business processes onto handheld or mobile devices. Instead, mobile systems should be tightly integrated with enterprise systems to streamline processes. Thus, technology innovation can be a stepping stone to achieve business process innovation. The ultimate objective of mobility in business can be seen as providing the power of a desktop computer in the hands of the mobile worker with the ability to connect to a network or the Internet from virtually anywhere. However, this may not be required or even feasible in all situations—technically or economically. Thus, it is important to identify and assess processes and aspects of processes that would benefit from mobile enabling to make a sound business case.

PROCESS REDESIGN PROJECTS

Most of the current interest and development in the area of business process change has its roots in the works done by Michael Hammer, James Champy, and Thomas Davenport in early 1990s. Since then the subject of business process change has evolved a great deal, with the practice growing across industries. Many different names have been given to such projects, for example, business transformation, business process improvement, and so on. Organisations change their processes with different objectives. All process change projects are different in a number of ways, mainly the objectives and the scope. Some projects have a strategic objective to achieve such as projects aiming to take advantage of strategic opportunities or address strategic problems. These include objectives such as to reduce cost, increase revenue, improve customer satisfaction, or strategically change business direction, while some projects have a specific goal of improving process efficiencies such as to implement an ERP/ CRM system, Web-enable existing applications, or introduce electronic commerce. Projects can also vary in scope, as some cut

across multiple functions while others are focused on very specific sub-processes or activities.

Depending on the focus and objectives of the change, different terms are being employed to describe process change projects. In addition to that, some terms evolved simply due to poor usage of terms, which resulted in the terms being unfairly disliked; for example, the use of "Business Process Reengineering" has often been seen as a pseudonym for staff cutbacks. According to Harmon (2003, p. 39):

If the process is relatively stable and the goal is to introduce incremental improvements, then the preferred term is process improvement. If the process is very large and seeks to redesign the process in a comprehensive manner, then the term used is process reengineering.

Reengineering relies on re-conceptualising how the business process should work and is radical in nature. Process improvement is mostly done on a gradual and ongoing basis, and its objective is tactical problem solving. Reengineering is strategically oriented and aims to either scize strategic opportunities or address strategic business problems. The key difference is that there are no boundaries with reengineering while there are boundaries with process improvement. These two types of projects form the two extremes. For most process-change projects that fall in between these two extremes, the term generally used is *process redesign*. Most of the projects that aim to mobile enable processes will fall under process redesign. However, mobile enabling large processes can be part of a reengineering project, or mobile enabling a few activities can be an integral part of a process improvement initiative. For the sake of simplicity, I will use the term process redesign to refer to process changes that aim to address mobility needs.

PROCESS REDESIGN METHODOLOGY

In order to undertake a process change or process redesign project that involves mobile enabling processes, it is important to understand the considerations for redesigning to support mobility. A process redesign project may include a number of changes that help in achieving the process goals or overall project objectives such as: all orders should be processed in a maximum of two business days or reduce cost of end-to-end purchasing by 10%. For simplicity, I am only considering the mobility objective of a process redesign project. I am not considering changes done to achieve any other objectives. Regardless of the objectives of the redesign project, the overall approach to mobility should essentially be the same. Redesigning for mobility can be distinguished as a specialised pattern for redesign. According to Harmon (2003, p. 236), there are two sets of redesign patterns—namely, *basic business process redesign patterns* and *specialised business process redesign patterns*. Basic redesign patterns are generic approaches used by almost every process redesign project such as value-added analysis and reengineering. Specialised redesign patterns are used to extend the basic patterns or to solve specialised problems, such as ERP-driven design, workflow automation, and Six Sigma. Extending the patterns definition to mobility implies that another specialised redesign pattern can be defined, the driver for which is the need for mobility of business processes. I prefer the term "mobile enabling" for such a pattern.

The methodology applied depends on the type of project, scope of project, and individual preferences of the program managers. The activities involved vary depending on the methodology. The activities involved in basic methodology are explained briefly as follows:

- **Understand Business Goals:** As for any major redesign project, it is important to understand the overall objectives of the business, any strategic directions the business is considering, and how mobility ties into it.
- **Define Project Objectives and Scope:** The objectives and constraints of the project should be clearly defined and documented, for example, reduce the average processing time for service request from four days to three days. The definition should clearly explain how the strategic business objectives would be achieved or enabled by mobile enabling processes. The scope of the project should clearly define processes that will be covered, define the boundary of the scope by explicitly describing out-of-scope elements, and define the activities that will be covered by internal teams and those that will be contracted to outside vendors or consultants. The project charter should also include the benefits expected after the successful completion of the mobile-enabling project.
- **Define Project Plan:** The project plan should include the time, effort, and resources required for each activity involved. The inputs and outputs of each activity should be clearly identified and defined.
- **Analyse Existing Processes, People, and Technology:** The existing process are analysed to understand problems, gaps, disconnects, and any improvement areas. As part of the process analysis, the roles of the people involved need to be understood. In addition, the current use of technology in the processes needs to be clearly understood.
- **Evaluate Redesign and Technology Options:** Explore the possible redesign options and technology options. The redesign options are evaluated in light of project objectives, constraints, benefits, and costs.

For instance, for service engineers completing jobs at customer sites, do they need handheld computers to feed job completion data online into the enterprise database server, or can they make a phone call to a call centre and provide information which can then be input into the system. The evaluation in such cases is primarily driven by cost-benefit analysis. When it comes to personal productivity, in most cases providing access to wireless e-mails and a calendar can result in dramatic productivity improvements.

- **Redesign Processes, People Roles, and Technological Architecture:** Based on the most suitable option, new designs of processes job roles and systems to be implemented are established.
- **Define Implementation Plan:** The implementation plan includes the plan for introducing all the changes in processes, people, and technology. The implementation plan answers detailed and specific questions such as: Who owns the new field service process? and What system changes will be done? It includes the detailed plans for technical design and developments, testing, and so on.
- **Define Transition Plan:** The transition plan ensures that the transition to new processes and systems is controlled and smooth. The plan should address cutovers, systems failures, back-ups, and so forth.

In the following sections, I have included detailed explanations on two key aspects of redesigning processes for mobility.

ANALYSING EXISTING PROCESSES

When redesigning processes for mobility, there are two key activities: analysing current processes and deriving the "to be" processes. In the analysis phase, analysing the distributed structure of a business process and identifying mobile sub-processes or activities assesses the need for mobility. Gruhn and Kohler (2003) suggest that when dealing with mobility of processes, we are dealing with a mobile partition or activity within a business process. A process can consist of one or more mobile activities, and as these activities affect the whole process, the complete business process is called *mobile business process.*

When analysing the distributed structure of the business processes in question, two critical questions need to be answered. First, where are the process activities to be executed, and second, what data is needed at which location of execution (Gruhn & Kohler, 2003). The other key consideration is the type of mobility of the user during the execution of the process activity. The different types of mobility of the worker executing the process activity can be classified into the following three modes (Kristoffersen & Ljungberg, 2000):

- **Wandering:** Worker performs activities while moving between different locations. The locations are locally defined within a building or local area.
- **Visiting:** Worker performs activities at different locations.
- **Travelling:** Worker performs activities while moving between different locations usually inside a vehicle.

I will use the terms user mobility and worker mobility interchangeably to refer to mobility of the person executing the process activity. Valiente and van der Heijden (2002) defined the concept of "location uncertainty", according to which the place of the execution of an activity can be different in different instances of the business process or the places can change during the execution of an activity.

The main characteristic of a mobile business process is that location uncertainty and worker mobility are externally determined, for instance: (a) a field service engineer inquiring for/requesting a repair part, and (b) a salesperson inquiring of the latest price and stock availability from a customer location. In both these examples, the location of activity execution is externally determined and worker mobility is of the type "visiting". In another example where a maintenance engineer is updating work order details during plant/asset maintenance, though location uncertainty is low due to limited geographical area, it is externally determined. Worker mobility is of the type "wandering" in this case. Location uncertainty and user mobility are the chief determinants of the choice of mobile connectivity (e.g., GPRS, WLAN, Bluetooth, etc.) and type of mobile solution required. A sound mobile solution will appropriately address these two aspects of the mobile business process. Some sub-processes or activities could be performed when a user is mobile—that is, travelling or away from the office—as location is irrelevant for their execution. Leveraging location freedom can help increase productivity. Business processes that include such activities or sub-processes can be termed mobile friendly processes. Time and travel management are typical mobile friendly processes (e.g., filling and submitting time sheets, travel expense forms, and leave requests). Similarly, a number of workflow activities such as approving requisitions and leave requests are also mobile friendly. These activities can be performed through a mobile system when the user is travelling or away from the office and make substantial productivity gains for the individual. This may not be true in all cases, as all productivity gains do not always lead to real productivity improvements. It is critical to note however that in certain roles

this may not lead to a productivity gain for the company; for example, workers who do unpaid overtime will frequently utilise overtime to perform time management activities, and therefore there may be no productivity gain. However, even in such cases the business can gain other benefits such as reduced transport costs, as the staff is not required to travel to the office for administrative purposes. Horizontal processes, such as those discussed here, are generally mobile friendly but do not directly add value to customer-oriented processes. They can however add indirect value in terms of productivity gains and more time for customer-oriented processes.

The activities in the business processes of an enterprise can be segregated based on location uncertainty and worker mobility. Figure 2 shows some typical processes in a large engineering business. The processes are mapped against varying degrees of location uncertainty and user mobility (Valiente & van der Heijden, 2002). This is useful for understanding the drivers of mobility and as a guiding tool for classifying processes for mobile enabling. After identifying the processes that need to be mobile enabled, managers then need to consider the value that mobile systems would add to these business processes. They need to consider how well the process goals are achieved and how best the processes can be streamlined with the support of mobile systems. The identified processes can then be prioritised based on the added business value. As part of the exercise, problems and bottlenecks in the processes are also analysed. For each of the business processes to be mobile enabled, the key is identifying mobile activities that need to be supported by mobile systems. The basic requirement to conceptualise a mobile system is to understand the location at which the mobile activity is to be performed, data required for the activity, and mode of user mobility. A more

Figure 2. Process mobility map

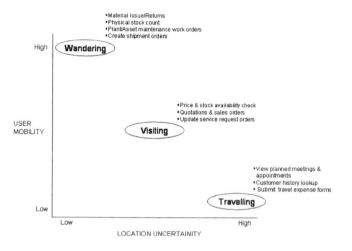

detailed analysis of activities can be performed to determine time taken to complete the activity and so on. I will focus only on specific areas of analysis that are relevant to mobility.

To summarise, the analysis of existing processes revolves around the following questions. The answers to these questions provide crucial direction in identifying candidate processes for mobile enabling.

- Is there any uncertainty of location of process execution?
- Which activities in the process have location uncertainty?
- What is the type of mobility of the person executing the specific activity?
- Will mobile enabling the process improve process efficiency or productivity of users or not?
- What is the estimated actual cost, revenue, or strategic impact on the business as a whole of enabling the user to work from multiple locations?

MODELLING EXISTING PROCESSES

In order to analyse the identified processes in further detail, process flow diagrams are drawn for the current processes, to show the details of activities and their flow in the process. Process diagrams use the concept of swimlanes to indicate the user, function, department, or technology responsible for the process or activity. From mobility perspective, we are interested in swimlanes drawn on process diagrams depicting activity level detail. Swimlanes are also known as actor lanes, as the lanes can be used to represent the worker who performs the activities in the respective lane. Swimlanes can be depicted as either vertical lanes or horizontal lanes. Technical analysts and authors tend to prefer vertical swimlanes due to their similarity to UML, while those from an organisational or pure process background tend to prefer horizontal; either can be reviewed so it can be left as a matter for individual companies to decide. Figure 3 is a simplified example of a horizontal

Figure 3. Conventional process diagram

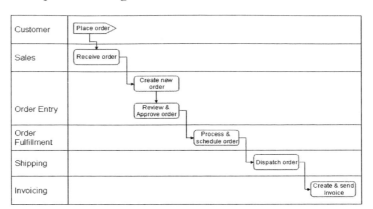

Figure 4. Process diagram with location information

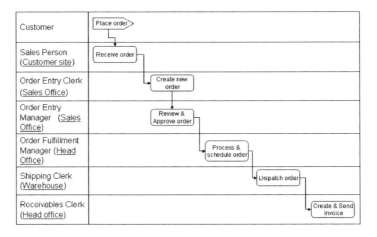

swimlane for the sales process in an engineering company.

Conventionally, locations are not specified on the swimlanes, as they are not relevant to modelling processes for traditional information systems. However, from a mobility perspective, location is key. In order to reflect the location aspect on process diagrams, swimlanes can be shown for a user and location combination (Valiente & van der Heijden, 2002). Thus the labels of lanes should indicate location along with the actor; such process diagrams can provide valuable insight into mobility needs. It is quite possible that a user is at more than one location to perform his or her activities, in which case that user can appear on multiple lanes, representing the multiple locations where the process activity is performed. However, there can be only one lane for the required user-location combination, on the process diagram. These diagrams can highlight activities that are performed at an externally determined location such as customer site. The diagrams can be enhanced to capture high-level details of the data required for the mobile activities. Figure 4 shows a simplified example of process diagram with mobility information for the same sales order process. Such a diagram can help identify the activities for potential mobile enabling and role changes; for example, the salesperson at the customer site can do the order entry instead of the order entry clerk doing it at the sales office.

DESIGNING "TO BE" PROCESSES

After the detailed analysis of the current processes, various redesign options are explored. There are more considerations and complexities involved as I move on to the next stage of redesigning mobile business processes. In order to arrive at the best redesign option, the objectives of mobility need to be clearly defined in order to drive the options at this stage. After the analysis is complete, it is useful to review the technology options and constraints at a broad level along with the generic redesign options. It is at this stage that new ideas for redesigning processes are generated and options established. The objectives can be increased efficiency, improved realisation of process goals, or simply productivity gains. The best option in general is the one that best achieves these objectives. However, there can

be a trade-off between the cost and benefits of the options. In such situations, the most suitable option is the one that adds maximum business value. In some situations, a detailed cost-benefit analysis is also conducted to decide on the best option. The new design is captured in "to be" process diagrams, as this represents the design "to be" implemented. "To be" process diagrams should be drawn in a similar way as the existing process diagrams. The "to be" process diagrams can provide necessary inputs to define high-level requirements of the mobile element of the information system. Figure 5 illustrates a process diagram for a process that is redesigned to support mobility.

There are additional considerations for defining requirements for mobile systems compared to that for traditional information systems. Some important ones are: (a) understanding which mobile systems model is to be used, (b) controlling the access of enterprise data on mobile device, and (c) the profile of the users. The user is really at the centre of the design. The mobile systems model can be understood as a basic framework for the design of the mobile information system. The model determines how the data in the enterprise systems will be mobile enabled to support process mobility, as well as the application design and deployment of the mobile system.

Figure 5. "To be" process redesigned for mobility

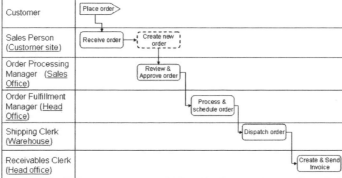

There are three basic mobile systems models that have evolved, based on which enterprise mobile systems can be conceptualised.

1. Build a new mobile application using existing enterprise data
2. Build a cut-down "mobile" version of the current application
3. Entire existing application is mobile enabled to run on mobile devices

The choice of model(s) is determined by a number of factors, including the type of users to be supported, the type of mobility to be supported, the mobile devices to be supported such as laptops and handhelds, the capabilities of mobile connectivity, and the target time for activity completion. Organisations can either use one of the models or any combination of the three models depending on their needs or constraints if any. In general, the first two models are preferable for a number of reasons. The primary reason is that hardware and connectivity requirements for desktop applications are very high and cannot be met on expensive and compact mobile devices with limited screen sizes. The other reason is the advantage of reduced data entry time on the mobile version of the application, which is pivotal to the efficiency of a mobile worker. Using the first two models is also an opportunity to address the need of users who generally prefer to deal with a minimal set of data for their system tasks. A separate "mobile" version of the application created for mobile users can fill this gap. In situations where large contracts have to be read and amended online, it becomes imperative to extend the existing application to laptops to provide larger screen sizes.

ENTERPRISE MOBILITY STRATEGY

Businesses considering enterprise-wide process mobility will require a mobility strategy. The mobility strategy should guide operations and technology employees through the process redesign, application design, and implementation of the mobile systems. Mobility strategies will depend on a number of factors, including the nature of the business, its strategic goals, need for process mobility, existing IT investments, and financial budgets. As a result, each organisation's mobility strategy will be unique in the same way their IT strategy is unique; one thing all sound mobility strategies will have in common, however, is that they will leverage existing IT investments and ensure maximisation of the return on investment on mobility projects.

There are a number of common factors that can be considered to devise an effective mobility strategy. I have identified a few key considerations for formulating an effective mobility strategy, which are as follows:

- **User Profiles:** Within each organisation there will be multiple "mobile worker" profiles, as the needs of mobile workers with different roles are likely to be very different from each other. Some mobile workers may require complex applications on a laptop, while others may need basic applications on handheld devices or a PDA.

- **Interoperability of Solution and Choice of Devices:** An effective mobile solution will be platform independent. The solution chosen should support multiple mobile devices and protect existing investments in IT infrastructure. The mobility strategy should clearly define the types of mobile devices that will be supported at different stages, such as handhelds and PDAs. The mobile devices to be used should be carefully selected in line with the applications and technologies to be supported and the needs of the users. For users who require reading large reports and contracts, the devices should have larger screen sizes.

- **Mobile Infrastructure:** The mobile infrastructure chosen should allow applications to be accessed by multiple channels and device types so that the implementation teams can focus on functional and operational issues, and not get bogged down with deployment complexities. Setting up and maintaining a stand-alone mobile infrastructure can be very costly. Thus, extending the existing infrastructure is a cost-effective approach. Mobile infrastructure and applications chosen should aim to leverage existing investments in IT infrastructure and enterprise systems.

- **Mobile Connectivity:** The choice of mobile connectivity is driven primarily by the type of mobile application and the geography to be supported. The connectivity medium chosen should be scalable for the future needs of existing application or newer mobile applications. If a combination of mobile connectivity options is to be used (e.g., Wi-Fi and GSM networks), it is critical to understand the hand-off capabilities of the service provider or the technology deployed. In addition to the above, when choosing mobile connectivity it is important to consider the regional and global support of the chosen standard depending on geographies covered by the business; for example, GPRS is more global than CDMA, which only works in a few countries like the U.S., Canada, Japan, and Korea.

- **Service Providers:** It is also critical to assess the selected solution providers in terms of their existing and future capabilities of products and services. The requirements should be clearly mapped with vendor capabilities, and the capabilities of the chosen solution providers should be scalable in line with the mobility strategy of the organisation.

- **Alignment with Business Strategy:** The mobility strategy should be aligned with the business strategy. In other words, the business strategy should drive the mobility strategy, and not the other way around. If, for example, the business strategy is to reduce customer service costs, then in such a scenario, mobile enabling customer service processes will take priority over the sales processes.

- **Maximise ROI:** Mobile enabling business processes could involve major investments, and such investments have to be justified in terms of their return. As a guiding principle, businesses should first consider mobile enabling processes that provide maximum return on investment on mobile enabling. In simple terms, processes that can provide maximum cost savings or new revenue generation potential should be top on the list.

- **Beginner or Experienced Organisation:** If an organisation is experimenting with mobile systems for the first time, then it is rational to begin with horizontal processes instead of vertical processes. Horizontal processes such as time management are internal processes that are low risk and span across functions and departments. The advantage in doing so is that there is wider acceptance of new systems within the organisation, and the return on investment is faster.

- **User Requirements:** Mobility strategies will need to include direction for providing support for users in using new mobile technologies and tools. Additional considerations included here could be maximum weight of combined mobile articles per person, insurance, and the equipment the user should have access to within the office (e.g., should a user with a laptop also be allocated a desktop?).
- **Systems Security:** As mobile systems have more risks than conventional systems, it is vital to have a mobile systems security strategy in place. For example, theft or loss of a handheld device or unauthorised access to wireless networks can pose a serious threat to critical enterprise data. The mobile security strategy should be a part of the IT security strategy as well as the enterprise mobility strategy. The mobile security strategy should determine the type and extent of data that can be accessed on a combination of mobile devices and user profiles.

BUSINESS CASE FOR MOBILE ENABLING PROCESSES

Organisations have to make significant investments in order to mobile enable business processes and implement mobile systems. In order to make investment decisions, decision makers need a case for investment supported by cost-benefit analysis, so managers who propose to mobile enable their processes should provide a detailed business case to justify the investments. The business case should provide an assessment of the value of benefits in comparison to the total costs incurred in achieving process mobility. Quantifying the monetary value of benefits, particularly intangible benefits such as improved customer satisfaction,

can be more complex. There are number of books published that discuss business case development and cost-benefit analysis in detail which can be used to assist less experienced analysts in this area. I have enlisted some of the main benefits of mobile enabling business processes.

- Increased speed of end-to-end process execution and increased accuracy
- Enhanced reputation of customer services and increased customer satisfaction
- Reduced costs such as inventory, travel, and other administrative activities
- Enhanced decision making
- Increased productivity

A more scientific approach to measuring those benefits that cannot be measured easily is process metrics or measurements, and these are frequently used as part of standard process management practice. Process metrics, also known as KPIs (Key Performance Indicators), provide insight into the performance of the processes and hence can be utilised to measure the benefits of mobile enabling business processes. Metrics used will vary from industry to industry and from business to business dependent on each organisation's strategy. These metrics can help assess the improvements in process execution and the value added thereof.

Some illustrative process measurements for an engineering services business are:

- Average order fulfilment time
- Average service request resolution time
- Timeliness and accuracy of delivery
- Percentage sales returns
- Customer satisfaction ratings
- Average service down time

Process measurements are gathered initially before the processes are changed and new systems are implemented. These measurements are captured again after the processes have been changed and new systems are put to use. The variance in measurements indicates the improvements in the process and can be attributed to the process and system changes. Gathering process measurements can be a very involved task, and it is recommended that initially these be kept to a small set of indicators until overall business value can be proven to prevent excessive expenditure on measurement, reducing the value of the actual improvements made. Organisations that collect process measurements on a regular basis are well placed to measure the benefits of the mobile-enabling processes. Process redesign projects aimed at mobile enabling business processes may also undertake some redesign that is not related to mobility. In such cases, the process measurements may not provide an accurate picture of benefits realised, due solely to mobility of processes if it is not possible to distinguish between benefits derived from mobility and those derived from other process changes. In this case, it is recommended that managers assign a proportion of improvement in measurements to each aspect of the process improvement including mobility.

REFERENCES

Computer Business Review Online. (2005a). *Free enterprise.* Retrieved January 22, 2005, from www.cbronline.com

Computer Business Review Online. (2005b). *Going mobile.* Retrieved January 22, 2005, from www.cbronline.com

Davenport, T.H. (1993). *Process innovation: Reengineering work through information technology.* Boston: Harvard Business School Press.

Gruhn, V., & Kohler, A. (2003). *Analysis of mobile business processes for the design of mobile information systems* (pp. 1-5). Chair of Applied Telematics/E-Business, University of Leipzig, Germany.

Harmon, P. (2003). *Business process change—A manager's guide to improving, redesigning and automating processes.* San Francisco: Morgan Kaufmann.

Henley, J. (2004). *Building the case for mobile business* (pp. 1-2). Redwood Shores, CA: Oracle Corporation.

IT Facts. (2005). Retrieved April 2005 from www.itfacts.biz

Mobile Advisor Magazine Online. (2005). Retrieved March/April 2005 from www.advisor.com

Pak, C. (2004). *Evaluating the business value of mobile enterprise systems* (pp. 1-5).

SAP–Global. (2005). *Mobile solutions.* Retrieved January 23, 2005, from www.sap.com

Valiente, P., & van der Heijden, H. (2002). *A method to identify opportunities for mobile business processes* (pp. 1-10). SSE/EFI Working Paper Series in Business Administration, Stockholm School of Economics, Sweden.

Chapter XLII
Evaluation of Mobile Technologies in the Context of Their Applications, Limitations, and Transformation

Abbass Ghanbary
University of Western Sydney, Australia

ABSTRACT

Emerging mobile technologies have changed the way we conduct business. This is because communication, more than anything else, has become extremely significant in the context of today's business. Organizations are looking for communication technologies and corresponding strategies to reach and serve their customers. Mobile technologies provide ability to communicate independent of time and location. Therefore, understanding mobile technologies and the process of transitioning the organization to a mobile organization is crucial to the success of adopting mobility in business. Such a process provides a robust basis for the organization's desire to reach a wide customer base. This chapter discusses the assessment of a business in the context of mobile technology, describes the application and limitations of mobile technology, presents a brief history of mobile technology and outlines an initial approach for transitioning an organization to a mobile organization.

INTRODUCTION

This chapter evaluates the effects of mobile technologies on business and outlines an initial process of transitioning to mobile business. In 1874, when Alexander Graham Bell invented the telephone, he could not have imagined the significant impact it would have on number of aspects of human life. Similarly today, ad-vancement in information and communication technology (ICT) has dramatically changed the way people live and conduct their businesses. One of the dramatic aspects of modern-day business activities is that these activities are conducted independent of location and time. For example, businesses are able to sell goods, facilitate customer enquiries, and coordinate their services through disparate geographical and time boundaries primarily due to the wonders of communications technologies. Alter (1996) describes the ICT as tools for doing things, rather than just for monitoring performance of yesterday or last week. Thus it is quite logical to conclude that ICT has changed the very nature of the workplace.

The basis for the communications technologies in most modern business applications is the Internet. Increasingly, the required access and connection to the Internet has become very simple and ubiquitous in most developed nations. This Internet access has opened up opportunities for organizations to revolutionize their business processes. Undoubtedly, improvement of the communication technology has impacted not only our business domain, but also our socio-cultural domain. This, as per Unhelkar (2004), has resulted in the "next wave" of technologies called mobile technologies:

Mobile technologies are becoming the next technology wave as the increasing popularity and the functionality captures many hearts. Riding on the back of traditional Internet, mobile networks ensure that information is available to its users independent of a physical location.

This ubiquitous connectivity accorded by mobile networks referred to above impact has facilitated the increased communication between people. Furthermore, this mobile connectivity has also improved the ability of business processes to exchange data and conduct transactions.

This transformation of businesses has been evolutionary rather than revolutionary. For example, at the beginning of the Internet age, with the aid of its communications capabilities, businesses were transferred to e-business, and we even had the opportunity to do our daily business activities from home. Ghanbary (2003) has described the Internet as the most powerful tool that brings information to our homes through communication lines, like water and electricity that come by power lines and pipes. Powerful search engines and the capability of sharing information are the great advantages of the Internet.

With the aforementioned strengths of mobile connectivity, it is also essential to work out a process that would outline "how" an organization can transition to such an m-enabled organization. However, this potential process of transitioning an organization to a mobile organization needs to incorporate all the major advances of mobile technologies of the past decade.

This is so because the philosophy of ordinary communication has given way to more advanced and efficient communications based on mobile and wireless technologies that enable business processes to be executed independent of time and location, resulting in a better, faster, and satisfactory response to the needs of the customer.

This impact of mobility is an important element of the mobile transition process that is felt at both business and personal levels. However, as of today, this process framework remains a challenge that needs to be further researched to enable businesses to transition successfully. This need for further investigation is also ratified by Ranjbar (2002), who correctly mentions: "It is not always possible to foresee all the implications of a new technology until it is adopted by the mass of population and used for a relatively long time."

With the increase in the number of mobile organizations, the service providers realize that they need to identify their strengths as well as their weaknesses in terms of providing mobile services that provide solutions as well as rectify the shortcomings. The analyses of the weaknesses and strengths will give them an advantage to provide a convenient service and increase their customer loyalty.

BACKGROUND OF MOBILE TECHNOLOGY

The known mobile phones used today are the extension of American mobile radiotelephone. However, the distinction between such phones and two-way radios is not clearly known. The advancement on mobile technology and the relevant gadgets have been very moderate due to the limitation of this technology and the government regulations on radio transmission. The major concern of the American Communication Commission was to decide who would get what frequencies and which emergency service should have the priority on air for transmitting first.

The ordinary usage of mobile took place in the 1980s. The first generation of cellular mobile phones was used widely only by transmitting analogue signals. The large size and very high prices of mobile devices in addition to the cost of the calls are the well-known facts of this period.

In the 1990s, the second generation of mobile phones was available in the market using advanced digital technology. Very fast signals, cheaper calls and handsets, more reliable services, and the smaller handset devices are the characteristics of this period.

The uncontrollable growth of mobile technology has created new culture in the business world. The use of mobile and Internet technology has passed their boundaries to a business and social revolution. The new technology has capabilities of text, voice, and videoconferences using wireless devices as well as the ability to connect to the World Wide Web.

According to Conners and Conners (2004), the application and the use of mobile and Internet technologies are to organize people into various common interest groups. These groups can vary from harmless fun to serious military or political operations.

Information and communication technology has enlightened the business activities. The use of mobile devices is going to be another crucial factor to remain in the competitive market. Vaghijiani and Teoh (2005) explain this phenomenon of mobile technologies as an enormous opportunity for players up and down the value chain, from the device suppliers to carriers to the end users.

BACKGROUND TO A TRANSITION FRAMEWORK

New process frameworks need to build on existing work on process transitions. Electronic transitions have been studied and experimented by Ginige et al. (2002). In the electronic transitions, there has been ample focus on the effect of a dynamic environment and the rapidly evolving technology on organizations. Undoubtedly, these changes cause organizations to restructure and would introduce a new suite of business processes for them to enable them to remain in the market as well as grow by dealing with a greater number of customers.

The new business model and the use of technology in the development of these changes were the cause of the new term e-business. The term e-business might mean trade on the Internet for some managers, however it is looking at the facts in deeper methodology. The Australian e-business guide (Philipson, 2001) translates e-business as any business transaction or activity that uses the Internet. This includes not only the sale of goods and services directly over the Internet, but also the use of the Internet to promote and facilitate the sale of goods and services.

By using mobile phones or any other mobile devices, we are able to make our e-business model more accessible. The improvement and the efficiency will create more benefits, hence

Figure 1. Electronic mobile business model

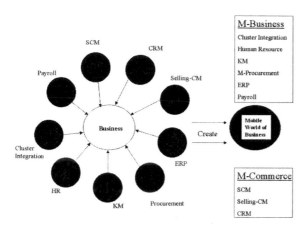

the productivity remains even when people are out of their offices.

The proper design of the e-business model is a necessary component for the success of the m-business model. M-business makes the practice of the e-business model easier, more effective, and more profitable since there is no restriction on approaching the required data. The share of internal data, providing better and more reliable customer service and better control over the organization in general, are other major benefits of mobile business.

Figure 1 shows the re-engineered and mobilized individual enterprise enabling the transformation of an ordinary business to a new and modern world of mobile business. The business and commerce sections are given in different boxes that provide a more comprehensive study of the transformation.

Figure 1 represents electronic mobile business activities in more detail.

- **Clusters Integration:** Our mobile business model must be able to integrate all the clusters of the organizations. This fact might look easy, however by careful analysis it is realised that to connect all the clusters of the organizations is a huge task. It needs more than technological

advancement since some clusters might hesitate to share their information.

- **Customer Relationship Management:** The organizations must create close relationships with their customers. Customers want reliable and fast service. If the organizations provide them with what they want, they gain more business, and more business basically means more revenue. The CRM could be classified as a crucial factor of a mobile business, as it is in direct contact with people who are outside of the company. These people practically do not care how things are running as long as the great service is provided to them. A combination of technology, software, people, and re-engineered business processes are the fundamental of great customer relationship management.

- **Selling Change Management:** The right mobile business model should provide information about the available product directly to the customers. Direct interaction with the customers eliminates the retailer, and this will enable the business officials to provide detailed information about the product to the clients. As there are fewer hands involved in a purchase order, the prices offered to the customers will be dramatically low.

- **Enterprise Resource Planning:** The organizations have realised the importance of having the knowledge about back-office systems which could improve their customer order, integration of their clusters, and provide them with more sufficient information on how to run day-to-day activities by reducing the cost.

- **Supply Chain Management:** The technology is enabling the organizations to eliminate unnecessary processes to save time as well as money. The supply chain management of a business is the plan for materials to be directed to the customers as quickly as possible by cutting the inessential retailers. The information about the delivery, financial matters, as well as order transmissions are provided on any mobile gadgets. The mobile business has more power of monitoring and control over the order status.

- **Procurement:** As per Kalakota (1999), purchasing refers to the actual buying of materials and those activities associated with the buying process. Procurement on the other hand has a broader meaning and includes purchasing, transportation, warehousing, and inbound receiving.

  Organizations spend millions of dollars for procurement every year. Mobile procurement gives them the opportunity to have a better control on their inventory, better control over the purchase approval, and so on. They have better control over the cost as well as knowledge regarding their assets.

- **Human Resources:** The functionality of human resources in the mobile world could be classified as providing the title of available jobs on mobile devices and providing more sufficient information about the positions and required level of desired criteria on the Internet. Enterprise Bargaining Agreement online and other responsibilities of HR could be done while using the new technologies.

- **Payroll:** The available technology is giving the opportunity to advance the payroll that saves a great deal of time and human resources. This strategy is more beneficial to the organizations with odd clusters such as shift workers. If their roster is also automated, the payroll office could automatically generate a payroll list based on the automated roster, while only a human supervisor is required.

- **Knowledge Management:** In today's competitive business world, knowledge plays the crucial rule. The knowledge must be reliable and accessible through all clusters of the organization anywhere and anytime. Knowledge created by the supplier is about available services and products, and knowledge about the users is in the form of profiles. Imagine if a customer calls the engineering department of his electricity company to get the approval for the extension of his house. At the end the inquiry, he asks for the amount of his bill; the system should be able to provide the necessary information rather than transferring his call to another operator. The available information should support the day-to-day running of the business.

By the aid of the emerging mobile technology and re-engineering the individual departments of the organization, the new mobile organization is created. In general, the purposes of the value-added services are to impress customers and create more control over the business. By replacing business with m-business, we can reduce the cost and create more revenue by having more satisfied customers.

APPLYING MOBILE TECHNOLOGIES

By correct application of mobile technologies into the business processes, the business enterprises are likely to gain advantages such as increased profits, satisfied customers, and greater customer loyalty. These customer-related advantages will accrue only when the organization investigates its customer behaviour in the context of the mobile environment.

In general, the application of mobile technologies could be classified in two different categories, online and off-line services. The applications of online services are the executed applications when the mobile gadgets are connected to the mobile Internet.

The applications of the online wireless mobile Internet depends on situation (walking, driving), place (remote area, city metropolitan area), goal (aim of the connection), immediacy (instant action and reaction to demand), and load (how occupied the Internet provider is at the time of the connection). The congestion of the network clearly depends on the usage, which varies at different times of the day. As expected during the business hours, the network load on the Internet service provider is heavy.

The major online mobile applications are:

- **Information:** General information about movies. Location of cinemas, hotels, hospitals, and universities. News, sports, travel, weather, and financial information.
- **E-Mail:** To send and receive mail while online using the mobile handheld.
- **Payment:** To buy the product and receive the service and pay by your mobile device and receive the payment on your mobile bill. There is trusted third party in mobile commerce regarding billing inquiries that increase the cost since another party is involved.
- **Mobile Internet Banking:** To complete the banking transaction using your mobile device while online. The participating banks decide what kind of transaction is allowed and how they provide the security for their clients.
- **Mobile Internet Shopping:** To shop online using the mobile gadgets. There are advantages and disadvantages in mobile shopping since the participants are not able to touch or smell the items they are buying unless it is first sale vs. the repeated sale.

- **Education:** Using mobile handset to download lectures, use library facilities (order book, search the library), and use laboratory.
- **Government Applications:** Election, government bulletin and broadcasting, disaster information system with the aid of location-based services, and automatically giving the priority to the broadcast.
- **Communication:** Videoconferencing, telephony, sending and receiving pictures, and international communication.
- **Leisure:** Download music, video, TV, and games, and for some particular people gambling could be classified as a leisure activity.
- **Telemetric:** Location-based services, global positioning services, and car navigation systems. Telemetric applications could be expanded to give the opportunity to your device to book a hotel room, purchase a ticket, gather your required information, and any related scenario just by a click of the button or voice order, assuming your personal mobile gadget is already holding all your personal and credit card details. These functions could be performed by connecting to your mobile Internet or just by connecting to your network provider.
- **Advertising:** Conjunction of mobile and Internet technology for advertising. Receiving an advertisement on the Internet (pop-up screens) is not something new; furthermore marketing organizations could use the same idea for mobile Internet.

The off-line applications are the services offered by related network providers and extra available features on the particular mobile devices. Networks must have infrastructure to support the fast transmission of the data, reliability of the data, the integrity of the data, and quality of service. The major off-line mobile applications are:

- **Communication:** Phone calls, SMS, messages, sending and receiving pictures.
- **Memory:** Phone book, music, different sounds, different effects (vibrate, volume), display, storing desired pictures and schedules and entertainment.
- **Expert System:** It would be possible for the organizations to use their mobile device as an expert system if their network provider has the capability to support it.
- **Remote Supervision:** To have control over the personnel while they are in an inaccessible area.
- **Traffic Information System:** Informing the drivers of traffic locations when they are driving close to the congested area.
- **M-Newspaper:** Subscribing to a newspaper if provided by the newspaper agency.
- **Advertising:** Based on the device's position, receiving the local advertisement.

Consumers' demands and corporate objectives could be different in the m-enabled world. While the application remains the same, expectation and usage are different. Usage in an m-enabled society is classified in three categories of interaction (voice, e-mail, chat, digital postcards, etc.), trading and business (banking, shopping, auctions, advertising, ticketing, etc.), and mobile-provided services (news, entertainment, driving direction, and much more).

It is very important to identify the sufficient information that is required to make the purchase decisions while the organizations are re-engineering their business processes. Ease of navigation and necessary links to other related Web sites are crucial factors for the software developers to consider while they are designing the new applications.

LIMITATION OF MOBILE TECHNOLOGY

Rising customer expectations have a direct connection to the advancement of technology. People's demand of the technology has not always been so realistic. The word "technology" has constantly fascinated human beings. Information and communication as the defining technology of the modern era have increased the expectations to an irrationally higher level. As Toffler (1980) predicted, people's dependence on technology has increased to a high level where technology has affected every aspect of human life. Mobile technology, which is an integration of communication and computer technology, has created such expectations in human behaviour that people cannot think of an era without such technology today.

It is clear that people rely on technology even when technology does not have the capability, or it is not robust enough, to support their task. There is no guarantee that I will not lose my work while writing this chapter on my computer, and the very same technology is used when human lives are involved. As an example, computers are used to take off and land airplanes.

However, mobile technology has its own characteristics and limitations which should be clearly identifiable to business enterprises. Of course these limitations will increase when mobile devices are connected to the mobile Internet.

Jamalipour (2003) explains that the access to the wireless mobile Internet is not just an extension of the Internet into the mobile environment giving users access to the Internet while on the move. However, it is about integrating the Internet and telecommunications technologies into a single system that covers all communication needs of people. He also believes that current network architectures used in either the wired Internet or the cellular networks would not be appropriate and efficient for future wireless mobile Internet, even if we assume that the cellular network will provide the major infrastructure of the mobile Internet. He concludes by saying that access to the mobile Internet is slow, expensive, and confusing.

Some limitations of mobile technology are as follows:

- **Cost:** The cost of restructuring the organization and personal devices.
- **Call Drops:** Disconnection while taking or downloading the data.
- **Connectivity:** Constant connectivity to a network is a big issue for mobile network providers. There are improvements in this area, but it should be taken into consideration when we are mobilizing the enterprise that the network must support the expected assignment of the enterprise.
- **Lost Work:** Losing the performed work due to disconnection or dead battery.
- **Managing Technology:** Consistent maintenance of the software and the hardware.
- **Security:** Payment online, user behaviour, rules and hassles, mobile virus protection, file encryption, access control, and authentication are the most important security factors in the mobile environment, considering that mobile devices are very personal; in case of loss or theft, who is accessing the corporate or personal data?
- **Integrity:** The transmitted data is actually going to the expected individual. The message received is actually the message sent, and also the sender is the real owner of the mobile handset.
- **Privacy:** Who is accessing the corporate database and where personal details of the individuals are involved.

- **Regulations:** Government roles and regulations regarding the mobile matters.
- **Standardization:** Technical standards and compatibility of the users (business-to-business, business-to-customer).
- **Health Hazards:** By encouraging people to use mobile gadgets, are we jeopardizing their health?
- **Data Transmission Speed:** Slow transmission is very costly and ineffective.
- **Coverage:** The coverage of the network in a remote area is an identified and unresolved problem.
- **Adaptation:** Some people are resistant towards technology, and it would take time for them to adapt to the new technology.
- **Training:** The cost of training, managing mobile workforce, and controlling their activity.
- **Marketing Issues:** It would be a new era of marketing issues in mobile age such as sex, age, and so on.
- **Social Aspects:** Technology is creating a new pressure for ordinary people. People resistant to changes should be the major concern while planning the mobile transformation. Perpetual contact is another issue that is changing the face of our society. Mobile users are communicating with some other person while driving, walking on the street, and when they are in different public places. It is becoming commonly acceptable in our society to give priority to the mobile caller even when personal face-to-face conversation is getting disrupted.

Limited processing powers of handsets' microprocessors, memory size, battery life, small screen of handheld devices and their resolution, replacement costs, required ongoing support, network charges, as well as mobile Internet charges and enhancements are the other critical shortcomings of the mobile technology.

TRANSITIONING TO A MOBILE ORGANIZATION

The transformation of the organization by introducing the new and re-engineered processes is a very crucial matter. Should the enterprise revolutionize and transform as soon as possible or use the evolutionary process? Should the enterprise adapt to the new technology as soon as it is available or delay the process to see the outcome by using another organization's experience?

Considering there is no suitable answer for the above questions, there may be a need for a new approach to clarify this uncertainty. This new approach is supposed to be the combination of the revolution and evolution—revolution since the organization should not fall behind by remaining competitive in the market, and evolution to reduce the risk of not having a successful e-transformation. According to Murugeson and Deshpande (2001), the development of an organization could be classified as the following:

The choice of a suitable development model, according to practitioners and researchers, is site (and applications), its document orientation, content and graphic design, budget and time constrains and the changing technology.

It could be quite risky if the organisations adapt to the new technology as soon as it is available in the market, as the system is definitely unknown and there might be hundreds of unresolved issues as well. Another factor at an early stage entry is the high cost involved in the introductory level of the new technology.

However, if they do not adapt during a specific period of time and their competitors do, there is a great chance of not being able to catch up with the advancements in the professional world. These are some issues that management is facing today. Their crucial decision making will determine the future of their companies. Serour (2005) clarifies that senior and middle management find it hard to proceed when there is (still) very little guidance available from real-world experience.

The organizations must allow internal and external parties involved to know that there are some changes that need to take place. In view of people's resistance to change, this will give them some time to prepare and adjust. The core of the training is for internal parties of the organization; however it is very important to provide sufficient information to external parties and advise them about the change.

The organizations must plan and manage change (cultural, technological, internal, and external) and understand the key areas associated to dangers related to their working environment that others have discovered and faced. The Australian e-business guide (Philipson, 2001) describes that implementations for e-business initiatives must be rapid and each project should be delivered in a maximum of three months. Build quickly and move to the learning stage, then build the next stage and fix the previous ones based on what you have learned.

Management must support the variation in business and market strategies, organizational restructure, and management strategies. The corporations must prepare all the existing clusters ready for change. Managing the transformation by having a reliable and calculated plan is the crucial factor for success. The transition must remain persistent alongside with detailed knowledge of the development of the individual clusters. According to Brans (2003), generally mobile transition takes place by distinguishing what kind of portable devices, networks, application gateways, and enterprise applications are required.

The benefits of mobilizing the organizations are: quick sale, closer communication within the internal departments, more strength and opportunities and less weaknesses and threats, professional façade for the organization, quick and reliable generation of customer data, and mobility at work.

CONCLUSION AND FUTURE DIRECTIONS

This chapter described some characteristics of m-business and offered a brief background of mobile phone technology.

When the Internet was introduced, nobody could imagine that this tool was going to make the next paradigm shift in all human interaction as well as business transactions. M-business is enabling organizations to increase global productivity. With the aid of mobile technology, the capability exists to operate in a very modern and extraordinary manner. The problems faced in the transformation of an organization to m-organization were identified and some solutions were recommended.

The domain of this chapter was to explain the particulars of a mobile business model, emphasising their significance, application, and the shortcomings of this technology in the world of business and trade. It is hoped that this chapter could convince the developers of the m-applications to spend more time in their design to fulfil the needs of the end users.

However, there are some critical issues that are unresolved in the world of mobile technology. These issues can be classified as security (integrity and privacy), national/international regulation, international standardization, security and integrity of databases on mobile devices, managing mobile workers and coordina-

tion of their activities, and the consistent maintenance of mobile hardware and software.

To create a robust and reliable mobile world, the developers could consider some other shortcomings of mobile technology. These issues are mobile payments, health hazards, the cost of restructuring the organization, damage to the handheld devices, legal liability of handheld devices (since the mobile gadgets are very personal and can keep confidential data related to the organization), and constant connectivity of mobile devices.

REFERENCES

Alter, S. (1996). *Information systems. A management perspective.* Benjamin/Cummings.

Brans, P. (2003). *Mobilize your enterprise.* Pearson Education.

Conners, J., & Conners, S. (2004). The impact of mobile technology on business planning. *Proceedings of IRMA 2004,* New Orleans, LA.

Deshpande, Y., & Ginige, A. (2001). Corporate Web development: From process infancy to maturity. In S. Murugesan, & Y. Deshpande (Eds.), *Web engineering managing diversity and complexity of Web application development* (p. 36). Germany: Springer-Verlag.

Ghanbary, A. (2003). *Effects of computers on family and leisure time.* Honour Thesis, University of Western Sydney, Australia.

Ginige, A. (2002). New paradigm for developing evolutionary software to support business. In S. K. Chang (Ed.), *Handbook of software engineering and knowledge engineering* (Vol. 2). World Scientific.

Jamalipour, A. (2003). *Wireless mobile Internet: Architectures, protocols and ser-*

vices. Hoboken, NJ: John Wiley & Sons.

Kalakota, R., & Robinson, M. (1999). *E-business roadmap for success.* Boston: Addison Wesley Longman.

Murugesan, S., & Deshpande, Y. (2001). *Web engineering (publication data).* Berlin/Heidelberg: Springer-Verlag.

Philipson, G. (2001). *Australian e-business guide.* McPherson's Printing Group.

Ranjbar, M. (2002). *Social aspects of information technology.* Sydney, Australia: University of Western Sydney.

Serour, M. K. (2005). The organizational transformation process to globalization. In Y. Lan (Ed.), *Global information society: Operating information systems in a dynamic global business environment.* Hershey, PA: Idea Group Publishing.

Toffler, A. (1980). *The third wave.* William Morrow and Company.

Unhelkar, B. (2005). Web services and their impact in creating a domain shift in the process of globalization. In Y. Lan (Ed.), *Global information society: Operating information systems in a dynamic global business environment.* Hershey, PA: Idea Group Publishing.

Vaghjiani, K., & Teoh, J. (2005). Comprehensive impact of mobile technology on business. In Y. Lan (Ed.), *Global information society: Operating information systems in a dynamic global business environment.* Hershey, PA: Idea Group Publishing.

Chapter XLIII
Policy–Based Mobile Computing

S. Rajeev
PSG College of Technology, India

S. N. Sivanandam
PSG College of Technology, India

K. V. Sreenaath
PSG College of Technology, India

ABSTRACT

Mobile computing is associated with mobility of hardware, data and software in computer applications. With growing mobile users, dynamicity in catering of mobile services becomes and important issue. Polices define the overall behavior of the system. Policy based approaches are very dynamic in nature because the events are triggered dynamically through policies, thereby suiting mobile applications. Much of the existing architectures fail to address important issues such as dynamicity in providing service, Service Level provisioning, policy based QoS and security aspects in mobile systems. In this chapter we propose policy based architectures and test results catering to different needs of mobile computing

INTRODUCTION

Policies are rules that govern the overall functioning of the system. *Policy computing* is used in a variety of areas. *Mobile computing,* with its ever-expanding networks and ever-growing number of users, needs to effectively implement a policy-based approach to enhance data communication. This can result in increasing customer satisfaction as well as efficient mobile network management.

POLICY COMPUTING AND NEED FOR POLICY-BASED MOBILE COMPUTING

Policies in society and organizations are often captured and enforced as laws, rules, procedures, contracts, agreements, and memorandums. Policies are rules that govern the choices of system behavior. A policy is defined as "a definite goal, course or method of action to guide and determine present and future decisions." Security policies define what actions are permitted or not permitted, for what or for whom, and under what conditions. Management policies define what actions need to be carried out when specific events occur within a system or what resources must be allocated under specific conditions. They are widely used for the mobile user whose requirements are dynamic.

Policy-based computing is the art of using policy-based approaches for effective and efficient computing; it is widely used because of its dynamicity. Hence in areas such as mobile computing, policy computing can be effectively used.

Much of the existing network systems' are configured statically (Fankhauser, Schweikert, & Plattner, 1999). In the present-day scenario, the number of mobile/wireless network users increases day by day. With the static systems being deployed, it is very difficult to achieve the needed dynamicity for mobile computing resulting from changing user base. In order to achieve efficient communication for fluctuating user base, policy-based systems need to be implemented in different areas of the existing wireless mobile network infrastructure.

POLICY IN MOBILE COMPUTING

Mobile computing is conducted by intermittently connected users who access network resources that need to escalate with increasing computing needs. Mobile computing has expanded the role of broadcast radio in data communication, and with increasing users, providing quality service becomes a challenging issue. The mobile users must be provided with the best possible service so that the service provider can stay in competition with peer service providers. In order for the best possible

service to be provided to the mobile users, there are certain criteria that should be met. They are:

- The quality of service should be guaranteed.
- There should be effective service-level agreement (SLA) between the mobile user and the service provider.
- Security should be foolproof.

With the existing system (without a policy-based approach), it becomes very difficult to achieve the mentioned criteria. It is very difficult to provide a guaranteed quality of service (QoS), which is also dynamic (not statically configured). Moreover SLA is a very static procedure. Because of the mobility and dynamicity of mobile networks, SLAs also must be made very dynamic. Similarly, security should also be made very dynamic and efficient. To overcome all these shortcomings of the existing system, *a policy-based approach should be used in mobile networks.*

Policy computing can be effectively implemented in mobile networks using policy compilers. Policies can be written in different ways. There are different languages for writing policies that are used for different purposes of specifying policies. In order that the "security policies" be specified, languages such as Trust Policy Language (TPL), LaSCO, and so forth are used. In a similar way, for specifying management-related policies, languages such as Ponder, Policy Maker, and so forth are used. Thus for different scopes of application of policies, specific languages are used.

Policy validation checks a solution's conformance to the policy file. The actual process of policy validation has three primary stages. First, a node or hierarchy change event in Solution Explorer (such as add, drag, or delete event) begins the validation process. Then the validation process maps items discovered in the solution (such as files, references, classes, or interface definitions) to a corresponding Template Description Language (TDL) policy ELEMENT node. Finally, for recognized ELEMENT nodes, the validation process checks the parent ELEMENT for policy compatibility with the child ELEMENT. When the policies are compatible, the validation process applies any ELEMENT-specific policy.

APPLICATIONS OF POLICY COMPUTING

Policy-based management is an over-arching technology for an automated management of networks (Lewis, 1996). Policy-based management is being adopted widely for different domains like quality of service, wireless networks, service-level agreement, virtual private networks (VPNs), network security, and IP address allocation. Therefore policy-based networking configures and controls the various operational characteristics of a network as a whole, providing the network operator with a simplified, logically centralized, and automated control over the entire network. In a wireless/mobile network, events are user and time based. These events are very dynamic in nature in order to provide the best service to the mobile users and also to maximize the profit of the service provider. But most of the existing systems are very static in nature. With the static systems being deployed, it is very difficult to achieve the needed dynamicity for mobile computing. In order to achieve this, policy-based systems need to be implemented in different areas of the existing wireless mobile network infrastructure such as QoS in wireless networks (especially differentiated networks), security, and SLAs.

Policy-Based Architecture for Security

Some of the key issues involved in providing services for wireless networks are (Sivanandam, Santosh Rao, Pradeep, & Rajeev, 2003):

1. **Bandwidth Cost:** Depending upon the number of users connected, the location (e.g., urban or remote), and the type of service (e.g., video, audio, etc.) being offered, dynamic allocation of bandwidth plays an important role.
2. **Limited Memory:** Today's wireless device places constraints on the amount of data that it can hold. Moreover, this limit depends upon the device being used and hence causes greater concern with low memory devices.
3. **Access Cost:** Optimizing the cost (Boertien, Janssen, & Middelkoop, 2001) of accessing and transferring data is more complex in wireless networks than in wired networks. If the number of servers used by a service or the number of services provided by an enterprise increases, then maintaining service consistency would turn out to be a cost factor in itself.
4. **Scalability Requirements:** These requirements force the service provider to think in terms of developing a solution that would support increasing and decreasing the number of services offered by the enterprise.

These constraints adversely affect the process of implementation of wireless/mobile services over the existing architecture. To overcome this, an identity management architecture for a wireless differentiated service schema that could be implemented using LDAP (lightweight directory access protocol) (Hodges &

Morgan, 2002), directory structures are constructed.

Policy Warehouse

The concept of differentiated services entitles the maintenance of a large amount of information pertaining to the user (e.g., user names, passwords, services registered, premium amount, etc.) and an efficient quick access mechanism to retrieve the relevant details. This overhead increases when it comes to wireless services. Here, the policy warehouse acts as the information backbone of the service provisioning system. The service provisioning engine contacts the policy warehouse whenever the service provider forwards to it a request from the user, after the right user has been authenticated.

1. **Id-Synch and P-Synch:** The synchronization of user identities and passwords pertaining to a single user is highly crucial in providing a hassle-free connection to services that require subscription to external back-end service providers. This architecture uses Id-Synch and P-Synch mechanisms for identity synchronization and password synchronization respectively.

2. **Meta Directories:** In general, service providers need to maintain a global user profile to uniquely identify a user over the various services provided to him. This information, which mainly comprises a collection of information pertaining to the user sign-on details of various services, is stored in meta directories. This profile makes it easy for the service provisions engine to authenticate the user. The architecture has a provision of compiling as well as retrieving meta directory information.

3. **LDAP Access Engine and Directory Structures:** Information pertaining to the user is stored in lightweight directory structures that can be retrieved using the LDAP. Directory structures are used to store user information because they provide a systematic mechanism for organizing data under a common head like user profiles, user services, user privileges, and so forth that are organized in a hierarchical manner on multiple workstations that are distributed over a network. This not only makes data retrieval fast, querying complexity less, and volume of data storage minimum, but it also makes easy implementation of policies that depend on the enterprise using the system.

The LDAP engine acts as an interface between the various servers like Id-Synch and P-Synch, and directories like meta directories and the underlying LDAP directory structures. They process the request for data from the higher layers and hand over the appropriate data to the requested application in the required format.

The LDAP directory services are part of the directory-enabled network services (DEN) that provide standard APIs for the access of network objects.

Service Provisioning Mechanism

The service provisioning system can be generally viewed in the following stages:

*   **User Login:** In this stage, the user sends his details like User Name, Password, and Chip Index Number to the service provider for authentication.
*   **Service Provider:** The Service Provider has to perform the following two actions on a service request:

Figure 1. Policy-based provisioning

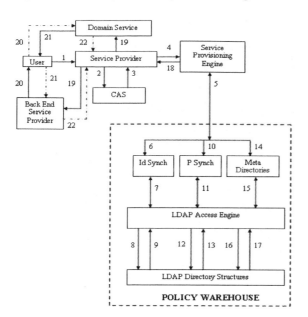

1. **Authentication:** The authentication is accomplished using the DSAP (distributed substring authentication protocol) (Sivanandam et al., 2003). This is done by fragmenting the user details into sub-strings and distributing them over a network which is monitored by a central authentication system. When the user is required to authenticate, the protocol fetches the appropriate sub-strings from the network and compares them to the user input. A match signifies a valid user. After this stage, the appropriate user policy is fetched from the policy warehouse using the service provisioning engine.

2. **Providing the Service:** This is the last step of the service provisioning. In this stage, the actual service that the user has requested is granted. The service could be from a back-end service provider or from the main service provider. This detail is an abstraction to the user who undertakes all transactions with the main service provider only. When the user disconnects from the service, intimation is sent to both the back-end service provider and to the main service provider. This has two implications: firstly, the main service provider's load is shared by the back-end service provider, and secondly, the intimation during the connection termination ensures that the main service provider gets the appropriate usage details. This can act as verification of the details that the back-end service provider will submit later.

Policy Based Architecture for QoS

The interface to the network device and the information models required for specifying policies are either standardized or being standardized in IETF and DMTF. An architecture for a policy-based QoS management system for Diffserv-based wireless networks, which are based on COPS for interfacing with the network device and on LDAP for interfacing with a directory server for storing policies, is constructed. The Diffserv policies are installed based on role combination assigned to the network device interfaces. The directory access could become a bottleneck in scaling the performance of the policy server, and it can be improved substantially by employing appropriate policy caching mechanisms. The framework considers various QoS parameters in the wireless network and proposes the policy-based architecture for QoS management in wireless networks.

Wireless Network QoS Parameters

The wireless/mobile network is affected by the following QoS parameters:

- **High Loss Rate:** Wireless/mobile networks are characterized by more frequent packet losses because of fading effects. The scheduler may think that a certain DSCP is being satisfied with the required number of packets scheduled, but the receiver is not receiving the packets at the required rate. It will be useful to have feedback from the receiver so that some compensation techniques can be employed. The base station (BS) can better handle compensation of lost bandwidth using this information.

- **Battery Power Constraints:** Current mobile battery technology does not allow more than a few hours of continuous mobile operation. Two of the major consumers of power in a mobile network are the network interface (14%) and the CPU/memory (21%). Therefore, network protocols should be designed to be more energy efficient (Agrawal, Chen, & Sivalingam, 1999). The mobile device can use the signaling mechanism to periodically send messages about its power level to the BS. The BS can then use this information to dynamically decide packet scheduling, packet dropping, and so forth.

- **Classification of Packets within a Flow:** Present Diffserv (Chan, Sahita, Hahn, & McCloghrie, 2003) mechanisms treat all packets within a flow identically. Even though a distinction can be made between packets as in-profile or out-of-profile, all in-profile packets are treated the same way. In many situations (e.g., while using layered video), it may be necessary to distinguish packets within a flow. This is because some packets from a flow level could be more important than the others, and a local condition like power level may lead to different treatments of these packets. Thus, the packets within a flow must

be made distinguishable, and bits in the TOS field may be used for this purpose. To summarize, the various possible factors needed to make the Diffserv architecture suitable for wireless networks were discussed in this section.

- **Low Bandwidth:** Wireless networks available today are mostly low bandwidth systems. Most of the current LANs operate at 2 Mbps with migration up to 11 Mbps available. However, the available wireless LAN bandwidth is still an order of magnitude less than the typical wired LAN bandwidth of 100 Mbps. This leads to two decisions. First, the signaling protocol should be very simple and highly scalable. It is also better to modify an existing protocol for compatibility with other existing network protocols. Second, the mobile should not be swamped with too much data from a wired sender with higher network bandwidth. This can be handled to a large extent by transport protocol control, but the problem can be alleviated by handling it partially at the base station. Therefore, mechanisms may be used at the BS to send data to the mobile devices based on current conditions such as channel condition, bandwidth available, and so forth.

Policy-Based QoS

The IETF Resource Allocation Protocol (RAP) working group has defined, among other standards, the policy-based admission control framework, and the common open policy service (COPS) protocol and its extension—COPS for provisioning (COPS-PR). COPS is a simple query protocol that facilitates communication between the policy clients and remote policy server(s). Two policy control models have been defined: outsourcing and provisioning. While COPS supports the outsourcing model, its ex-

Figure 2. Policy-based management system architecture

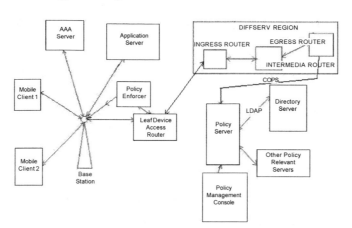

tension COPS-PR integrates both the outsourcing and provisioning models. The outsourcing model is tailored to signaling protocols such as the resource *reservation protocol* (RSVP) (Braden, Zhang, Berson, Herzog, & Jamin, 1997), which requires traffic management on a per-flow basis. On the other hand, the provisioning or configuration model is used to control aggregate traffic-handling mechanisms such as the Differentiated Services (Diffserv) architecture. In the outsourcing model, when the PEP receives an event (e.g., RSVP reservation request) that requires a new policy decision, it sends a request (REQ) message to the remote policy decision point (PDP). The PDP then makes a decision and sends a decision (DEC) message (e.g., accept or reject) back to the PEP. The outsourcing model is thus PEP driven and involves a direct 1:1 relation between PEP events and PDP decisions. On the other hand, the provisioning or configurations model (Chan et al., 2001) makes no assumptions of such direct one-to-one correlation between PEP events and PDP decisions. The PDP may proactively provision the PEP reacting to external events, PEP events, and any combination thereof (N: M correlation). Provisioning thus tends to be PDP driven and may be

performed in bulk (e.g., entire router QoS configuration) or in portions (e.g., updating a Diffserv marking filter).

Architecture of a Policy-Based Management System for a Diffserv-Based Wireless Network

Figure 2 illustrates the architecture of the policy-based management system for Diffserv-based wireless networks. The policy server is responsible for interpreting higher-level policies and translating them into device-specific commands for realizing those policies. For allocating resources on inter-domain links and for implementing SLAs, the policy server (especially the bandwidth broker component) has to communicate with the policy server in the provider.

The policy server is mainly responsible for the following:

- retrieving relevant policies created by the network administrator through the policy console after resolving any conflicts with existing policies;
- translating the policies relevant for each PEP into the corresponding policy information base (PIB) commands;

- arriving at policy decisions from relevant policies for policy decision requests, and maintaining those decision states; and
- taking appropriate actions such as deletion of existing decision states or modification of installed traffic control parameters in the PEP for any modifications to currently installed policies.

All the policies are stored in the LDAP server. The policy editor (PE) is the entity responsible for creating, modifying, or deleting policy rules or entries in the LDAP server. LDAP protocol provides access to directories supporting the X.500 models, while not incurring the resource requirements of the X.500 directory access protocol (DAP). It is specifically targeted at management applications and browser applications that provide read/write interactive access to directories. It does not have the mechanism to notify policy consumers of changes in the LDAP server. Therefore, it is the responsibility of the policy editor to indicate the changes in the LDAP server, as and when required, using an internal event messaging service. The policy server, in addition to querying the LDAP server, queries other policy-relevant servers such as Certificate server, Time server, and so on.

The policy management client—also referred to as the policy editor—provides a high-level user interface, for operator input translates this input into the proper schema for storage in the directory server and pushes it out to the directory for storage. The authentication, authorization, and accounting (AAA) server is responsible for authentication, authorization, and accounting of the user after the relevant policies have been picked and enforced in the policy enforcers (routers).This AAA server is used by the base station to check if the user is authenticated and authorized for the resource he requests, and to check if he is accounted.

The policy enforcer nearer to the base station enforces the policy decisions taken from the policy server. The base station then requests the nearest application server (after policies are enforced) and waits for the response from the application server.

The base station first sends the request to the leaf access router, which then sends it to the ingress router in the region. The ingress router then passes on the requests to the intermediate router. The request passes through the other intermediate routers and reaches the egress router, which sends the request to the policy server through COPS.

Policy-Based Architecture for SLA

Mobile ad hoc networks (MANETs) are autonomous networks operating either in isolation or as "stub networks" connecting to a fixed infrastructure. Depending on the nodes' geographical positions, transceiver coverage patterns, transmission power levels, and co-channel interference levels, a network can be formed and unformed on the fly. Ad hoc networks have found a growing number of applications: wearable computing, disaster management/relief and other emergency operations, rapidly deployable military battle-site networks, and sensor fields, to name a few. The main characteristics of ad hoc networks are:

- **Dynamic Topological Changes:** Nodes are free to move about arbitrarily. Thus, the network topology may change randomly and rapidly over unpredictable times.
- **Bandwidth Constraints:** Wireless links have significantly lower capacity than wired links. Due to the effects such as multiple accesses, multi-path fading, noise, and signal interference, the capacity of a wireless link can be degraded over time and

the effective throughput may be less than the radio's maximum transmission capacity.

- **Multi-Hop Communications:** Due to signal propagation characteristics of wireless transceivers, ad hoc networks require the support of multi-hop communications; that is, mobile nodes that cannot reach the destination node directly will need to relay their messages through other nodes.

- **Limited Security:** Mobile wireless networks are generally more vulnerable to security threats than wired networks. The increased possibility of eavesdropping, spoofing, and denial-of-service (DoS) attacks should be carefully considered when an ad hoc wireless network system is designed.

- **Energy Constrained Nodes:** Mobile nodes rely on batteries for proper operation. As an ad hoc network consists of several nodes, depletion of batteries in these nodes will have a great influence on overall network performance. Therefore, one of the most important protocol design factors is related to device energy conservation.

To support mobile computing in ad hoc wireless networks, a mobile host must be able to communicate with other mobile hosts that may not lie within its radio transmission range. Therefore in order for one mobile host in the ad hoc network to communicate with the other not lying in its transmission range, some other hosts in its transmission range should route the packets from the source to the destination host. The conventional routing protocols used in wired networks cannot be effectively used in ad hoc networks. Hence new routing mechanisms are suggested which may be used for routing in ad hoc networks. Routing issues in ad hoc networks are beyond the scope of this chapter and are not considered here.

Since many mobile hosts may be within transmission range of each other, there may be multiple routes for a packet to reach a destination. Therefore the source host should decide which route to use to send the packets to reach its destination. Obviously, the sending host has to decide on the best optimal route before sending its packets towards the destination. Thus, there should be a service level agreement between the source mobile host and the host which routes the packets to the destination host. Moreover there are certain constraints based on the characteristics of the ad hoc network which play a major role in deciding which route is optimal, given there are more routes to reach the destination.

Architectural Framework of the Policy-Based Mobile Ad Hoc Network

MANET is a collection of mobile hosts forming a temporary network without the aid of any centralized administration or standard support services. The architecture for the policy-based SLAs in ad hoc networks is given below. The architecture is designed where at least one host has connectivity with the wired network. In the architecture shown in Figure 3, the policy server is placed in the wired network. Polices are stored in the directory server. The ad hoc 'host1' is within the vicinity of both 'host2' and 'host3'. 'Host4' is not within the transmission region of 'host1'. So when 'host1' wants to send a packet to 'host4', intermediatory hosts, 'host2' and 'host3', help 'host1' with connection establishment.

Assuming that both the host services satisfy the constraints of 'host1', 'host1' must choose a service level agreement among the two. In this case since 'host1' is connected to a base station, which in turn is connected to the wired network having the policy server, 'host1' can query the policy server through the base station

Figure 3. Architecture of policy-based MANET

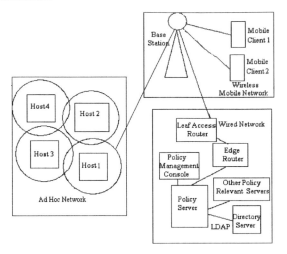

and then the leaf access router and edge router. For simplicity we have shown the policy server being connected to the base station through only a few hops. But in practice it may be many hops away from it. Once the request reaches the policy server, it takes appropriate policies from the directory server through the LDAP.

The policy server also communicates with other relevant policy servers such as Time Servers, Certificate Servers, and AAA servers, and validates the host providing service by means of certificates and AAA. The policy server makes the decision on whether the host providing the service is an authenticated one, and his services are authorized with accountability and certificates. Then the policy server based on the higher level polices stored in the directory server chooses an agreement among the available agreements. The decision to choose an agreement from among the available ones may be done giving more weight to those performance metrics which affect the overall performance the most. Over a period of time, the history of the hosts providing the service will be stored; solutions based on a neural network model may be used for finding an optimal solution. Once the policy server chooses an

agreement, it sends its reply back to 'host1', which agrees for the service with the appropriate host.

Policy Based E-Supply Chain Management Architecture

Internet-based e-purchases and e-supply chain management are now being widely used. This however has a major disadvantage of very limited mobility and the absence of a dynamic policy that will efficiently manage the entire supply chain. The activities that are required to provision mobile users comprise a surprisingly large number of steps that cross an entire enterprise. Policy setting and implementation, approval workflows, physical resource setup/teardown (provisioning), account maintenance, reconciliation of actual resource assignments with approved user lists, audit, and overall service management are some examples. They are together called policy-based provisioning. An extension to e-purchases through mobile phones using policy-based e-supply chain management is constructed.

Architecture of Policy-Based E-Supply Chain Management

The architectural framework of the e-purchase through policy based e-supply chain management is shown in Figure 4. Mobile customer, policy server, nodes (1-4), and suppliers (1 and 2) form the key elements of the proposed architecture. The mobile consumer requests the policy server of the service provider through the base station. The mobile user's authentication, authorization, and accounting rights are then verified by the AAA server. If the mobile user is found to be an authenticated one and his request for service is an authorized one, then the policy server fetches the corresponding policies for the user from the directory server through LDAP. The policy server of the service provider is connected to other nodes (Node 1 to Node 4) in the Internet. The nodes of the suppliers such as 'supplier1' and 'supplier2' are also connected to the nodes through the public network (Internet). When the relevant policies are fetched from the directory server, the user's request is sent to an efficient supplier who will supply the product to the mobile customer. The supply chain is made electronic as discussed earlier and is policy based. Once the user's request is sent to an efficient supplier relatively nearby to the customer, the ordered products will be delivered to the mobile customer through the shipping department. The billing of the products purchased is taken into the credit account of the mobile user and is charged along with the mobile phone bill, simplifying user billing and payment. The customer at the end of the month would pay the bill through the electronic account facility available with his existing bank account. Thus the whole process of placing the purchase order, delivering the product through the supply chain, and paying for the product is made electronically, thereby facilitating the customers, who are mostly travelers and tourists.

Figure 4. Framework of e-purchase through policy-based e-supply chain management

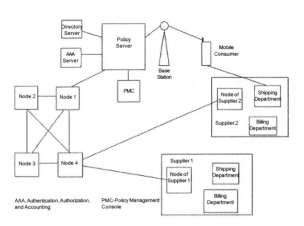

CASE STUDY

A simulation for the policy-based MANET shown in Figure 3 was performed using the QualNet Network Simulator, using the simplex method to solve the linear program model given below, and using the Ponder Toolkit. The mathematical model is given in the following section.

Mathematical Model

A mathematical model considered for policy-based MANET uses Linear Programming and Simplex Method to solve it. The following performance metrics that are crucial for effective SLA trading and choice of route are considered in our model: (a) Bandwidth, (b) Delay, (c) Demand, (d) Packet Loss, (e) Congestion, (f) Queuing Delay, (g) Throughput, (h) Buffer Capacity, (i) Battery Consumption, and (j) Mobility. Let,

T_{ij} = Total (maximum) Bandwidth (channel capacity) available from host i to host j.

U_{ij} = Bandwidth being used for traffic flow between host i to host j at instant 't'.

R_{ij} = Reserved bandwidth from host i to host j.

Hence the bandwidth that can be leased to other hosts G_{ij} is given by $G_{ij} = T_{ij} - U_{ij} - R_{ij}$.

Let the required bandwidth—that is, the bandwidth consumed by the host k to reach host j through host i—be RB_{ij}.

And,

D_{ij} = Delay from host i to host j.

C_{ij} = Cost of reaching host j through host i.

F_{ij} = Fraction of bandwidth bought from host i to reach host j.

The objective here is to minimize the cost of reaching host *j* through other hosts.

$$Minimise \sum_{i,j} F_{ij} C_{ij} \qquad (1)$$

As stated earlier, in the above equation represents the cost hostcharges to reach host through host.

Constraints

There are a set of constraints that define the model. The first constraint is the demand for bandwidth to reach host *j* through host *i*; De_{ij} should be less than or equal to the amount of bandwidth host *i* is ready to offer for cost to reach host *j*, G_{ij}.

$$DE_{ij} \le \sum_{i,j} G_{ij}$$

(2)

The following constraints check if the service performance metrics in the service offered by the host *i* to reach host *j* fall within the predetermined and pre-calculated boundaries as expected by host *k* which needs the service. These boundary constants for the performance metrics can also be set dynamically and SLA negotiated accordingly.

Buffer Capacity B_{ij} should not be less than a bearable value given by the constant N=Number of packets that can be buffered.

$$B_{ij} \ge N \qquad (3)$$

The time delay *D* should be set to a limit expressed by a constant '*p*1'as expected by the 'ISP k' which needs the service. The constant '*p*1' is arrived as derived as follows:

'*p*1' = Propagation Time + Transmission Time + Queuing Delay (+ Setup Time)
Propagation Time: Time for signal to travel length of network
= Distance/Speed of light
Transmission Time = Size/Bandwidth

Therefore, we have

$$D_{ij} \le p1 \qquad (4)$$

Queuing Delay Q_{ij} should not exceed an allowable limit '*p*2' expressed as

$$p2 = \frac{D}{2} \times (N-1)$$

where, *D* = the time delay, *N* is the Buffer Capacity

$$Q_{ij} \le p2 \qquad (5)$$

The Packet Loss P_{ij} for the service provided should not exceed a maximum limit set as constant '*p*3', and Congestion in the channel offered for service Co_{ij} should also be within the acceptable limits represented by the constant '*p*4', both of which are arrived at as shown as follows:

T_{min} = Minimum Inter-Arrival Time observed by the receiver.
P_0 = Out of order packet.
P_i = Last in-sequence packet received before P_0.
T_g = Time between arrival of packets P_0 and P_i.
n = Packets missing between P_i and P_0.

If $(n+1) T_{min} \le T_g < (n+2)T_{min}$, then *n* missing packets are lost due to transmission errors and hence '*p*3'='*n*'and

$$P_{ij} \leq p3 \tag{6}$$

Else n missing packets are assumed to be lost due to congestion and hence '$p4$'='n' and

$$Co_{ij} \leq p4 \tag{7}$$

Throughput TH_{ij} should be greater than or equal to '$p5$', which is given by

$$p5 = \{MSS / RTT\} \times C / (\sqrt{p})$$

where,

MSS = Maximum Segment size in bytes, typically 1460 bytes.
RTT = Round Trip Time in seconds, measured by TCP.
p = Packet loss.
C = Constant assumed to be 1.

$$TH_{ij} \geq p5 \tag{8}$$

The jitter J_{ij} should be within the acceptable limit '$p6$' given by

$$p6 = p6 + (|D(i-1,i)| - p6)/16$$

given

where, S_i, S_j are sender timestamps for packets i, j and R_i, R_j are receiver timestamps for packets i, j.

Therefore

$$J_{ij} \leq p6 \tag{9}$$

The Battery Consumption BC_{ij} for the offered service should be within the boundary constant '$p7$',

$$BC_{ij} \leq p7 \tag{10}$$

The Mobility Factor M_{ij} which gives the idea of how long the host j will be in the transmission range of host for which packets need to be routed should not be smaller than a particular constant represented by '$p8$',

$$M_{ij} \geq p8 \tag{11}$$

This mobility factor M_{ij} plays a crucial role in ad hoc networks because the hosts are all mobile. It may be minutes or in any preferred time unit as the case may be. We generally assume that a mobile which has joined the ad hoc has more probability of staying in the network than the ones which came earlier than that. But the exact nature of the mobility of a host can be predicted only based on past performances of the mobile.

Non-Negativity Constraints

The following are the non-negativity constraints applied in the model:

Cost C_{ij} should always be positive,

$$C_{ij} \geq 0 \tag{12}$$

Fraction of bandwidth bought from host i to reach host j, F_{ij} should also be positive,

$$F_{ij} \geq 0 \tag{13}$$

The bandwidth that can be offered for cost to other hosts by host i should be positive,

$$G_{ij} \geq 0 \tag{14}$$

Given the objective, for example, to minimize the agreement cost along with the performance metrics constraints, the proposed linear programming model solved using simplex method suffices for arriving at a suitable agreement for service with other hosts. There are always cases that the above model will fetch more than one solution if other solutions exist. Hence in such cases the decision of choosing the most appropriate of the available solutions should be taken which is described in the next section.

The test environment has four ad hoc hosts from 'host1' to 'host4', as shown in Figure 3. The total bandwidth, used bandwidth, reserve bandwidth, battery consumption, mobility factor, and other performance metrics of the hosts are tabulated below.

Table 1. Performance metrics and other parameters of the hosts

Performance Metrics	Host 1	Host 2	Host 3
Total Bandwidth Allocated (MBps)	3	6	5
Bandwidth Used at Instant (MBps)	2	2	1
Reserve Bandwidth (MBps)	0	1	1
Remaining Bandwidth G_y (MBps)	1	3	3
Demand for Bandwidth to Reach 'host4' (MBps)	1	0	0
Delay (x 10-3/sec)	7	8	10
Packet Loss Factor	7	5	6
Congestion Factor	30	20	25
Queuing Delay (x 10-4sec)	8	7	10
Throughput (x 103 Bits/sec)	100	100	90
Buffer Capacity (No. of Packets)	9	10	8
Battery Consumption (mWh)	-	8	9
Mobility Factor—Minutes	-	25	18

In the simulation test environment, 'host1' needs to communicate with 'host4', which is not in its transmission range. So both 'host2' and 'host3' offer the service to 'host1'. Using the mathematical model proposed, 'host1' decides upon the suitable service among the offers using the SLA trading algorithm (Rajeev, Sivanandam, Sreenaath, & Bharathi Manivannan, 2005) and the mathematical model given previously. Since only the service offered by 'host2' adheres to the performance metric constraints, 'host1' chooses the service offered by 'host2'. All the simulation is done with respect to the packet flow from 'host1' to 'host4'.

The trade for the service is decided by using the simplex method to solve the linear programming model and SLA trading algorithm, by which a feasible solution is obtained. The performance constraints and other parameters of the hosts are given in Tables 1 and 2. According

Table 2. Performance metrics and other parameters of 'host2' and 'host3'

Performance Metrics	Host 2	Host 3
Delay (x 10-3/sec)	2.9	3.1
Packet Loss Factor	0.2	0.3
Congestion Factor	0.3	0.3
Queuing Delay (x 10-4sec)	0.2	0.2
Throughput (x 103 Bits/sec)	4.2	4.2
Buffer Capacity (No. of Packets)	20	15
Battery Consumption (mWh)	8	9
Mobility Factor (minutes)	25	18
Jitter (x 10-4sec)	3.9	4.1
Fraction of Bandwidth that Can Be Given F_y (MBps)	1	1
Cost C_y ($)	2	6

Table 3. Original and final value of the objective

Objective	Original Value	Final Value
$\sum_{i,j} F_{ij} G_{ij} C_{ij}$ of 'host2' (\$)	12	6
Cost of 'host2' (\$)	4	2

to the constraints given by 'host1' for the required service, the simplex method and SLA trading algorithm are used, and the best bid among the bids offered by the two hosts ('host2' and 'host3') is selected. Since only the bid for the service offered by 'host2' satisfies the constraints of 'host 1', SLA between 'host1' and 'host 2' takes place. The LP model is solved by using the simplex method. As only the trade provided by 'host2' satisfies all the constraints with the objective of minimum cost, the Service offered by 'host2' is agreed upon for trade.

From the performance metrics and the constraints on performance metrics, the objective of minimizing cost is arrived at (see Figure 5 and Table 3). Thus an effective SLA is traded between 'host1' and 'host2', satisfying the constraints on the performance metrics which affect the service.

Figure 5. Objective

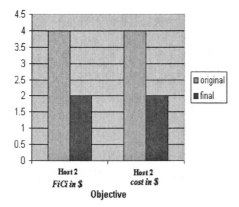

CONCLUSION AND FUTURE DIRECTIONS

Policy computing can be effectively used in mobile computing in various arenas such as QoS, security, SLA, and e-purchase. The architectural framework demonstrated in the case study gives insight as to how QoS, SLA, and security can be implemented in mobile networks. Policy-based architectures for billing in mobile networks are currently being constructed which could bring transparency in mobile billing with added dynamicity.

REFERENCES

Agrawal, P., Chen, J. C., & Sivalingam, K. M. (1999). *Energy efficient protocols for wireless networks.* Norwell, MA: Kluwer Academic Publishers.

Braden, R., Zhang, L., Berson, S., Herzog, S., & Jamin, S. (1997). *Resource ReSerVation Protocol (RSVP)—version 1 functional specification.* IETF RFC 2205.

Chan, K. et al. (2001). *COPS usage for policy provisioning (COPS-PR).* IETF RFC 3084.

Chan, K., Sahita, R., Hahn, S., & McCloghrie, K. (2003). *Differentiated Services quality of service policy information base.* IETF RFC 3317.

Fankhauser, G., Schweikert, D., & Plattner, B. (1999). *Service level agreement trading for the Differentiated Services architecture.* Technical Report No. 59, Computer Engineering and Networks Lab, Swiss Federal Institute of Technology, Switzerland.

Hodges, J., & Morgan, R. (2002). *Lightweight Directory Access Protocol (v3): Technical specification.* IETF RFC 3377.

Lewis, L. (1996). Implementing policy in enterprise networks. *IEEE Communications Magazine, 34*(1), 50-55.

Rajeev, S., Sivanandam, S. N., Sreenaath, K. V., & Bharathi Manivannan, A. S. (2005). Policy-based SLA for wireless ad hoc networks. In *Proceedings of the International Conference on Services Management,* India.

Sivanandam, S. N., Santosh Rao, G., Pradeep, P., & Rajeev, S. (2003). Policy-based architecture for authentication in wireless Differentiated Services using Distributed Substring Authentication Protocol (DSAP). In *Proceedings of the International Conference on Advanced Computing,* India.

Section IX

Customer

Chapter XLIV

Investigation of Consumer Behavior in Using Mobile Payment Services—A Case Study of Mobile Recreational Services in Taiwan

Maria Ruey-Yuan Lee
Shih Chien University, Taiwan

Yi-chen Lan
University of Western Sydney, Australia

Hsiang-ju Su
Shih Chien University, Taiwan

ABSTRACT

The growing popularity of the mobile phone and the diverse functionality of mobile services have forced mobile service providers to enter into a highly competitive business arena. In digital life today, mobile phone services are not restricted merely to communicating with people but more and more value-added services have emerged to amalgamate disparate industries/businesses and open up greater market opportunities. These disparate industries/ businesses may include recreational and travel services, mobile learning services, mobile banking services, and many others. Nevertheless the service providers must understand the consumer behaviour in value-added services in order to enhance their product design. The key objectives of this research is to investigate and analyze the relationships between the consumer behaviour, consumer personality and lifestyle in adopting mobile recreational services; and provide recommendations to the service providers for increasing competitiveness—in the context of Taiwan.

INTRODUCTION

The rapid evolution of mobile phone technologies and services provide consumers with enormous interests in using mobile phones for many other daily activities. The service providers are racking their brains to develop increasingly value-added services to attract consumers. Today, consumers anticipate the products and services they purchased are in personalized form. Consequently mobile phone services cannot be restricted merely in communication function, and should be customized and characterized in accordance with a consumer's personality and individuality.

According to the statistic report announced by the Directorate General of Budget, Accounting and Statistics, Taiwan[1] in March 2004:

1. Total mobile phone accounts (25 million accounts) at the end of 2003 increased 5% from the previous year.
2. GPRS (General Packet Radio Service) accounts (2.68 million accounts) at the end of 2003 increased six times from the previous year.
3. There were an average of 111 accounts per 100 users at the end of 2003.
4. The total mobile phone communications duration at the end of 2003 was 23.3 billion minutes, which was a 16.6% increase from the previous year.

Based on the above figures, there is no doubt that the mobile telecommunication market is continuously growing. Furthermore, swift growth of GPRS accounts shows that the usage of mobile phones has expanded to other value-added services such as the ability to access and operate Internet applications, and the ability to remotely access and control in-house appliances and machines. For this reason, the study focuses on the following two aspects:

1. the investigation and analysis of consumer behavior of adopting value-added mobile services; and
2. the relationships with specific consumer characteristics such as personality, lifestyle, and corresponding demographic parameters (age, gender, education level, and occupation) to identify their implications.

This research concentrates on consumers adopting value-added mobile services in terms of downloading ring tones and images. Through an online questionnaire, data related to personality, lifestyle, and experiences of downloading ring tones and images can be collected and further analyzed for service providers in decision making and strategic planning.

The above research objectives are converted into the following specific research activities, which have been addressed during this study:

1. Analyze value-added mobile services users' (VAMS user) personalities, lifestyles, and their relationships with demographic parameters.
2. Evaluate the associations between VAMS users' personalities and product/service
3. Evaluate the associations between VAMS users' lifestyles and product/service categories.

Figure 1. Research framework

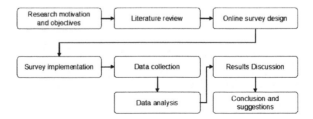

Table 1. Comparison of four generations of mobile phone technologies (based on Yu, 2002)

Generation	1G	2G	3G	4G
Main Technologies	AMPS, TACS, NMT	GSM, CDMA	UTRAN	Under development
Type of Transmission	Analogue	Digital	Digital	Combination of network and communication infrastructure
Support	Basic mobility	Roaming services	Integrated roaming services	Integrated roaming services
Bandwidth	Low, voice only	Low, voice and digital signals	Hi, digital signals and multimedia	Hi, large and real-time multimedia
Transmitting Speed	Very slow	GSM-9.6Kbps CDMA-14.4Kbps	Up to 2Mbps	Up to 100Mbps

Table 2. Comparison of characteristics between mobile business and e-business (based on Liu, 2002)

	Mobile Business	E-Business
Communication Technologies	Connecting with clients via mobile and wireless communication technologies	Connecting with clients via traditional telecommunication technologies
Scope of Services	Provides location-based information	Limited Web site-based information
Equipment for Services	Mobile phone, PDA	Desktop, laptop
Locations of Acquiring and Using the Services	Clients can use mobile business services at any time and any location	Due to fixed-point access, clients can only use the services in a single location at any one time

Furthermore a research framework is illustrated in Figure 1 to enable readers to capture the overall picture of this study.

The rest of the chapter is organized into four main sections. Firstly, the literature survey section provides a comprehensive review on the evolution of mobile phone technologies, mobile business operations, and consumer behavior on mobile services. Secondly, the research method and framework section highlights the research conceptual model, hypotheses, limitation, and questionnaire design. Thirdly is the discussion of analyzed survey results, and finally the chapter concludes with suggestions and future directions.

LITERATURE SURVEY

In order to understand and enhance mobile phone value-added services, it is imperative to review the fundamentals and evolution of mobile phone technologies, mobile business services, value-added mobile services, and mobile consumer behavior. This section has been subdivided into various corresponding sub-sections to accommodate the aforementioned components.

Fundamentals and Evolution of Mobile Phone Technologies

To give readers an overall picture of the fundamental mobile phone technologies, the authors have summarized the comparison of four generations (1G, 2G, 3G, and 4G) of mobile phone technologies in Table 1.

Mobile Business Services

Although both mobile business and e-business services are based on the Internet platform, there are still quite a few dissimilarities between them. Table 2 summarizes the differences between mobile business and e-business in terms of their communication technologies, scope of services, equipment for services, and locations of acquiring and using the services.

Mobile business services may include the following features:

1. Users can access the services regardless of time and location limitation—that is, "ubiquity."
2. Easy and convenient—users are not required to go through a sophisticated procedure to complete transactions.
3. Mobile business services can provide users with real-time information, such as such as traffic report, weather conditions, and locations of ATMs, based on the users' location.
4. Mobile business services allow users to select functions based on their personal preferences and setup the services details—that is, "personalization."
5. Users can access the services network (e.g., mobile banking) without complicated procedures.

Value-Added Mobile Services

At present, the mobile service providers deliver value-added mobile services in two main categories. One is aimed at end users for consuming services, another is targeted at enterprise users for systems integration services. In general, value-added mobile services can be

Table 3. Value-added mobile service categories

Service Categories	End Users	Enterprise Users
Information Services	• Reference information (e.g., maps) • Real-time updated information • Location-based information	• Data collection • Data monitoring • Alerting and reminding mechanisms
Communication Services	• Text messaging services • Multimedia messaging services	• Calendar • Messaging services • E-mail services • Group messaging
Mobile Business Services	• Retailer services • Finance and banking services • Transaction and payment mechanisms	• Customer relationship management • Sales and marketing services • Customer services
Entertainment Services	• Games, images, and ring tones' downloading • Image transmissions	

grouped into four main categories, including information services, communication services, mobile business services, and entertainment services (see Table 3).

Mobile Consumer Behavior

According to Pearson Education's online glossary (2000), "consumer behavior" refers to the process by which people determine whether, what, when, where, how, from whom, and how often to purchase goods and services. In other words, it is to investigate why people purchase commodities. Once sales and marketing staff recognized the reasons people purchased particular brands or products, enterprises could rely on such information to fine-tune their business strategies to enhance customer services and satisfaction.

In general, consumers make purchasing decisions through six phases. Nevertheless consumers may withdraw purchase decisions at any point. The following six phases delineate the pattern of Internet consumers' decision procedures (Lin et al., 2002).

Phase 1— Requirement Confirmation

This phase initializes the entire purchase decision process. The purchase requirement is confirmed when consumers think, feel, see, and understand how the featured products or services will help make their life or job easier.

Phase 2—Information Gathering

There are two types of information gathering. The first type is "internal search"—consumers search their past experiences. The second approach is "external search"—consumers search for related information via external channels to assist their purchase decision making. In the traditional purchase environment, gathering information requires time and effort; however, in the Internet environment, information is searched and gathered through powerful search engine facilities and various information providers and agents. The time and effort required for Internet-based information gathering are far less than the traditional approach.

Phase 3—Evaluation of Acquired Information

After gathering sufficient information, consumers will evaluate the information to identify the appropriate solution (products or services).

Phase 4—Purchasing

At this point consumers may or may not proceed with their purchases. It is important that online merchants should recognize the unsuccessful purchase factors.

Figure 2. Conceptual model

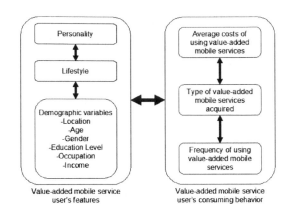

Phase 5—After-Purchase Evaluation

After the successful purchase, consumers would compare the purchase with their expectations. After-sales services are an important evaluation element and a key factor to determining the returning consumer.

Phase 6—Returning and Repurchasing

If the after-purchase evaluation does not meet consumers' satisfaction, the consumers may return their purchase and restart their repurchasing activities.

RESEARCH METHOD AND CONCEPTUAL MODEL

After the review of mobile phone technologies, mobile business and value-added services, and consumer behavior, this section identifies an appropriate research method and develops a conceptual model (see Figure 2) to carry out the study.

Hypotheses

The following hypotheses are derived in accordance with the literature review, research framework, and research objectives:

H1: There is a significant correlation between the VAMS user's personality and lifestyle.

H2: There is a significant correlation between the VAMS user changing ring tones frequency and demographic variables.

H3: There is a significant correlation between the VAMS user changing display images frequency and demographic variables.

H4: There is a significant correlation between the VAMS user's average spending on value-added services and demographic variables.

H5: There is no significant correlation between the VAMS user's personality and the costs of using value-added services.

H6: There is a significant correlation between the VAMS user's lifestyle and the type of acquired value-added services.

H7: There is no significant correlation between the VAMS user's lifestyle and changing display images frequency.

H8: There is no significant correlation between the VAMS user's lifestyle and changing ring tones frequency.

H9: There is no significant correlation between the VAMS user's lifestyle and the costs of using value-added services.

Research Design

An online survey was designed to allow any Internet users to answer the survey questionnaires. The survey was written in ASP with SQL backend database system and is located at http://home1.usc.edu.tw/M9196012/main.asp. The survey system has the capability of avoiding repeated participants and uses entered e-mail addresses to remove suspicious samples.

Constructing Questionnaires

The survey is divided into three main sections. The first section contains 35 questions relating to the VAMS user's personality. This section is based on Cohen's (1967) CAD personality measurement. The second section adopts from SRI Company's VALS2 survey (2005). It measures people's lifestyles of managing time and money. The section contains 33 questions. The third section is aimed at collecting demographic data (age, gender, education level, occupation, and income) of the VAMS user, and it contains eight questions. The survey can be found in the Appendix.

RESULTS ANALYSIS AND DISCUSSION

The duration of survey data collection was four weeks after it was launched on the Internet. A total of 138 samples derived after a validation process. A sample of 138 respondents yields a sampling error of just under +/-10% at the 95% confidence level. Therefore, for example, if half the respondents in the sample agree with an item and half disagree, then in 19 out of 20 cases the true percentage agreeing in the wider population would range between 40% and 60% (i.e., +/-10%). That would be the maximum variance potentially arising due to sampling effects.

In addition to basic (descriptive) statistics analysis, the study also embraces Reliability Analysis (Cronbach's Coefficient Alpha), Factor Analysis, and Bivariate Correlation Analysis.

A summary of data analysis results is divided into two main groups and presented as follows.

Mobile Phone User Profile in Taiwan

1. *Age Distribution*—61.6% of the respondents are in the age group of 23-30.
2. *Gender Distribution*—There is not much difference between male (44.9%) and female (55.1%) respondents.
3. *Education Level*—89.9% of the respondents hold college degrees or higher.
4. *Monthly Allowance*—The majority of the respondents (55.8%) have a monthly allowance in the following two groups:
    a. NT$1,001-5,000—26.8%
    b. NT$5,001-10,000—29%
5. *Occupation Distribution of the VAMS User*—Student is the largest group (30.4%), followed by IT industry (13.0%). The communication industry also holds 11.6% of the VAMS users.
6. Nearly two-thirds of value-added mobile services users (69.6%) live in the northern part of Taiwan (close to the capital city of Taipei).
7. 43.5% of VAMS users have used mobile phone services for more than five years.
8. A total of 96.4% of mobile phone accounts are monthly subscription and pre-paid services.
9. The primary concern of adopting value-added mobile services of the mobile phone users is the "cost" of the services (31.9%).
10. About 42.0% of the mobile phone users learned the value-added mobile services from TV advertisements.

Relationships between Personality, Lifestyle, and Mobile Phone Services Consuming Behavior

1. There are significant correlations between the consuming behavior and the VAMS user's age, education level, occupation, and the available monthly allowance.
2. A significant negative correlation between the VAMS user's personality and the frequency of changing mobile phone ring tones.
3. A negative correlation (not significant) between the VAMS user's personality and the frequency of changing mobile phone display images.
4. A positive correlation between the VAMS user's personality and the average costs of subscribing value-added mobile services.
5. A positive correlation between the VAMS user's lifestyle and the types of subscribed value-added mobile services.
6. A positive correlation between the VAMS user's lifestyle and considering adopting the types of value-added mobile services.
7. No correlation between the VAMS user's lifestyle and the frequency of changing mobile phone display images.
8. No correlation between the VAMS user's lifestyle and the frequency of changing the mobile phone ring tones.
9. A positive correlation between the VAMS user's lifestyle and the average costs of subscribing value-added mobile services.

Overall, the characteristics of the VAMS user such as age, gender, education level, personality, and lifestyle have positive influence on the user's consuming behavior.

Discussion

Derived from the above analysis, we identify and highlight the following recommendations for sales and marketing strategy of mobile services providers.

From the demographic variable perspective, balanced gender distribution illustrates that

*   the sales and marketing campaign should not focus on either the female or male domain;

- age distribution clearly indicates the younger generation (23-30 age group) makes up the majority of mobile phone users, hence the types of value-added mobile services should be designed to suit these age group's needs; and
- costs of adopting value-added mobile services is another major hurdle to attract new customers, consequently subscription fees of popular value-added services should be positioned within the mobile phone user's monthly allowance range.

The VAMS users' lifestyles are influenced by their personalities, and the differences of lifestyles will influence their decisions in choosing value-added mobile services. Therefore, the mobile service providers need to identify the characteristics of each kind of lifestyle, and design and package the value-added services to satisfy the requirements.

Furthermore, the results evident indicate that students are the prominent mobile phone user group. However, many service providers possess a misperception of consuming capability of the student group and fail to benefit from such a great market. The authors would like to suggest that service providers investigate the trends and needs among the student domain and design suitable value-added mobile services to attain such a market.

CONCLUSION AND FUTURE DIRECTIONS

This study aims at investigating the VAMS users' consuming behavior and identifying the correlations between the consuming behavior and various characteristics such as personality, lifestyle, and a number of demographic variables. The research findings indicate that there are significant correlations between the VAMS users' consuming behavior and various factors.

These factors have imperative impact on the decisions of VAMS users adopting and choosing value-added mobile services—namely, personal characteristics (personality) and lifestyle. A number of demographic variables are embedded into the research design to further analyze the implications of the factors from the demographic perspective. From the results discussion, the authors have recommended that mobile phone providers explore the experiences and issues faced by the existing customers, and revisit their sales and marketing campaigns and strategies.

The investigation of VAMS users' consuming behavior has led naturally to many ideas that could be further pursued by the authors, as well as other researchers who might be interested in these ideas and who embrace a similar research paradigm. Furthermore, as mentioned in the introduction, the value-added mobile services are continuously changing and evolving. Therefore, mobile phone service providers are likely to face many other issues and challenges. The following points suggest future research topics that are emerging from this study:

1. Various environmental factors may be included in future study to represent more accurate and realistic situations.
2. Value-added mobile services embrace a much wider scope than the current study. Many other services such as mobile Internet access, mobile shopping, small amount payment, and real-time and on-demand multimedia services should be considered and integrated for further analysis.

REFERENCES

Cohen, B. J. (1967, August). An interpersonal orientation to the study of consumer behavior. *Journal of Marketing Research, 4,* 270-278.

Frolick, M. N., & Chen, L. (2004). Assessing m-commerce opportunities. *Journal of Information Systems Management, 21*(2), 53-61.

Lin, et al. (2002). *Internet marketing.* Taipei, Taiwan: Flag Publishing.

Market Intelligence Center. (2005). Retrieved March 3, 2005, from http://mic.iii.org.tw/intelligence/

Patrick, E., & William, A. (1979). A modernized family lifecycle. *Journal of Consumer Research, 6*(1), 16.

Pearson Education. (2000). *Online glossary.* Retrieved April 29, 2005, from http://www.prenhall.com/rm_student/html/glossary/c_gloss.html

SRI Consulting Business Intelligence. (2005). *The VALS survey.* Retrieved March 30, 2005, from http://www.sric-bi.com/VALS/presurvey.shtml

Stafford, T. F., & Gillenson, M. L. (2003). Mobile commerce: What it is and what is could be. *Communications of the ACM, 46*(12), 33-34.

Yu, F. (2002). *Mobile telecommunications.* Taipei, Taiwan: Kings Information Co.

ENDNOTE

[1]   http://eng.dgbas.gov.tw/mp.asp?mp=2

APPENDIX

1. Have you ever used any value-added mobile services (e.g., SMS, downloading music, ring tones, or images)?
   Yes (please continue)   No (please stop, thank you)

2. Have you answered this survey previously?
   Yes (please stop, thank you)   No (please continue)

Part 1: About your personality
Please choose the answers that most suit you.

No.	Question	1	2	3	4
1	Being free of emotional ties with others				
2	Giving comfort to those in need of friends				
3	The knowledge that most people would be fond of me at all times				
4	To refuse to give in to others in an argument				
5	Enjoying a good movie by myself				
6	For me to pay little attention to what others think of me				
7	For me to be able to own an item before most of my friends are able to buy it				
8	Knowing that others are somewhat envious of me				
9	To feel that I like everyone I know				
10	To be able to work hard while others are elsewhere having fun				
11	Using pull to get ahead				
12	For me to have enough money or power to impress self-styled "big shots"				
13	Basing my life on duty to others				
14	To work under tension				
15	If I could live all alone in a cabin in the woods or mountains				
16	Punishing those who insult my honor				
17	To give aid to the poor and underprivileged				
18	Standing in the way of people who are too sure of themselves				
19	Being free of social obligations				

		1	2	3	4
20	To have something good to say about everybody				
21	Telling a waiter when you have received inferior food				
22	Planning to get along without others				
23	To be able to spot and exploit weakness in others				
24	A strong desire to surpass others' achievements				
25	Sharing my personal feelings with others				
26	To have the ability to blame others for their mistakes				
27	For me to avoid situations where others can influence me				
28	Wanting to repay others' thoughtless actions with friendship				
29	Having to compete with others for various rewards				
30	I knew that others paid very little attention to my affairs				
31	To defend my rights by force				
32	Putting myself out to be considerate of others' feelings				
33	Correcting people who express an ignorant belief				
34	For me to work alone				
35	To be fair to people who do things which I consider wrong				

Part 2: Your lifestyle

No.	Question	1	2	3	4
1	I am often interested in theories.				
2	I like outrageous people and things.				
3	I like a lot of variety in my life.				
4	I love to make things I can use everyday.				
5	I follow the latest trends and fashions.				
6	I like being in charge of a group.				
7	I like to learn about art, culture, and history.				
8	I often crave excitement.				
9	I am really interested only in a few things.				
10	I would rather make something than buy it.				
11	I dress more fashionably than most people.				
12	I have more ability than most people.				
13	I consider myself an intellectual.				
14	I must admit that I like to show off.				
15	I like trying new things.				
16	I am very interested in how mechanical things, such as engines, work.				
17	I like to dress in the latest fashions.				
18	There is too much sex on television today.				
19	I like to lead others.				
20	I would like to spend a year or more in a foreign country.				
21	I like a lot of excitement in my life.				
22	I must admit that my interests are somewhat narrow and limited.				
23	I like making things of wood, metal, or other such material.				
24	I want to be considered fashionable.				
25	A woman's life is fulfilled only if she can provide a happy home for her family.				
26	I like the challenge of doing something I have never done before.				
27	I like to learn about things even if they may never be of any use to me.				
28	I like to make things with my hands.				
29	I am always looking for a thrill.				
30	I like doing things that are new and different.				
31	I like to look through hardware or automotive stores.				
32	I would like to understand more about how the universe works.				
33	I like my life to be pretty much the same from week to week.				

Part 3: Mobile phone useage

1. How long have you been using mobile phone services?

   a. < 6 months

   b. 6-12 months

   c. 1-2 years

   d. 2-3 years

   e. 3-4 years

   f. 4-5 years

   g. > 5 years

2. What type of mobile phone account do you subscribe to?

   a. Monthly subscription

   b. Pre-paid

3. What is your main concern when you consider adopting images, ring tones, and call waiting tones download services?

   a. The contents are new and fashion

   b. The contents represent my personal style

   c. Costs of the services

   d. Easy to set up

4. Where did you [hear about] the value-added mobile services?

   a. TV advertisements

   b. Magazine and newspaper

   c. Friends

   d. SMS advertisements

   e. Internet

   f. Other, please specify _____

5. On average, how often do you change your mobile phone display image (or wallpaper)?

   a. More than twice a week

   b. Once a week

   c. Once every fortnight

   d. Once a month

   e. Twice a year

   f. Once a year

   g. Never change

6. On average, how often do you change your mobile phone ring tones (or call waiting ring tones)?

   a. More than twice a week

   b. Once a week

   c. Once every fortnight

   d. Once a month

   e. Twice a year

   f. Once a year

   g. Never change

7. How much do you pay for value-added mobile services every month?

   a. < NT$100

      b. NT$100-NT$200

      c. NT$200-NT$300

      d. NT$300-NT$400

      e. NT$400-NT$500

      f. NT$500-NT$1,000

      g. > NT$1,000

8. Who is your mobile phone services provider?

      a. China Telecom

      b. FETNet

      c. Taiwan Mobile

      d. KGT

      e. MOBITAI

      f. TransAsia

      g. PHS

      h. APBW

      i. Other, please specify _____

Part 4: Personal details

1. Gender

      a. Male

      b. Female

2. Age

      a. < 12

      b. 13-18

      c. 19-22

      d. 23-25

      e. 26-30

      f. 31-35

      g. 36-40

      h. 41-50

      i. 51-60

      j. > 60

3. Education Level

      a. Primary School

      b. Junior High School

      c. Senior High School

      d. College and University

      e. Master and above

4. Occupation

      a. Student

      b. IT Industry

      c. Manufacturing (exclude IT Industry)

      d. Teacher/Professor

e. Public Sector

f. Business and Trade

g. Finance and Real Estate

h. Public Broadcasting/Advertisement

i. Building

j. Telecommunication

k. House Administration

l. Retired

m. Other, please specify _____

5. Monthly Allowance (available for any purposes)

a. < NT$1,000

b. NT$1,001-NT$5,000

c. NT$5,001-NT$10,000

d. NT$10,001-NT$15,000

e. NT$15,001-NT$20,000

f. NT$20,001-NT$50,000

g. > NT$50,001

6. Where do you live (which part of Taiwan)?

a. North

b. Centre

c. South

d. East

e. Islands

Chapter XLV
Mobile CRM:
Reaching, Acquiring, and Retaining Mobility Consumers

Chean Lee
Methodscience.com, Australia

ABSTRACT

This chapter provides an introduction of using Mobile CRM to reach, acquire, convert and retain consumers. Firstly, a definition of the term CRM is provided and the author also gives an insight on extending CRM to the wireless world. Having presented the benefits of mobile data services and their benefits to businesses in terms of customer relations and marketing, however, businesses still faced the challenges on delivering the promise to consumers. More importantly, the adoption of mobile services is still low in business and consumer segments. The author identifies content appropriateness, usability issues, personalization, willingness to pay, security and privacy as major challenges for businesses, and then, recommends businesses to start segmenting their mobile consumers into: Mobile Tweens, Mobile Yuppro and Senior Mobile users and the understanding of demographics, social and behavioural issues of these three consumer groups as initial step in Mobile CRM, before finally recommending the use of viral marketing as a mechanism to market mobile services. This is followed by matching relevant services to consumers create positive usability experience and always build a critical mass but develop a customer at one time.

INTRODUCTION

In recent years, we have seen an explosion in mobile entertainment, mobile B2B applications, mobile devices, and the wireless Web access packages from the telcos. This chapter discusses the utilization of CRM (customer relationship management) in reaching, acquiring, converting, and retaining mobile consumers. This chapter, therefore, includes discussions on mobile CRM strategy, market segmentation, and applying mobile CRM in the customer engagement process.

The term customer relationship management encompasses many descriptions which depend on different purposes. In the early days of CRM, business application providers often related CRM to sets of technology modules which included: marketing automation (i.e., campaign management, Web analytics, market forecasting); sales force automation (i.e., opportunity management, quotation generation, sales analytics, etc.); call centre application; order management system; and partner relationship management (PRM) that automates the entire customer management cycle.

The CRM is a business strategy that turns customer data into insights and provides a profitable process in handling client relationships. CRM technology packages act as tools to enhance the entire business strategy and processes.

The core objective of this chapter is to extend the CRM approach to mobility and the Wireless Web. In short, mobile CRM could consist of the following core components:

- Short Message Service-based advertising;
- mobile opt-in for customer data acquisition;
- mobile coupon or redeemable m-voucher;
- personalized mobile portal that offers content ranging from ring tones, news, horoscope, m-payment, and others;
- mobile alert function; and
- Web-based mobile campaign management engine.

THE POSSIBILITY OF MOBILE CRM

With the abundant availability of mobile commerce and technology, the questions arise on the benefits of mobile application in shaping the new customer relationship experience. According to research conducted by the management consulting firm McKinsey & Company in Europe, SMS is an effective mechanism to boost ratings and advertising sales to TV broadcasting. In the study, McKinsey pointed out that by linking TV broadcasting with SMS platform to certain shows, it enables cable TV broadcasters to boost audience loyalty. In some cases, the addition of SMS boosted the viewership of popular free-to-air television shows by up to 20% (Bughin, 2004).

In addition, SMS has also become a chatting medium for audiences ranging from ages 16-30. With the evolution of mobile technology, multimedia messaging services (MMSs), Java Games, mobile portals, and so forth, many opportunities have been created for carriers, marketers, telecoms, and businesses in building customer loyalty. Last but not least, SMS marketing from carrier and portal provider has become an acceptable advertising format compared to magazines, direct mail, or telesales, according to the mobile marketing solutions provider Enpocket (2005).

Despite the bright future of mobile technology, the adoption of mobile CRM as a form of relationship building tool is still facing obstacles. We will present some challenges faced by carriers, content providers, middleware companies and mobile marketing agencies, or even businesses in implementing mobile CRM.

CHALLENGES OF GENERATING DEMAND AND BUILDING RELATIONSHIPS WITH MOBILITY CUSTOMERS

It is reported that over $180 billion has been spent on 3G licenses for developing the next generation of mobile data services, which is believed to have attracted 10 million regular users in the UK alone, according to research conducted by the management and IT consulting firm, Accenture (2001a).

These services offered by the mobile service provider seem hard to impress and capture the attention of mobility consumers. So what could be classified as the potential obstacles and challenges in offering a range of wireless Web and data solutions to customers?

There are several factors that caused mobility customers to dropout from a range of wireless services:

- content appropriateness and freshness;
- usability issues;
- personalization;
- willingness to pay; and
- privacy, spam, and security.

Content Appropriateness and Freshness

Appropriateness refers to the relevancy of mobile content and data services to the consumers. Freshness means the frequency of mobile content updates. For example, if a person is attracted to horoscopes, he/she must receive alerts of career, fortune, and luck of the Scorpio to his/her mobile device every morning before going to work. The quality and the daily update of the content are very important for these individuals to continue their subscription to horoscope mobile content.

Usability

Mobile application or services are often limited by screen size, and mobile users are often not very technical people to operate the PC and Internet browser in a perfect manner. As a result, simplicity in design limits the layer of mobile phone keypads operations, the completion of a WAP page navigation cycle, or even

the reduction of the number of scrolls to complete an air ticket reservation using SMS.

Personalization

Can I create my own mobile "universe" in a mobile portal? This refers to personalization whereby a mobility consumer creates his/her favourite contents like news alert, stock market updates, ESPN sports on the mobile, and others.

Willingness to Pay

Questions often arise whether consumers are willing to continuously subscribe to mobile content such as games, songs downloads, and other related leisure and entertainment. This is often a challenge to the conversion and retains the cycle of a mobile content provider.

Privacy, Spam, and Security

Do mobile marketers, ad agencies, mobile network operators, and content providers respect the privacy in obtaining opt-in data from consumers for a range of mobile services? Nowadays, consumers are bombarded by junk mail and SMS promotions; these will lead to dropouts of consumers for mobile marketing messages. In addition, security is an important issue to be solved in offering m-payment and m-banking.

In the next section, we provide a framework for mobile service providers in building the relationship in a wireless world. Practitioners are encouraged to customize the framework that suits their business objectives, marketing, and CRM strategy. Because the write-ups in this chapter serve as a blueprint in building a mobility relationship, they need to be adjusted and amended when dealing with different scenarios.

Table 1. Mobility consumers segments

User Groups	Demographic	Character	Lifestyle	Mobility Motif
Mobile Teens	Aged between 9-20	Tech-savvy innovators who are willing to try on new mobile gadgets and services.	Enjoy variety in entertainments, live and breathe on own communities, love social and network games.	Entertainment and online community participation.
Mobile Yuppro	Aged between 26-40	Rational users used the Internet and mobile frequently in improving job productivity as well as entertainment.	Work hard and play hard. Seeking balances between leisure and career.	As a productivity tool for office administration, work scheduling, save time, and also entertainment.
Senior Mobitizen	Aged 50+	Realistic, used mobile and Internet for researching finance package, welfare, and e-mail. Cautious on trying new technologies.	Have time to spare, enjoy quality lifestyle after retirement.	Mobile is a tool for me to receive target ads on finance, best bargains, etc. for improving my life and family lifestyle.

MOBILITY CUSTOMER SEGMENTATION AND POSITIONING

Before implementing mobile CRM to acquire and build customer relationships in the wireless world, it is important for us to look at the customer segments and mobility motives. Based on findings from various researches (typically, Nokia and Sony Ericsson), we can divide mobility consumers into three major groups—*Mobile Teens* (teenagers and youngsters who used mobile phones frequently), *Mobile Yuppro* (young, urban professionals that used Mobile Internet), and *Senior Mobitizen* (senior consumer groups). Table 1 provides a summary of the characteristics, usage, motives, and sets of mobile CRM strategy and applications to be offered. The summary is based on the Scenario Grid by Lindgren, Jedbratt, and Svensson (2002).

Figure 1 segments the mobility consumer group we identified in Table 1 into five major adoption phases:

- **Innovators:** This segment refers to consumers described as techies. They are pioneers in technology adoption throughout the marketing cycle. They are influencers that help mobile application, content, and service providers in developing new products.
- **Early Adopters:** This segment refers to consumers willing to try out new options, as well as visionaries that believe mobile services will boost their work productivity as well as entertainment lifestyle.
- **Early Majority:** This segment refers to consumers who possessed a "wait and see" buying behaviour. They are sceptical of mobile services and applications which really create benefits to them. They often conduct research and ask for referral before the adoption of new services.

Figure 1. Phases of mobile services adoptions (based on Accenture, 2001a)

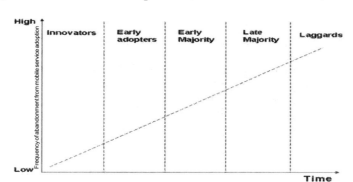

- **Late Majority:** This segment refers to consumers that are conservative towards new mobile technologies.
- **Laggards:** This segment refers to consumers that are hard to change and resist changing. It takes time to convert them into new mobile technology adopters.

From this analysis, we can conclude that mobile services adoption is seen as an interactive relationship between suppliers and consumers. Thus, there is urgency for service providers to provide the relevant product to the users. It is important to understand the consumers' need in the wireless world, and we are now entering a "pull marketing" era. The following are CRM strategies that mobile marketers and service providers adhere to in order to reach, acquire, and retain targeted audiences:

1. **Understand Customers and Make Use of Viral Marketing:** Get to know your customers often affected by the overall CRM campaign used to communicate and interact with them. First, understand the overall needs of consumers in relation to their demographics, age, lifestyle, and adoption behaviour as mentioned in Table 1; Figure 1 helps organizations implement a successful mobile CRM in acquiring customers. Secondly, do not ignore the

power of viral marketing, which is the creation of positive word of mouth or chain reaction in mobile eco systems. This means that we need to build a critical mass of product innovators, early adopters, or even early majority user groups, as we defined in Figure 1. This is due to the fact that innovators, early adopters, and the majority often are opinion leaders; they often act as a reference for the late majority or even conservative customers. Figure 2 illustrates the power of viral marketing plays as a mobile CRM tactic.

2. **Product Matching:** Offering relevant mobility products and services that matched the needs of the consumers.

Figure 2. Viral marketing in mobile ecosystem (based on Accenture, 2001a)

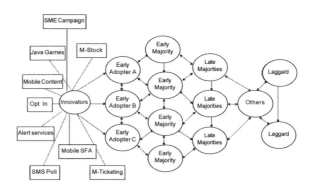

Table 2. Example of mobile services offering based on behavioural mode

Behavioral Mode	Characteristics	Example of Services	Mobile CRM Methods for Reaching Consumers
Productivity	Increase work productivity and save time.	Mobile personal organizer, e-mail, mobile sales force automation, etc.	Mobile Portal, mobile opt-in application for gathering client information regarding their interest; permission-based marketing, campaign.
Social and Entertainment	Entertainment and socialize with friends.	Mobile Java games, ring tones.	Mobile game engines for tracking the number of game downloads, winning points tracking, viral campaign.
Mobility	Get information at any time and anywhere.	Locations and positioning services.	Alert services subscription. Example, notification on bank account balance via SMS

Before mobile service providers offer more personalized products to consumers, they can always start with basic features like bulk SMS, mobile e-mails, personal organizers, and others. In the second step, we can further divide our segments we identified in Table 1 in terms of "behaviourgraphic" of mobility consumers. Table 2 illustrates the possible services as well as mobile CRM methods in reaching consumers with different behaviourgraphics.

We have seen the possibility of CRM tactics in reaching, acquiring, and retaining customers based on their behaviour. Remember, it is recommended that network operators, content developers, technology companies, or even marketing agencies should provide an open platform that enables partners to offer their services to consumers—for example, the Nokia developer networks, Telenor Mobile open platform and NTT Do Como open platform.

3. **Create User Experience:** When I was an e-business analyst at a Web development company, I always spoke to my clients about usability design and positive user experience on Web sites and Web-based applications. With the emergence of Mobile Internet and application as a mechanism in improving work productivity, entertainment options, and m-commerce, it is important for marketers to design their mobile services in usability standard. Creating positive user experience is an important CRM strategy in mobility space. Think about the available characteristics of connection protocols like WAP, GPRS, EDGE, and UMTS: screen size issues; differences on mobile gadgets like smart phones, Pocket PCs, Palm Pilots, and mobile notebooks. As a result, design with standards is crucial to create a positive experience. User experience is about accessible, content freshness, relevance, and personalization.

4. **Develop One Customer at a Time:** In the mobile universe, reaching consumers with SMS campaigns and mobile coupons, followed by relevant content to acquire consumers, is not the end of the journey. The key to mobile success is retaining the consumers. The following are the CRM strategies that help to develop one-to-one marketing for mobility consumers:

   - Understand the consumers by segmenting the users based on demographics, behaviours, lifestyle, and motivations.
   - Practice permission-based marketing to gather consumer information

on preferences about mobile services. Respect privacy and do not create mobile spam.

- Offer customer education and relationship building with relevant content and personalization.
- Retain the consumer by influencing and changing their spending behaviour. Build a mobility Internet community to create a mobile value chain.

Now that we have a clear blueprint and framework about pull marketing for our mobile CRM strategy to reach, acquire, and retain consumers, we will provide an in-depth walkthrough of the consumer engagement process.

THE CUSTOMER ENGAGEMENT PROCESS

The focus on the customer engagement process in the mobile world and understanding of the process will enable us to apply the pull marketing concept to each phase of the process. We can also further analyse the dropout risks in the process and provide examples of applying mobile CRM in each phase with the provision of user scenario or persona. Measurement metrics on the effectiveness of mobile CRM will be provided.

So how do we define the customer engagement process? It refers to the series of stages where a customer converts to a purchase of a product and services offered by an organization. It consists of four phases: Reach, Acquire, Convert, and Retain. Figure 3 shows these phases, modified based on the online consumer engagement process by Sterne (2002).

Reach

Reach in a mobile world refers to an informative stage where wireless marketers, mobile network operators, content providers, software companies, and mobile manufacturers utilized communication mediums to provide information for consumers—for example, an SMS marketing campaign to alert customers to participate in mobile game contests to win free movie tickets. In a mobile world, reach also must be built based on customer permission.

Acquire

This stage involves the effectiveness of the outcomes of the first stage, where mobility consumers are alerted by the SMS marketing campaign. Further interest is built between marketers and consumers.

Convert

This refers to the persuasion stage of the SMS campaign or even MMS content. It involves the participation of users in response to the SMS campaign. For example, the download of mobile Java games, the use of mobile a phone to book an airline ticket, or even the purchase of movie tickets using SMS.

Retain

When the mobile consumers have the experience and familiarity with the applications, they will become frequent users of the application. For example, a consumer will continue to use the SMS as a tool for movie ticket purchasing. In this stage, a vital factor is where the satisfied customer will spread positive messages to their friends regarding the mobile application or cam-

Figure 3. Customer engagement process in mobile world

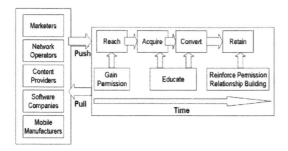

Figure 4. Mobility customers conversion funnel

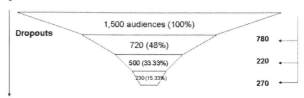

paign. It is also important for marketers to reinforce the permission and retain customers by offering fresh and relevant mobile content and services.

We conclude from Figure 3 that in wireless world, reaching and acquiring customers utilizing CRM methods like SMS or MMS ad messages and content are not the destiny; relationship building is the key to success in the customer engagement cycle. Let us apply the famous 80/20 rule into the wireless world. We can say 80% of the mobile content provider sales revenue is generated from 20% of the customers. Thus, we can see how important ongoing customer retention is in the mobile world. Again, customer acquisition is about reaching mass audiences, but developing selected target consumer groups over time. When I was involved in Pre Sales in an ERP solutions company, it took much cold calling from my telesales colleague to generate leads, and then I had to determine that the leads were strong during my client engagement. From there I prepared proposals to turn them into my prospects, and I offered demos to turn them into hot prospects. I closed a few of them from the hot prospect lists. So we can see the cycle is like a funnel: it involves some key metrics and CRM strategies we apply in each stage, is like playing with numbers, and builds up targeted long-term customers (see Figure 4).

Figure 4 illustrates an example of a mobile marketing campaign. During the process, the marketer successfully sends 1,500 SMS marketing messages to 1,500 audiences, and during the entire consumer engagement cycle, we expect leakage and dropout of consumers during each stage, which could be due to irrelevant mobile content, accessible issues, negative user experience, and so on. In the final stage, we successfully converted 230 out of 1,500 users into our customers. Thus, we can see that conversion in a wireless world is also a numbers game; however, it is important for marketers to practice the concept of relationship building in the CRM context.

DROPOUT RISKS OF CUSTOMER ENGAGEMENT CYCLE

We have already discussed the issues that cause the dropout of consumers from wireless

Figure 5. Dropouts in customer engagement process

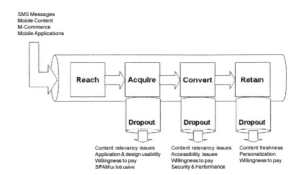

services. Figure 5 shows the various leakage or dropout risks that happen in the engagement process. This figure is modified based on the customer retention process addressed by Sterne (2002).

The challenges and factors for the adoption of mobile services are already identified. Figure 5 shows how to apply these factors to each phase of the customer engagement process. This figure summarizes that overall dropout in several phases are due to content relevance and freshness, application design usability, willingness to pay for mobile services, and security and privacy issues.

In the next section, we will provide a real-life user scenario, best industry practices, and measurement metrics for each phase in the customer engagement process.

APPLYING THE FRAMEWORK: DEFINED MARKETING OBJECTIVES AND CRM CAMPAIGN DESIGN

It is important to have a clear objective when we carry out a specific task or project. It could be the objective of building a bridge, flying to the moon, developing new products, or even implementing a project. In a mobile world, it is important to have sets of marketing objectives available when we decide to apply CRM. Table 3 summarizes the relationship between marketing objectives and mobile CRM implementation to acquire and strengthen relationships with consumers.

Table 3 shows the six key marketing objectives of businesses and examples of CRM applications that enable development of one customer at one time.

Table 3. Relationship between marketing objectives and mobile CRM application

Marketing Objectives	Description	Mobile CRM Application for Amy
Awareness and brand involvement.	Create brand and product reach and awareness by encouraging consumers to interact with the brand to strengthen the benefit and emotional feel of consumers to a product.	Mobile and interactive TV convergence, voting, and program contest via SMS.
Sales leads generation and drive physical store traffic.	Generate subscription for mobile content, opt-in and drive customers to a retail store.	Dynamic offering by sending SMS promotion to Amy's mobile when she is in a shopping mall.
M-commerce.	Similar to e-commerce, the main objective is creating sales.	M-commerce application such as M-Ticketing.
Self-service/productivity.	Improve work productivity, reduce service request, and streamline process.	Property industry-specific mobile sales force automation.
Content and communities	Deliver premium mobile content to subscribers. Personalization and content freshness are keys for success.	Astrology content delivered via SMS. It includes horoscope, games that relate to Amy's horoscope, and movie star news that shares the same horoscope with Amy.
Customer retention	Retain customers and build long-term relationship for cross-selling opportunities.	Bank interest rate, forex update, account balance, SMS alerts, etc.

Figure 6. Mobile CRM, and integrated approach

THE INTEGRATION APPROACH

We have shown readers mobile CRM examples. To be successful in reaching, acquiring, converting, and retaining mobile consumers, we recommend that businesses align strategy, marketing objectives, and the customer engagement process with the appropriate technical platform and application. Figure 6 summarizes an integrated CRM approach in winning loyal consumers in a wireless world.

The spiral model in Figure 6 shows an abstract of combining strategy, marketing objectives, process, implementation, and measurement for a mobile CRM project. It is strongly believed that there is no priority on whether the entire marketing strategy drives the creation of mobile CRM campaign or technical design since all the proposed elements are equally important. For example, you might have a great vision of using a mobile device to acquire customers; however in terms of technical aspect, it would be not realistic or vice versa.

Furthermore, the use of a multi-channel approach and integration of different technology are also crucial for reaching, acquiring, and retaining wireless consumers. A mobile CRM

application needs to be integrated with an application server, middleware, partners' interface, and an other Web-based CRM and call centre system. Figure 7 shows the mobile CRM architecture, which is based on a leading mobile marketing solution vendor, Enpocket's (2005) mobile engine.

Based on the proposed mobile CRM architecture, end users are connected and receive mobile content using standards such as SMS, MMS, J2ME, and so forth. The core components that sit in the application engine are:

- **Messaging Module:** This module handles functions of setting up permission for campaign delivery, as well as a campaign manager that enables SMS blasting services and an information manager.
- **Mobile Internet Module:** This module deals with mobile content production, for example, astrology, news, sports updates, and other mobile content. It consists of content creation, the targeting function to match relevant content for end users, as well as reporting analysis to measure the overall content delivery performance such as number of subscriptions, content viewing, impression, response rate, and so on.
- **Handset Module:** This module deals with B2C-based services. The m-coupon redeems and loyalty-based sweep takers where we received in promotional leaflets lies in this module. It also functions as a download engine for ring tones and mobile network games. The interaction between user registration and games participation activity data collection also sits in this module.

Finally, the mobile application engine also needs to be integrated with backend systems and other CRM systems to support all CRM activities, from marketing campaign management to sales closing and after-sales services.

Figure 7. Mobile CRM architecture

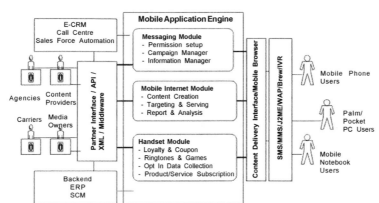

CONCLUSION AND FUTURE PREDICTION

In conclusion, businesses and organizations are still facing obstacles in acquiring customers and delivering consistency data services to wireless consumers considering many possibilities and promises on mobile technology. Successful mobile projects need to identify the challenges of a wireless world such as screen size, bandwidth, usability, security, and content relevance. Therefore, segmenting the consumers, choose a positioning strategy, and effectively using the viral marketing concept are keys for success. In addition, the understanding of multi-channel conversion and matching CRM campaign design with marketing objectives is equally important. In a final word, we can build a critical mass, but establish one customer at a time in the mobile venture.

What are the implications of a wireless world in five years' time? I can say they are full of surprises and hopes. The following are some of the predictions:

- The boundary between carriers, agencies, content providers, and mobile marketers will become blurring in the future. Major mergers and acquisitions between mobile service providers as well as smart strategic alliances cause this.

- New mobile solution players will emerge in the market. They will focus on delivering niche applications such as real estate management and map-based systems to vertical industry.

- A consumer-driven process means that consumers will require on-demand mobile data services and content that matches their needs and daily lives.

- The availability of new technology and delivery of promises will be important, for example, the complete 3G infrastructure will create a high bandwidth user experience where mobile devices will function as entertainment gadgets in receiving high-quality broadcast movies and live sports events.

- An emerging market is that of mobile teens and children.

Telstra, Australia's largest telecommunication company, has just introduced iMode services to Australian mobile users. iMode guarantees an optimistic future of mobile businesses, and will soon shape and improve productivity for our work lives and increase fun in business and entertainment.

REFERENCES

Accenture. (2001a). *Moving customers: Building customer relationships and generating demand in the mobile data world.* Accenture.

Accenture. (2001b). *In lieu of interactive skin: Getting a grip on mobile commerce in the entertainment industry.* Accenture.

Bughin, J. R. (2004). Using mobile phones to boost TV ratings. *The McKinsey Quarterly.*

Enpocket. (n.d.). Retrieved from www.enpocket.com

Lindgren, M., Jedbratt, J., & Svensson, E. (2002). *Beyond mobile: People, communications and marketing in a mobilized world.* New York: Palgrave.

Sterne, J. (2002). *Web metrics: Proven methods for measuring Web site success.* New York: John Wiley & Sons.

Chapter XLVI
Factors Influencing Segmentation and Demographics of Mobile-Customers

Anne-Marie Ranft
University of Technology, Australia

ABSTRACT

This chapter addresses important factors for consideration when readying a mobile commerce business for global business, addressing both regional differentiation in demographics that influence classifications of customer segments, and differentiation in demographics within a region. Globally, not all customer segments have regular access to mobile commerce facilities, and even for those that do, other demographic factors can impede their potential as mobile-customers. When starting from an Anglo-centric perspective, it is vital to have awareness of global differences in culture, language, payment options, time zones, legal restrictions, infrastructures, product needs, and market growth that could either improve or inhibit mobile-customer uptake, and in the worst case, result in unexpected litigation.

INTRODUCTION

Mobile-customers should be considered as one of the most significant elements for a mobile commerce enterprise. Mobile-customers of the enterprise are those customers that use mobile devices—the most common ones being mobile phones, personal digital assistants (PDAs), and notebook PCs. Mobile commerce products can include: physical devices, applications, and ac-cessories; access to the mobile infrastructure; and unrelated products and services marketed, bought, and sold using a mobile device as the communication tool.

Internet-based e-commerce interactions are generally categorised by the broad segments of Consumer (C), Business (B), and Government (G), and then decomposed into the relevant market segments. However, when undertaking global commerce, regional factors providing

Table 1. Potential access to mobile commerce and e-commerce—summary

	OECD 1990	1998/2000	Non-OECD 1990	1998/2000
Fixed and mobile telecommunications access paths per 100 inhabitants	41.1	72.1	2.7	7.8
Internet hosts per 1,000 inhabitants	23	82	0.21	0.85
Data source: OECD, 2001				
OECD Countries—There are 30 member countries, mainly in the European and North American regions, as well as the United Kingdom, Australia, and New Zealand.				

differentiation in demographics can alter classifications of customer segments, and differences in demographics can occur within a region. A market segment that exists in Australia, the United States, or the United Kingdom may not exist in some regions. It should also be noted that market segments based on Internet e-customer demographics may not necessarily be directly applicable for mobile-customers.

Before targeting a product or service to a particular market segment and location, these issues should be considered to maximise mobile-customer uptake and prevent unexpected litigation.

FACTORS INFLUENCING GLOBAL DIFFERENCES

First, the question of regional mobile-customer segment sizes will be discussed with reference to the *digital divide*, then other differentiating factors will be listed, followed by a list of possible strategies to consider when designing global mobile commerce products and marketing.

Digital Divide—Historical Factors

The first issue to be addressed is one of whether potential mobile-customers for a segment even exist in the targeted regions.

"Visions of a global knowledge-based economy and universal electronic commerce, characterised by the 'death of distance' must be tempered by the reality that half the world's population has never made a telephone call, much less accessed the Internet" is the caveat noted by the Organisation for Economic Cooperation and Development (OECD, 1999). The OECD uses the term "digital divide" to describe the inter- and intra-country inequalities in access to information and communication technologies by both individuals and businesses due to socio-economic and geographic differences (OECD, 2001). They provided statistics that highlight the differences between OECD and non-OECD countries (see Table 1).

They further noted that the higher growth rate in telecommunication access for non-OECD countries is especially due to rises in China, but there was insignificant African growth during that period.

Within a geographic region, different demographic factors also contribute to a reduction in potential mobile-customers.

Uptake of mobile commerce in some regions is still biased towards the business and professional consumer sectors, especially mobile phone ownership in the Asian region.

It should be noted that many *developing* nations suffer from lack of suitable telecommunication infrastructure; access to a reliable electrical source for re-charging of mobile devices and permanent housing can be limited for lower socio-economic groups. For instance, in my experience I have found that many Indian businesses have access to high-speed Internet lines and mobile connectivity, but require their own generators to back up the state power supply.

Overall, it can be concluded that there are two major groups of potential mobile-customer segments not currently available due to this *digital divide* factor—consumer and business segments whose geographical demographics are characterised by lack of telecommunication and other infrastructure, and consumer and business segments whose socio-economic demographics make mobile commerce unviable.

In many regions, especially Asia and Africa, consumer and small business sectors in lower socio-economic groups have the double barriers of no infrastructure and un-affordability, with the result that much of their population cannot today be counted as potential mobile-customers for C2C, B2C, and G2C segments.

Digital Divide—Transition Factors

The last few years have seen an enormous increase in the number of mobile phone connections in all global regions. This is shown in Table 2.

A common trend noted globally is the increase in the proportion of mobile subscribers to fixed telephone line customers. Some customer segments, especially youth segments in rental accommodation, may no longer see the necessity for a fixed line. Logistically, the resources required for installation of new mobile infrastructures in rural or undeveloped regions may be less than that required for new fixed-line infrastructures.

In Australia, the Australian Communications Authority's "Telecommunications Performance Report 2003-04" tabled that the number of mobile phone services had exceeded the number of fixed telephone services operating by June 2004. The number of mobile phone services grew by 15.4% over the period, with a growth in prepaid services, which by then made

Table 2. Mobile phone connections—summary

Region	1998 (1000s)	2003 (1000s)	CAGR (%) 1998-03[a]	Per 100 Inhabitants 2003	As % of Total Telephone Subscribers 2003
Africa	4,156.9	50,803.2	65.0	6.16	67.3
Americas	95,066.8	288,219.9	24.8	33.80	49.8
Asia[b]	108,320.6	543,153.4	38.1	15.03	52.4
Europe[c]	104,382.0	441,234.9	33.4	55.40	57.5
Oceania	5,748.5	17,256.3	24.6	54.45	57.2
World	317,674.8	1,340,667.7	33.4	21.91	53.9

Notes:
[a] The compound annual growth rate (CAGR) is computed by the formula:
$$[(P_v / P_0)^{(1/n)}]-1$$
where P_v = Present value
P_0 = Beginning value
n = Number of periods
The result is multiplied by 100 to obtain a percentage.

[b] By the end of 2003, Hong Kong and Taiwan had exceeded a rate of 100% phones per inhabitant.
[c] By the end of 2003, Italy and Luxembourg had exceeded a rate of 100% phones per inhabitant.

Data Source: International Telecommunication Union, 2004 up 43% of mobile services (Australian Communications Authority, 2004).

Customer segments in regions with limited fixed-line infrastructures may now, for the first time, have access to modern telecommunications. Of particular interest is the increase in the size of the potential mobile-customer segments in regions that until 2002 were limited in the infrastructure required to support mobile commerce, thus enabling the creation of an emerging market segment.

The UN's International Telecommunications Union industry report, "Trends in Telecommunications Reform 2004-2005," has been reported by the press to state that globally, 2004 revenue from mobile services is expected to be higher than revenue from fixed telephone line services. China, India, and Russia were stated to have the highest rate of increase (Australian IT, 2004).

In India, a press release from the Telecom Regulatory Authority of India stated that during 2004, approximately 19.5 million mobile subscribers were added, giving a total of 48 million mobile subscribers (an increase of 68%). The number of mobile subscribers now exceeds that of fixed-line subscribers, who only experienced a small increase in numbers over the same period (Telecom Regulatory Authority of India, 2005).

Influence on Demographic Factors for This New Segment

The demographics of this emerging segment, especially of those located in less developed regions, may differ from early adopters of mobile commerce and Internet users in these regions by factors including:

- more likely to use a pre-paid account and less likely to own a credit card or have access to other e-commerce payment methods;

- mobile devices more likely to be limited to mobile phones, rather than business-oriented devices such as PDAs;
- wider geographic location—that is, rural areas without fixed-line telephony and Internet may now have access to mobile telephony infrastructure;
- wider age spread—that is, may be used for communication between many generations of a family structure; parents may purchase a mobile phone for their children to enable a sense of security, and conversely, adult children may purchase a mobile phone for their elderly parents to satisfy the same objective;
- may have attained lower levels of education and literacy;
- less likely to speak or read English, or even to be fluent in their own national language;
- may be less familiar with current communication technologies; and
- small businesses, especially in the rural sector, may now have access to mobile telephony, thus facilitating the potential for the deployment of new business and agricultural techniques.

Location Differentiation

Some differences affect all potential mobile-customers in a specific location, be it geographical region or individual country/province.

Geography

- Time differences in mobile-customers' time zones, established business hours, and public and religious holidays could affect peak and off-peak system processing loads, with implications for the scheduling of system downtimes for maintenance or upgrades, and the staffing of call centres and other customer services.

- Seasonal and climate differences affect the marketability and usability of some products.
- Metropolitan vs. rural locality can impact the availability and quality of communications and product delivery infrastructure, unless the product can be delivered via the mobile device. Many Asian and African rural areas lack communication and other infrastructures, and even remote locations well serviced by satellite communications, such as the Australian outback or Antarctic bases, can have poor or expensive product delivery services.

Products and Services

- Suitability for use in global locations must be considered. Is there a need for the product or service? What use is a service to send payment details to a parking meter if few customers in the region own a car? Will the product actually work? This is especially an issue for electrical goods such as chargers for mobile devices or other items purchased via mobile commerce which may not be compatible with local equipment.
- Accuracy and knowledge of locality is important for some products, especially location-based services that interact with and require a global positioning system (GPS) infrastructure in the region.
- Social acceptance of products needs to be understood. Is the product attractive to the locations' typical mobile-customer needs, social values, and religious beliefs, or even legal?
- Equipment and availability for mobile commerce may differ for some customer segments. A business traveller expecting global availability of Wi-Fi "hotspots" for PDA or PC connection may be disappointed when travelling in less technically developed regions, and there are some regions that are not yet reachable by commercial GPS satellites.

Handset types required depend on whether the local networks offer Global System for Mobile communication (GSM), Code-Division Multiple Access (CDMA) of which there are many variations, Personal Digital Cellular (PDC), or Third-Generation/Universal Mobile Telecommunications System (3G/UMTS) services. The Japanese network types are fairly unique to Japan; few commercially available handsets can be used both in Japan and other countries. Despite the availability of Japans' NTT DoCoMo iMode service in many countries, including Australia, the applications available and handsets required do differ between the individual countries of implementation.

- The number of potential customers who are visiting a region affect the viability of services that are aimed at the visitor, for example local directories, tourism guides, or special communication roaming deals such as SingTel's "Local Direct Dial" in Singapore (SingTel, 2005).

Product Content and Interface Presentation

- Language and keyboard/screen character sets differ. This is especially important to remember if mobile-customers are sought in China or Japan. Emerging mobile-customer segments may require mobile devices and applications to be designed using the local language for the interfacing component.
- Marketing promotions should be sensitive to customers' varying social backgrounds and local legislation regarding content.

Financial

- Credit card ownership is not ubiquitous in some regions. While customers can use other forms of payments such as invoicing, COD, or local debit cards for national purchases, credit cards are the most widely acceptable payment method for international mobile commerce. In many Western European countries, Spain and Germany in particular, most consumers use debit rather than credit cards, limiting their global mobile-customer potential (Barclays, 2001; Forrester Research Technology, 2004). Some Asian countries such as South Korea, Japan, and Hong Kong have high credit card ownership (Lafferty Cards International, 2004a), while most others do not.

  On an optimistic note for global mobile commerce, data shows that credit card ownership is growing/is projected to grow strongly in the Asia Pacific (Visa, 2004a), especially Indian credit card use (Gupta & Dasgupta, 2004), as well as Central and Eastern Europe, and Middle East regions (Visa, 2004a). To overcome this limitation, billing options that integrate with the customers' mobile account should be considered, whether it is a pre-paid or post-paid account.

- Cash payments are preferred by consumers in some regions. Visa notes that over 90% of transactions in the Asia Pacific region are made in cash (Visa, 2004b). Some European countries such as Greece are still cash oriented (Lafferty Cards International, 2004b). Again, these customers could be catered for by billing options that integrate with the customers' mobile account, which may well be a pre-paid account.

- Currencies for transactions—can customers pay in their own currency, only in the major currencies, or only in the currency of the mobile commerce business?

- Taxes—VAT, GST, state, and other sales taxes may or may not be payable on transactions depending on where the mobile commerce site is located and the location of the customer.

Legal

- Forbidden products both create and limit mobile commerce opportunities in some regions. There may be a large potential market for prohibited goods, especially in countries such as Saudi Arabia where alcohol and a range of other goods are forbidden (Department of Foreign Affairs and Trade, 2004) for a mobile commerce enterprise willing to engage in a high-risk venture. Otherwise, such products should not be included when targeting consumers in those regions to avoid causing offence, litigation, or censorship.

  Mobile services and content that does not meet local legislative requirements could cause the loss of a mobile operator's license.

- Privacy regulations differ greatly across the world in regard to data collection and management, and unsolicited marketing. In Australia the Privacy Act applies to businesses with an annual turnover of more than $3 million and all businesses of certain types (Office of the Federal Privacy Commissioner, 2005). There are no significant data protection laws in the U.S. at this point. Member countries in the European Union have some of the strictest data protection laws in the world which attempt to control their citizens' data stored in non-member countries too (European Commission, 2005).

Customer Differentiation

Within a location, individual customer demographic and lifestyle differences may alter the identification and classification of customer segments from standards in the mobile commerce's home location.

Demographic

- Age group usage may differ especially in locations where older groups have limited literacy. Younger groups may embrace internationalism and be confident using a wide range of services, including those marketed in the English language, while older groups may be more conservative and prefer using brands and services that reflect their own culture. Younger groups may be more confident using their mobile telephone for more than just telephony and are enthusiastic users of Short Message Service (SMS).
- Education is especially important in developing locations, where generally only the better educated have an opportunity to earn sufficient income to acquire the neccessary infrastructure.
- Gender may affect customer segments in locations where females in lower socio-economic groups are less educated.
- Family lifecycle stage groups may differ in relative segment sizes. For instance, the relative size of the European "adult with no dependents" demographic is larger than that in many Asian countries.
- Metropolitan vs. rural locality differentiation is covered above. In some regions, education and financial infrastructures may also be limited in rural and remote areas.
- Language used may be different to the national language. Many regions comprise many ethnic language groups, espe-

cially India. English is more likely to be understood by the higher socio-economic groups.

Lifestyle

- Time consciousness—mobile devices are more likely to be used in the course of performing business functions when timing of communications is critical, or of a personal nature when the customer has limited time for family and social activities. Different cultures experience a difference in expectations of what is considered "on time" or not.
- Moral attitudes vary greatly, especially for sexuality. Various "adult" services of a sexual nature are marketed heavily to mobile customers in some regions, but could cause the loss of an operator's license if marketed or offered in a region with strict legislation controlling mobile content.
- Personal values differ between cultures, which should be taken into account when marketing and designing features. Is the target society one that values concepts of individuality or social and family group membership? Is there prestige associated with acquisition of new mobile and other technologies?
- Attitude to adoption of new technology may differ between different segments within a region. Japanese youth are well known for their enthusiastic embrace of mobile telephones, individualizing accessories, and mobile services offered in particular by their iMode system.

Firmographics

- Size does matter. Globally, smaller businesses are less likely to use the latest technologies (OECD, 2001). Small businesses in developing locations are even less likely due to infrastructure issues listed above.

- Industry sector is shown to affect Internet use (OECD, 2001). Predominately subsistence-level agricultural communities may not require mobile commerce.

STRATEGIES FOR THE DESIGN OF PRODUCTS AND MARKETING

Strategies for the products, services, and marketing delivered by the mobile commerce business require tailoring for the targeted mobile-customer segments.

First, the customer segments should be identified according to the global differentiations outlined. Next, a decision should be made whether to create individual products, services, and marketing for different segments, or create a common suite to be used for all.

Factors indicating individual suites include:

- Significant differences in deliverable products and services, and customer differentiation, especially in legal restrictions, currencies, language, and social values.
- Economic justification for developing multiple products, services, and marketing campaigns.

Factors indicating a common suite include:

- Uniformity in products, services, and customer demographics.
- Uneconomic to develop multiple products, services, and marketing campaigns.

Then, the targeted mobile-customer segments should be guided to the appropriate site by strategies such as:

- Marketing and linking the mobile commerce product or service from an established mobile commerce portal, perhaps run by the telecommunication operator in the region or from a relevant Internet site in the region.
- Marketing via traditional channels such as print advertising in the region.

And finally, the mobile-customers should be provided with good "quality of service" regardless of their time zones and other differences. Consistent service availability and customer support should be provided to the most profitable customer segments at least, and ideally, to all.

CONCLUSION

While the recent arrival of mobile telecommunications infrastructures in most regions of the world has created a vast number of potential mobile-customers, mobile commerce businesses should be aware of the many geographical, legal, and demographic differences summarised in the following diagram before attempting to trade internationally, or deliver products and services developed outside their region to the local market.

Shrinkage of the digital divide for business and medium-high socio-economic groups across international boundaries, especially in the Asian region and within developed countries, is enabling the potential for even more growth in the size and variety of mobile-customer segments.

The recent emergence of new potential mobile-customers outside the established socio-economic and urban-located demographic groups requires more careful tailoring of products, services, and billing options than for the more established segments.

Benefiting from this expected growth can only be achieved by ensuring the mobile

Figure 1. Summary of influencing factors

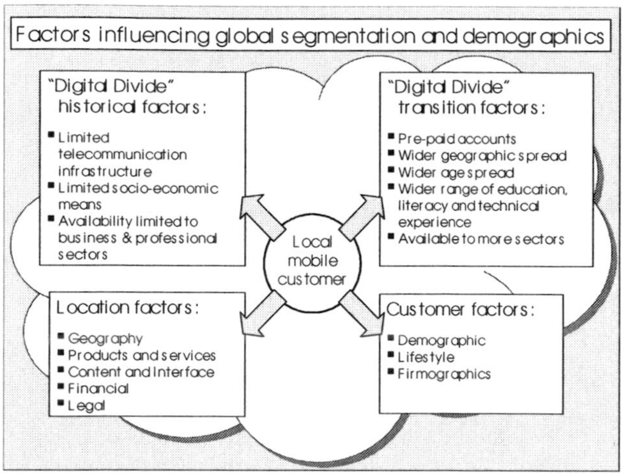

commerce's products and services, interface design, and marketing; customer service is tailored to satisfy the targeted market segment, being either the established or emerging mobile-customer segments.

REFERENCES

Australian Communications Authority. (2004, December). *Media release 95: Growth in mobiles and wireless broadband highlight year in telecommunications.* Retrieved January 22, 2005, from http://internet.aca.gov.au

Australian IT. (2004, December 14). *Mobile revenue to outstrip landlines.* Retrieved January 10, 2005, from http://www.australianit.news.com.au

Barclays. (2001). *International growth.* Retrieved September 10, 2004, from http://www.investor.barclays.co.uk

Department of Foreign Affairs and Trade. (2004). *Department of Foreign Affairs and Trade, Saudi Arabia country brief.* Retrieved September 10, 2004, from http://www.smartraveller.gov.au

European Commission. (2005). *Information society—Telecommunications, privacy protection.* Retrieved January 23, 2005, from http://europa.eu.int

Forrester Research Technology. (2004, August). *Forrester's consumer technographics.* Retrieved September 10, 2004, from http://www.forrester.com

Gupta, N. S., & Dasgupta, S. (2004). Dragon fire's no match for India's credit card club. *The Economic Times* (April 8). Retrieved September 10, 2004, from http://economictimes.indiatimes.com

International Telecommunication Union. (2004). *Mobile cellular, subscribers per 100 people 2003.* Retrieved January 21, 2005, from http://www.itu.int

Lafferty Cards International. (2004a, August). *Korean card use declines.* Retrieved September 10, 2004, from http://www.lafferty.com

Lafferty Cards International. (2004b, August). *Olympian leap forward for Greek cards.* Retrieved September 10, 2004, from http://www.lafferty.com

OECD (Organisation For Economic Cooperation and Development). (1999). *The economic and social impact of electronic commerce: Preliminary findings and research agenda.* Retrieved September 11, 2004, from http://www.oecd.org

OECD. (2001). *Understanding the digital divide.* Retrieved September 11, 2004, from http://www.oecd.org

Office of the Federal Privacy Commissioner. (2005). *Private sector—business.* Retrieved January 22, 2005, from http://www.privacy.gov.au

SingTel. (2005). *Visiting Singapore.* Retrieved January 23, 2005, from http://home.singtel.com

Telecom Regulatory Authority of India. (2005, January 9). *Press Release no. 6/2005.* Retrieved January 22, 2005, from http://www.trai.gov.in

Visa. (2004a). *Visa Asia Pacific.* Retrieved September 10, 2004, from http://corporate.visa.com

Visa. (2004b). *CEMEA.* Retrieved September 10, 2004, from http://corporate.visa.com

Section X
Social

Chapter XLVII

Mobile Camera Phones—Dealing with Privacy, Harassment, and Spying/Surveillance Concerns

Christopher Abood
Australian Computer Society, Australia

ABSTRACT

This chapter discusses the growing inappropriate use of mobile camera phones within our society. There are two areas of concern that are dealt within this chapter. The first concern deals with individual privacy and the use of mobile camera phones as a tool of harassment. The second concern deals with organizations seeking to prevent industrial espionage and employee protection. This chapter outlines how these devices are being used to invade individuals' privacy, to harass individuals, and to infiltrate organizations. The author outlines strategies and recommendations that both government and manufacturers of mobile camera phones can implement to better protect individual privacy, and policies that organizations can implement to help protect them from industrial espionage.

INTRODUCTION

During 2004, Samsung ran a television advertisement depicting a young man sitting in a café. He was taking photos of a young girl walking across the promenade with his mobile phone with in-built camera. The young girl was not aware that she was having her photo taken as she walked by. However, she turned to look over at the café and realized she was being photographed. She walked over to the man sitting in the café, took the mobile camera phone from him, and began to take pictures of herself rolling over a car.

During the 2004 Olympics, LG ran a television advertisement where a girl on a beach phones her friend in a shop to show her a live video feed of a muscular man applying suntan lotion next to his surfboard. It is obvious that the man is unaware that he is being videoed. The girl in the shop receiving the images passes the phone to a male companion to view. The catch phrase for the advertisement was: "LG, Official Sponsors of Eye Candy."

In both cases, these television advertisements depict voyeurism as a legitimate activity. In the Samsung advertisement, it goes further to suggest that people like having their photos taken with or without their knowledge or consent.

It is therefore ironic that both Samsung and LG have banned the use of mobile camera phones within their operations due to concerns these devises will be used for industrial espionage (BBC News, 2003). Also ironic is that Sydney, Australia, resident Peter Mackenzie was fined $500 for using his mobile phone to photograph women sunbathing topless at Coogee Beach (The Australian, 2004). Appearing in Waverley Local Court, Mackenzie pleaded guilty to behaving offensively in a public place. He told reporters later that he regretted his actions and realized they had been inappropriate. However, his behaviour was actively encouraged by mobile phone vendor advertising campaigns. Mackenzie's actions on Coogee Beach were entirely consistent with advertising campaigns for mobile camera phone technology, but the fact that he was arrested, charged, and subsequently fined makes it clear that these campaigns are out of step with reality and public standards.

Also during 2004, Virgin ran a television advertisement depicting a game called Ming Mong. The game essentially involves sending a picture to someone's mobile phone with a caption. One such example in the television advertisement was a picture of a toilet with the caption, "your breath." This advertising campaign is out of step with community concerns over the increasing use of mobile phones to bully and intimidate others, especially within the school environment.

The past few years have seen rapid convergence within various technologies, none more so than the mobile phone. The mobile phone now has PDA functionality and the ability to send and receive e-mails, view Web pages, listen to the radio and MP3 songs, and play games. Mobile phones are now coming onto the market with one or more gigabytes of storage and of course digital camera facilities with 3-plus mega-pixel resolution. We are now starting to see the adoption of videoing facilities (enabling real-time chat), and it will not be long before these devices start incorporating global positioning system (GPS) mapping technologies (which raises all sorts of surveillance/stalking issues). In short, the mobile phone is morphing into the everyday must-have mobile information and communications centre.

This convergence in technology, while providing many benefits, also raises issues dealing with privacy and surveillance/spying. Although we have had digital photography for a number of years, people generally tend not to carry their digital cameras with them all the time, whereas people tend to carry their mobile phones with them constantly. With a digital camera, you need to go home, connect it to your PC, and transfer the images from the camera to a storage medium. You then possibly e-mailed the images or uploaded them to a Web site. However, with a mobile camera phone, you can immediately send the image to an e-mail address, another phone, or to a computer server (for display on a Web site).

This can be great if you have one of those photo magic moments that you wish your friends to share. But it is not so great if the photo being forwarded is one that has been taken without the subject's knowledge. It is not readily obvious if someone is using their mobile camera phone to take photographs, as they may appear to be just chatting on the phone. Most mobile camera phones have the lens on the back, so when you are talking, the lens has a clear view to your side. Some mobile camera phones have a swivel lens, which makes it easier to conceal the fact that you are taking a photograph. You can now purchase a mobile camera phone relatively cheaply, walk into a competitor's premises, take photos of a sensitive industrial nature (while pretending to be talking to someone), immediately send on the image, dump the phone into a bin, and walk out clean.

Individual privacy and industrial espionage are becoming two major concerns dealing with mobile camera phones. Governments and organizations are grappling with how to deal with the growing misuse of these devises, and there does not seem to be a clear answer. How we protect an individual's and an organization's privacy while still allowing people to enjoy the benefits that mobile camera phones provide will be a difficult juggling act.

In this chapter, I discuss how to juggle protecting individuals' and organizations' rights, while at the same time enabling users to gain the benefits of mobile camera phones. Much of the discussion will be derived from the Australian Computer Society's policy on mobile camera phones (of which I led the development). This policy will be discussed at the end—but first, protecting individual privacy.

INDIVIDUAL PRIVACY

A number of Web sites have appeared that cater for images taken by mobile camera phones. There is www.mobog.com, which is a Web blog (an online journal) for images taken by mobile camera phones. People can create their own section to upload, store, and display photos taken with their mobile camera phone. This is particularly useful if you are travelling and wish friends back home to view where you have been. www.phonepiks.com is a central

place for people to upload and view pictures taken with their mobile camera phone. Unlike mobog, photos are uploaded to predefined categories. Both these sites contain images that range from the relatively mundane through to the pornographic. Another site, www.mobileasses.com, specializes in images taken of people's backsides by mobile camera phones. The photos are displayed with information about the type of mobile phone used, where it was taken, and when. The site even has a "backside of the day" competition, where viewers vote on the picture of a backside they like best. The winner wins a t-shirt. In most instances of images involving people displayed on these sites, it is obvious that the picture has been taken without the subject's knowledge or consent. They are probably unaware that their picture resides on these sites, let alone what to do to have the image removed. Mobileasses.com even has a section on tips on how to take pictures of people's backsides without their knowledge.

It is inappropriate to take a photo of someone without his or her knowledge or consent. It is even more so to then upload that image where all control over the image is lost. Even if the victim were successful in having his or her image removed, it would be of little consequence, as the image can be re-downloaded and forwarded on many times. Such is the nature of the Internet. Once the image has left the geographical boundaries of a nation, an individual has little chance of having the image removed, especially as they will now have to deal with a number of different legal systems in various countries. The difficulty with dealing with different legal systems is highlighted by the recent legislation passed by the Australian government. It would be illegal for a site like mobileasses.com to be hosted in Australia. A complaint can be made and a take down notice issued by the Australian Broadcasting Agency.

This means the Web site must be removed from the server and access denied to the Web site (in Australia). Many such sites in Australia have been issued with such take down notices. All of these sites simply relocated to servers located in other countries where the Australian law has no jurisdiction and have continued business uninterrupted. These sites still use the same Web address as before with the .au domain.

Like e-mail in its early days, it is difficult to predict what other ways people will use mobile camera phones in the future. Examples include "up-skirting" (prevalent on mobileasses.com) and digital shoplifting. Up-skirting involves using your mobile camera phone to take photos up a lady's skirt. On mobileasses.com, there are plenty of examples of a photograph being taken under a table. Holding a mobile camera phone in your hand looks inconspicuous and not readily obvious that your photo is being taken. Digital shoplifting involves someone taking a photo of an article in a magazine in a newsagent without having to buy the magazine. Some newsagents have banned the use of mobile camera phones from within their premises, but again, unless you are going to search each person as they enter and make them leave their phone at the counter (as some make you leave your bag at the counter), it is going to be hard to eliminate this practice.

Probably the most concerning of inappropriate use of mobile camera phones is using images taken of people in awkward situations for bullying and intimidation. Indeed, the biggest problem of using mobile phones for bullying is via the use of Short Message System (SMS). However, I believe that Multimedia Messaging Service (MMS) will replace SMS as the preferred means of bullying and harassment (as encouraged in the Virgin Ming Mong television advertisement). MMS will allow you to send images, text, video, and audio. So without your knowledge, your conversation can be recorded, your movements videoed, and images taken of you. These can then be used against you, especially if what has been captured is of an embarrassing nature. The video, image, and audio that is captured can also be uploaded immediately to a Web site for the entire world to view. This is what happened to a 17-year-old Indian boy who used his mobile camera phone to record his girlfriend giving him oral sex (Sydney Morning Herald, 2004). The video clip somehow made its way onto the Internet. This is something that they probably did not want to happen.

So how can we protect our privacy and control over our image without infringing on the benefits that these devices provide, such as documenting an accident? Simply banning mobile camera phones from places such as beaches, swimming pools, gyms, and other public places will not work. Not all mobile phones have in-built cameras, and people have a genuine need to have their mobile phones with them. To effectively ban these devices, you would need to search everyone, which would be impractical. As stated earlier, it is not always obvious that someone talking on his or her mobile phone is actually taking a photograph. You may also open yourself up for legal action if you confront someone who is actually just talking on the phone.

Governments are also grappling with the problem of misuse by passing various laws; however, it is difficult to enforce these laws. In New South Wales, Australia, it is against the law to drive and talk on your mobile phone at the same time, but I see people driving and chatting everyday. The United States is looking at passing a law to make it an offence to photograph people in situations where they would expect to have a reasonable amount of privacy. So you can take their photo on a public beach (which Peter Mackenzie found out you cannot do in Australia), but not in their backyard. However,

I believe this law would be open to interpretation; for instance, if I take a photo of you in your backyard at a party, is it still reasonable to expect a high level of privacy?

Again, laws will not help if you see a picture of yourself on a Web site and you have no idea who took it or how you could track down the person who uploaded it to the Web site. The local law would be even less powerful if that Web site resides in a different country with its own set of laws.

I believe the best way to reduce the amount of misuse of mobile camera phones is though education. For starters, mobile camera phone manufacturers should stop advertising their products in a way that encourages owners to use mobile camera phones in an inappropriate (and sometimes illegal) way. A mobile camera phone etiquette guide should be developed by mobile camera phone manufacturers (with appropriate input from the community) to be distributed with all new mobile camera phones. A campaign should be undertaken to educate users as to their responsibilities and the appropriate use of mobile camera phones. Both government and manufacturer should undertake this campaign.

The community must also be fully informed about their rights and what they should do if they suspect that they have had their photo taken without their consent. When taking a photo, you should make it clear that you are doing so and seek permission of the person being photographed, especially if they are unknown to you. If you do take a photo, you should take all reasonable steps to ensure that the photo is not uploaded to a Web site or e-mailed to others without the photographed person's permission.

For public venues such as gyms and swimming pools, visible signs should be displayed that indicate use of mobile camera phones, indeed any camera, is prohibited. A sign designed similar to the non-smoking sign would be ideal. People who visit these public places should be made fully aware of their responsibilities and any penalties that may apply if they are caught photographing within the premises. Penalties could include banishment or suspension from the premise.

It is hoped to that when governments are passing laws into inappropriate use of mobile camera phones, and indeed any organization passing rules, that they do not infringe upon a person's right to use these devices legitimately. Mobile camera phones can enhance your safety and prevent crime. In Yokohama, Japan, an 18-year-old female store clerk used her camera phone to take a photo of a 38-year-old man who was fondling her on a commuter train. She called police during the train ride and presented her phone shots as evidence. The man was arrested at the next stop (CBS News, 2003).

It should be noted that if you suspect that someone has taken your photo without your consent, you should contact the appropriate authority. Try and avoid confronting the person unless you are sure that it safe to do so; your physical safety is far more important.

As stated earlier, it is difficult dealing with Web sites that are hosted in a different country. Such is the nature that the Internet knows no borders. Perhaps it is time to look at an international approach to dealing with complaints against Web sites that publish images of you without your permission.

ORGANIZATIONAL PRIVACY

Business has also started to show concern regarding the use of these devices on their premises. Industrial espionage and spying is of real concern, namely because of the ease of use of mobile camera phones and the fact that it is not readily obvious that a person is capturing images while using them. It is difficult to tell whether a mobile phone is equipped with a built-in camera. A number of organizations will not issue staff with a mobile phone that has a built-in camera. Apart from spying and espionage concerns, there is the likelihood that if photos are taken of employees inappropriately either by employers or other employees, the organizations may face legal action for invasion of privacy or workplace harassment.

Mobiles with a built-in camera are not their organizations' only concern. Mobiles equipped with General Packet Radio Service (GPRS) may allow the downloading of data without your knowledge. Once downloaded, it can immediately be forwarded on to any number of destinations.

Raghu Raman of Mahindra Special Services Group says:

Some corporates are becoming sensitive that a camera phone is a potential snooping or espionage device and should not be treated lightly in high-risk areas. But most companies do not realize this and are blissfully unaware of leakage of sensitive information through this route. (India Times Infotech, 2004)

Some businesses have tried to deal with the problem by requiring sticky tape to be placed over the lens or banning mobiles from their premises. Neither of these options works. Sticky tape can easily be removed, and banning would require a search to be effective. It would also be difficult to ban mobile camera phones when employees need to be contactable, whether for business or personal reasons. There is also the prospect of using jamming technologies, but these are expensive and again would prevent vital communications, such as emergency calls. However, jamming technologies would be useful in sensitive areas.

Organizations will need to start looking at a variety of solutions to protect both sensitive information and personal privacy. Organizations will need to look at two separate and simultaneous sets of solutions, those that deal with employees and those that deal with non-employees interacting with your organization.

For employees, an organization should first begin with reviewing and incorporating acceptable use of mobile camera phones within their Human Resources policies. Employees should be made aware (on a regular basis) of the issues involved with using a mobile camera phone. It is an education step, the same as the education process that we went through for safe work practices and sexual harassment. Employees need to be made aware that taking a photo of a workplace colleague without their consent can constitute harassment. Taking a photo of corporate documents or downloading data without permission can constitute theft. Education of employees to their responsibilities will be important in avoiding future problems where employees may take legal action because of harassment and privacy violations.

Where companies issue an employee a mobile phone, they should have in place policies that limit those who are issued with a built-in camera only if it is required as part of their work. Many companies now only issue to employees mobile phones without built-in cameras. Samsung, manufacturer of mobile phones, stated that it had been asked by Telstra and Optus (two Australian communications companies) to continue to offer conventional phones to corporate clients.

The general manager of mobile phones for Samsung Australia, Josh Delgado, said:

We have spoken to our carrier customers, Optus and Telstra, who sell to the corporate market, and they have mentioned that camera phones and this area of privacy are an issue. They would not mention who those customers were, but we've been told that we should continue to manufacture both. (Lee, 2004)

Employees who bring their own mobile phones to work need to be aware of appropriate use within the organization. It will be difficult and impractical to ban private mobiles from the workplace, as people often need to be contactable.

Organizations that have non-employees interacting within their premises need to look at other solutions to protect sensitive information and personal privacy. As stated earlier, requiring sticky tape to be placed over the lens or banning them from their premises would be counterproductive, as sticky tape can easily be removed and banning would require a search to be effective. Also, these people would need their mobile phones to contact others, especially if they have to contact someone from their own organization for information or advice.

Organizations can look at requiring all entities interacting with the organization and be required to sign a non-disclosure agreement—that is, that all images and data of an organization remain the copyright of the organization. The agreement should also state that they would not use their mobile camera phone to take photos or download data without prior written permission. The agreement should also outline penalties that will be levied if the agreement is broken. This can be incorporated into existing agreements that your organization may already require. Again, making non-employees aware of the issues will greatly help in avoiding future problems.

Organizations need to also assess future changes in this technology and the impact this will have. One such change likely to come about is the incorporation of GPS mapping technologies within mobile phones. This will provide enormous benefits to organizations and their employees. Employees will be better able to find clients' premises, and employers who need to deploy staff to a client's premises (for example, their computer has gone down) will be able to find the closest employee easily and readily.

However, organizations and employees need to be aware that privacy issues can arise. For instance, does an organization tracking an employee outside of business hours, or during breaks, constitute a privacy invasion? Also, tracking employees with the view to enhance productivity may expose the organization to potential legal problems. For example, an organization tracks a courier to ensure that they are not taking an excessive amount of time to make individual deliveries. If the courier knows that they are being monitored and can face disciplinary action for taking too long, they may start to take risks to shorten their delivery times to avoid disciplinary action. If these risk lead to an accident, the organization may find that they are liable if it can be shown that the monitoring led to the risk to shorten the delivery time.

The point is, when an organization implements the use or allows the use of new technologies such as camera phones or GPS-equipped phones, they need to think through the implications of these technologies and how people may use them. Clear guidelines need to be established, and employer and employee education needs to be undertaken to ensure that the organization's policies and guidelines are being met.

AUSTRALIAN COMPUTER SOCIETY POLICY ON MOBILE CAMERA PHONES

The Australian Computer Society has taken a leadership position on this topic with the release of its mobile camera phone policy in July 2004 (Abood, 2004). The policy has been well received, with both government and the media showing much interest in it.

Many have asked me why the Australian Computer Society is getting involved in this subject. Information and communication technologies have become prevalent within our lives. From school to work to home, we are becoming more reliant on these technologies. The benefits these technologies are providing us are many. However, these technologies also can be used to the detriment of others. As the Australian Computer Society is the guardian of ethics within our industry in Australia, it is our responsibility to alert society to inappropriate use of these technologies, and help guide society to better understand and use these technologies appropriately.

At the time that the policy was released, there was (and still is) a lack of clarity with regards to the protocols, policies, and laws regarding the use and misuse of camera phones. The policy recommends the following:

- Raising the level of awareness and potential impact of camera phones with Australian companies and the Australian public, including an education campaign to advise of rights, guidelines, and etiquette on the appropriate use of mobile phone cameras.
- Assessing the full implications of this technology, including relevant overseas developments.
- Introducing appropriate guidelines by businesses to address use of phone cameras as a "tool of trade".
- Developing responsible guidelines by manufacturers, retailers, and promoters of this technology to be distributed with all phone cameras sold/issued and to accompany advertising of these products.
- Examining the use of technological solutions to enhance privacy including the use

of sound or a flashing light when a picture or video is being taken and the use of jamming technology.
- Developing a code of conduct for the use of mobile camera phones in sensitive areas such as change rooms.

The Australian Computer Society is also calling for a more transparent debate on this topic among the Australian federal and state governments and the telecommunications industry. This debate should be inclusive to ensure policy making in this area is not influenced solely by those with a commercial vested interest. Greater clarity and consistency in the approach taken by government(s), and by Australian corporations and venues to policy making in this area need to be undertaken.

The Australian Computer Society is also calling for greater awareness and education regarding:

- The protocols and legal responsibilities of the holders/users of this technology.
- The potential for misuse of this increasingly pervasive technology.
- The civil rights and protections that are in place should a member of the public feel they have been subject to abuse or misuse of this technology.
- The appropriate security precautions Australian companies can take in relation to this new technology.
- The appropriate use of this technology in public advertising—in light of the personal privacy concerns.

CONCLUSION AND FUTURE DIRECTION

While mobile camera phones provide many benefits, these devices are increasingly being used for inappropriate use. Areas of inappropriate use include invasion of privacy and use as a tool for harassment and spying/surveillance within organizations. With appropriate policies and education of the public with regards to the appropriate use of mobile camera phones, we will be all able to enjoy the benefits of this technology without the unwanted side effects.

REFERENCES

Abood, C. (2004, July). *The Australian Computer Society policy on mobile camera phones.* Retrieved from http://www.acs.org.au/acs_policies/docs/2004/mobilespolicyfinal.pdf

The Australian. (2004). Fine over mobile topless pictures. *The Australian,* (December 1). Retrieved from http://www.theaustralian.news.com.au/commonstory_page/0,5744,11555056%255E1702,00.html

BBC News. (2003, July 7). *Samsung bans "spy" phones.* Retrieved from http://news.bbc.co.uk/1/hi/world/asia-pacific/3052156.stm

CBS News. (2003, July 9). *Camera phone etiquette abuses.* Retrieved from http://www.cbsnews.com/stories/2003/07/09/tech/main562434.shtml

India Times Infotech. (2004, December 23). *Camera phone: A security threat.* Retrieved from http://infotech.indiatimes.com/articleshow/968730.cms

Lee, J. (2004). Is that a camera in your briefcase? *The Sydney Morning Herald* (June 12). Retrieved from http://www.smh.com.au/articles/2004/06/11/1086749894185.html?from=storylhs&oneclick=true

The Sydney Morning Herald. (2004). Indian schoolboy's phone sex prank reverberates around the world. *The Sydney Morning Herald* (December 22). Retrieved from http://www.smh.com.au/news/World/Indian-schoolboys-phone-sex-prank-reverberates-around-the-world/2004/12/21/1103391774573.html?oneclick=true

Chapter XLVIII
Social Context for Mobile Computing Device Adoption and Diffusion:
A Proposed Research Model and Key Research Issues

Andrew P. Ciganek
University of Wisconsin-Milwaukee, USA

K. Ramamurthy
University of Wisconsin-Milwaukee, USA

ABSTRACT

The purpose of this chapter is to explore and suggest how perceptions of the social context of an organization moderate the usage of an innovative technology. We propose a research model that is strongly grounded in theory and offer a number of associated propositions that can be used to investigate adoption and diffusion of mobile computing devices for business-to-business (B2B) interactions (including transactions and other informational exchanges). Mobile computing devices for B2B are treated as a technological innovation. An extension of existing adoption and diffusion models by considering the social contextual factors is necessary and appropriate in light of the fact that various aspects of the social context have been generally cited to be important in the introduction of new technologies. In particular, a micro-level analysis of this phenomenon for the introduction of new technologies is not common. Since the technological innovation that is considered here is very much in its nascent stages there may not as yet be a large body of users in a B2B context. Therefore, this provides a rich opportunity to conduct academic research. We expect this chapter to sow the seeds for extensive empirical research in the future.

INTRODUCTION

What causes individuals to adopt new information technologies (ITs)? How much influence do the perceptions of the social context of an organization have on the acceptance of new ITs? These questions are significant because systems that are not utilized will not result in expected efficiency and effectiveness gains (Agarwal & Prasad, 1999), and will end up as unproductive use of organizational resources. Academic research consequently has focused on the determinants of computer technology acceptance and utilization among users. Some of this research comes from the literature on adoption and diffusion of innovations (DOI), where an individual's perceptions about an innovation's attributes (e.g., compatibility, complexity, relative advantage, trialability, visibility) are posited to influence adoption behavior (Moore & Benbasat, 1991; Rogers, 2003). Another stream of research stems from the technology acceptance model (TAM), which has become widely accepted among IS researchers because of its parsimony and empirical support (Agarwal & Prasad, 1999; Davis, 1989; Davis, Bagozzi, & Warshaw, 1989; Hu, Chau, Sheng, & Tam, 1999; Jackson, Chow, & Leitch, 1997; Mathieson, 1991; Taylor & Todd, 1995; Venkatesh, 1999, 2000; Venkatesh & Davis, 1996, 2000; Venkatesh & Morris, 2000).

Individual differences indeed are believed to be very relevant to information system (IS) success (Zmud, 1979). Nelson (1990) also acknowledged the importance of individual differences in affecting the acceptance of new technologies. A variety of research has investigated differences in the perceptions of individuals when using TAM (Harrison & Rainer, 1992; Jackson et al., 1997; Venkatesh, 1999, 2000; Venkatesh & Morris, 2000); however, the perceptions and influences of the social context of an organization have not been widely examined in the literature. Hartwick and Barki (1994) suggest

that it is imperative to examine the acceptance of new technologies with different user populations in different organizational contexts.

Although mobile computing devices have existed for several years, strategic applications of this technology are still in their infancy. Mobile computing devices (in the context of business-to-business—B2B) is treated as a technology innovation in this chapter due to their newness and short history. An investigation into the usage of mobile computing devices within a B2B context, which we define as two or more entities engaged within a business relationship, is of value because of its increasing popularity (March, Hevner, & Ram, 2000). As an emergent phenomenon, relatively modest academic literature has examined the nature of adoption and use of this technology. Mobile computing devices, which have been described as both ubiquitous (March et al., 2000) and nomadic (Lyytinen & Yoo, 2002a, 2002b), offer a stark difference from traditional, static computing environments. A good characterization of these differences is provided in Satyanarayanan (1996). New technology innovations typically require changes in users' existing operating procedures, knowledge bases, or organizational relationships (Van de Ven, 1986). Such innovations may even require users to develop new ways of classifying, examining, and understanding problems. The domain of mobile computing devices has the potential to become the dominant paradigm for future computing applications (March et al., 2000), and topics of such contemporary interest are recommended to be pursued in IS research (Benbasat & Zmud, 1999; Lyytinen, 1999).

The primary objective of this chapter is to examine whether and how perceptions of the social context of an organization moderate the adoption, use, and infusion[1] of mobile computing devices for B2B transactions. We extend TAM to include individuals' perceptions of the social context of their organization, which in-

Figure 1. Technology acceptance model (Adapted from Davis, Bagozzi, & Warshaw, 1989)

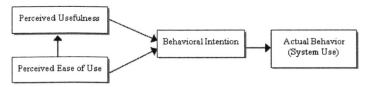

corporates aspects of both culture and climate research as recommended in the literature (Denison, 1996; Moran & Volkwein, 1992). Aspects of the social context of an organization are suggested as having a significant role in the introduction of new technologies (Boudreau, Loch, Robey, & Straub, 1998; Denison & Mishra, 1995; Legler & Reischl, 2003; Orlikowski, 1993; Zammuto & O'Connor, 1992), particularly with the introduction of mobile computing devices (Jessup & Robey, 2002; Sarker & Wells, 2003). Only a handful of studies in the past have specifically looked at the micro-level connections of these relationships (Straub, 1994); unfortunately, even this has not been within a mobile computing context. We argue that an organization's social context will have a significant moderating effect on the perceptions of employees considering adoption and use of mobile computing applications for B2B purposes.

The chapter proceeds as follows: the next section presents the background research in the domains (adoption and diffusion of technology innovations within the context of TAM, DOI, and social context) underlying this research. This will be followed by the presentation and discussion of our proposed model and accompanying propositions. A brief discussion of the types of B2B application domains that are relevant to mobile-computing and would be of (future) interest to our investigation is then presented, accompanied by one methodological approach to how such research can be conducted. This chapter concludes with some po-

tential implications for research and practice, limitations of the book chapter, and potential future directions.

BACKGROUND RESEARCH

In this section, we first discuss the extant research connected with the technology acceptance model followed by research related to social context.

Technology Acceptance Model

The technology acceptance model proposed by Davis (1989) has its roots in the theory of reasoned action (TRA) of Fishbein and Ajzen (1975). As earlier alluded to, it is one of the most widely used models of IT acceptance. This model accounts for the psychological factors that influence user acceptance, adoption, and usage behavior of new IT (Davis, 1989; Davis et al., 1989; Hu et al., 1999; Mathieson, 1991; Taylor & Todd, 1995). The TAM model is displayed in Figure 1.

As is fairly well known in the IT literature, TAM specifies two beliefs—*perceived usefulness* (PU) and *perceived ease of use* (PEOU)—to be determinants of IT usage. It incorporates behavioral intention as a mediating variable in the model, which is important for both substantive and sensible reasons. In terms of substantive reasons, the formation of an intention to carry out a behavior is thought to be a necessary precursor to actual behavior

(Fishbein & Ajzen, 1975). In terms of sensible reasons, the inclusion of intention is found to increase the predictive power of models such as TAM and TRA, relative to models that do not include intention (Fishbein & Ajzen, 1975). Perceived usefulness is defined as "the degree to which a person believes that using a particular system would enhance her/his job performance"; perceived ease of use is defined as "the degree to which a person believes that using a particular system would be free of effort" (Davis, 1989, p. 320).

The TAM model and other subsequent IT models of acceptance have largely ignored the influence that continued usage has on the acceptance of an IT. For example, Karahanna, Straub, and Chervany (1999) found differences in the determinants of attitudes between potential adopters and actual users of an IT. In particular, they found that perceived usefulness continued to play an important role in the attitudes of IT users, while ease of use ceased to be important over time. Consequently, the relationship between actual/demonstrated usefulness and continued use is added by us to the original TAM model. Once the actual/realized usefulness of an IT is confirmed by a potential adopter, it is likely to continue to play a significant role in the overall infusion of the technology.

Based upon conceptual and empirical similarities across eight prominent models in the user acceptance literature, Venkatesh, Morris, Davis, and Davis (2003) developed a unified theory of individual acceptance of technology (the unified theory of acceptance and use of technology, or UTAUT). The UTAUT theorizes four constructs as having a significant role as direct determinants of acceptance and usage behavior: performance expectancy (subsuming perceived usefulness), effort expectancy (subsuming perceived ease of use), social influence, and facilitating conditions. In addition, it considers four moderators—age, gender, voluntariness of use, and experience of the users to influence the relationship between the four direct antecedents and intentions to use (and in the case of facilitating conditions on actual use behavior). Although the UTAUT model explains a significant amount of variance in the intention to adopt an IT, the model lacks the parsimony and empirical replication of the TAM model. In this light, the modified TAM model that we propose may be considered a viable and prudent alternative to the UTAUT model. An empirical comparison between these two models is, of course, necessary.

Recent research employing the TAM model had identified individual differences as a major external variable (Agarwal & Prasad, 1999; Jackson et al., 1997; Venkatesh, 2000; Venkatesh & Morris, 2000). Individual differences are any forms of dissimilarity across people, including differences in perceptions and behavior (Agarwal & Prasad, 1999). For example, Agarwal and Prasad (1999) found that an individual's role (provider or user) with regard to a technology innovation, level of education, and previous experiences with similar technology were significantly related to their beliefs about the ease of use of a technology innovation. Agarwal and Prasad also found a significant relationship between an individual's participation in training and their beliefs about the usefulness of a technology innovation. Jackson et al. (1997) examined variables such as situational involvement, intrinsic involvement, and prior use of IT by users, and Venkatesh (2000) considered individual specific variables such as beliefs about computers and computer usage, and beliefs shaped by experiences with the technology in the traditional TAM. Both these studies found significant relationships among these individual differences and TAM constructs. Further, Venkatesh and Morris (2000) argue from their findings that "men are more driven by instrumental factors (i.c., perceived usefulness) while women are more motivated by process (perceived ease of use) and social (subjective norm) factors" (p. 129). Thus, while the various above-noted research studies have investigated the differences in the perceptions of individuals using TAM as the underlying theoretical basis, as noted earlier, perceptions of the social context of an organization is not common in the literature. Most of these refinements to TAM and findings are accommodated in the earlier-noted overarching UTAUT model proposed by Venkatesh et al. (2003).

Social Context of an Organization and Innovativeness

As noted in the introduction, although the social context of an organization has been suggested as having a significant role in the introduction of new technologies (Boudreau et al., 1998; Denison & Mishra, 1995; Legler & Reischl, 2003; Orlikowski, 1993; Zammuto & O'Connor, 1992), particularly with the introduction of mobile computing devices (Jessup & Robey, 2002; Sarker & Wells, 2003), it has not been widely examined in the literature. In this chapter we extend the TAM to incorporate an individual's perceptions of the social context of their organization. The perceptions of the social context are of value to consider since they are likely to be fairly stable in the mind of the potential adopter and less subject to change than other perceived factors or the underlying technological innovation. As recommended in the literature, we examine the social context of an organization to incorporate aspects of both culture and climate (Denison, 1996; Moran & Volkwein, 1992). We take the stand that a study of organizational culture and organizational climate actually examine the same phenomenon—namely, the creation and influence of social contexts in organizations—but from

different perspectives (Denison, 1996). Following the recommendation of prior research, we examine the broader social context in order to improve our understanding of the organizational phenomenon (Astley & Van de Ven, 1983; Denison, 1996; Moran & Volkwein, 1992; Pfeffer, 1982).

Organizational climate can be described as the shared perceptions of organizational members who are exposed to the same organizational structure (Schneider, 1990). Zmud (1982) suggests that it is not the structure of the organization that triggers innovation; rather, innovation emerges from the organizational climate within which members recognize the desirability of innovation, and within which opportunities for innovation arise and efforts toward innovation are supported. As summarized in Schneider (1990) and in Moran and Volkwein (1992), a number of different conceptualizations of organizational climate have been suggested over the years. Pareek (1987) advanced the idea that climate and culture can only be discussed in terms of how it is perceived and felt by individual members/employees of the organization, which is a perspective that is supported in the literature (Legler & Reischl, 2003). Thus, we are interested in capturing the perceptions of individuals within organizations. Since the unit of analysis (during empirical evaluation) in this chapter is the individual employees within organizations, appropriate measures of examining social context can be derived from psychological climate literature.

Rather than focusing on how the psychological climate of an organization gets formed and can be influenced (certainly important), of interest in this chapter is how the prevailing climate of an organization moderates the relationship between individuals' perceptions of an innovation's usefulness and ease of use, and their intentions to adopt and use the innovation. Psychological climate is a multi-dimensional

Table 1. Dimensions of psychological climate (Adapted from Koys & DeCotiis, 1991)

Dimension Name	Definition
Autonomy	Employee's perception of their own sovereignty with respect to work procedures, goals and priorities.
Cohesion	Employee's perception of sharing and togetherness within their organization.
Trust	Employee's perception of freedom to communicate openly with members at higher organizational levels about sensitive or personal issues with the expectation that the integrity of such communications will not be violated.
Pressure	Employee's perception that time demands are incongruent with respect to task completion and performance standards.
Support	Employee's perception of the tolerance of their behavior by superiors, including the willingness to let employees learn from their mistakes without fear of reprisal.
Recognition	Employee's perception that their contributions to their organization are acknowledged.
Fairness	Employees' perception that their organization's practices are equitable and non-arbitrary.
Innovation	Employee's perception that change and originality are encouraged and valued within their organization, including risk-taking in domains where the individual may have little to no prior experience.

construct that can be conceptualized and operationalized at the individual level (Glick, 1985; Legler & Reischl, 2003). In an attempt to integrate several different measures of psychological climate, Koys and DeCotiis (1991) derived eight summary dimensions—*autonomy*, *cohesiveness*, *fairness*, *innovation*, *pressure*, *recognition*, *support*, and *trust*. A brief definition/description of each of these dimensions is provided in Table 1.

In the next section, while presenting our research model and associated propositions, we will discuss how each of these dimensions would be expected to moderate the relationship between an individual's perceptions (of an innovation) and behavioral intention (to adopt and use it). Briefly, however, we will take a couple of these climate dimensions (*support* and *autonomy*) and discuss the relevance of these dimensions of organizational climate for the adoption of technological innovations.

Senior management's attitude toward change (consequential to the introduction of technology innovations) and thus the extent of their *sup-*

port impacts the adoption of these technology innovations (Damanpour, 1991). Senior management teams may be very conservative, preferring the status quo and using current or time-tested methods innovating only when they are seriously challenged by their competition or by shifting consumer preferences (Miller & Friesen, 1982). By contrast, they may be risk prone, actually encouraging and actively supporting the use of innovative techniques to move the organization forward, usually trying to obtain a competitive advantage by routinely making dramatic innovative changes and taking the inherent risks associated with those innovations (Litwin & Stringer, 1968). The potentially disruptive features typically associated with the adoption of (radical) innovations require an organizational context where managers encourage individual members of the organization to take (prudent levels of) risk, support adoption of technology innovations, and be supportive of changes in their organizations (Dewar & Dutton, 1986). Organizations should be wary, however, that a follower approach taken by employees

Figure 2. Research model

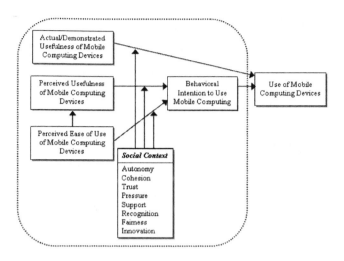

may promote a "mindless" environment resulting in undesirable levels of risk-taking, which can cause significant problems (Swanson & Ramiller, 2004).

Organizational context/climate also reflects the extent of focus on autonomy/empowerment vs. control of its members. An organic organization as contrasted with mechanistic organization is typically associated with open and free-flowing communication, sharing of necessary information and knowledge, flexibility, and absence of rigid rules and regulations; such an organization context is usually positively related to innovation (Aiken & Hage, 1971; Kimberly & Evanisko, 1981). Furthermore, an organizational climate that is geared toward and has built-in expectation of high levels of achievement and high standards of excellence nurtures a vibrant base of challenges posed to its members who have the freedom to apply innovative technologies, techniques, and procedures to effectively accomplish the tasks (Rosenthal & Crain, 1963). Such an organizational context will be more prone to encouraging its members to adopt technology innovations to accomplish high levels of performance.

RESEARCH MODEL AND TENTATIVE PROPOSITIONS

Based on the foregoing brief discussion of the extant research, we extend the standard TAM model with social context dimensions as shown in Figure 2.

Traditional TAM Propositions

An individual's intention to adopt/use technology is driven by his or her perceptions of the usefulness of the technology (Davis et al., 1989). This contention, as noted in the background research section, has been supported extensively in previous research (Agarwal & Prasad, 1999; Davis et al., 1989; Hu et al., 1999; Jackson et al., 1997; Venkatesh, 1999, 2000; Venkatesh & Davis, 2000; Venkatesh & Davis, 1996; Venkatesh & Morris, 2000). A primary reason why individuals would intend to adopt/use mobile computing devices for B2B transactions is that they believe that this technology will provide them the flexibility to perform their job and enable their job performance enhancement (Davis, 2002; Intel, 2003). Furthermore, following the findings of Karahanna et al. (1999), the perceived usefulness of an IT influences the attitudes of both potential adopters and users of an IT. However, we contend that when an IT has demonstrated its usefulness over time, it is likely to play a significant role in the overall infusion of the technology. Therefore, we propose:

- **Proposition 1:** Perceived usefulness will have a positive effect on organizational members' intention to adopt/use mobile computing devices for B2B transactions.
- **Proposition 2:** Actual/demonstrated usefulness will have a positive effect on organizational members' continued usage of mobile computing devices for B2B transactions.

As noted earlier, the second major determinant of behavioral intentions in the TAM model, *perceived ease of use,* has been observed to have both a (somewhat weak) direct influence on behavioral intention as well as a (strong) indirect influence through its effect on perceived usefulness (Davis, 1989; Davis et al., 1989; Hu et al., 1999; Jackson et al., 1997). This is understandable since a person who believes that a technology innovation is (relatively) easy to understand and use, and is less demanding of efforts, would likely believe that using such a technology is also more useful. While perceived ease of use may trigger users' intention to adopt/use the innovation (mobile computing devices for B2B), it is unlikely to play a key role in the spread/infusion since users would likely become more familiar with all the features of the innovation and gain significant expertise with time following the initial use. Hence, we propose the two following propositions:

- **Proposition 3a:** Perceived ease of use will have a positive effect on organizational members' intention to adopt/use

mobile computing devices for B2B transactions.
- **Proposition 3b:** Perceived ease of use will positively influence organizational members' perceptions of the usefulness of mobile computing devices for B2B transactions.

Extended TAM Propositions

One of the key objectives of this chapter is to examine what role, if any, social context plays in the link between individuals' perceptions of usefulness/ease of use and behavioral intentions of the TAM model. We pointed out that social context, when conceptualized in terms of climate/culture, is a multi-dimensional construct composed of eight dimensions (Koys & DeCotiis, 1991). Since there has been no attempt to examine this additional set of dimensions within the context of TAM, many of the arguments and much of the rationale that we provide in the rest of this section while developing the propositions are likely to be tentative.

Autonomy

At one end of the spectrum, an organization can be extremely control and compliance oriented (*mechanistic* organizational context) in formulating, administering, and closely monitoring and enforcing a set of policies and procedures that guide employee work activities. At the opposite end of the spectrum, an organization can be performance and achievement oriented (*organic* organizational context) by empowering their employees to determine their task priorities and schedule, providing them the autonomy to make use of any and all techniques, tools, and technologies that they deem best for getting the work done, and being flexible with respect to adherence on the standard policies and procedures. Thus, organizations where the members perceive greater *autonomy* and flexibility being provided to them in making decisions and choices on their task-related activities are likely to more quickly exploit (any) opportunity that technology innovations offer. While this is fairly obvious when the technology is perceived to be useful and easy to use, even in instances where such perceptions (of ease of use and usefulness) may not be completely true, the organizational members may still be more willing to make informed decisions that they are responsible and accountable for (Aiken & Hage, 1971; Kimberly & Evanisko, 1981). To become better informed, they may actively seek out knowledge from various pockets of the (internal) organization as well as from external sources (e.g., consultants, vendors, trade literature, etc.). Therefore:

- **Proposition 4a:** The relationship between employees' perceptions (of usefulness and ease of use of the technology) and their intentions to adopt/use mobile computing devices for B2B transactions will be stronger in organizational contexts that provide greater autonomy to their employees.
- **Proposition 4b:** The relationship between the actual/demonstrated usefulness and continued usage of mobile computing devices for B2B transactions will be stronger in organizational contexts that provide greater autonomy to their employees.

Cohesion

As would be noted from the brief description provided in Table 1, *cohesion* refers to an organizational context/climate that fosters a sense of sharing, caring, accommodation, and togetherness among the members/employees (Koys & DeCotiis, 1991). Communication, sharing, and exchange of information and knowledge amongst the members is bound to be much more open in such a context. Employees would more willingly share their experiences and sup-

port one another when attempting to make decisions on complex and unknown topic areas (e.g., relevance and mastery of new technologies). It is, therefore, reasonable to expect that potential adopters of new technology innovations (mobile computing devices) would be more willing and prepared to assume any challenges posed by the new technology environment in view of the potential support that they can expect from their colleagues in their work environment. Therefore, we propose the following:

Proposition 5a: The relationship between employees' perceptions (of usefulness and ease of use of the technology) and their intentions to adopt/use mobile computing devices for B2B transactions will be stronger in organizational contexts that foster a greater sense of cohesion/cohesiveness among their employees.

Proposition 5b: The relationship between the actual/demonstrated usefulness and continued usage of mobile computing devices for B2B transactions will be stronger in organizational contexts that foster a greater sense of cohesion/cohesiveness among their employees.

Trust

The third dimension of organizational climate, *trust*, refers to the extent to which employees within the organization can openly communicate with their superiors, seek their guidance and expertise, and be confident that the integrity of sensitive information will not be compromised (Koys & DeCotiis, 1991). It is easy to visualize that such expectations of trust work in both directions—from subordinate to superiors and vice versa. Trust also involves an expectation of confidence in the goodwill of others in the organizational context/environment, as well as the prospects for continuity of the relationship entered into (Hart & Saunders, 1997). It is normal to expect that in these trusting organizational contexts, employees will be more prepared to share their difficulties and concerns (work related and even personal), propose potential technology-based solutions, and seek approval/guidance/advice from their superiors and peers. This can be quite important as in the case of introduction of mobile computing devices where the work arrangements and workflows are bound to be disrupted and changed quite radically (e.g., employees may not have to be always present on site and could increasingly work from off-site locations, at home, or on the move). Trust is a significant determinant of a stable relationship (Mayer, Davis, & Schoorman, 1995; McKnight, Choudhury, & Kacmar, 2002). Therefore, we propose:

- **Proposition 6a:** The relationship between employees' perceptions (of usefulness and ease of use of the technology) and their intentions to adopt/use mobile computing devices for B2B transactions will be stronger in organizational contexts that promote and reinforce trust between employees and the organization.
- **Proposition 6b:** The relationship between the actual/demonstrated usefulness and continued usage of mobile computing devices for B2B transactions will be stronger in organizational contexts that promote and reinforce trust between employees and the organization.

Pressure

The fourth dimension of organizational climate, *pressure*, refers to the fact that the work context may not provide adequate time for the employees to accomplish their task-related activities and achieve the required standards of performance and goals (Koys & DeCotiis, 1991). Typically, it would be reflective of a situation of significant stress, perhaps hasty

decisions and actions resulting in suboptimal results, and generally chaos. However, such a stressful environment may also be one that could spur the organizational members to creatively look for (technologically) innovative solutions to alleviate the difficulties and infuse some order. To the extent that the performance of tasks is not geographically constrained (e.g., assembly-line work in automotive manufacturing, patrons being serviced in a restaurant or a bank), it is possible that mobile computing devices may indeed alleviate the time pressure that is so rampant in the work context. For example, employees may become skillful in time management through the convenience of mobile computing devices in coordinating work and personal tasks (Davis, 2002; Intel, 2003). Therefore, surprising and counter-intuitive as it might sound, we propose:

- **Proposition 7a:** The relationship between employees' perceptions (of usefulness and ease of use of the technology) and their intentions to adopt/use mobile computing devices for B2B transactions is likely to be stronger in organizational contexts that reflect one of (time) pressure for employees to accomplish their task and realize the set performance standards.
- **Proposition 7b:** The relationship between the actual/demonstrated usefulness and continued usage of mobile computing devices for B2B transactions is likely to be stronger in organizational contexts that reflect one of (time) pressure for employees to accomplish their task and realize the set performance standards.

Support

The fifth dimension of organizational climate, *support*, reflects an organizational context that is tolerant of errors and mistakes that employees may commit, and is supportive of them as long as they learn from these (Koys & DeCotiis, 1991). An environment that is permissive and lets its members learn from mistakes without fear of punishment and reprisal could engender deep-rooted learning, a "can-do" attitude to problem solving, and (reasonable) risk-taking orientation (Litwin & Stringer, 1968). As noted earlier, management's attitude toward change (often triggered by the introduction of technology innovations) and thus the extent of their support impacts the adoption and successful implementation of these technology innovations (Damanpour, 1991; Sanders & Courtney, 1985). The potentially disruptive features typically associated with the adoption of (radical technology) innovations require an organization context where managers encourage individual members of the organization to take (prudent levels of) risk, support adoption of technology innovations, and be supportive of changes in their organizations (Dewar & Dutton, 1986). Supportive organizational context is also conducive to successful IT implementation (Ramamurthy, Premkumar, & Crum, 1999). Caron, Jarvenpaa, and Stoddard (1994) chronicle how CIGNA Corporation, due to its supportive and tolerance-for-failure environment, facilitated significant learning to accrue in the context of major disruptive and radical changes triggered by business process reengineering projects. Therefore, we propose:

- **Proposition 8a:** The relationship between employees' perceptions (of usefulness and ease of use of the technology) and their intentions to adopt/use mobile computing devices for B2B transactions is likely to be stronger in organizational contexts that are tolerant and supportive of employees in accomplishing their work.
- **Proposition 8b:** The relationship between the actual/demonstrated usefulness and

continued usage of mobile computing devices for B2B transactions is likely to be stronger in organizational contexts that are tolerant and supportive of employees in accomplishing their work.

Recognition

The sixth dimension of organizational climate, *recognition*, reflects an organizational context where employee achievements and accomplishments are acknowledged and recognized (Koys & DeCotiis, 1991). *Human relations management* and *job enrichment* literature (Hackman & Oldham, 1980) points out that intrinsic rewards (e.g., employee-of-the-month recognition) at times are more important than extrinsic rewards (e.g., salary raises, promotion). Extrinsic and intrinsic motivation literature has also been used significantly to explain adoption and use of innovations (Davis, Bagozzi, & Warshaw, 1992). Resource-based theory also acknowledges the vital role human assets/resources play in contemporary hyper-competitive external environments where progressive organizations strive to keep their employees satisfied and thus retain top talent. It is, therefore, natural to expect that organizations should strive to create a climate that spurs their employees to constantly look out for creative solutions (including new technology innovations) that foster excellence in achievement. Obviously, this is unlikely when such efforts and accomplishments go unrecognized. Thus, we would propose:

- **Proposition 9a:** The relationship between employees' perceptions (of usefulness and ease of use of the technology) and their intentions to adopt/use mobile computing devices for B2B transactions is likely to be stronger in organizational contexts that are open to acknowledge and recognize the accomplishments of their employees.

- **Proposition 9b:** The relationship between the actual/demonstrated usefulness and continued usage of mobile computing devices for B2B transactions is likely to be stronger in organizational contexts that are open to acknowledge and recognize the accomplishments of their employees.

Fairness

The seventh dimension of organizational climate, *fairness*, reflects an organizational context where employees believe in equitable and non-arbitrary treatment (Koys & DeCotiis, 1991). This reinforces the notion that hard, sincere, and smart work pays off. Individuals that believe an inequity exists, for example, are likely to resent and resist organizational changes (Joshi, 1989, 1991). Clearly an organization that does not design its workplace context with work/job assignments that are perceived to be fair and rewards that are perceived to be equitable for similar accomplishments would trigger significant discontent and distrust. Such an environment is hardly likely to evoke any voluntary or enthusiastic response to work-related organizational challenges, including searching for new technology innovations. Therefore, we would propose:

- **Proposition 10a:** The relationship between employees' perceptions (of usefulness and ease of use of the technology) and their intentions to adopt/use mobile computing devices for B2B transactions is likely to be stronger in organizational contexts that are deemed to be fair in the treatment of their employees.
- **Proposition 10b:** The relationship between the actual/demonstrated usefulness and continued usage of mobile computing devices for B2B transactions is likely to be stronger in organizational contexts that

are deemed to be fair in the treatment of their employees.

Innovation

The last (eighth) dimension of organizational climate, *innovation*, reflects an organizational context where employees believe change from status-quo can be good, that originality is valued, and risk taking will be encouraged (Koys & DeCotiis, 1991). As noted earlier, management's attitude toward change (often triggered by the introduction of technology innovations) impacts the adoption of these technology innovations (Damanpour, 1991). Some senior management teams may have conservative attitudes toward innovation and associated risk, preferring the status quo and using current or time-tested methods; such organizations innovate only when they are seriously challenged by their competition or by shifting consumer preferences (Miller & Friesen, 1982). By contrast, other senior management teams may be risk prone, actually encouraging and actively supporting the use of innovative techniques to move the organization forward. Such organizations usually try to obtain a competitive advantage by routinely making dramatic innovative changes and taking the inherent risks associated with those innovations. The potentially disruptive features typically associated with the adoption of (radical technology) innovations require an organization context where managers encourage individual members of the organization to take prudent levels of risk, support adoption of technology innovations, and be supportive of changes in their organizations (Dewar & Dutton, 1986). Thus, we would propose:

- **Proposition 11a:** The relationship between employees' perceptions (of usefulness and ease of use of the technology) and their intentions to adopt/use mobile computing devices for B2B transactions is likely to be stronger in progressive/innovative organizational contexts.

- **Proposition 11b:** The relationship between the actual/demonstrated usefulness and continued usage of mobile computing devices for B2B transactions is likely to be stronger in progressive/innovative organizational contexts.

B2B APPLICATION DOMAIN AND SUGGESTED RESEARCH METHODOLOGY

Some of the broad domains of B2B application areas that are relevant for mobile-computing and of interest to us for this research would be inventory management, customer relationship and service management, sales force automation, product locating and purchasing, dispatching and diagnosis support to, say, technicians in remote locations, mobile shop-floor quality control systems, as well as those applications and transactions in supply chain management (SCM) that facilitate the integration of business processes along the supply chain (Rao & Minakakis, 2003; Turban, King, Lee, & Viehland, 2004; Varshney & Vetter, 2001). An example of B2B transactions in the SCM context includes data transmission from one business partner to another through the typical enterprise resource planning (ERP) interactions. Other scenarios may involve the ability to continue working on projects while in transit or the ubiquitous access to documents via "hot spots" or wireless network access (Intel, 2003). Consequently, in light of the fact that a number of application domains have preexisted the Internet, the choice of application areas could be either Internet or non-Internet based.

As noted before, mobile computing is still in a very early stage of its evolution and use within organizations in a B2B context. Although a large-scale field survey would be required to test the research model that we presented, such an approach may not be appropriate in this context due to the exploratory nature of the inquiry proposed here. Therefore, the research methodology that we suggest and propose that researchers use at this stage is a combination of both qualitative and quantitative research for data collection. Rather than a large national random sample, we propose a purposive convenience-based sample of a few (say, 8-12) large and medium-sized corporations with almost equal composition of manufacturing and service sectors. Furthermore, based on secondary information and personal contacts, we would prefer that researchers select an equal mix of corporations that do not (yet) use and those that currently use mobile computing so that we can capture their "intention" and subsequently their "continued use." Although the "social context" or "climate" prevailing within each of these organizations may be a "given reality" at least at a point in time, as observed in most past research, it is the interpretations of this social context/climate that would drive individual actions, especially when the intended/actual behavior (in this case, adoption and use of mobile computing) is not mandatory (Moran & Volkwein, 1992). Thus, in-depth interviews coupled with a questionnaire survey from a number of focal members (about 20 to 25), sampled from multiple functional areas (that are amenable for use of mobile computing devices such as sales and marketing, purchasing, and operations) within participant organizations, should be used to capture individual perceptions of the mobile computing devices and their organization's social context. As argued above, since the rate of diffusion for mobile computing devices for B2B transactions is still relatively small, a convenient sampling approach among organizations that have and have yet to adopt these technologies is appropriate. To ensure relevance and reasonable generalizability of the study findings of the convenience-based sampling suggested by us, participants from each organization should be chosen randomly. A number of statistical techniques such as logistic regression (for the "intention to adopt" stage) and structural equation modeling or hierarchical moderated regression analysis (for the "infusion" stage) would be candidates for data analyses.

CONCLUSION

In this chapter we incorporated the social context of an organization into TAM and proposed an extended model to investigate adoption/use of mobile computing devices for B2B transactions as a technological innovation. We believe that such an extension is appropriate because aspects of social context have in general been found significant with the introduction of new technologies. In particular, a micro-level analysis of this phenomenon for the introduction of new technologies is rare. Since the unit of analysis of this chapter is individual employees, we utilized dimensions of psychological climate to represent the social context of an organization. The primary objective of this chapter was to posit how perceptions of the social context of an organization would moderate the intention to adopt/use and infusion of a technology innovation.

A key feature of this study is that we examined an information technology that has the potential of becoming a dominant paradigm and platform for future computing applications. As we noted, although mobile computing devices have existed for several years, their use for business-to-business transactions or operating context has not been adequately or systematically explored in academic research. We drew upon theories from the diffusion of inno-

vation, information systems, and organizational behavior literature, among others, to develop our research model and the associated 10 propositions. The model we proposed could serve as a foundation for one stream of IS research that integrates social context of an organization into TAM to examine the vital role of mobile computing devices in electronic commerce.

IMPLICATIONS, LIMITATIONS, AND FUTURE RESEARCH DIRECTIONS

Since the empirical segment of this research has not yet been conducted, we can only conjecture several potential research contributions for researchers and practitioners. One implication that this work has for future research is the exploration of how the social context of an organization may influence the acceptance and spread of an information technology innovation. The social context of an organization has not been applied to TAM, and an extension focusing on the micro-level aspects of the social context have not been widely examined in the literature. By explicitly investigating the social context of an organization, this study extends the innovation adoption and TAM literature base. Our model may be considered a viable and prudent alternative to the UTAUT model. Utilizing a (valid and popular base) model and measures that have become widely accepted among IS researchers allows for researchers in future research to replicate our study and examine other factors of interest. This chapter also addresses the need to explore technology that is close to the "leading edge" (Lyytinen, 1999, p. 26), which is recommended for maintaining the relevance of IS research (Benbasat & Zmud, 1999; Lyytinen, 1999; Orlikowski & Iacono, 2001). Obviously, considerable care and precautions (in the design of the study, operationalization, and evaluation of the measurement properties) will be needed in translating the theoretical model proposed in this chapter into a large-scale empirical investigation that can establish validity and reliability of its results.

The potential implication that this work has for IS practice is that it identifies a number of contextual factors that may influence the acceptance of a technological innovation that an organization wishes to introduce. Mobile computing devices can enhance employee productivity by granting them flexibility in work location and time management (Intel, 2003). Organizations that covet such gains in productivity are likely cognizant of the investments typically at stake when implementing IT innovations. Given that aspects of the social context of an organization are suggested as having a significant role in the introduction of mobile computing devices (Jessup & Robey, 2002; Sarker & Wells, 2003), it is desirable to understand the influence that the social context of an organization plays. Moran and Volkwein (1992) state that focusing on the micro-level aspect of the social context is appealing because it is relatively accessible, more malleable, and the appropriate level to target short-term interventions aimed at producing positive organizational change. This study helps to uncover several future opportunities for organizations since mobile computing devices have the potential to become the dominant paradigm for future computing applications (March et al., 2000; Sarker & Wells, 2003).

Although this chapter offers several potential contributions, several limitations exist. The social context of an organization is operationalized through psychological climate dimensions. The definition of social context that we adopted takes a much broader view than focusing on the individual incorporating traditions from research in the organizational culture literature as well. We feel that it is

appropriate to use the social context of the organization to begin the integration of culture and climate literature. It is our opinion that the psychological climate research is the most appropriate theory to support the research model, which presents opportunities in future work to examine other aspects of the social context of an organization that may be influential in the acceptance of a technological innovation. Another limitation of this study (when an empirical investigation is conducted) is that it may obtain retrospective accounts/information from (current) users of mobile computing devices. Retrospective accounts are an issue because individuals may not be able to accurately recall the past. It would be necessary to consider preventive measures on this front to ensure validity and reliability of the results.

REFERENCES

Agarwal, R., & Prasad, J. (1999). Are individual differences germane to the acceptance of new information technologies? *Decision Sciences, 30*(2), 361-391.

Aiken, M., & Hage, J. (1971). The organic organization and innovation. *Sociology, 5*, 63-82.

Astley, W., & Van de Ven, A. (1983). Central perspectives and debates in organizational theory. *Administrative Science Quarterly, 28*, 245-273.

Benbasat, I., & Zmud, R. W. (1999). Empirical research in information systems: The practice of relevance. *MIS Quarterly, 23*(1), 3-16.

Boudreau, M., Loch, K., Robey, D., & Straub, D. (1998). Going global: Using information technology to advance the competitiveness of the virtual transnational organization. *Academy of Management Executive, 12*(4), 120-128.

Caron, J., Jarvenpaa, S., & Stoddard, D. (1994). Business reengineering at CIGNA Corporation: Experiences and lessons learned from the first five years. *MIS Quarterly, 18*(3), 233-250.

Damanpour, F. (1991). Organizational innovation: A meta-analysis of effects of determinants and moderators. *Academy of Management Journal, 34*, 555-590.

Davis, F. (1989). Perceived usefulness, perceived ease of use, and user acceptance of information technology. *MIS Quarterly, 13*(3), 319-340.

Davis, F., Bagozzi, R., & Warshaw, P. (1989). User acceptance of computer technology: A comparison of two theoretical models. *Management Science, 35*(8), 982-1003.

Davis, F., Bagozzi, R., & Warshaw, P. (1992). Extrinsic and intrinsic motivation to use computers in the workplace. *Journal of Applied Social Psychology, 22*, 1111-1132.

Davis, G. (2002). Anytime/anyplace computing and the future of knowledge work. *Communications of the ACM, 45*(12), 67-73.

Denison, D. (1996). What is the difference between organizational culture and organizational climate? *Academy of Management Review, 21*(3), 619-654.

Denison, D., & Mishra, A. (1995). Toward a theory of organizational culture and effectiveness. *Organization Science, 6*(2), 204-223.

Dewar, R., & Dutton, J. (1986). The adoption of radical and incremental innovation: An empirical analysis. *Management Science, 23*, 1422-1433.

Fishbein, M., & Ajzen, I. (1975). *Belief, attitude, intention, and behavior: An introduction to theory and research.* Reading, MA: Addison-Wesley.

Glick, W. (1985). Conceptualizing and measuring organizational and psychological climate: Pitfalls in multilevel research. *Academy of Management Review, 10*(3), 601-616.

Hackman, J., & Oldham, G. (1980). *Work redesign.* Reading, MA: Addison-Wesley.

Harrison, A., & Rainer, R. (1992). The influence of individual differences on skill in end-user computing. *Journal of Management Information Systems, 9*(1), 93-111.

Hart, P., & Saunders, C. (1997). Power and trust: Critical factors in the adoption and use of electronic data interchange. *Organization Science, 8*(1), 23-42.

Hartwick, J., & Barki, H. (1994). Explaining the role of user participation in information system use. *Management Science, 40*(4), 440-465.

Hu, P., Chau, P., Sheng, O., & Tam, K. (1999). Examining the Technology Acceptance Model using physician acceptance of telemedicine technology. *Journal of Management Information Systems, 16*(2), 91-112.

Intel. (2003). *Effects of wireless mobile technology on employee productivity.* Intel Information Technology White Paper (pp. 1-20), USA.

Jackson, C., Chow, S., & Leitch, R. (1997). Toward an understanding of the behavioral intention to use an information system. *Decision Sciences, 28*(2), 357-389.

Jessup, L., & Robey, D. (2002). The relevance of social issues in ubiquitous computing environments. *Communications of the ACM, 45*(12), 88-91.

Joshi, K. (1989). The measurement of fairness or equity perceptions of management information systems users. *MIS Quarterly, 13*(3), 343-358.

Joshi, K. (1991). A model of users' perspective on change: The case of information systems technology implementation. *MIS Quarterly, 15*(2), 229-242.

Karahanna, E., Straub, D., & Chervany, N. (1999). Information technology adoption across time: A cross-sectional comparison of pre-adoption and post-adoption beliefs. *MIS Quarterly, 23*(2), 183-213.

Kimberly, J., & Evanisko, M. (1981). Organizational innovation: The influence of individual, organizational and contextual factors on hospital adoption of technological and administrative innovations. *Academy of Management Journal, 24,* 689-713.

Koys, D., & DeCotiis, T. (1991). Inductive measures of psychological climate. *Human Relations, 44*(3), 265-283.

Legler, R., & Reischl, T. (2003). The relationship of key factors in the process of collaboration. *The Journal of Applied Behavioral Science, 39*(1), 53-72.

Litwin, G., & Stringer, R. (1968). *Motivation and organizational climate.* Boston: Harvard University Press.

Lyytinen, K. (1999). Empirical research in information systems: On the relevance of practice in thinking of IS research. *MIS Quarterly, 23*(1), 25-28.

Lyytinen, K., & Yoo, Y. (2002a). Issues and challenges in ubiquitous computing. *Communications of the ACM, 45*(12), 63-65.

Lyytinen, K., & Yoo, Y. (2002b). Research commentary: The next wave of nomadic computing. *Information Systems Research, 13*(4), 377-388.

March, S., Hevner, A., & Ram, S. (2000). Research commentary: An agenda for information technology research in heterogeneous and distributed environments. *Information Systems Research, 11*(4), 327-341.

Mathieson, K. (1991). Predicting user intentions: Comparing the Technology Acceptance Model with the Theory of Planned Behavior. *Information Systems Research, 2*(3), 173-191.

Mayer, R., Davis, J., & Schoorman, F. (1995). An integrative model of organizational trust. *Academy of Management Review, 20*(3), 709-734.

McKnight, D., Choudhury, V., & Kacmar, C. (2002). Developing and validating trust measures for e-commerce: An integrative typology. *Information Systems Research, 13*(3), 334-359.

Miller, D., & Friesen, P. (1982). Innovation in conservative and entrepreneurial firms: Two modes of strategic momentum. *Strategic Management Journal, 3,* 1-25.

Moore, G., & Benbasat, I. (1991). Development of an instrument to measure the perceptions of adopting an information technology innovation. *Information Systems Research, 2*(3), 192-222.

Moran, E., & Volkwein, J. (1992). The cultural approach to the formation of organizational climate. *Human Relations, 45,* 19-47.

Nelson, D. (1990). Individual adjustment to information-driven technologies: A critical review. *MIS Quarterly, 14*(1), 79-98.

Orlikowski, W. (1993). Learning from notes: Organizational issues in groupware implementation. *The Information Society, 9,* 223-250.

Orlikowski, W., & Iacono, C. (2001). Research commentary: Desperately seeking the "IT" in IT research—A call to theorizing the IT artifact. *Information Systems Research, 12*(2), 121-134.

Pareek, U. (1987). *Motivating organizational roles*. New Delhi, India: Oxford and IBH.

Pfeffer, J. (1982). *Organizations and organizational theory*. Boston: Pitman.

Ramamurthy, K., Premkumar, G., & Crum, M. (1999). Organizational and inter-organizational determinants of the EDI diffusion: A causal model. *Journal of Organizational Computing and Electronic Commerce, 9*(4), 253-285.

Rao, B., & Minakakis, L. (2003). Evolution of mobile location-based services. *Communications of the ACM, 46*(12), 61-65.

Rogers, E. (2003). *Diffusion of innovations* (5th ed.). New York: The Free Press.

Rosenthal, D., & Crain, R. (1963). Executive leadership and community innovation: The fluoridation experience. *Urban Affairs Quarterly, 1*, 39-57.

Sanders, L., & Courtney, J. (1985). A field study of organizational factors influencing DSS success. *MIS Quarterly, 9*(1), 77-93.

Sarker, S., & Wells, J. (2003). Understanding mobile handheld device use and adoption. *Communications of the ACM, 46*(12), 35-40.

Satyanarayanan, M. (1996). Fundamental challenges in mobile computing. *Proceedings of the ACM Symposium—Principles of Distributed Computing*, Philadelphia.

Schneider, B. (Ed.). (1990). *Organizational climate and culture*. San Francisco: Jossey-Bass.

Straub, D. (1994). The effect of culture on IT diffusion: E-mail & fax in Japan and the U.S. *Information Systems Research, 5*(1), 23-47.

Swanson, E. B., & Ramiller, N. C. (2004). Innovating mindfully with information technology. *MIS Quarterly, 28*(4), 553-583.

Taylor, S., & Todd, P. (1995). Understanding information technology usage: A test of competing models. *Information Systems Research, 6*(2), 144-176.

Turban, E., King, D., Lee, J., & Viehland, D. (2004). *Electronic commerce: A managerial perspective*. Upper Saddle River, NJ: Prentice-Hall.

Van de Ven, A. (1986). Central problems in the management of innovation. *Management Science, 32*, 590-607.

Varshney, U., & Vetter, R. (2001). A framework for the emerging m-commerce applications. *Proceedings of the 34th Hawaii International Conference on Systems Sciences*.

Venkatesh, V. (1999). Creation of favorable user perceptions: Exploring the role of intrinsic motivation. *MIS Quarterly, 23*(2), 239-260.

Venkatesh, V. (2000). Determinants of perceived ease of use: Integrating control, intrinsic motivation, and emotion into the Technology Acceptance Model. *Information Systems Research, 11*(4), 342-365.

Venkatesh, V., & Davis, F. (2000). A theoretical extension of the Technology Acceptance Model: Four longitudinal field studies. *Management Science, 46*(2), 186-204.

Venkatesh, V., & Davis, F. D. (1996). A model of the antecedents of perceived ease of use: development and test. *Decision Sciences, 27*(3), 451-481.

Venkatesh, V., & Morris, M. (2000). Why don't men ever stop to ask for directions? Gender, social influence, and their role in technology acceptance and usage behavior. *MIS Quarterly, 24*(1), 115-139.

Venkatesh, V., Morris, M. G., Davis, G. B., & Davis, F. D. (2003). User acceptance of information technology: Toward a unified view. *MIS Quarterly, 27*(3), 425-478.

Zammuto, R., & O'Connor, E. (1992). Gaining advanced manufacturing technologies' benefits: The roles of organization design and culture. *Academy of Management Review, 17*(4), 701-728.

Zmud, R. (1979). Individual differences and MIS success: A review of the empirical literature. *Management Science, 25*(10), 966-979.

Zmud, R. (1982). Diffusion of modern software practices: Influence of centralization and formalization. *Management Science, 28*, 1421-1431.

ENDNOTE

[1] We use the term *infusion* to refer to diffusion and spread of the innovation within an organization's internal environment.

Chapter XLIX
A Socio–Cultural Analysis of the Present and the Future of the M–Commerce Industry

Ritanjan Das
University of Portsmouth, UK

Jia Jia Wang
University of Bradford, UK

Pouwan Lei
University of Bradford, UK

ABSTRACT

With high optimism, the third generation mobile communication technologies were launched and adopted by telecommunication giants in different parts of the globe—Hutchison 3G in the UK, Verizon in the USA and NTT DoCoMo in Japan. However, with an uncertain and turbulent social, economic and political environment, and the downturn in the global economy, difficult conditions are pronounced for the initial promises of m-commerce technologies to be fully realized. The causes for this, determined so far, have been largely of a technical nature. In this chapter, we shift the focus of analysis from a pure technical approach to a socio-cultural one. The basic premise of the chapter is that cultural variations do play a very important part in shaping potential consumers' choice, belief and attitude about m-commerce services. We believe that to be an important way for the m-commerce industry to fulfill its potential.

INTRODUCTION

This chapter discusses the impact of socio-cultural aspects on mobile commerce (m-commerce). While m-commerce heralds the next revolutionary phase in the advent of digital technology, still the digital industry can be considered in its infancy. This makes its specific categorization difficult. However, as Mahatanankoon, Wen, and Lim (2004) point out, the 1980s can be roughly classified as the age of PCs, the '90s as the "decade of the Internet, and…the first decade of the 21st century as the decade of mobile computing and mobile commerce." By the end of 2004, the number of mobile phone subscribers was expected to be 1.5 billion—about one-quarter of the world's population (Evans, 2004). The ITU (International Telecommunication Union) said the growth in mobile phone subscribers outpaced growth in the number of users of fixed lines (1.185 billion) today and is outstripping the rate of increase in Internet users. Emerging markets such as China, India, and Russia contribute to the growth. The current state of the digital age that we live in convinces us of the remarkable rate that wireless data communication (WDC)/mobile computing/mobile commerce services are penetrating the market with. It is predicted to be one of the main driving forces for the computing industry, as well as a substantial revenue-generating platform for businesses. Recent major findings by the research firm IDC (Mahatanankoon et al., 2004) predicted a growth in the mobile commerce revenues from US$500 million in 2002 to US$27 billion by 2005. Predictions by Forrester Research (Mahatanankoon et al., 2004) estimate an average of 2.2 wireless phones per U.S. household by 2007, with up to 2.3 million wired phone subscribers making a switch to wireless services. Worldwide, there were 94.9 million users of m-commerce in 2003; this is expected to grow to 1.67 billion in 2008, resulting in estimated global revenue of US$554.37 billion (Wireless Week, 2004).

UNDERSTANDING M-COMMERCE

Mobile-commerce can be defined as the commercial transactions conducted through a variety of mobile equipment over a wireless telecommunication network in a wireless environment (Barnes, 2002; Coursaris & Hassanein, 2002; Gunsaekaran & Ngai, 2003). Currently these wireless devices include two-way pagers/SMS (short message systems), WAP-

(wireless application protocol) equipped mobile phones, PDAs (personal digital assistants), Internet-enabled laptop computers with wireless access capacity, and consumer premise IEEE 802.11 (a/b) wireless network devices (Leung & Antypas, 2001). The range of applications that characterize m-commerce activities can be largely divided into:

- **Entertainment:** Includes online TV broadcasts, online mobile games, and downloaded music or ring tones.
- **Content Delivery:** Includes reporting, notification, consultation, and so forth.
- **Transactions:** Includes data entry, purchasing, promotions, and so forth (Balasubramanian, Peterson, & Jarvenpaa, 2002; Leung & Antypas, 2001).

Wireless cellular technology (third- and fourth-generation wireless cellular networks) areas have witnessed exciting innovations in recent years. 3G cellular networks offer broadband transmission with speeds up to 2Mbps, allowing for high-speed wireless access to the Internet, e-commerce transactions, and other information services from any location across the globe. Shim and Shim (2003) describe the not-so-far future of the industry as

...a true wireless broadband cellular system (4G), which can support a much higher bandwidth, global mobility, and tight network security; all at a lower cost. 4G systems should be able to offer a peak speed of more than 100 Mbits per second in stationary mode and an average of 20Mbits per second when in motion. The deployment of 4G technologies will allow the dream of a unified wireless Internet to become a reality.

On the other hand, Wi-Fi (wireless fidelity), wireless area local networks that allow users to surf the Internet while moving, are proliferating

Figure 1. M-commerce applications

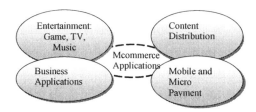

at astonishing speed on a global scale. Worldwide retail chains like Starbucks and McDonalds offer wireless Internet access to their customers. Wi-Fi offers a fast and stable connection; the data rate is several times faster than 3G. Wi-Fi is an important, new, and disruptive technology to mobile telephone technology, and it may be a watershed for all other m-commerce investment by telecom and content providers in the world of the mobile Internet (Lamont, 2001). In making use of this technology, a mobile phone manufacturer (Nokia) and wireless network manufacturer (Cisco) have been working together closely to produce the Wi-Fi phone. A U.S. mobile telecommunication operator has integrated a network of Wi-Fi hotspots with the existing mobile network systems. In such a way, the tariff of accessing the mobile Internet will be reduced to a budget price. More handheld device users will surf the Internet through their compact mobile phones or PDAs when on the move. Furthermore, WiMax (Worldwide Interoperability for Microwave Access), a low-cost wireless broadband connection in wide area network (WAN), will be rolled out (Cherry, 2004). As the wireless cellular and wireless technologies are converging, the tariff of mobile Internet will be affordable in the future. This rapid development of m-commerce technologies, which was earlier considered to be a mere extension of e-commerce activities, has opened up hitherto unseen business opportunities. It has increased an organization's ability to reach its customers regardless of location and

distance, and has also been successful to a certain extent in creating a consumer demand for more advanced mobile devices with interactive features. While e-commerce is characterized by e-marketplaces, an explosion in m-commerce applications has presented the business world with a fresh set of strategy based on personalized and location-based services. Many new business models have been established around the use of mobile devices which typically have the characteristics of portability, low cost, more personalization, GPS (global positioning system), voice, and so on. These new business models include micro-payment and mobile payment, business services, entertainment, and content distribution services (see Figure 1). Because of their existing customer base, technical expertise, and familiarity with billing, mobile telephone operators are the natural candidates for the provision of mobile and micro-payment services, the latter involving small purchases such as vending and other items.

The mobile phone has become a new personal entertainment medium. A wide range of entertainment services are available on it. These include playing online games, downloading ring tones, watching football video clips, broadcasting live TV, downloading music, and so on. Unsurprisingly, adult mobile services and mobile gambling services are among the fast growing services. According to Juniper research, the total revenue from adult mobile services and mobile gambling services could be worth US$1 billion and US$15 billion respectively by 2008 (Kowk, 2004). Law regulators have to stay ahead of the fast-growing development. Content distribution is concerned with providing real-time information, notification, and positioning systems for personalized information by location and advertising. Real-time information such as news, traffic reports, stock prices, and even weather forecasts can be distributed to mobile phones via the Internet. The information can also be personalized to a user's interests thus achieving a greater degree of personalization and localized services. For example the user's profile such as past behaviour, situation, and location can all be used to determine the relevant information to be supplied at any given time. This in turn can lead to such services being effectively provided through a mobile portal. The mobile network operators (MNOs) can prove to be the central player in providing such services, as they have a number of advantages over other portal players (Tsalgatidou & Veijalainen, 2000). Firstly, they have an existing customer relationship that will lead them, with sufficient ease, to identify the location of the subscriber. Secondly, they already have an existing billing relationship with the customers, while the traditional portals do not. These MNOs can also act as a trusted third party and hence play a dominant role in m-commerce applications. Business applications can be (and to some extent, have been) greatly influenced by these m-commerce applications, especially for companies with remote staff. Extending the existing enterprise resource planning (ERP) systems with mobile functionality will provide remote staff such as sales personnel with real-time corporate and management data. Time and location constraints are reduced and the ability of mobile employees is enhanced. The paperless office becomes a reality. M-commerce offers tremendous potential for businesses to respond quickly in supply chains.

CHARACTERISTICS OF M-COMMERCE

With increasing Internet users across the globe, the m-commerce industry definitely holds great promises for the future. However, to realize the complete potential of this industry, it is ex-

tremely important to understand the unique set of features that guide its development, apart from the general advantages of using e-commerce—that is, efficiency, convenience, broader selections, competitive pricing, rich information, diversity, and so on (Wu & Wang, 2004). While all of these apply to m-commerce applications as well, there is the additional requirement of such services blending into the *world of mobility* by focusing on the moments users will spend on the wireless Internet. Distinctive characteristics of the mobile environment make m-entertainment qualitatively different from fixed-line online entertainment (Baldi & Thaung, 2002). These characteristics are as follows:

- **Accessibility:** The services are available irrespective of location and time. As Mizukoshi, Okino, and Tardy (2001) observe, users of m-commerce services have unique usage patterns and niche access timings in contrast to PC-based Internet users.
- **Ubiquity:** It is possible for the users to enjoy m-commerce services whenever they might feel the need of it. The ubiquity of these services also increases the anonymity of the user in comparison to the wired Internet. Uses of services like gambling and adult entertainment, which are to a large extent socially unexpected, also get promoted as a result of this ubiquitousness.
- **Localization:** The entertainment services can be customized to the user's location, thus allowing for better targeted information and transaction-based entertainment services (e.g., mobile coupons).
- **Reachability:** With the user's permission, it is possible to reach him/her anywhere and anytime, thus allowing for immediate interactions in communication ap-

plications. From a service provider's point of view, this enables the transmission of time-critical information (stock prices, game results, etc.) to the interested clients.

- **Personalization:** As wireless devices are a very personal item, it is possible to customize the interface as per its user's preferences, thus creating an individual relationship with the customer, which in turn will encourage transactions and will also prevent the user from switching between service providers (Granger & Huggins, 2000).

PROMISE VS. REALITY

In spite of such promises and expectations, the situation in which the m-commerce industry finds itself today does not present much scope for such high optimism. As Jarvenpaa, Lang, Takeda, and Tuunainen (2003) write:

The advent of m-commerce has fuelled much anticipation of future possibilities. However, predictions that Internet technologies and wireless communication would greatly benefit both firms and individuals have now come under increasing scrutiny. Uncertain technology standards, the complexities of interactive multimedia applications, and the threat of governmental regulation have all contributed to a deflated vision of mobile commerce.

With tremendous expectations of 3G services, mobile phone operators spent billions of dollars in obtaining 3G licenses, marking their first successful launch in Japan in October 2001. But since then, the global economy has taken an unexpected turn towards the worse. In addition, the interest in 3G being pioneered by the mobile phone operators has confronted unforeseen low ebb in customer response. In effect, the much anticipated global launch of 3G services has been postponed. There are also inherent criticisms of the 3G mechanisms, especially on the security front.

As far as government policies go, Japan, Korea, a number of European nations, and the United States were moving efficiently towards establishing an m-commerce market. But so far, Japan and Korea are the only markets where the wireless Internet has exceeded expectations, while the European markets did not live up to the hype (Baldi & Thaung, 2002). The U.S. has also found itself not so successful in the race to establish a "functional and interoperable infrastructure for m-commerce" (Shim & Shim, 2003).

The portrayed picture in the early days of m-commerce also finds itself in stark contrast with the reality today, amidst a time of turbulent and uncertain political environment, especially in the post-9/11 world. The financial industry's dependence on telecommunications is well known, and 9/11 provided a clear demonstration of how disruptions to the nation's critical infrastructure can and will close markets and disrupt payment flow (Ferguson, 2002). Hence the adoption of m-commerce applications has definitely suffered a setback after a series of disastrous and malicious events, epitomising themselves in the 9/11 and Enron incidents. The telecommunication sector is to be among the worst-affected ones in terms of job loss and stock prices. Many telecommunication companies—Global Crossing, WorldCom, AT&T, and Level 3 Communications—are already gone or are struggling through bankruptcy (Borland, 2002). The downturn in the global economy has indeed spelt a difficult situation for further developments in the digital world to be widely and freely accepted.

However, although cautiously, the world economy is definitely gaining momentum again; concurrently, mobile technologies have advanced in both hardware and software, resulting in affordable handsets with reasonable cost of 3G mobile services. Furthermore, there is growing interest in 3G from the existing mobile phone users, especially in Asia. While in Korea, the number of 3G users has surpassed 1.7 million; in Japan it grew from 150,000 in January 2003 to 2 million in January 2004 (3G Newsroom.com, 2003; NTT DoCoMo, 2004). In Europe however, the growth is slow. There are only 361,000 3G users in the UK and 453,000 in Italy, despite a huge price lowering campaign (Hutchison Whampoa, 2004). The growth rate in other countries is not promising either.

A CRITICAL DIMENSION OF M-COMMERCE: CULTURAL VARIATIONS

Given this varied distribution of interest towards m-commerce applications, it is perhaps justifiable to shift the focus of analysis of the situation from an application-based rigorous technical discussion to a more soft but relevant socio-cultural one. Several researchers have pointed out that the perceptions and usage patterns of m-commerce differ from one customer segment to the another in terms of age, gender, professional orientation, economic standing, and most importantly, their cultural backgrounds. Consumers from different cultural backgrounds considerably vary in their perceptions, outlooks, and beliefs (Hofstede, 1980), which undoubtedly have a great influence on their choices of mobile technologies/ applications. Hence *culture* becomes a highly relevant and critical dimension behind the successful realisation of the initial predictions regarding the m-commerce industry. As Shim

and Shim (2003) write, we need to understand, appreciate, and "leverage subtle but important cultural differences exhibited by individuals."

Such cultural differences exist across societies, geographical locations, age groups, and so forth. It is beyond the scope of the current study to present a complete description and analysis of all the factors that such variations originate from, which is a large and fascinating field of study in itself. However, to show how such variations influence the diffusion of IT-based communication in general (Straub, 1994) and acceptance of mobile entertainment applications in particular, it will be suitable to draw the reader's attention to a set of factors identified by Baldi and Thaung (2002) which they see as the determining factors behind the relative success of m-commerce in Japan and other eastern societies as compared to the western societies.

- **Attitude towards Time (Killing Time vs. Saving Time):** People from western societies, especially aged ones, would rather prefer to wait for mature services. As Simon, Brilliant, and Macmillan (2000) note, in a society where saving time is a highly relevant part of daily life, the users must first be shown the value, how mobile entertainment applications "kill time".
- **Commuting Habits (Public Transport vs. Car):** While the public transport system is a more common way for daily commuters in the east, the phones represent an amusing way to spend the time on the way to work or school. On the other hand Europeans/Americans use less public transportation, but use their cars instead. And it is not possible to use interactive entertainment services while driving.
- **Role of the Mobile Phone (Life Style Component vs. Tool):** The mobile phone is fast becoming an important part of the

eastern society, making a fashion statement (using different handsets for different purposes), adding a lifestyle component, and almost becoming a part of the owner's identity. As a result, using a phones for different entertainment purposes is a common practice in the eastern societies. But the scene is not similar in the western world, where people perceive a mobile phone not a fashion statement, but merely a communication tool.

The basic premise of the world of mobile technology is that it would present its users with their own individual space so that they can enjoy a large degree of freedom, but at the same time increase their connectedness with both the physical and cyber world. But as Jarvenpaa et al. (2003) point out, this increased sense of freedom and connectedness at the same time might be a potential harbour for social disorder:

Excessive mobile use encourages superficiality, indifferent behaviour toward one's surrounding, the privatisation of lifestyle, and increased opportunity of control of others' lives. Similarly, independence can lead to abuses of freedom, compulsive and self-destructive behaviour, isolation and depression. (Jarvenpaa et al., 2003)

What degree this conflict between interconnectedness and individual freedom might reach once again depends on the particular user's social surrounding, cultural background, and even historical context. While certain cultures naturally emphasise freedom, individual needs, values, and goals, others hold social connectivity as central to their lives. However, although a study of a group's cultural dimensions over individual choices (and vice versa) opens fantastic possibilities, we will have to restrict ourselves from steering our topic to such a direction, as that lies outside the purview of the topic under discussion. We would conclude this section with recognising individual cultural elements as one of the most (if not the most) important factors influencing people's attitude towards mobile applications/entertainment, and this recognition will be the main basis of the course that our argument will take in the next few sections.

CONSUMER GROUPS AND MARKET STRUCTURES

The variety of consumer groups is another important element that influences the diffusion of m-commerce applications in the world. Although highly related to the cultural issues as discussed in the previous section, the different consumer groups deserve a special mention. The importance behind recognising different groups of consumers lies in the fact that as far as mobile entertainment is concerned, each such group has its own unique set of preferences. These preferences differ according to the varying social environments that each consumer group largely finds itself situated in. Baldi and Thaung (2002) distinguishes the following three major customer groups:

- **Kids and Teenagers:** Are the most prolific users of both PC-based wired Internet as well as the wireless environment. Many kids across different societies are regular and efficient users of mobile devices, and are familiar with services such as games and ring tone downloads. These kids probably will also be the first to link their consoles to wireless phones to play games with their peers, given the amount of free time they enjoy and the importance of socializing in a kid's/teenager's life (Hall,

2001). Hence, undoubtedly they will be the most regular, dedicated, and efficient users of the different services mobile technologies can offer.

- **Young Adults:** Especially high school and university students, who have minimal expenses but strong demand for brands, are highly likely to use their mobile phones for daily entertainment services. This group also includes employed people in their mid- to late twenties, who tend to spend a considerable amount on their lifestyle. Familiarity with the digital technology wave, perceptive to new trends, and high rate of education make them a highly likely group to increase the demand for wireless data usage.

- **Business Users:** Although they are unlikely to have a considerable amount of time to gain familiarity with the latest technologies and entertainment services, will use their phones to a large extent for infotainment and communication. Real-time news applications—such as stock market information, share prices, political incidents, weather forecasts, and so forth—will be frequently used and popular services among this group.

The overall market structure is another important determining factor of the mobile industry. Once again, the western and eastern worlds show a number of differences in the way their markets operate. Baldi and Thaung (2002) have also enumerated these as follows:

- **Originator for Phone Specifications (Carrier vs. Manufacturer):** Compatibility amongst different handsets is a problem in European and American markets, as each manufacturer specifies the phone attributes themselves, and there is no central guideline. But the scenario is exactly the opposite in the eastern world, where central regulations exist regarding handset features. For example, in Japan, NTT DoCoMo gives and monitors handset specifications to each of the four major handset manufacturers of the country.

- **Wired Internet Experience (Low vs. High):** The acceptance of mobile technologies as an alternative business and entertainment paradigm in itself depends on how experienced the public is with wired Internet services. In Japan, due to high costs for wired telephone and Internet connections, wireless Internet has become the primary means of communication. But in other parts of the world, such as Western Europe, the wireless Internet has been marketed as only an alternative means of accessing the Web, not as the next natural progression of the cyber world.

- **Billing Scheme (Centralized vs. Decentralized):** In Japan, there exists a central iMode site (the Japanese equivalent of European WAP services) that provides a centralized billing facility, where the users are charged for value-added services via their phone bills. The WAP users, on the other hand, have to pay beforehand with their credit cards for services they want to use, as no such centralized billing system is available in Europe.

- **Pricing (Moderate vs. Medium):** Pricing differences between wired Internet and wireless services also have led to the successful acceptance of mobile phones as the primary means of communication and entertainment in the eastern world. Fixed-line connections are priced at a much higher rate than wireless connections there. The situation is just the opposite in Europe where the mobile industry is yet to be established as the primary mode of communication due to the cheap landline connections.

Figure 2. Factors determining choice of mobile services

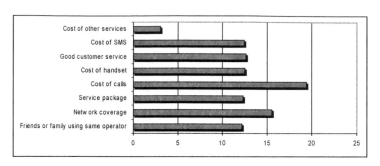

STUDYING M-COMMERCE USAGE PATTERNS

The importance of cultural factors affecting and/or determining patterns of m-commerce usage has been emphasised in the earlier sections. Keeping these in mind, the authors conducted a research study in order to investigate the perception and acceptability rates of different mobile services among current/potential users. The study was based primarily in Europe, with about 200 subjects from different communities, age groups, professions, and so forth. However, the teenagers/young adult age group (19-30 years) was chosen as being the major focus group, since this generation is known to have the most familiarity with and access to the latest digital technologies. Some of the research findings are briefly presented below. Although it is hardly a complete discussion of the different dimensions that affect the m-commerce market in Europe, the issues discussed do provide insight into the users' perception of the industry and the services provided.

Factors Determining Choice of Mobile Services

The survey revealed that while the telecommunication giants in the UK are heavily campaigning and gearing up for a 3G mobile services launch on a global scale, the users still do not see the service package offered as the one major criteria that might influence their choice of operators. While the cost of calls still remains the most important factor behind choosing an operator, other factors such as network coverage are also considered to be more vital than the range of services offered. Figure 2 shows the users' preferences.

Customer Saturation

Another factor that has often been described as a major concern of the m-commerce industry not having taken off as promised is the aspect of *customer saturation*. While mobile companies launch new service packages and upgrades on a frequent basis, a very large segment of the customer population just does not feel inclined to possess the latest in the market. The reasons for this are twofold.

Firstly, mobile phone usage has a very high penetration rate in European markets. A substantial percentage of the customer base already possess a mobile phone of fairly high capacity, and they are satisfied with the same. This prevents a constant lookout for latest upgrades.

The second reason, which is more fundamental, is that unlike the Japanese society, a mobile phone, although highly essential, is not perceived as a fashion statement in Europe.

Figure 3. Customer demand for latest services offered/upgrades

(**Question:** Are you on the constant look out for better service package/upgrade opportunities?)

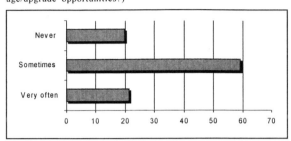

The possession of the latest in the market is therefore rendered unnecessary.

Only about 21% of the users in the research answered in favour of a constant lookout for latest upgrades. Given that the focus of this research was the *techno savvy* younger generation, this is an alarmingly low rate. The reason for this can be traced back to the how current/potential users *perceive* the mobile industry. Albeit the latest, the services offered are still considered to be *additional/extra.* Unless this changes to *necessary,* it is difficult for the mobile industry to realise its full potential. Figure 3 illustrates the related findings from the research.

Education as a Determining Factor

Another interesting fact that emerged out of the research was the co-relation between a person's experiences of mobile technologies with his/her education level. More than 90% of the users who use their mobile phones on a regular basis (not just for making calls, but also availing other services offered) are educated to at least "A" levels or equivalent. The importance of this also lies in the fact that it might be a useful strategy for generating personalized contents. An individual's educational background is likely to have a high impact on his/her choices, tastes, and preferences. Thus it might be possible to customize the content in order to reflect the choices of people with different educational backgrounds. It can also be inferred that an individual's educational background would affect his/her social status as well, which in turn is an important determinant of personalized, customized contents. Figure 4 illustrates this co-relation, as revealed in the research.

A Comparative Evaluation of the Different Services Provided

The range of different services that m-commerce offers also presented an interesting theme for the study. These services can be broadly divided into the following categories:

- Information Services
- Entertainment Services
- Financial Transactions
- Location-Based Services

Figure 4. Education levels of m-commerce users

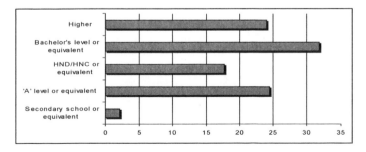

Table 1. Ranking m-commerce services (1—most important; 4—least important)

Rank Value	1		2		3		4	
	Response Frequency	Percentage	Response Frequency	Percentage	Response Frequency	Percentage	Response Frequency	Percentage
Information services	134	69.79	33	17.46	13	6.95	12	6.35
Entertainment service	22	11.46	53	28.04	50	26.74	64	33.86
Financial Transactions	15	7.81	43	22.75	67	35.83	63	33.33
Location based services	21	10.94	60	31.75	57	30.48	50	26.46

While this is an important input to design content strategy formulation, it might also be an important issue to consider for developing marketing strategy.

The research findings show that in spite of all the marketing hype about the range of services that m-commerce promises to provide, almost 70% of people still ranked traditional information services as most important (see Table 1). Both entertainment services and financial transactions have been ranked as the least important by one-third of the respondents.

Cost of Services

In spite of the advanced technologics, user-friendly nature, and exciting applications provided, costs still remain the primary criteria behind the users subscribing to such services. While some services such as ring tone/music download have gained quite some popularity among the public due to relatively cheap rates,

other large applications are priced at a comparatively higher rate. Moreover, in case of some applications or service providers, certain hidden costs are also involved. This includes paying an additional per-use fee, registration charges, and so forth. The cheaper alternative of the wired PC-based Internet thus gains preference over the wireless Internet. So it was not surprising when more than 70% of the subjects interviewed said that they are likely to consider the services offered if the costs involved are less (see Figure 5).

In relation to the above, it was also interesting to note the subjects' estimated monthly budget for their mobile phone services. While on one hand, the young adults have been determined as the most prolific users of m-commerce, it was also found that they have a not-so-high budget, thus preventing them from subscribing to the services available. More than 50% of the people interviewed had a monthly budget of £15 or less, while another 20% had a

Figure 5. Determining the likelihood of users subscribing to cheaper services

Figure 6. Estimated monthly budget for your mobile phone service (including both voice telephony and data service) (1GBP=1.89USD on March 2005)

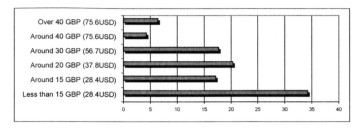

budget of £20. Thus marketing different s-commerce services to the young-adults user group, in spite of them being the most technology enthusiastic generation, is placed under certain constraints. The relevant research findings are shown in Figure 6.

CONCLUSION AND FUTURE DIRECTIONS

As the above statistics and discussions show, the market for m-commerce in Europe is yet to fulfil its initial promises. The picture in other parts of the world is not perfect either. The United States, as discussed earlier, has not been too successful in its effort to establish a successful m-commerce infrastructure and industry. The only country to have been able to meet the expectations is perhaps Japan, where telecommunications market works in close contact with their equipment makers, as well as the government—that is, the communications market in Japan, as well as in Hong Kong and Korea, is highly regulated by the government.

It is difficult to identify a unique set of reasons acting as an impediment to the successful realisation of the promises made in the early days of m-commerce. While the technology is definitely in place and ever improving, the focus has rightly shifted to a socio-cultural analysis of the situation over the past few years.

In this chapter, an effort has been made to present a complete overview of the m-commerce industry. While the initial sections have outlined the technological infrastructure and the applications/services offered, the later ones have discussed the market structure and its determinants. However, the most important aspect of this exercise has been the attempt to shift the focus of analysis from a technology-driven perspective to a socio-cultural one. However, even in a socio-cultural mode, it is important to make a conscious effort to refrain from a reductionism way of analysis. It must be remembered that the objective is not to decipher a certain number of explanations, but rather endeavouring to draw the reader's attention to a large plethora of possible social-cultural, political, and economic dimensions that might have contributed towards the not-so-successful m-commerce industry in Europe.

It is also very difficult, if not impossible, to determine the future of the industry. While it is almost certain that the industry did go through a period of over-hype in its early days, there is also no denying of the fact that the opportunities presented to the human civilisation in this m-commerce age were hitherto unseen, and they indeed mark the beginning of a new chapter in the world of digital technologies. As Jarvenpaa et al. (2003) write: "M-commerce lets the Web come to the user, at any time, any location."

However, the m-commerce industry still lacks the vital quality of being *necessary* rather than its present status of just an *additional* set of services. The main challenge that the industry faces, which in turn also determines future research directions in this area, lies not in improving and refining the technology, but rather in changing the users' perception of the services provided as the services being something that will positively affect their lifestyle and surroundings. Such a change in perception, which will take m-commerce growth to a new level, can only be brought by an appreciation of the various socio-cultural dynamics that influence the consumers' choices, beliefs, and attitudes.

REFERENCES

3G Newsroom.com. (2003, October 9). *South Korea claims success with 3G*. Retrieved March 30, 2005, from http://www.3gnewsroom.com/3g-news/oct_03/news_3831.shtml

Balasubramanian, S., Peterson, R. A., & Jarvenpaa, S. L. (2002). Exploring the implications of m-commerce for markets and marketing. *Journal of the Academy of Marketing Science, 30*(4), 348-361.

Baldi, S., & Thaung, H. P. (2002). The entertaining way to m-commerce: Japan's approach to the mobile Internet—a model for Europe? *Electronics Market, 12*(1), 6-13.

Barnes, S. J. (2002). The mobile commerce value chain: Analysis and future developments. *International Journal of Information Management, 22*(2), 91-108.

Borland, J. (2002). *WorldCom piles on telecom collapse*. CNET News.com. Retrieved March 30, 2005, from http://news.zdnet.com/2100-1009_22-939488.html

Cherry, S. M. (2004, March). WiMax and Wi-Fi: Separate and unequal. *IEEE Spectrum, 43*(3), 16.

Coursaris, C., & Hassanein, K. (2002). Understanding m-commerce. A consumer-centric model. *Quarterly Journal of Electronic Commerce, 3*(3), 247-271.

Evans, R. (2004, December 9). *Mobile phone users double since 2000*. Retrieved March 30, 2005, from http://www.computerworld.com/mobiletopics/mobile/story/0,10801,98142,00.html

Ferguson, R. W. Jr. (2002, May 9). Implications of 9/11 for the financial services sector. *Proceedings of the Conference on Bank Structure and Competition,* Chicago, IL (pp. 46-52).

Granger, V., & Huggins, K. (2000, June). *Wireless Internet—more than voice: The opportunity and the issues*. Report, Merrill Lynch Global Securities Research and Economics Group, USA.

Gunasaekaran, A., & Ngai, E. (2003). Special issue on mobile commerce: Strategies, technologies and applications. *Decision Support Systems, 35*(1), 187-188.

Hall, J. (2001, February 17). Chatty teens seen as growth market for wireless. *Reuters Technology News.*

Hofstede, G. (1980). *Culture's consequences*. London: Sage.

Hutchison Whampoa. (2004). *Audited results for the year ended 31 December 2003*. Retrieved March 30, 2005, from http://www.hutchison-whampoa.com

Jarvenpaa, S. L., Lang, K. R., Takeda, Y., Tuunainen, K. V. (2003). Mobile commerce at crossroads. *Communications of the ACM, 46*(12), 41-44.

Kwok, B. (2004, January 3). Watershed year for mobile phones. Companies and finance in South China. *Morning Post* (Hong Kong).

Lamont, D. (2001). *Conquering the wireless world: The age of m-commerce.* Oxford: Capstone.

Leung, K., & Antypas, J. (2001). Improving returns on m-commerce investment. *Journal of Business Strategy, 22*(5), 12-14.

Mahatanankoon, P., Wen, J., & Lim, B. (2004). Consumer-based m-commerce: Exploring consumer perception of mobile applications. *Computer Standards and Interfaces.* Retrieved from www.sciencedirect.com

Mizukoshi, Y., Okino, K., & Tardy, O. (2001). Lessons from Japan. *Telephony, 240*(3), 92-95.

NTT DoCoMo. (2004). *Results for the third quarter of the fiscal year ending March 31, 2004.* Retrieved March 30, 2005, from http://www.nttdocomo.co.jp

Shim, J. P., & Shim, J. M. (2003, September/October). M-commerce around the world: Mobile services and applications in Japan, Korea, Hong Kong, Finland, and the U.S. *Decision Line, 34*(5), 9-13.

Simon, S., Brilliant, P., Macmillan, R., et al. (2000, September). *Wireless Internet report: Boxing clever.* Morgan Stanley Dean Witter.

Straub, D. (1994). The effects of culture on IT diffusion: E-mail and fax in Japan and the U.S. *Information Systems Research, 5*(1), 23-47.

Tsalgatidou, A., & Veijalainen, J. (2000, September). Mobile electronic commerce: Emerging issues. *Proceedings of EC-WEB 2000, the 1ˢᵗ International Conference on E-Commerce and Web Technologies,* London, Greenwich, UK (pp. 477-486).

Wireless Week. (2004). *Buying numbers,* p. 30.

Wu, J., & Wang, S. (2004). What drives mobile commerce? An empirical evaluation of the revised technology model. *Information and Management.* Retrieved from www.sciencedirect.com

Chapter L

The Mobile Network as a New Medium for Marketing Communications:
A Case Study

Heikki Karjaluoto
University of Oulu, Finland

Matti Leppäniemi
University of Oulu, Finland

Jari Salo
University of Oulu, Finland

Jaakko Sinisalo
University of Oulu, Finland

Feng Li
University of Newcastle upon Tyne, UK

ABSTRACT

This chapter discusses the mobile network as a new medium for marketing communications. It illustrates that the mobile medium, defined as two-way communications via mobile handsets, can be utilized in a company's promotion mix by initiating and maintaining relationships. First, by using the mobile medium companies can attract new customers by organizing SMS (short message service) -based competitions and lotteries. Second, the mobile medium can be used as a relationship building tool as companies can send information and discount coupons to existing customers' mobile devices or collect marketing research data. The authors explore these scenarios by presenting and analyzing a mobile marketing case from Finland. The chapter concludes by pondering different future avenues for the mobile medium in promotion mix.

INTRODUCTION

In the repercussion of the mobile hype around wireless access protocol (WAP), followed by the launch of third-generation (3G) networks/ Universal Mobile Telecommunications System (UMTS), the debate over the role of the mobile medium in promoting goods and services has emerged as a topic of considerable magnitude that echoes across different academic disciplines. The burst of the telecommunications bubble in 2000 eventually led telecommunications companies and information technology firms to change their way of thinking, from a technology-driven viewpoint to a more user-oriented perspective. In Europe, only a few mobile services have prospered, while others like many WAP-based services have proved to be unpopular (e.g., Williams, 2003). In fact, only ring tone downloading, logo services, and Short Message Service (SMS) can to date be considered as successful mobile services. The reasons underlying the success of these services fundamentally lie with the strong market demand and easy-to-use technology. When thinking about future mobile services, the *Mobile Internet* is often seen as a messiah of the 3G. Third-generation mobile telephony protocols support higher data rates, measured in kbps (kilobits per second) or Mbps (megabits per second), intended for applications other than voice-centric (3GPP, 2005; Symbian Glossary, 2005). The underlying idea of the 3G/ UMTS networks is that mobile phones are always connected to the best available network ranging from 2G GSM networks to EDGE (General Packet Radio Service), HSCSD (High-Speed Circuit-Switched Data) to WLAN (Wireless Local Area Network), and 3G networks. However, many companies operating in the telecommunications field are facing the same challenge when thinking about the right mobile services to the right mobile users. Recently, a project led by Nokia and a couple of other Finnish companies announced that television will find its way on mobile phone screens. Consumer acceptance of mobile TV services as well as the underlying technology will be tested and developed with 500 users in Finland (Nokia, 2005).

Since the future of mobile services is still unpredictable, this chapter will not speculate on new mobile services that might take off in the next few years. Instead, we will focus on technologies and applications that are already here and in use, which allow us to examine the utilization of text messaging (SMS) in managing customer relationships in the business-to-consumer markets. In this chapter, we will present a mobile marketing case in a Finnish general store that integrated mobile media in its marketing communications mix as shown in Figure 1.

BACKGROUND TO THE RESEARCH PROJECT

This research is based on a project called PEAR, Personalized Mobile Advertising Services (www.pear.fi), which aims at developing a multi-channel mobile marketing service system for planning, implementing, and analyzing mobile marketing that utilizes value-added features such as personalization, user grouping, presence, profile, and location information. The service system will be tested and developed with end users in real-life settings. The results are expected to contribute to the invention of new customer-oriented service concepts and business models, which can open up potential new business opportunities in global markets. Mobile marketing is in this project defined as marketing communications sent to and received on smart phones, mobile phones, or personal digital assistants (PDAs).

The Campaign Logic

The basic idea of the advertising campaign was to redirect customers to the company's Web page and to get them to register on the company's electronic marketplace. The campaign was advertised in various media (print media, Web pages, and at the store). Advertisements contained instructions of how to participate in the lottery that offered a prize worth 200EUR for registered users. Users were requested to send a text message to a short number and receive a text message back from the company that contains a five-digit short code. The mobile marketing service system generated 100,000 different five-number digits so each participant received an individual code. With the use of this "lucky number"—the five-digit code—customers were able to register with the online shop and thus participate in the lottery.

INTEGRATING THE MOBILE MEDIUM INTO THE MARKETING COMMUNICATIONS MIX

Generally speaking, marketing communications refer to the promotion of both the organization and its offerings (Fill, 2002). The marketing communications mix, also called the promotional mix, comprises a set of tools that can be

Figure 1. Mobile marketing campaign integrated with other marketing channels

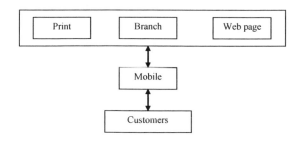

used in different combinations and in different degrees of intensity in order to communicate with a target audience. In recent years, the traditional way of thinking about how firms communicate with their customers has changed (e.g., Duncan & Moriarty, 1998; Kim, Han, & Schultz, 2004). With the help of new technologies, companies nowadays have a variety of digital channels allowing assorted ways to both send and receive information. Broadly speaking, we have been witnessing a change from mass communications to more direct and personal communications in which the messages are highly targeted and personalized. This has happened especially in digital communication channels (Kitchen & Schultz, 1999).

Mobile marketing (m-marketing) communications, defined as all forms of marketing, advertising, or sales promotion activities aimed at consumers (MMA, 2003), is one of the most modern digital channels in the promotion mix. Its role in advertising campaigns has not been studied widely, and relatively little is known of its role in the overall communications mix (Karjaluoto, Leppäniemi, & Salo, 2004). M-marketing can be either *push* based, which refers to communications such as SMS alerts sent to wireless devices requiring user permission, or *pull* based, which refers to information a user requests from a provider or advertiser (Barnes & Scornavacca, 2004; Carat Interactive, 2002). The mobile medium has to date mainly been used in promotions such as lotteries and various competitions (e.g., Pura, 2002). However, the market seems to be ready for more sophisticated two-way mobile marketing campaigns such as mobile customer relationship management (Finnish Direct Marketing Association, 2004), as customers are more and more using mobile data services such as text messaging and multimedia messaging in buying purposes and in providing feedback.

Figure 2. Special features of mobile media

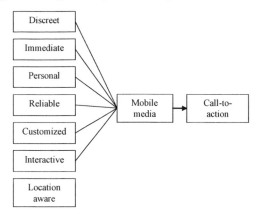

SPECIAL FEATURES OF THE MOBILE MARKETING MEDIUM

The mobile medium has some unique features that other direct marketing channels lack. In general terms, the mobile medium is favoured by marketers for its broad reach, low cost, and high retention rates (Clickatell, 2002). For mobile phones, several features are particularly relevant: the mobile phone is seen as an extremely personal, immediate, and interactive medium allowing marketers an effective way to reach customers in a fresh manner (Koranteng, 2001; Peters, 2002). As a marketing communication channel, the mobile (especially text) messaging is seen as immediate, automated, reliable, personal, discreet, and customized, allowing an efficient way to reach markets directly and providing mobile phone users a direct call-to-action, which would be almost impossible via other channels (Barnes & Scornavacca, 2004; Clickatell, 2004; Leppäniemi & Karjaluoto, 2005). These special features are illustrated in Figure 2.

As mobile phones are extremely personal in nature, advertising to mobile devices has to be very discreet in the sense that unwanted messages are easily perceived as spam. Messaging services (e.g., SMS and MMS) can be considered a very reliable way to distribute information, not only due to the fact that messages almost always arrive in time, but also because the majority of consumers usually read all messages they receive. In relation to mobile marketing campaigns, studies reported that in over 90% of cases, respondents read mobile advertising messages they receive (Enpocket, 2003). Moreover, the mobile media can also be regarded as an interactive media. As Enpocket's (2003) study indicated, of 5,000 consumers participating in SMS campaigns, 15% of them responded to mobile marketing messages. Finally, mobile devices have the ability to identify the location of users through the use of various technologies such as network-based positioning (or remote positioning), accurate local area positioning techniques, and satellite positioning (Kumar & Stokkeland, 2003; Zeimpekis, Giaglis, & Lekakos, 2003). Location-based advertising (LBA) is based on the idea that in a certain location, and additionally at a certain time, consumers receive advertisements based on their location (Salo & Tähtinen, 2005; Tsang, Ho, & Liang, 2004). However, as the rules of protecting consumers' privacy, including the use of location-based information for marketing purposes, are becoming stricter, the development and diffusion of location-based advertising have many obstacles to overcome.

Research has shown that the primary role of mobile marketing in a company's promotion mix has to date been promoting call-to-action (e.g., Paananen, 2003; Clickatell, 2004; Karjaluoto et al., 2004). In providing a direct call-to-action, location awareness and time open up the possibility to personalize messages in the manner that provide straight call-to-action. For instance, if a consumer arrives at a store, he or she might receive a personalized advertisement from that store based on his or her profile. It is important to note that to receive the benefits, for instance a discount to the store, consumers need to be in a certain location.

CUSTOMER RELATIONSHIP MANAGEMENT WITH MOBILE PHONES

Over the past decade it has become increasingly difficult to differentiate from competitors in serving the general product needs of customers. Therefore, companies have had to shift their focus to customer orientation and to search for novel ways to create value to customers. As a result, customer relationship management (CRM) is currently gaining widespread popularity in several disciplines and industries (e.g., Ryals, 2003; Zablah, Bellenger, & Johnston, 2004). On the one hand, the objective of CRM is to build and maintain customer relationships, and on the other to provide value for customers. Despite the potential of traditional CRM to provide value for customers, customers are expecting more and more individual attention. From the viewpoint of marketing communications, new digital marketing channels such as the Internet and mobile phones are considered to be powerful new media to reach consumers by allowing personalization and interactivity of both the content and the context of the message (Heinonen & Strandvik, 2003; Kim et al., 2004). Furthermore, as mobile marketing can combine the capacities of both direct marketing and ever-present nature and power of mobile digital technology, this form of communication is seen to provide synergy that will increase the potential of direct marketing (Mort & Drennan, 2002). Although electronic customer relationship management (e-CRM), defined broadly as CRM through the Internet, has received much attention among practitioners and academics (Bradshaw & Brash, 2001; Feinberg, Kadam, Hokama, & Kim, 2002; Fjermestad & Romano,

2003), the mobile medium as an element of CRM has gained far less attention.

Customer relationship management with mobile phones—in other words mobile customer relationship management (we use the term m-CRM)—can be defined as an ongoing process that provides seamless integration of every area of business that touches the customer, for the purpose of building and maintaining a profit-maximizing portfolio of customer relationships, by taking advantage of the mobile medium.

Because customer relationships evolve with distinct stages (Dwyer, Schurr, & Oh, 1987), companies should also interact with customers and manage relationships differently at each stage (Srivastava, Shervani, & Fahey, 1998). The CRM process outlines three key stages, namely the initiation, maintenance, and termination phases (Reinartz, Krafft, & Hoyer, 2004). In this study, it is implicitly assumed that m-CRM consists of these three stages as well. Because the main interest of this study is about how to redirect customers to the company's Web page and to get them register to the company's electronic marketplace, we focus primarily on the initiation stage.

USING MOBILE CRM IN INITIATING RELATIONSHIPS

Integrating the mobile medium as an element of CRM involves three key aspects: technology, implementation, and customers. This has been illustrated in Figure 3.

Figure 3. Dimensions of the initiation stage of m-CRM

The first key aspect to consider during the initiation stage of m-CRM is the technology. The company has to decide whether to rent the hosting mobile marketing server or to acquire it. Then gateways have to be built to connect with the mobile operators whose customers are allowed to be contacted.

The implementation consists of two decisions: first, through what marketing medium that the customers are acquired; and second, how much it will cost customers to use the messaging service—that is, sending the SMS message.

Given the personal nature of mobile phones, customers are often unwilling to use the mobile medium for marketing purposes due to fears of unsolicited marketing messages or even spam. Therefore, the third aspect of the initiation stage comprises understanding customers; to discover what are the key factors that lure the customer to use the mobile medium; and in this case, to redirect them to the company's electronic marketplace.

CASE STUDY

The central idea of the case study was to analyze the role of mobile media in generating traffic to company A's e-commerce Web page. Moreover, we wanted to investigate the company's (an ironmonger store) consumer responses to mobile marketing. The marketing campaign began at the end of November 2004 with a full-page advertisement in local newspaper. The newspaper advertisement contained instructions on how to participate in the SMS lottery. Consumers were asked to send a simple text message to a five-digit short number code. Immediately after sending the message, the mobile service system generated a four-digit code that the consumer received on his or her phone. This code was then entered on the

Figure 4. Marketing communications mix in company A's case

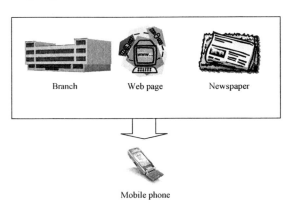

company's Web site. After entering the code to the Web site, consumers were guided to the registration page and asked to fill in the registration form. In case a registered person is participating in the campaign, they are guided to update their registration information. All registered consumers would be entered into a drawing for a digital video camera. The logic of the campaign is illustrated in Figure 4.

MAIN RESULTS

Altogether 232 consumers took part in the lottery during the period from November 24 to December 20, 2004. The newspaper advertisement on November 24, 2004, generated approximately 100 new registered customers in two days. After the newspaper advertisement, the company's Web site was, practically speaking, the only advertising channel for the lottery. The amount of new registrations per day varied from a few to around 10, with a final number of 232 registered consumers. Of these, only five were existing customers of the case company's online store. There were 47 female (20.3%) and 185 men (79.7%) in the sample. Approximately 33% of the participants were between ages 36 and 49, 22% were between 26 and 35, and 12% were between 50 and 64. The remaining categories had less than 11% of responses per category. With respect to careers, most respondents were blue-collar workers (34%), followed by white-collar workers (24%); 13% belonged to top or middle management and 11% to lower management, and students were 15%. Additionally, the registration form inquired about respondents' areas of interests. A total of 118 respondents reported interest in electronics and photography, followed by computers, leisure time, and motor vehicles. Of the total 21 interest areas, pets (26 responses), toys (29 responses), and watches and jewelry (30 responses) were considered least interesting.

CONSUMER WILLINGNESS TO RECEIVE DIRECT MARKETING

The registration form finally inquired about respondents' willingness to receive digital marketing communications from the company either via SMS or e-mail. While most respondents wanted to opt out of text messaging service (73%), the majority opted in to e-mail marketing (67%). This finding might relate to the fact that the case in question was relating to online commerce, and thus respondents regarded the e-mail channel as more suitable for them to receive marketing communication. In addition, approximately 24% of the respondents opted out of both marketing communication channels.

We next compared willingness to receive SMS and e-mail marketing with demographic variables with the use of a series of chi-square analyses. In respect of gender, 25% of males welcomed SMS marketing from the company, whereas the corresponding number for females was 34%. Concerning willingness to receive e-mail marketing, around 70% of males and 55% of females opted in to e-mail marketing. However, the difference between genders was neither statistically significant for SMS (chi-square=1.612, p=0.204) nor for e-mail (chi-square=3.510, p=0.061), although in e-mail it was relatively close to significant (p<0.05 level). With regard to respondents' age and willingness to receive digital marketing communications, a statistically significant difference was found in welcoming e-mail marketing (chi-square=15.725, p=0.008). This finding indicates that the younger age categories were not so willing to receive e-mail marketing compared to the older ones. Age category 26 to 35 most eagerly welcomed e-mail marketing (86% opted in), followed by age group 36 to 49 (70% opted in), and age group 50 to 64 (64% opted in). Approximately 50% of those under 18 opted in to e-mail marketing, as the corresponding number for age group 19 to 25 was 48%. No relationship was found between age and willingness to receive SMS communications (chi-square=4.752, p=0.451).

In terms of careers and willingness to receive marketing communications, no differences were found with the original classification of professions. However, after recoding the profession variable into a two-class variable, in which 1 was equal to blue-collar worker and 2 equal to white-collar worker, a statistically significant difference was found between profession and willingness to receive SMS communications (chi-square=5.520, p=0.019). It seems that blue-collar workers more eagerly welcomed SMS communications (33% opted in) than white-collar workers (15% opted in). No differences were found between profession and willingness to receive e-mail marketing (chi-square=0.236, p=0.627).

Finally, there is some evidence that respondents' place of residence has an impact on willingness to receive both SMS and e-mail marketing. The place of residence variable was divided into three categories, in which one was equal to "I live within ten-mile radius of the company's physical store," two was equal to "I live within a hundred-mile radius from the company's physical branch," and three was equal to "I live elsewhere in Finland." Interestingly, with respect to willingness to receive SMS marketing, most respondents living elsewhere in Finland did not want to receive SMS marketing (80% opted out). A total of 71% living near the store opted out of SMS marketing, and 65% of those living in a hundred-mile radius opted out. The difference between groups was close to statistical significance (chi-square=5.664, p=0.059). Moreover, those living the farthest away from the store were most willing to welcome e-mail marketing (75% opted in), followed by those living within a hundred-mile radius (60% opted in), and those living near the store (54% opted in). The chi-square value (chi-square=6.860, p=0.032) indicates a statistically significant relationship between place of residence and willingness to receive e-mail marketing. On this basis, it seems that the further a person lives, the more likely he or she is to opt in to e-mail marketing. However, in regard to SMS marketing, there is no linear relationship between location and willingness to receive SMS marketing.

To conclude, our results indicate that blue-collar workers are more willing to receive SMS marketing from the case company than white-collar workers. This fresh finding should be interpreted in light of the empirical case setting, which might indicate that blue-collar workers are not in general so familiar with e-mailing as white-collar workers, and thus they regard the SMS channel as more suitable for them in receiving marketing communications. However, validation of further work is needed. Moreover, with regard to welcoming e-mail marketing, it seems that age groups 26-35, 36-49, and 50-64 offer the most potential for e-mail marketing

campaigns, as more than 50% of them welcomed e-mail marketing. Furthermore, it seems that age itself has no influence on willingness to receive SMS marketing communications. Finally, our results showed that place of residence has an impact on willingness to receive e-mail marketing. The further a person lives from the physical store, the more likely he or she was to opt in to e-mail marketing communication.

CONCLUSION AND FUTURE DIRECTIONS

This chapter responds to the call for research on the use of mobile media in the marketing communications mix by investigating its specific features and role in integrated marketing communication mix and by describing the use of mobile media in amassing mobile marketing customer database. Our objective was to discuss the ways to integrate the mobile medium into the promotion mix of companies. By using a single case study from the retail sector, we showed that by combining the mobile medium and e-commerce store, it is possible to build a customer database in an efficient and cost-effective manner. Although our empirical case mainly contributes to the discussion of how to initiate customer relationships, it also gives some insights into the maintenance process of relationships by asking respondents about their digital channel preferences.

In light of the main results, several conclusions can be drawn. First of all, with relatively small promotional activity, the case company gained close to 250 new registered customers to their online store mainly by the use of a newspaper advertisement and online advertising on the company's own Web site. As most respondents opted in to e-mail marketing (67%), and most opted out of SMS marketing (73%), in the next stage of relationship building, the e-mail channel might be the right avenue to continue. However, some specific customer segments (blue-collar workers especially) welcomed SMS marketing. This finding should be interpreted as: those customers who opted in to SMS channel either do not use e-mail or for other reasons want marketing communications via SMS. By giving customers a choice, it is supposedly also contributing to the overall customer satisfaction and thus driving those registered potential customers into purchasing online customers as well.

This study has some limitations that present opportunities for further research. First, the study is among the first ones examining the use of a mobile device as a marketing communication channel, and thus the results obtained should be considered tentative. Second, we study only one retailer and its marketing communication, and despite the fact that its communication mix is in line with other companies operating in the same field nationwide, it would be valuable to scrutinize other retailers as well. In sum, we assume that these limitations do not endanger the reliability and validity of the findings, yet they do place bounds on the conclusions and implications that can be drawn from the study.

While mobile marketing is today almost entirely SMS based, the diffusion of MMS-enabled phones will presumably shape the industry in the future (Paananen, 2003; Barwise & Strong, 2002). Also the mushroom of devices with larger screens will guide mobile marketers to new avenues (Yonos, Gao, & Shim, 2003). These should be taken into consideration when planning future studies. Furthermore, a natural extension of our study would be the investigation of the role of mobile media in the marketing communications mix with other retailers. By doing so we might get valuable insights into how companies nationwide use, or plan to use, mobiles as a media in marketing communications.

REFERENCES

3GPP. (2005). *Terms & abbreviations.* Retrieved March 17, 2005, from http://www.3gpp.org

Barnes, S., & Scornavacca, E., (2004). Mobile marketing: the role of permission and acceptance. *International Journal of Mobile Communications, 2*(2), 128-139.

Barwise, P., & Strong, C. (2002). Permission-based mobile advertising. *Journal of Interactive Marketing, 16*(1), 14-24.

Bradshaw, D., & Brash, C. (2001). Managing customer relationships in the e-business world: How to personalize computer relationships for increased profitability. *International Journal of Retail & Distribution Management, 29*(12), 520-529.

Carat Interactive. (2002). The future of wireless marketing. *Research Report.* Retrieved from http://www.mmaglobal.com/resources/Wireless_WhitePaper.pdf

Clickatell. (2004). Business mobility guide. *Research Report.* Retrieved from http://www.clickatell.com

Duncan, T., & Moriarty, S. E. (1998). A communication-based marketing model for managing relationships. *Journal of Marketing, 62*(2), 1-13.

Dwyer, R., Schurr, P., & Oh, S. (1987). Developing buyer-seller relations. *Journal of Marketing, 51*(2), 11-28.

Enpocket. (2003, February). The response performance of SMS advertising. *Research Report, 3.* Retrieved from http://www.enpocket.com

Feinberg, R. A., Kadam, R., Hokama, L., & Kim, I. (2002). The state of electronic customer relationship management in retailing. *International Journal of Retail & Distribution Management, 30*(10), 470-481.

Fill, C. (2002). *Marketing communications— Context, strategies and applications.* Englewood Cliffs, NJ: Prentice-Hall, Harlow.

Fjermestad, J., & Romano, N.C. (2003). Electronic customer relationship management. Revisiting the general principles of usability and resistance—An integrative implementation framework. *Business Process Management Journal, 9*(5), 572-591.

Heinonen, K., & Strandvik, T. (2003, May 22-23). Consumer responsiveness to mobile marketing. Paper presented at the *Proceedings of the Stockholm Mobility Roundtable*, Stockholm, Sweden.

Karjaluoto, H., Leppäniemi, M., & Salo, J. (2004). The role of mobile marketing in companies' promotion mix. Empirical evidence from Finland. *Journal of International Business and Economics, 2*(1), 111-116.

Kim, I., Han, D., & Schultz, D.E. (2004). Understanding the diffusion of integrated marketing communications. *Journal of Advertising Research, 44*(1), 31-45.

Kitchen, P., & Schultz, D. (1999). A multi-country comparison of the drive for IMC. *Journal of Advertising Research, 39*(1), 21-38.

Koranteng, J. (2001). ZAP! There's no escaping the mobile ad. *Ad Age Global, 1*(5), 9.

Kumar, S., & Stokkeland, J. (2003). Evolution of GPS technology and its subsequent use in commercial markets. *International Journal of Mobile Communications, 1*(½), 180-193.

Leppäniemi, M., & Karjaluoto, H. (2005). Factors influencing consumer willingness to accept mobile advertising. A conceptual model. *International Journal of Mobile Communications, 3*(3), 197-213.

MMA (Mobile Marketing Association). (2003). *MMA code for responsible mobile marketing.* Retrieved December 14, 2003, from www.mmaglobal.co.uk/

Mort, G., & Drennan, J. (2002). Mobile digital technology: Emerging issues for marketing. *Journal of Database Marketing and Customer Strategy Management, 10*(1), 9-23.

Nokia. (2005). *Mobile TV pilot begins in Finland.* Retrieved March 17, 2005, from http://press.nokia.com/PR/200503/983665_5.html

Paananen, V. M. (2003, February 12). *European mobile marketing: Case Add2Phone.* Presentation at the NETS Seminar, Helsinki, Finland.

Peters, B. (2002, November 6). The future of wireless marketing. *Carat Interactive Study.* Retrieved from http://www.caratinteractive.com/resources/articles.html

Pura, M. (2002). Case study: The role of mobile advertising in building a brand. In B. E. Mennecke & T. J. Strader (Eds.), *Mobile commerce: Technology, theory and applications* (pp. 291-308). Hershey, PA: Idea Group Publishing.

Reinartz, W., Krafft, M., & Hoyer, W. (2004). The customer relationship management process: Its measurement and impact on performance. *Journal of Marketing Research, 41*(3), 293-305.

Ryals, L. (2003). Making customers pay: Measuring and managing customer risk and return. *Journal of Strategic Marketing, 11*(3), 165-175.

Salo, J., & Tähtinen, J. (2005). Retailer use of permission-based mobile advertising. In I.

Clarke, III, & T. B. Flaherty (Eds.), *Advances in electronic marketing* (pp. 139-155). Hershey, PA: Idea Group Publishing Group.

Srivastava, R., Shervani, T., & Fahey, L. (1998). Marketing based assets and shareholder value: A framework for analysis. *Journal of Marketing, 62*(1), 2-18.

Symbian Glossary. (2005). Retrieved March 17, 2005, from http://www.symbian.com/technology/glossary.html

Tsang, M. M., Ho, S.-C., & Liang, T.-P. (2004). Consumer attitudes toward mobile advertising: An empirical study. *International Journal of Electronic Commerce, 8*(3), 65-78.

Williams, C. (2003, March 17). Can GPRS get back on track? *Lucent Technologies: 3G Solutions for Operators, 12.* Retrieved from http://www.lucent.com/livelink/0900940 380033c60_Newsletter.pdf

Yonos, H. M., Gao, J. Z., & Shim, S. (2003). Wireless advertising's challenges and opportunities. *IEEE Computer, 36*(5), 30-37.

Zablah, A. R., Bellenger, D. N., & Johnston, W. J. (2004). An evaluation of divergent perspectives on customer relationship management: Towards a common understanding of an emerging phenomenon. *Industrial Marketing Management, 33*(6), 475-489.

Zeimpekis, V., Giaglis, G. M., & Lekakos, G. (2003). Towards a taxonomy of indoor and outdoor positioning techniques for mobile location-based applications. *ACM SIGecom Exchanges, 3*(4), 19-27.

Chapter LI
Overview and Understanding of Mobile Business in the Age of Communication

Joseph Barjis
Georgia Southern University, USA

ABSTRACT

This chapter provides an introduction, review and study of mobile businesses with emphasis on its supporting mobile technologies and wireless networking. The chapter first discusses the concept of mobile business where opportunities, motivations and needs for this type of business are studied. Following this discussion, the chapter studies the current status of mobiles business, key hardware and software solutions (business applications) available on the market. The chapter also discusses different mobile devices, communication infrastructure, supporting networks and other crucial components that make businesses mobile and able to be conducted anytime and anywhere. Finally, an extended discussion is focused on issues and future developments of mobile businesses along with some recommendations, and suggestions regarding mobile business.

INTRODUCTION

Mobile business is considered an offspring of the Advanced Communication Age and a driving force of the new economy, therefore its discussion is strongly aligned with the discussion of underlying information and communication technologies (ICTs).

The amazing pace of innovations in ICTs during recent years has opened a wide spectrum of new opportunities and challenges for the business industry. These opportunities demand a dramatic shift towards mobility in almost every aspect of life such as education, entertainment, health care, and business. Rapid developments in wireless communication technologies, mobile devices, high-speed transmission facilities, and broad bandwidth prepared and paved the way for transforming human activities towards mobility. The most noticeable impact of these evolving technologies can be seen in business, which is preparing for another revolutionary change. First, business has gone through transformation from traditional business to electronic business, and now it has been adapting towards mobile business, or m-business for short.

In the coming years, m-business will be a fashion for industry, researchers, enterprise managers, and society as a whole. This will be the business style of the Age of Communication, inspiring managers and enterprises for serious shift. Perhaps even, enterprises will experience another era of Business Process Reengineering, or maybe this time, m-Business Process Engineering (mBPR)!

Before diving deep into the topic, it is worthwhile to mention some example where mobility has been making its initial breakthrough. These different examples aim to provide an idea about the breadth, depth, and diversity of mobility and mobile business.

Education

Educational institutions have been pioneering implementation of the wireless networking environment, providing students with flexibility of accessing campus resources and downloading academic applications at their convenience and desired location (lab, classroom, library, cafeteria, campus garden, or while watching campus games). Not being tied to lab hours and classrooms, students are given more flexibility and opportunity to pursue their education, which in turn increases quality of education. So, campuses are going mobile within the campus area.

Health Care

Hospitals in general and modern medical practices in particular are adapting towards mobile health care delivery. Computer-based patient records, also referred to as electronic medical records, are part of a system that provides a mobile working environment for physicians, staff, and managers of medical practices. Each physician carries a handheld computer that access patients records, X-rays, and surgery videos, allowing sharing and discussing of images with specialists from other hospitals, coordinating remote operations, and so forth. If a physician wants immediate information about a particular medicine prior to issuing a prescription, his handheld computer allows him to access the relevant Internet page for such information. Productivity and quality of health care service is impressive with mobile facilities.

Sales and Marketing

Retail, wholesale, mass distribution centers are using a mobile business environment for goods delivery, shelf refill, inventory control, warehouse management, transport and logistics, and working with branches in different locations. Mobile devices help to track goods delivery and movement of products.

These three small examples illustrate different ranges of mobility within a building, within an enterprise, and within a town, used for a wide range of activities. These examples help analysts extract some important characteristics of m-business including range of functionality and types of mobile devices (wireless laptops, tablet PCs, smart phones, etc.). According to some authors, the application of m-business can be distinguished as "macro" applications in outdoor settings or "micro applications in indoor environments—for example, hospital, libraries, hypermarkets. Like the underlying wireless networks supporting it, m-business may be distinguished by its span as a local, regional, or global m-business.

With this brief introduction, this section is concluded and the rest of the chapter will discuss different aspects of m-business and its technological components, and elaborate on various facets of m-business.

CONCEPT OF M-BUSINESS

In general, the main concept of m-business is moving enterprises' critical business to the point of sale and service, or even closer to the point of consumers.

Like its predecessor, electronic business, the concept of mobile business has been used in many applications from communication to consumer transactions and corporate services (Vos & de Klein, 2002). However, the real potential of m-business is much broader than merely providing service, conducting sales, or delivering goods. A well-engineered, well-designed, and well-integrated m-business supports not only conducting business, but also adds collaboration, coordination, instant communication, and management features in the business. Being based on most advanced ICTs, m-business aims to be more productive than traditional business or business supported by networked computers, and productivity means in all aspects.

Evolving from mainframe and wired network eras, m-business is the leading edge of the new generation (3G or third generation) of wireless networking that aims to adapt the best business and management practices, standards, and styles.

Motivations for M-Business

On the business horizon, the main motivations that have been pushing business towards m-business are competition for flexibility in conducting business, extending functionality and service to the business point, convenience of employees and comfort of consumers, better

satisfaction, quality improvements, personalization, and localization of business. And obviously, revenue increase and market gain are at the very core of these driving forces.

On the technology horizon, the revolutionary progress in wireless networking, information technologies, and mobile devices making high-speed wireless communication everywhere has provided new infrastructure for business that resulted in more mobility.

Definition of M-Business

Mobile business is the business of the future which is based on a wireless infrastructure, using mobile devices, bringing critical business to the point of service and sale, aiming for more productivity in wide economic sense. Depending on a focal point and perspective, m-business can be defined in quite different ways.

Mobile business study and application attracted the attention of many outstanding authors, scholars, and researchers (e.g., Deitel, Deitel, & Steinbuhler, 2001; Deitel, Deitel, Nieto, & Steinbuhler, 2003; Vos & Klein, 2002; Paavilainen, 2002). In the abundance of available definitions used by different authors, it is quite challenging to find a unique definition, however the definition given by Kalakota and Robinson (2001) could be cited here as an example: m-business is "the application infrastructure required to maintain business relationships and sell information, services, and commodities by means of the mobile devices." This is just one of definitions that characterizes m-business. Different definitions are given from different perspectives. However, regardless of definition perspective, what is generic about m-business is that m-business encompasses three essential components: wireless networking technology (3G Networks, WLAN, WWAN), mobile devices, and improved business practices (procedures). The last component is a key component of m-business. If the first two are facilitators, the third one is the main objective of m-business.

M-BUSINESS TODAY

Although application of m-business started not too long ago, this opportunity very quickly attracted enterprise managers, industry leaders, and researchers and authors. Today a number of periodicals are adapted or founded on m-business, numerous monographs are published (e.g., Paavilainen, 2002; Vos & Klein, 2002; Kalakota & Kurchina, 2004; Sadeh, 2002), annual conferences such as ICMB (International Conference Mobile Business) are held, and a wealth of Internet-based resources reporting studies, results, examples, and models of m-business are available. In addition, tens of IT and business consulting companies refocused their activity from electronic commerce and electronic business towards m-business.

Studying the overwhelming opportunities and increasing demands in m-business, Kalakota (2005), in his work "Mobile Business: Vision to Value," emphasizes how rapidly the emerging technologies change the way enterprises conduct their business and how dramatically m-business is replacing traditional business. As a result of these changes, the author states, questions have shifted from "Should I do mobile business?" to "How can mobile create business value?"

Enterprises and businesses worldwide are implementing mobile business solutions to accelerate business cycles, increase productivity, reduce operating costs, and extend their enterprise infrastructure.

The need to go mobile turned into serious competition between leading companies that provide wireless infrastructure, application solutions, and mobile devices for m-business.

Today, the following market leaders are among the top providers of applications and wireless networking infrastructure for mobile businesses.

- In the *software market,* SAP as the world's largest inter-enterprise software company took a pioneering initiative in providing software packages for different types of m-business as listed below. In gaining a leading position in mobile business solutions, Microsoft is adding m-business features to Windows.
- In the *wireless market,* Cingular and Verizon are pioneering in providing modern wireless networking services by introducing and expanding 3G networks in major cities and metropolitans (Segan, 2005).
- In the *mobile devices market,* Siemens, Nokia, and other leading providers are introducing mobile technologies that significantly boost competitiveness of businesses.

The SAP mobile business solution set includes ready-made applications that provide access to the corporate information and processes—anytime, anywhere—allowing use of a variety of mobile devices. Among various software packages, SAP provides mobile business applications (SAP, 2005):

- **SAP Mobile Time and Travel:** This package gives mobile workers access to timesheets and travel management functionality.
- **SAP Mobile Sales:** This package provides a solution for salespeople who need to perform their tasks quickly and productively.
- **SAP Mobile Service:** This package enables field service engineers to react quickly to customer needs.

- **SAP Mobile Asset Management:** This package allows in-house service engineers to access relevant business processes anywhere, anytime.
- **SAP Mobile Procurement:** This package enables mobile workers to manage the entire procurement process, from price comparison to ordering.

INFRASTRUCURE OF M-BUSINESS

Think of m-business as a two-level framework (as shown in Figure 1), where the upper level is the business level and the bottom level is the IT infrastructure that supports business to carry out its mission and tasks, and sets business in motion.

Just in a narrow sense, let us say, from a high-level perspective, one can consider that the business level is a variable and the IT level is relatively a constant. Business level is variable because it represents any type of business, but the IT infrastructure, in general functionality, may be the same for almost any type of m-business, with some differences such as range, size, structure, configuration, potential, and complexity.

The IT infrastructure of m-business includes two major components—mobile devices and wireless networks.

Mobile Devices

One of the main characteristics of mobile devices used in e-business is the communication or networking facility and capacity to receive,

Figure 1. M-business IT infrastructure

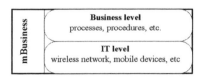

transmit, and process different types of data (text, audio, video) at a high rate. In addition to high-speed data exchange, ease of handling, portability, and size are important in these devices. Furthermore, these devices must be able to download essential business, office, and enterprise applications and have sufficient memory to run them.

Currently, devices used in m-business are wireless handheld computers, laptops, PDAs, tablet PCs, smart phones, Blackberry, and so on. These devices should be capable to run high-performance business, office, and enterprise applications such as multimedia, full-motion video; wireless teleconferencing; and use connection over wireless networks using Wi-Fi, GPRS, Bluetooth, or other advanced connections.

For a better idea and distinction between ordinary mobile devices and business-quality devices, and a better idea about business quality mobile devices, have a look at the features provided by the Nokia 9500, having a full set of critical business tools, full keyboard, with a large memory capacity and versatile network connections:

Browse the Internet in rich full color, on a wide, easy-to-read screen. Work with office documents—not just e-mail and memos, but presentations and databases too. Get them from your corporate network via Wireless LAN or EDGE for fast mobile access. Keep your Personal Information Management data in sync and up-to-date calendar and contacts—with PC Suite and SyncML, you can easily exchange data between your Nokia 9500 Communicator and a compatible PC. (Nokia, 2005)

Wireless Networks

Although not yet in mass application, in the future m-business will be operating on third-generation (3G) wireless networks that provide high-speed download and upload rate. The speed of transmission in these networks using mobile devices is at the level of DSL connections. Because data costs on 3G networks are lower than on traditional networks (Solheim, 2005), more and more enterprises will shift into using 3G networks as a main infrastructure.

Currently, m-businesses are based on different types of networking technology:

- **Personal Area Network:** Using Bluetooth technology enables short-range device-to-device wireless connections within a small space (office, a desktop, a personal space).
- **Wireless Local Area Networks:** Using Wi-Fi technology, based on IEEE 802.20, 802.11 standards, support a wireless connection to a network from inside a home or from a hotspot in a building, campus, or airport.
- **Wireless Metropolitan Area Networks:** Using WiMax, based on IEEE 802.16 standard, will enable any remote worker to make a wireless connection anywhere in a range up to 50 kilometers.
- **Wireless Wide Area Networks or 3G Networks:** Provide the highest available bandwidth for mobile devices. Although theoretical rate of transmission is 2Mb, for practical purposes the transmission speed is like DSL letting users download text, audio, video, Web contents, and e-mail while in motion.
- **4G Technologies:** Promise to integrate different modes of wireless communications, from indoor networks such as wireless LANs and Bluetooth, to cellular signals, to radio and TV broadcasting, to satellite communications.

ISSUES OF M-BUSINESS

In the previous sections we discussed advantages and benefits of m-business; however, enterprise managers and business owners ought to be aware of issues and challenges before undertaking the transformation initiative.

It would be not less than an illusion, if one considered the transition from traditional or e-business to m-business as just shifting from a wired environment to wireless communication and networking. Transformation into mobile business is not about moving from wired environment to wireless, from desktops to handheld computers, or from office to field. And the challenge is not only about acquisition and implementation of best wireless technology or awareness of emerging mobile technology; it is rather a multi-dimensional issue where technology is only one facet of it. Challenges that may require more profound study are transformation of the business and enterprise, shift in the mindset from thinking of m-business as a different way of doing business to considering m-business as an implementation of best business practices, improved procedures, higher quality, and so forth. The challenge of going mobile is much more complicated than application and implementation of mobile technology.

Although each category of business may have its own peculiarities, some of the common issues for enterprises to carefully study while embarking on competing for faster implementation of m-business are as follows

- From business's perspective
  - How well are m-business opportunities studied?
  - How much Business Process Reengineering, ERP system changes, and customer relationship management is required to go mobile?

- What is expected from m-business, direct profit, or quality of service? How can qualitative values be turned into quantitative values?
- How much patience is needed before harvesting the first fruits of benefit?
- Are the employees ready to go mobile, or will they resist against?
- Also important to remember: Mobile business should not be considered as an immediate way for profits.
- From technology perspective
  - What will happen with the existing IT infrastructure?
  - How carefully are the problems of interfacing, integration, and legacy systems studied?
  - While going mobile, an important challenge is security of m-business—is this issue studied?
  - How well are the connectivity and management of mobile devices, security, and updates issues studied?
- From consumer's perspective:
  - What is the impact of m-business on consumers?
  - Is the transition for consumers straightforward or painful?
- Transition cost
  - Should the business mobilize a few employees or the whole enterprise?
  - How much it will cost for an enterprise with tens of thousands of employees?
  - What is the price and benefit of a mobile employee (m-employee)?

For each of the mentioned categories, the list of such questions can be much longer than shown here. These are just some of the issues not including public, political, and legislative issues.

CONCLUSION

This chapter provided a brief overview of m-business, opportunities that m-business opens, and challenges accompanying these opportunities.

The chapter also discussed m-business in connection with its underlying technology. In this part different technological components of m-business were introduced and discussed. Along with technology, the chapter provided information about some m-business software solutions.

Future of M-Business

With the arrival of 3G wireless networks in the market and development of powerful mobile devices, shift in business, and need for more ubiquity, in few years we will be participating in and watching TV-quality business meetings of your corporation on your laptop while in the air (flying back home), coordinating and managing your business while enjoying the beach, evaluating and managing a project and assigning new tasks while interacting with nature, processing loan applications while on the road, and conducting other serious business activities on the streets. What is most amazing is that neither a manager, nor employees or colleagues will realize that all these times you were miles away from the office, because m-business will provide you with facilities to work as if in the office all the time.

In short, you will be carrying your office or enterprise in your briefcase because you will be doing m-business. Your office and business will reside at your fingertips, and you will not be tied to your office. Then, we also may not need to maintain large office buildings. And all this will be in the near future because m-business is the business of the future. At present, however, there are more than enough challenges and issues in adapting m-business and transforming the traditional way of doing business. We will see its full realization in the next few years.

REFERENCES

Deitel, H., Deitel, P., & Steinbuhler, K. (2001). *E-business and e-commerce for managers.* Upper Saddle River, NJ: Prentice-Hall.

Deitel, H., Deitel, P., Nieto, T., & Steinbuhler, K. (2003). *The complete wireless Internet and mobile business programming training course.* Englewood Cliffs, NJ: Prentice-Hall.

ICMB. (2005, July 11-13). *Proceedings of the 4th International Conference on Mobile Business,* Sydney, Australia. Retrieved from http://www.mbusiness2005.org/

Kalakota, R. (2005). *Mobile business: Vision to value.* Retrieved from http://www.kalakota.com/speakingtopics/mbusiness.htm

Kalakota, R., & Kurchina, P. (2004). *Mobilizing SAP: Business processes, ROI and best practices.* Mivar Press.

Kalakota, R., & Robinson, M. (2001). *M-business: The race to mobility.* New York: McGraw-Hill.

Nokia. (2005). Retrieved March 2005 from http://www.nokia.com/mobilebusiness/americas/

Paavilainen, J. (2002). *Mobile business strategies.* Reading, MA: Addison-Wesley.

Sadeh, N. (2002). *M-commerce: Technologies, services, and business models* (1st ed.). New York: John Wiley & Sons.

SAP. (2005). Retrieved March 2005 from http://www.sap.com/solutions/mobilebusiness/index.epx

Segan, S. (2005). Wireless without borders: Networks for those on the go. *PC Magazine, 24*(5), 90.

Solheim, S. (2005). 3G gets real. *eWeek, 22*(3), 28.

Vos, I., & de Klein, P. (2002). *The essential guide to mobile business.* Upper Saddle River, NJ: Prentice-Hall.

Section XI

Case Study

Chapter LII
Successful Implementation of Emerging Communication Technologies in a Mobile-Intense Organization: A Case Study of Sydney Airport

Keyurkumar J. Patel
Box Hill Institute, Australia

ABSTRACT

Wireless Technology is growing at a phenomenal rate. Of the many present challenges highlighted by the author, increased security is one of the main challenges for both developers and end users. This chapter presents this important security aspect of implementing a mobile solution in the context of Sydney International airport. After tackling initial challenges and issues faced during the implementation of wireless technology, this chapter demonstrates how security issues and wireless application were implemented at this mobile-intense airport organization. The decision to deploy and manage the wireless spectrum throughout the Airport campus meant that the wireless LAN had to share the medium with public users, tenants and aircraft communications on the same bandwidth. Therefore, this case study also demonstrates invaluable approach to protect unintended users from breach of existing security policies adopted by their corporate network. Authentication and data privacy challenges, as well as complete WLAN connectivity for tenants, public and corporate usage is presented in this case study.

Copyright © 2006, Idea Group Inc., distributing in print or electronic forms without written permission of IGI is prohibited.

Figure 1. Typical wireless LAN topology

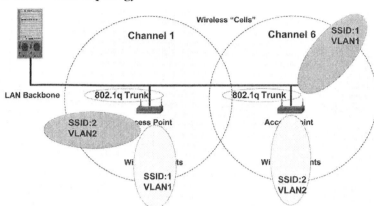

INTRODUCTION

Sydney's International Airport forms the hub of aviation in the Pacific region. It is an 85-year-old site, approximately 8 kilometers from Sydney CBD. With 5 terminals with 34 international, 31 domestic, and 5 airfreight gates, it is the largest airport catering to 8.7 million international and 15.5 million domestic passengers per year (McCubben, 2003). As such, an acute need was felt to ensure a high level of timely and quality service to the entire infrastructure of the airport. Mobile technologies were considered as a crucial ingredient in provision of this service. This need continues to be corroborated worldwide; for example, at the Airport Council International (ACI) World Assembly in Santiago in November 2000, the airport community expressed the importance of a wireless infrastructure at airports managed by the airport authority (Sydney Airport Corporation Limited, 2000). The following resolution was agreed upon:

Airport Operators should assert control over the use of Wireless Infrastructure at Airports, both inside and outside terminal buildings. Tenants, concessionaires and others should use a common infrastructure for wireless managed by the Airport Operator. In return

for this exclusivity, Airport operators should constantly evaluate competing technologies, so as to maintain low costs, increased capacity and security in line with demand for the benefits of all tenants, concessionaires and others.

Meanwhile, in 1998, with the impending Sydney Olympics 2000, Sydney Airport Corporation Limited (SACL) was formed. SACL took it upon itself to embark on the challenge of becoming the sole provider of wireless infrastructure at the International Terminal and Airfield. Past experience indicated that business customers preferred to install their own networks, and wireless—still an evolving technology with no ratified security standards and ease of deployment—gave SACL a unique challenge. This chapter discusses in detail the successful deployment of mobile applications at the Sydney International Airport.

WLAN Architecture and Security Challenges

With a typical wireless LAN (WLAN; see Figure 1), transmitted data is broadcast over the air using radio waves. With a WLAN, the boundary for SACL's network has moved and is now located in many airfield remote sites. In

Figure 2. Network architecture and security policy

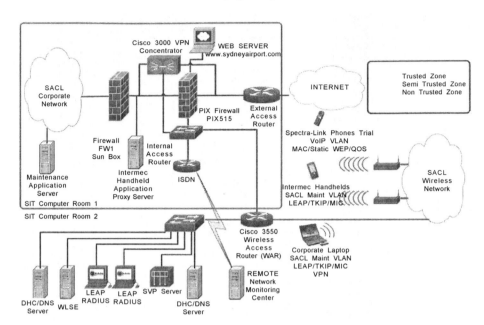

early 2001, SACL deployed some 120 access points within the International Terminal (Terminal 1) and at various sites on the airfield. Sydney Airport WLAN implementation in early 2001 deployed Cisco Aironet 350 Series Access Points. The IEEE 802.11b standard adopted uses the unlicensed 2.4x gigahertz frequency band, providing only three non-overlapping channels (1, 6, and 11) with data-rates of 1, 2, 5.5, and 11Mbps.

However, without stringent security measures in place, the wireless infrastructure is equivalent of putting Ethernet ports everywhere. Thus, SACL's wireless deployment challenge was to ensure that the implementation of the wireless network did not breach its existing security policies for the corporate network. SACL regards the wireless network infrastructure in much the same fashion as the Internet, an untrusted zone. Even with this view, SACL has still ensured that wireless network security protects Sydney Airport's wireless VLAN. The following outlines the type of wireless and network security utilized at Sydney Airport.

Network Architecture and Security Policy

An important decision when deploying a WLAN is how it will interface back into the corporate infrastructure. The Wireless LAN (WLAN) at Sydney Airport has been designed so that the WLAN infrastructure access is located outside the corporate firewalls (see Figure 2). This approach creates more administrative overhead, because of the need for configuration of the External access network, consisting of router access lists and firewall rules. Furthermore, Sydney Airport does not need to maintain multiple WAN (Wide Area Network) remote sites. This is due to the fact that SACL's WLAN network is not deployed in order to replace the wired LAN in the office and is a network not solely accessed by SACL users. All external access from the wireless network is via the Wireless Access Router (WAR). The WAR is a Cisco 3550 Router configured to perform access-list filtering on all traffic based on source and destination IP address, protocol, and port numbers (Cisco Systems, 2002).

Many regard wireless technology as insecure (Arbaugh, 2001). SACL regards the wireless infrastructure at Sydney Airport as an untrusted zone. There are only two ways that a SACL wireless client can gain access to data from SACL's corporate network. The first option is to install an application proxy server. The proxy server allows data to move from the wireless untrusted zone to a semi-trusted zone located in the De-Militarized Zone (DMZ) outside SACL's Corporate Network, Firewall 1. This application proxy server located in the semi-trusted zone accesses the corporate network or trusted zone via Firewall 1 on the clients' behalf. The second option and only way resources can be accessed directly from an untrusted to a trusted zone is via the use of a Virtual Private Network (VPN). A VPN is established between SACL's wireless client and the Cisco VPN concentrator connected in parallel to the PIX firewall located in the DMZ. Both of these methods are described in detail further on.

SACL Wireless Authentication

Originally, SACL only offered a Cisco Propriety Wireless Security solution of Light-weight Extensible Authentication Protocol (LEAP) for all tenants and concessionaires utilizing SACL's infrastructure. Cisco's LEAP utilizes a 128-bit dynamic Wireless Equivalent Privacy (WEP) key, along with radius username and password authentication (Geier, 2002). The WEP key is dynamically assigned from one of two Remote Authentication Dial-In User Service (RADIUS) servers (Cisco Secure ACS 2.6 or greater) when an authenticated user associates with an access point. This WEP key is again negotiated between the client and the RADIUS server after a pre-configured period set on the RADIUS server. At present this period is set at less than 10 minutes (see Figure 3).

The new firmware software introduced by Cisco for their access points supports the termination of 802.1q trunk. This allows a trunk to be provisioned between the access point and an Ethernet switch, the end result allowing users in a wireless VLAN cell to belong to different VLANs. With the use of different VLANs, user traffic is segmented per group (i.e., per VLAN) with the use of differentiated security policy per VLAN. The Service Set Identifier (SSID) is used to map the client to the wireless VLAN. RADIUS attributes passed in between the access point can also override this mapping if the users are not authorized for that SSID. Up

Figure 3. SACL wireless authentication—LEAP

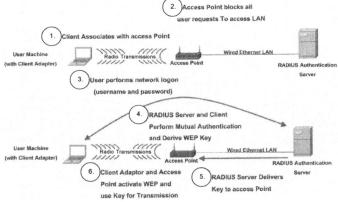

to 16 wireless VLAN's can be supported on each access point.

The introduction of wireless VLAN's allows the use of non-priority client cards, along with different security models. Although SACL believes that the Cisco LEAP solution is still the most secure and manageable solution presently available, when used in conjunction with Temporal Key Integrity Protocol (TKIP) and MIC, (TKIP and MIC security are explained in the next section), it is possible for a static WEP, Media Access Control (MAC), and or EAP security options to be used on a separate wireless VLAN. This enables the use of all vendors' 802.11b wireless clients adaptor and gives greater flexibility for products not yet supporting LEAP, such as most voice over IP wireless phones. At present SACL continues to use the LEAP solution with TKIP and MIC, as all their devices make use of Cisco client adapters.

Wireless Data Privacy Enhancements

While WLAN security that relies on Service Set Identifiers (SSIDs), open or shared-keys, static WEP keys, or MAC authentication is better than no security at all, it is not sufficient or truly manageable for the size of the Sydney Airport wireless network.

Sydney Airport, like all wireless network administrators, eagerly awaits the wireless IEEE 802.11i security standard that will allow vendor interoperability and still solve all known vulnerabilities of WEP (Stubblefield, 2002), the basic mechanism to date for interoperable security of Wireless 802.11b products. The IEEE 802.11i standards were published at the end of 2003. The Wi-Fi Alliance represented by many of the wireless vendors and in conjunction with the IEEE, has driven an effort to bring strongly enhanced, interoperable Wi-Fi security to market in the first quarter of 2003. The result of this effort is Wi-Fi Protected Access (WPA). Wi-Fi Protected Access is a specification of standards-based, interoperable security enhancements that strongly increase the level of data protection and access control for existing and future wireless LAN systems. Designed to run on existing hardware as a software upgrade, WPA is derived from and will be forward compatible with the upcoming IEEE 802.11i standard. One 802.11i component not required in WPA is Advanced Encryption Standard (AES) support. AES will replace 802.11's WEP initialization Vector RC4-based encryption under 802.11i specifications. Migrating to AES encryption, though, will require hardware changes, so this has been deferred by the Wi-Fi Alliance until the formal standard is in place to give vendors and customers some breathing room. The bad news is that 802.11i will require hardware changes regardless of whether WPA gets deployed over the next year or not.

Cisco has already been given WPA certification on the new IOS software available for both the 1100 and 1200 series Access Point range, with firmware for the 350 Series Access Points installed at Sydney Airport scheduled for the third quarter of 2005. Until the release Sydney Airport utilized Cisco's proprietary solution that features a subset of the 802.11i draft. As mentioned previously, Cisco has developed an 802.1X authentication type called EAP Cisco Wireless, or Cisco LEAP. Access points at Sydney Airport can be configured to support Cisco LEAP and all 802.1X authentication types, including EAP Transport Layer Security (EAP-TLS provides for certificate-based mutual authentication that relies on client-side and server-side digital certificates). With 802.1X authentication types such as LEAP and EAP-TLS, mutual authentication is implemented between the client and a RADIUS server. The credentials used for authentication, such as a log-on password, are never transmitted in the clear, or without encryption, over the wireless medium. Another benefit of 802.1X authentication is centralized management of WEP keys. Once

Figure 4. SACl wireless deployment (map © Sydway Publishing Pty. Ltd. Reproduced with permission)

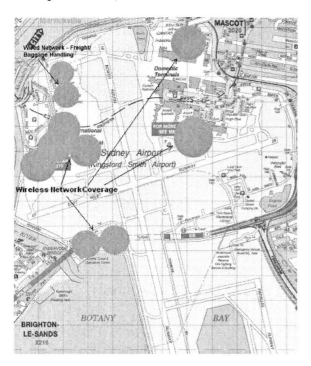

mutual authentication has been successfully completed, the client and RADIUS server each derive the same WEP key, which will be used to encrypt all data exchanged. The result is per-user, per-session WEP keys. AP software running at the airport provides several enhancements to WEP keys that have formed part of the Wi-Fi Protected Access. These WEP enhancements include Cisco's pre-standard Temporal Key Integrity Protocol and support for Message Integrity Check (MIC).

When TKIP, also known as key-hashing support, is implemented on both the AP and all associated client devices, the transmitter of data hashes the base key with the IV (Initialization Vector of RC4 Key Scheduling Algorithm) to create a new key for each packet (Fluhrer, 2001). By ensuring that every packet is encrypted with a different key, key hashing re-

moves the predictability that an eavesdropper relies on to determine the WEP key by exploiting IVs.

When MIC support is implemented on both the AP and all associated client devices, the transmitter of a packet adds a few bytes (the MIC) to the packet before encrypting and transmitting it. Upon receiving the packet, the recipient decrypts it and checks the MIC. If the MIC in the frame matches the calculated value (derived from the MIC function), the recipient accepts the packet; otherwise, the recipient discards the packet. Using MIC, packets that have been (maliciously) modified in transit are dropped. Attackers cannot use bit-flipping or active replay attacks to fool the network into authenticating them, because the MIC-enabled client and access points identify and reject altered packets.

Mobile Maintenance

The initial deployment of the wireless network within Terminal 1 and the airfield coincided with the Mobile Maintenance Project. The project utilized the wireless network to track and complete maintenance work in the field. Maintenance staff at Sydney Airport uses a Computerized Maintenance Management System (CMMS) known as MAXIMO (see Figure 5). The one limitation of MAXIMO was its inability to follow staff to the job. Previously "work orders" or job sheets were printed directly off the system and then taken into the field by the relevant trade staff or technician. Once the work was completed, the sheet was completed manually and the written data then entered in MAXIMO, either on return to the workshop or at the end of a shift. The mobile solution utilizing an industrialized handheld personal digital assistant (PDA) eliminates re-entry of data and has a positive follow-on effect in allowing more accurate reporting and a paperless system. With the limitation of VPN

Figure 5. Computerized maintenance management system

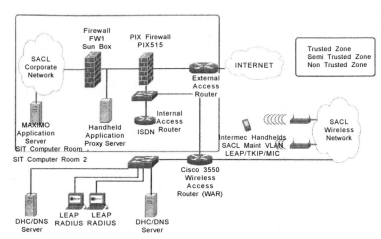

Figure 6. VPN authentication process

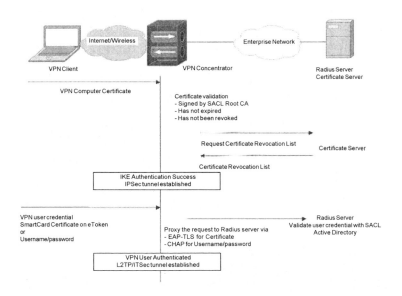

clients for PDAs at the time of delivery for the project, an application proxy server was installed within SACL's semi-trusted DMZ area to proxy wireless client requests to the MAXIMO application server. The PDA used was fitted with a Cisco Client Adaptor card and set up to use LEAP. The only changes from the initial installation are the upgrade of wireless client driver to utilize Cisco's TKIP and MIC

security enhancements configured on the access points.

Wireless VPN Remote Access Solution

In December 2002 SACL embarked on a corporate remote access solution. The solution enables SACL corporate users to access the corporate network with all desktop applications from broadband Internet access, corporate dialup, and wireless access. By placing the wireless infrastructure access outside the corporate firewalls, it allows SACL to best utilize its remote access VPN solution. The implemented solution integrates a Cisco VPN 3000 Concentrator, Microsoft 2000 Certificate Server, Microsoft VPN Client, USB Port Token, and Centralized remote PC firewall (ZoneLabs Integrity) to provide strong security, ease of use, and centralized management. The original goal was to utilize the Cisco VPN Client that allowed cooperative reinforcement with the remote PC firewall. The ZoneLabs VPN Enforcement feature ensures that the VPN users can only connect to and remain connected to the SACL network as long as the client is running a verified version of the ZoneLabs firewall agent and the client is enforcing the most up-to-date security policy.

The Cisco VPN client used was incompatible with a few of the corporate applications which required the client to have a virtual adaptor IP address in order to ftp data from a server back to the remote PC. Cisco released a VPN client that has a virtual adaptor which is still being piloted by SACL. In the interim period, SACL chose to utilize a Microsoft VPN client that does not provide cooperative reinforcement of the remote PCs. Remote PCs have utilized the Microsoft XP operating system to lock down the ZoneLabs application to ensure that the user cannot shutdown the firewall. As all the wireless users within SACL are utilizing Cisco wireless adaptors, the same Wireless VLAN utilizing LEAP is used. This will change in the near future, as the new laptops have inbuilt wireless adapters. As the VPN solution does not require the additional wireless security, as it is geared for broadband and hotspot users, another wireless VLAN will be created with a different security model. Figure 2 shows the present VPN establishment

and logon process to the corporate network. Strong authentication is required to secure the VPN connections (see Figure 6). VPN users must have a computer with a valid SACL digital certificate, valid Windows account, or USB eToken with SmartUser certification to successfully establish a VPN connection. After the use of valid Digital Certificate for Internet Key Exchange (IKE) authentication, the Microsoft VPN uses Internet Protocol Security (IPSec) Encryption ESP (encapsulation protocol)—Layer Two Tunneling Protocol (L2TP)—Transport IPSec SA.

Wireless Voice Over IP Pilot

The third device that has been trialed by SACL across its wireless LAN is a pilot of wireless IP phones. During the last quarter of 2002 SACL, undertook a voice over IP trial. This pilot included the deployment of wireless handsets that ran over the terminal's live wireless network. The wireless IP handsets used were spectra-link phones. These phones did not support LEAP, so a dedicated WLAN with static 128-bit WEP was set up across the international terminal. This wireless VLAN was also given the highest quality of service (QoS) on each access point to ensure phone calls would not drop out if a wireless access point was supporting multiple clients. Knowing the security vulnerabilities of static WEP, SACL combined MAC-level security on the wireless VLAN. MAC-level security on the Cisco access points can be centrally managed by the Cisco Secure ACS RADIUS Servers. This is performed by entering the MAC address of the phone as a user and password in the RADIUS server. While the security solution is not necessarily ideal, it was the only means available at the time of the pilot. Cisco's latest release of software for the 1100 and 1200 access points has a feature known as fast secure roaming.

This feature allows EAP authentication to be used for Wireless Voice Over IP.

Tenant Wireless Connectivity

As the wireless infrastructure is a shared medium for both SACL and its tenants, it is necessary to establish connections back into the tenants' own corporate network. Figure 8 shows how this connection is implemented, indicating demarcation points. A range of IP addresses is assigned to each tenant who will either hard code or utilize SACL's DHCP servers. Every tenant utilizing SACL's wireless infrastructure will be given a dedicated 100 Mbps UTP routed connection to the Wireless Access Router. Within the WAR, access-lists filtering on IP address, protocol, and port numbers are configured. It is advised that the 100 Mbps Ethernet port be connected to the tenants' network via their own firewall. The tenant is responsible for any additional security measures such as VPNs with their own wireless clients. The tenant is set up with their own Wireless VLAN that can be given a security policy that best meets their needs or device capabilities.

Until, the introduction of Wireless VLANs, there was only one option—to use Cisco client adaptors utilizing LEAP. This interoperability proved to be challenging, with one of the International Airline Lounges already entered into a commercial agreement with an ISP to provide wireless Internet access to their Frequent Flyer Members. As SACL could not control the public users' client adaptor card, an interim solution was put in place. Other tenants wanting to utilize symbol barcode scanning devices were told the warranty on the device would be void if they replaced the wireless adaptor with a Cisco Card. With the release of wireless VLANs, two organizations utilizing SACL's wireless network can now use symbol devices with symbol wireless client adaptors. Both are using static WEP with Symbols own VPN solution AirBEAM Safe.

Public Internet Connectivity

Sydney Airport deployed a public wireless network (see Figure 7) to allow high-speed Internet connectivity to public users. SACL is strategically well placed to target this market of mobile professionals. SACL has partnered with Internet service providers to offer their existing customers and new subscribers all the Internet ser-

Figure 7. WLAN public Internet connectivity

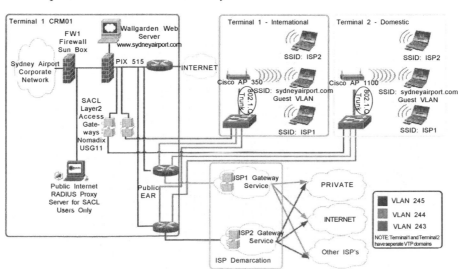

Figure 8. Total wireless connectivity diagram

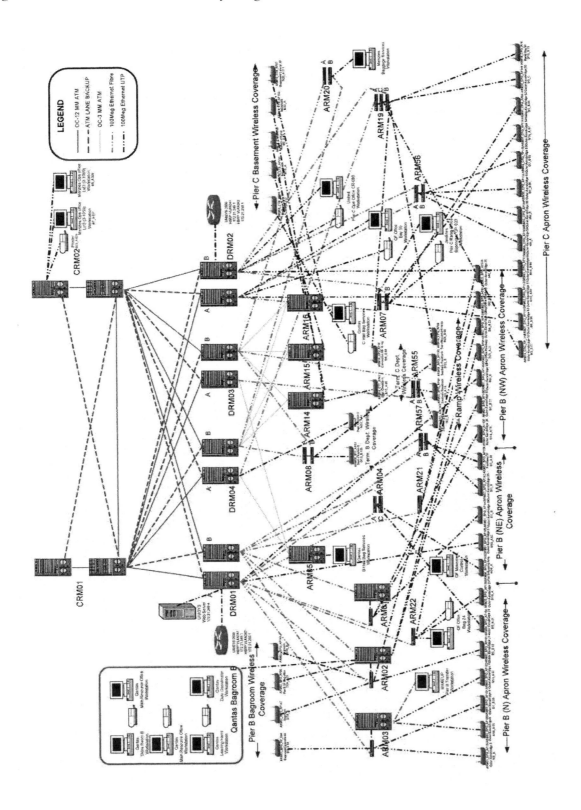

vices including e-mail, Web browsing, and connection back into their corporate networks via secure VPN. Like most hot spots, the wireless public LANs are set to open authentication with no WEP key encryption configured. The portal page login for the subscriber is made by opening an SSL-encrypted session hypertext transfer protocol over Secure Socket Layer (HTTPS). The difference between most hotspots is that Sydney Airport hosts three unencrypted wireless LANs on each of its access points. As Sydney Airport already had the wireless infrastructure installed, it made commercial sense to cater for multiple service providers instead of the usual "ISP-grabbing-real-estate" approach. The commercial model with each of the ISPs is based on revenue sharing of subscriber usage. The subscriber can be classified as retail and wholesale. A wholesale customer refers to a subscriber who is signed with a roaming partner of the ISP. To cater for multiple ISPs proved quite challenging. The technical challenges of routing users down different Internet service connections, government regulations on IP interception rules in Australia, and the airport not wanting to lawfully become an Internet service provider led to the decision of running multiple wireless VLANs. The latest 350 IOS firmware for Cisco access points allows for multiple unencrypted VLANs. SACL therefore maintains control of the access point infrastructure and WLAN services on the airport site, being responsible for ensuring the WLAN technical capability of the network. Each ISP is provided with a non-broadcast unencrypted wireless LAN, with their nominated Wireless Network Name (SSID) as set up in their other hotspot sites. This WLAN will be connected to the ISP's gateway service infrastructure. The Internet service provider will ensure compliance with the relevant government regulations. This compliance will mean that they will solely be responsible for interception of all IP traffic pertaining to their WLAN on the airport site.

The Internet service provider shall provide to SACL a non-repudiable system for reporting and auditing of the Sydney Airport hotspot site. The system shall be accessible to SACL in real time or close to it. Sydney Airport requires online data such as billing or RADIUS accounting records that detail each user's IP session time along with data byte usage.

CONCLUSION AND FUTURE DIRECTION

The security interoperability challenges are being addressed by Wi-Fi Protected Access range of Wi-Fi products based on the upcoming IEEE802.11i standard. With the WPA and ultimately 802.11i standard implementation in place, the need for add-on solutions such as VPNs may be deemed unnecessary in some enterprise wireless LAN environments (Patel & McCubben, 2004a). Other future enhancements not referenced in this chapter include the upcoming release of 802.11g Wireless Access Points. 802.11g can ideally deliver 54 Mbps maximum data rate and offer an additional and compelling advantage—backward compatibility with 802.11b equipment. This means that 802.11b client cards will work with 802.11g access points, and 802.11g client cards will work with 802.11b access points. Because 802.11g and 802.11b operate in the same 2.4 GHz unlicensed band, migrating to 802.11g will be an affordable choice for Sydney Airport with existing 802.11b wireless infrastructures (see Figure 8).

One drawback is that 802.11b products cannot be "software upgraded" to 802.11g because 802.11g radios will use a different chipset than 802.11b in order to deliver the higher data rate (Patel & McCubben, 2004b). However, like Ethernet and Fast Ethernet, 802.11g products can be co-mingled with 802.11b products in the same network. Sydney Airport will continue to provide solutions to business needs by utilizing innovative and leading-edge technology. Future applications include Mobile Self-Service, Check-In Kiosks, RF Bag Tags, Wireless Point of Sale, Wireless Stock Take, Wireless VoIP, and the end-of-the-year government requirements for Baggage Reconciliation. To-date total wireless connectivity diagrams are shown in Figures 4 and 8. So far it is safe to say that the wireless technology will play an important part in the future of Sydney Airport's total journey experience.

REFERENCES

Arbaugh, W. A., Shankar, N., Justin Wan, Y.C. (2001). *Your 802.11 wireless network has no clothes.* Retrieved June 20, 2003, from http://www.cs.umd.edu/~waa/wireless.pdf

Cisco Systems. (2002). Retrieved June 20, 2002, from http://www.cisco.com/en/US/netsol/ns340/ns394/ns348/ns337/networking_solutions_package.html

Fluhrer, S. R., Mantin, I., & Shamir, A. (2001). *Weaknesses in the key scheduling algorithm for RC4.* Retrieved November 11, 2001, from http://downloads.securityfocus.com/library/rc4_ksaproc.pdf

Geier, J. (2002). *802.11 security beyond WEP.* Retrieved July 12, 2002, from http://www.wifiplanet.com/tutorials/article.php/1377171

McCubben, S. (2003). *Trim document reference: M2003/06745.* Sydney: Sydney Airport Corporation Limited.

Patel, K. J., & McCubben, S. (2004a). Addressing wireless security issues during implementation of wireless applications in a highly mobile organization. In *Proceedings of the International Conference on Computing, Communications and Control Technologies* (Vol. 7, pp. 13-18).

Patel, K. J., & McCubben, S. (2004b). Implementation of wireless technology in a highly mobile organization: Challenges and issues. In *Proceedings of the 8th World Multi-Conference on Systemics, Cybernetics and Informatics* (Vol. 11, pp. 43-48).

Stubblefield, A., Ioannidis, D., & Rubin, A. (2001). *Fluhrer, Mantin, Shamir attack to break WEP.* Retrieved June 20, 2002, from http://www.isoc.org/isoc/conferences/ndss/02/proceedings/papers/stubbl.pdf

Sydney Airport Corporation Limited. (2000). Retrieved June 24, 2000, from http://www.sydeyairport.com

Chapter LIII
The Next Big RFID Application:
Correctly Steering Two Billion Bags a Year Through Today's Less-Than-Friendly Skies

David C. Wyld
Southeastern Louisiana University, USA

ABSTRACT

This chapter examines the adoption of radio frequency identification (RFID) technology in the commercial aviation industry, focusing on the role of RFID systems for improved baggage handling and security. The chapter provides a timely overview of developments with regard to the implementation of RFID technology in commercial aviation, which promises distinct advantages over the currently used bar-code system for baggage handling. The chapter focuses on how RFID technology can improve customer service through better operational efficiency in baggage handling, which has been demonstrated to be an integral component of the airline's customer service equation. Developments with RFID technology can dramatically improve the accuracy of baggage handling, which can enable air carriers to close an important service gap among customers in an increasingly turbulent operating environment. Other service industries can certainly benchmark the airline industry's use of RFID technology in luggage tracking as a way to improve their own operational capabilities.

INTRODUCTION

To put this chapter in perspective, consider this scenario: You have just landed in Alexandria, Egypt, or Alexandria, Louisiana. You are standing at the baggage carousel, having flown in on the last flight arriving that night. A constant stream of bags of all shapes, sizes, and colors circle past you, disappearing one by one as your "lucky" fellow passengers claim their prizes. After about 15 minutes, the carousel stops spinning. At that point, you realize that your checked roller-bag has not arrived on the same flight as you.

Now, you are in "lost luggage hell," and while the airline may do its best to accommodate you, no amount of compensation from the air carrier—whether in money, miles, or drink coupons—can change one simple fact: How are you going to make that winning presentation to a major new client at 8:00 the next morning? You realize that the only clothing you have in your possession is the warm-up suit you wore to be comfortable all day as you traveled; your "killer suit" and "confidence tie" are likely sitting on an airport tarmac thousands of miles away, with no clothing store in the city that will open before the meeting (unless you happen to be in Las Vegas).

The system that you are dependent upon to correctly track your checked luggage to either the Memphis in Tennessee or in Egypt, or wherever else it may be, is based on correct readings along the line of a bar-coded label, bearing a 10-digit IATA (International Air Transport Association) number. Gartner's Research Director, Jeff Woods, commented that "bags are very well tracked right now" by the airlines and their bar code-based systems (cited in Morphy, 2004). Yet, this is little consolation when it is *your* bag that is lost. The baggage tracking systems of the world's airlines are mature, and even under the best of conditions, bar code technology works in correctly reading only eight or nine bags out of every 10. This means that the airlines continue to devote considerable time and energy to manually intervene to correctly direct the right bags onto the right flights, while spending great amounts of money to reunite passengers with their bags when the system breaks down.

Today, savvy airlines, even in their precarious financial positions, are seeing the shift to RFID (radio frequency identification)-based baggage tracking systems as a solid operational investment that can produce significant cost savings and demonstrated return on investment (ROI). Airports as well are taking the initiative to shift to RFID-based systems, sensing the opportunity to produce greater traveler satisfaction with their experience at a specific airport. In a deregulated world of airline and airport choices, these entities are combining forces to enhance customer service and give them a competitive advantage, perhaps for a significant window of time until such RFID-based systems are made mandatory.

In this chapter, we will examine the mechanics of how RFID-based baggage tracking works and the benefits it can provide. After a brief overview of RFID technology, we will look at the experience of Delta Air Lines, which is the first airline to publicly commit to taking the technological leap forward to implementing RFID-based baggage tracking. We will then examine the confluence of technology, terrorism, and yes, marketing, that will likely drive the adoption of RFID-based tracking of checked baggage throughout the world. The RFID movement is also being spearheaded by the U.S. government. It is clearly interested in securing the safety of the traveling public and with it, what financial viability the airline industry has left in the wake of the after-effects of September 11, 2001, and the decline in travel spurred by that awful tragedy, an economic recession, and record fuel prices. We will examine the government push in this area and concerns over passenger privacy. Finally, we will look at an alternative vision of the future of airline customer service, which may preclude the need for baggage service as part of the air passenger experience altogether.

WHAT IS RFID?

In brief, radio frequency identification uses a semiconductor (microchip) in a tag or label to store data. Data is transmitted from, or written to the tag or label when it is exposed to radio waves of the correct frequency and with the correct communications protocols from an RFID reader. Tags can be either *active* (using a battery to broadcast a locating signal) or *passive* (using power from the RFID reader for location). A firm may use a combination of fixed and handheld readers for reading RFID tags to gain as complete a picture as has ever been possible on exactly what is where in their operations. Reading and writing distances range up to 100 feet, and tags can be read at high speeds (Booth-Thomas, 2003). For a detailed explanation of the technology, see Jones and Wyld (in press), McFarlane (2002), Kambil and Brooks (2002), and Reed Special Supplement (2004).

The advantages of RFID over bar code technology are summarized in Table 1.

RFID tags have been described as being a "quantum leap" over bar codes. *Inc. Magazine* characterized RFID vs. bar codes as "like going from the telegraph to the Internet" (Valentine, 2003).

As noted in an interview last year with the *Harvard Business Review*, William Copacino, group chief executive officer for Accenture's Business Consulting Capability Group, interest in RFID is picking up significantly throughout the global business community today. This is due not only to the fact that prices are rapidly dropping for both the RFID tags themselves and for the readers to sense them, but more importantly, the technology is providing significant improvements in operations and efficiency over traditional methods, while affording companies the concomitant opportunity to improve their customer service strategies (opinion cited in Kirby, 2003). From the perspective of Deloitte Consulting (2004), if RFID is viewed as simply an alternative means of identification and labeling to bar code technology, then businesses will have a "lost opportunity" on their hands. This is because RFID technology potentially offers wide-ranging opportunities for transformative change (a change of the highest magnitude) in internal business processes, supply chain management, security threat management, and customer service. Innovative applications of RFID technology are being seen in myriad industries today, including such critical areas as pharmaceuticals (Wyld & Jones, in press) and livestock tracking (Wyld, Juban, & Jones, 2005).

BAGGAGE AND AIRLINE CUSTOMER SERVICE

The critical link in customers' minds between seeing their luggage on the baggage carousel upon arrival and their perception of the quality of an airline's service offering has been empirically proven. Each year, professors Brent D. Bowen (University of Nebraska Omaha) and

Table 1. RFID vs. bar code technology

Bar Codes	RFID Tags
Bar codes require line of sight to be read.	RFID tags can be read or updated without line of sight.
Bar codes can only be read individually.	Multiple RFID tags can be read simultaneously.
Bar codes cannot be read if they become dirty or damaged.	RFID tags are able to cope with harsh and dirty environments.
Bar codes must be visible to be logged.	RFID tags are ultra thin, and they can be read even when concealed within an item.
Bar codes can only identify the type of item.	RFID tags can identify a specific item.
Bar code information cannot be updated.	Electronic information can be overwritten repeatedly on RFID tags.
Bar codes must be manually tracked for item identification, making human error an issue.	RFID tags can be automatically tracked, eliminating human error.

Table 2. 2004 airline quality ratings (adapted from Bowen & Headley, 2004)

RANK	AIRLINE	AQR SCORE
1	JETBLUE AIRWAYS	-0.64
2	ALASKA AIRLINES	-0.74
3	SOUTHWEST AIRLINES	-0.89
4	AMERICA WEST AIRLINES	-0.89
5	US AIRWAYS	-0.96
6	NORTHWEST AIRLINES	-1.02
7	CONTINENTAL AIRLINES	-1.04
8	AIRTRAN AIRWAYS	-1.05
9	UNITED AIRLINES	-1.11
10	ATA AIRLINES	-1.17
11	AMERICAN AIRLINES	-1.24
12	*DELTA AIR LINES*	*-1.24*
13	AMERICAN EAGLE AIRLINES	-2.10
14	ATLANTIC SOUTHEAST AIRLINES	-5.76
	INDUSTRY AVERAGE	-1.14

Dean E. Headley (Wichita State University) produce their *Airline Quality Rating* report. These researchers' analytical methodology ranks airline performance in the United States, based on a weighted average of four key performance measures. These benchmarks have been validated as key in determining consumer perceptions of the quality of airline services. The four measures, drawn from data that the airlines are mandated to report to the U.S. Department of Transportation, include:

1. on-time arrivals,
2. mishandled baggage,
3. involuntary denied boardings, and
4. 12 areas of customer complaints.

Several airlines in the U.S. that have performed well in the quality survey, including Southwest, JetBlue, and Midwest Express, have touted their rankings in Bowen and Headley's (2004) report in their advertising campaigns.

Such has not been the case with Atlanta, Georgia-based Delta Air Lines. Based on the recently released *Airline Quality Rating 2004* report (as seen in Table 2), Delta has now fallen to *last* among the 12 major U.S. airlines in consumer perceptions of service quality. To put this in perspective, while the airline's composite quality rating has actually *improved* over

time since 2000, in that same timeframe, Delta's competition has been making marked improvements in the service components that matter most to airline customers.

Today, Delta is a firm embroiled in the turmoil that makes up the airline industry in America. Facing rising fuel costs, a downturn in business travel, an uncertain economy, and discount competition, *all* the established, legacy carriers in the U.S. are struggling financially and operationally today, with prominent carriers such as US Airways and United barely surviving (e.g., see Tully, 2004). Delta itself has been the subject of bankruptcy rumors, and it has conducted layoffs and closed its major hub at the Dallas/Fort Worth International Airport to stave off its demise (Perez, 2004). In September, CEO Gerald Grinstein announced a comprehensive overhaul plan, including laying off thousands of employees, and received initial agreement from its pilots' union to the recall of retired pilots on a limited basis (Fein, 2004; Weber, 2004). The airline industry is finding that without the ability to raise fares or to spend lavishly to improve customer service, it must improve its operational efficiencies and performance to survive today.

One particular area of weakness for Delta has been its handling of air travelers' checked-in luggage. In fact, according to the recently released 2004 report (which uses annual data as of the close of 2003), Delta's mishandled baggage rate increased from 3.57 in 2002 to 3.84 in 2003. As can be seen in Table 3, Delta still remains below the industry average rate of four lost bags per 1,000 passengers. However, Delta's own performance is impacted by that of Atlantic Southeast Airlines (ASA), Delta's regional partner throughout much of the United States. ASA "earned" *the* lowest quality rating of *all* airlines operating in the United States, regardless of size. Luggage service is a particularly sore point for Delta's code-sharing partner, as ASA's rate of 15.41 mishandled bags per 1,000 passengers is almost *four times* the industry average.

Despite years of trying to improve the quality of its baggage-handling systems, Delta has seen the performance of its current bar code-based system flat-line, with bar-coded labels being successfully read by scanners only 85% of the time. According to Delta spokesman Reid Davis, the airline faced the fact that it had "reached the end of the improvements that could be accomplished without new technology" (cited in Rothfeder, 2004). Of course, just because a bag is not scanned correctly does not mean that your bag will end up in Wichita Falls when you were heading to Wichita. In the end, Delta estimates that only 0.7% of all checked luggage is actually "lost." However, the airline spends upwards of $100 million each year to return these bags to their rightful owners and provide compensation to passengers whose luggage is never found (Collins, 2004).

Delta's top management has decided to tackle its "bag problem" head-on, looking to RFID technology as the means to an end of

Table 3. Mishandled baggage reports for U.S. airlines—June 2004 (adapted from U.S. Department of Transportation, Air Travel Consumer Report, August 2004; http:// airconsumer.ost.dot.gov/reports/2004/ 0408atcr.doc)

RANK	AIRLINE	REPORTS PER 1,000 PASSENGERS
1	JETBLUE AIRWAYS	2.81
2	AIRTRAN AIRWAYS	3.02
3	SOUTHWEST AIRLINES	3.16
4	HAWAIIAN AIRLINES	3.18
5	ALASKA AIRLINES	3.32
6	CONTINENTAL AIRLINES	3.32
7	AMERICA WEST AIRLINES	3.55
8	NORTHWEST AIRLINES	3.80
9	ATA AIRLINES	3.80
10	UNITED AIRLINES	3.83
11	US AIRWAYS	4.10
12	*DELTA AIR LINES*	*4.23*
13	AMERICAN AIRLINES	4.66
14	EXPRESSJET AIRLINES	5.29
15	AMERICAN EAGLE AIRLINES	9.00
16	COMAIR	10.21
17	SKYWEST AIRLINES	10.71
18	ATLANTIC COAST AIRLINES	13.42
19	ATLANTIC SOUTHEAST AIRLINES	13.97

providing far-better luggage service to its passengers. In the fall of 2003, Delta implemented a pilot test of an RFID tracking system for checked luggage on flights between Jacksonville, Florida, and its hub in Atlanta, Georgia. In this testing program, Delta tracked 40,000 passenger bags equipped with radio frequency identification (RFID) tags from check-in to loading on an aircraft. As can be seen in Table 4, the RFID-enabled system provided far superior reading accuracy than the legacy bar code-based system. In the spring of 2004, Delta implemented another pilot RFID baggage-tracking system at its Cincinnati, Ohio, hub, producing similar results (Murray, 2004).

Through the two test programs, Delta learned several valuable lessons. It saw that tag antennas could be damaged by the static electricity generated along the conveyor systems (Collins, 2004). It also found that the lowest scanner accuracy rate (96.7%) was found when attempting to scan bags inside the unit load devices (ULDs), the large containers pre-loaded with checked luggage that are then loaded onto the plane. The ULDs are made of metal with canvas doors, and the metal housing impeded the radio signals. Delta plans to coat the ULDs with a material that can better reflect the radio waves (Brewin, 2004a). While the test programs were conducted in rather neutral weather environs, concerns were raised over the ability of the tagging systems to function in harsher environments, such as at Delta's western hub in Salt Lake City, Utah (Murray, 2004). Finally, there is a famous American commercial from Samsonite that shows a gorilla in his cage, tossing the bag around and eventually stomping on the suitcase. The obvious message is that checked bags are not always handled "delicately" by the humans or the machinery as it passes through baggage systems. Thus, it must be noted that baggage handling itself can damage or detach labels/tags, and concerns over

Table 4. Results from Delta Air Lines pilot RFID test in Jacksonville (adapted from AIM Global, 2004)

Errors per 40,000 bags	RFID	Bar Code
Worst Case	1,320 (96.7%)	8,000 (80.0%)
Best Case	80 (99.8%)	6,000 (85.0%)

the durability of the RFID tag are genuine.

Even with limited capital to invest in IT projects, in July 2004 Delta became the first airline to commit to having RFID-enabled baggage tracking in place system-wide by 2007. Delta plans to use passive tags, which will cost the airline 25 cents each initially. However, the airline hopes that the cost of the tags will drop to approximately 5 cents a unit by the time the system is fully implemented in 2007 (McDougall, 2004). Delta estimates that the full implementation cost of its RFID-based tracking system will ultimately fall somewhere between $15 and $25 million for its 81 airport locations. Delta has not yet announced plans for deploying the RFID-based system with its code-sharing partners, which would greatly raise the number of airports worldwide for implementation and the cost and complexity of the overall project (Murray, 2004). While this represents a significant investment, the ROI equation shows that this cost can be recouped in far less than a single year. This makes Delta unique, as it is one of the few examples to date in *any* industry where the decision to invest heavily in automatic identification technology is based on the desire to dramatically improve customer service.

Delta's RFID-enabled baggage system will give the company the ability to track a bag from the time a passenger checks it in at his/her departing airport till the time the bag is claimed at the baggage carousel at the arrival airport. At check-in, the RFID tag's serial number will be associated with the passenger's itinerary. Delta will position fixed readers at check-in counters and on conveyor belts where the bags are sorted. The airline will also equip baggage handlers with portable readers and outfit aircraft cargo holds with readers built into them. RFID readers can also be positioned to scan bags as they are loaded and unloaded from the unit load devices (ULD—the large containers that are loaded onto the plane). Through this surveillance system, Delta should be able to all but eliminate the problem of misloaded and misdirected checked luggage, and the attendant costs of reuniting the lost bag with the passenger. Ramp and flight crews will be able to make certain that the right luggage is on board before an aircraft takes off. And, in the event a passenger's bag is misdirected, Delta can instantly locate the bag through its RFID reader and more quickly route it to the passenger's destination. Pat Rary, a Delta bag systems manager, illustrated the fact that RFID will allow the airline to take proactive customer service steps on baggage problems. He observed that:

With this technology, we won't have to wait for the customer to come tell us that the bag is lost. We can tell the customer it's on the wrong plane and start responding before it's a crisis. Eventually, RFID should be able to signal an arriving passenger's cell phone with news of how long it will be before the bag is on the carousel. (cited in Field, 2004, p. 61)

Rob Maruster, Delta's director of airport strategy, recently commented in *Airline Business* that RFID tracking "will transform the airline on the ramp as much as radar did to transform air traffic control. When that happened, it was as if a light was turned on and people said, oh, so that's where the planes are.

This technology will do that for bags. People will say, oh, so that's where the bags are" (cited in Field, 2004, p. 60). Delta's ultimate goal is to have a baggage tracking system that will have a "zero mishandling rate" (Brewin, 2004a).

RFID AND BAGGAGE SECURITY

Unfortunately, in our post-September 11 world, there are worse things that can happen in the air or at the airport than losing one's luggage or even eating the "Chef's Surprise" at the airport restaurant. The twin, nearly simultaneous jet crashes in Russia in August 2004 have now been attributed to in-flight bomb detonations by Chechen female suicide bombers, raising fears that suicidal terrorists could use similar methods to attack the West (Hosenhall & Kuchment, 2004). While enhanced physical passenger screening, such as that just announced by the Transportation Security Administration (TSA) in the U.S., can deter such would-be suicidal terrorists, since September 11, the airline industry and national governments have placed renewed vigilance on screening both carry-on and checked bags for explosives and on making sure that all checked bags are matched to passengers who have actually boarded the aircraft. Writing in *Management Services*, Collins (2004) observed that one of the very real near-term applications for RFID technology is the prospect that a passenger's checked bag will be able to tell security personnel and the airline if it has not been properly screened.

The need for matching passengers with checked luggage has been at the forefront of anti-terrorism concerns ever since the in-flight bombings in the 1980s that took down a Pan American 747 over Lockerbie, Scotland, and an Air India jumbo jet over the Atlantic. Out of this concern, airlines must routinely remove bags from aircraft when a passenger fails to board,

out of fear that a homicidal, rather than suicidal, terrorist would attempt to down an airliner with a bomb in an unaccompanied, checked suitcase (AIM Global, 2004).

Often, this is a time-intensive, laborious process, which can delay flight departures indefinitely, as ramp workers face the daunting task of finding the bags in question out of the hold of an aircraft or from the unit load devices. Airport operations managers and airline flight crews will often tell horror stories of how the inability to find the one or two targeted bags of a non-boarding passenger in and amongst the bags of 300-400 passengers on a jumbo jet has caused flights to be delayed for hours, costing the airline countless amounts of goodwill amongst the passengers, even if such measures are done precisely to safeguard their transit and their very lives. Thus, airports are also very interested in providing better baggage tracking as part of their customer service equation.

In Florida, the Jacksonville International Airport installed an RFID-based system in 2003 to direct checked luggage through their newly installed baggage handling system. The city's airport authority and the TSA jointly funded the Jacksonville system. The contractors for the Jacksonville Airport project included FKI Logistex and SCS Corporation. The Jacksonville system was designed to only handle outbound luggage, directing checked bags from the check-in counter through explosive detection screening and on to the correct terminal serviced by the respective airlines. All checked bags have a bar code label affixed to them, with approximately 12% receiving an additional RFID tag, due to their being selected for special screening attention by a computer-assisted passenger profiling system (CAPPS) (Trebilcock, 2003).

The Jacksonville pilot program tested the effectiveness of both disposable and reusable tags. Passengers checking in on the north side of the airport who were selected by the CAPPS had a disposable tag attached to their luggage, while those checking in on the airport's south side had a reusable, credit card size tag affixed to their checked bags. Each reusable tag costs $2.40, and each disposable tag costs 63 cents. Van Dyke Walker, Jr., director of planning and development for the Jacksonville Airport Authority, believes that his airport's system is a precursor of what is to come. He commented that "RFID is the future of airline baggage tracking, and we want to be ready" (cited in Trebilcock, 2003, p. 40).

Las Vegas' McCarran International Airport is considered to be an ideal proving ground for RFID baggage tracking. This is due to the fact that the vast majority of the passengers using the airport either begin or end their journeys there. In fact, as Las Vegas sees only 8% of its passengers connecting to other flights at its airport, a rate that is only second to Los Angeles International Airport (Anonymous, 2003a), Las Vegas' system is designed to track all checked luggage, routing bags through bomb detection screening and on to the proper aircraft. From the perspective of Randall H. Walker, McCarran International Airport's director of aviation, the RFID-enabled baggage handling system "becomes a win for all concerned: the traveler, the airport, the TSA and the airlines" (cited in Anonymous, 2004a). In 2005, the TSA is slated to have similar systems in place at both LAX and Denver International as well (AIM Global, 2004). Alaska Airlines also uses the tags on its international flights out of San Francisco International Airport (Woods, 2004).

Internationally, RFID-based baggage tracking systems are being tested in Narita, Japan, Singapore, Hong Kong, and Amsterdam (CNETAsia, 2004; Atkinson, 2002). In fact, the RFID baggage tracking system being installed at Hong Kong International Airport is regarded as the largest automatic identification system to be developed and deployed to date in Asia. Hong Kong's airport is one of the busiest in the world, handling approximately 35 million passengers each year. Y. F. Wong, who heads Technical Services and Procurement at the Airport Authority of Hong Kong, believes that the airport's investment in RFID technology is essential, as it addresses the need for improved customer satisfaction, while also enabling increased levels of security assurances (cited in Anonymous, 2004a). According to John Shoemaker, senior vice president of corporate development at Matrics, which will supply the airport with upwards of 80 million smart labels over the next five years, "What is key about Hong Kong International is that it is deploying this system to also save money" (quoted in Collins, 2004).

RFID baggage tracking is thus a means to an end for airports—with the end being improved baggage security. Simon Ellis, a supply-chain futurist at Unilever, recently observed that: "Security is just a sub benefit of visibility. Knowing exactly what is where gives you better control...and if you have better control you have better security" (cited in Atkinson, 2002).

RFID IN THE NEAR FUTURE AT THE AIRPORT

In a widely read article in *Scientific American*, Roy Want predicted that airline baggage tracking would be one of the first commercially viable RFID applications (Want, 2004). The potential market size is outstanding, as the world's airlines currently handle approximately two billion checked bags annually (Anonymous, 2003b). In the view of AIM Global (2004), with the proven accuracy and effectiveness of RFID-enabled baggage tracking, it may just be a matter of time before the TSA mandates that such automatic identification technology-based systems be employed in the U.S. However, such mandates, whether in the U.S. alone or in conjunction with other civil aviation authorities worldwide, would raise a multitude of issues. These include who will bear the costs of such systems, the need for standards, and the need for international airlines that fly to the U.S. and/ or interconnect with U.S.-based carriers to employ such RFID tagging.

Paul Coby, chief information officer of British Airways, believes that members of the airline industry need to work together to ensure that investments in technologies such as RFID will yield the fullest possible ROI and customer service benefits. He suggested that the International Air Transport Association (IATA) should play a leading role in driving this technology, so as to ensure that the industry adopts common information systems standards. Coby commented, "For technology to fully bring business change, the whole industry needs to move forward" (cited in Thomas, 2004). In June, 2004, Delta and United jointly proposed an RFID-specification for baggage to the IATA (Collins, 2004).

The need for a unique air transport standard is obvious for the not-so-distant future, looking to the day when luggage will contain items with their own RFID tags, say on Gillette razors, Benetton shirts, and items purchased from Target, Wal-Mart, Metro, or countless other retailers. Wal-Mart mandated the use of tags on merchandise it purchases from key suppliers by January 2005 (RFID Journal, 2003), thus prompting other retailers to follow suit, or at least begin investigating the technological investment such a move to RFID will require. For example, the retailer Boscov's (U.S.) sees customer service benefits in terms of reduced stock-outs, yet worries about the tag and infrastructure costs (Sullivan, 2004a). Tesco (UK)

announced plans to expand its RFID test project to include eight big-name packaged-goods manufacturers like Proctor & Gamble (Sullivan, 2004b). Others, like Federated's Lazarus store in Columbus, Ohio, and restaurants in Texas have seen improvements in customer service in terms of improved sales transactions (Coupe, 2003; Dunne & Lusch, 2005, p. 405). However, retailers will need to address the issue of consumer privacy, much like the airlines must do (Dunne & Lusch, 2005, p. 306; Lacy 2004).

It is even more important when one considers that the aircraft itself will likely have key parts tagged with RFID sensors in the near future. Boeing and Airbus are taking the lead in outfitting their new passenger jets, the 7E7 and A380 respectively, with RFID-tagged parts to provide a new level of historical and performance information on the key components. The two dominant commercial aircraft manufacturers are cooperatively working to produce industry standards, which is especially important since they share 70% of their supplier base (Tegtmeier, 2004). Likewise, Federal Express and Delta have pilot tested, equipping both flight deck electronics and engine parts, with RFID sensors (Brewin, 2004a). Thus, in only a matter of a few years, commercial airliners will be perhaps one of the most concentrated locations for RFID tags, making standards a necessity for avoiding problems with signal collision and information overload.

Tracking luggage with RFID may not be the only automatic identification technology we will see in use at the airport. By 2015, the International Civil Aviation Organization (ICAO) has proposed putting RFID chips in the over a billion passports worldwide. This move, while drawing fire from civil rights groups around the globe, may become a reality, all in an effort for the airlines and civil authorities to have better insights into who exactly to let on their aircraft (Jones, 2004). Likewise, the U.S. Transportation Security Administration has begun looking at how to use RFID-tagged boarding passes to improve airline security. The goals would be both to enhance airport security by giving facility security the ability to track passengers' movements within the airport and to speed passengers through airport security lines. The latter would be accomplished by linking the issuance of boarding passes to the proposed "registered traveler" program. This would allow frequent fliers who have been through a background check to be given specially tagged boarding passes, which they could then use to be directed through special "fast lanes" at security checkpoints (Brewin, 2004b).

The TSA is investigating the RFID-enabled boarding pass concept in concert with a number of other airport security initiatives in the United States. However, working in conjunction with the Federal Aviation Administration's Safe Skies for Africa Initiative, RFID-tagged boarding passes are already being deployed in an undisclosed number of African states (Brewin, 2004b). There are concerns however as to how this data will be utilized in airport security. From a practical standpoint, critics have scoffed at the jumble of data that would be created by trying to track thousands of passengers simultaneously in an airport. Privacy advocates also object to the invasiveness of the tracking, leading one to ask, "Are they going to track how long I spend in the ladies' room?" (cited in Brewin, 2004b).

There is also concern that airports, in their push to provide wireless access for patrons, may find that such Wi-Fi systems can conflict with RFID tracking innovations. In fact, in mid-2004, Northwest Airlines discovered that the wireless communication system used by its baggage handling operators was overwhelmed by a new Wi-Fi antenna installed by AT&T Wireless Services. The problem was alleviated after AT&T agreed to adjust its power levels, but it seriously impinged on Northwest's own wireless systems for a time (Schatz, 2004).

A mid-September survey in 2004 by software supplier Wavelink found that approximately four out of five Frontline Conference and Expo attendees were currently piloting the technology or planned to do so in the next two years. Key concerns of the company executives included cost, lack of standards, and an early, untested market. Yet they expect adoption of the technology to grow, as it matures and benefits become reality (Gonsalves, 2004). However, as supply chain consultant Scott Elliff argues, all the new technology "simply isn't a substitute for superior business practices" (Elliff, 2004). The airline industry needs to remember this and, better yet, implement better business practices.

CONCLUSION AND FUTURE DIRECTIONS

This chapter has discussed the application of RFID in numerous airline applications across the world. The chapter has particularly discussed the advantages in using such applications for the benefit of all parties concerned. As has been shown, there is much promise for RFID to be applied in the airline industry to produce competitive advantage for airlines that are willing to implement the technology in a time of great competitive and economic turmoil in the industry. Both air carriers and airports themselves can leverage the technology to provide better customer service and heighten the security of air travel.

On a final note, moving to RFID may be the only way airlines may be able to even continue handling checked luggage in the future, both from a security and a cost standpoint. In fact, one company, the British-based low-fare carrier, Ryanair, has announced its intention to eventually stop providing checked baggage service altogether. Michael O'Leary, Ryanair's maverick CEO, believes that by banning checked luggage, the cost of flying each passenger could be cut by at least 15%. This would be due to the elimination of the staff needed at check-in counters and in baggage handling operations. Not only would there be a direct cost savings for Ryanair, but there is the very real prospect for improved service, as passengers would get through the airport much faster and that their aircraft could be utilized more productively. The latter would be due to the quicker turnarounds that the airline could achieve by not having to load outbound and unload inbound aircraft luggage holds (Noakes, 2004).

Over the next few years, Ryanair has planned to take steps to modify its passengers' mindsets regarding their baggage to encourage them to carry more of their baggage with them on board. The airline has already raised the weight limits for carry-on bags, while hiking its fees (up 17%) for overweight checked luggage. O'Leary even intended for the airline to begin giving passengers a small rebate if they choose to not check a bag sometime in 2005. While Ryanair's competition scoffs at O'Leary's luggage-ban plan, he notes that other innovations in the airline industry, including the elimination of paper tickets and Web-based travel booking, drew similar derision when they were first introduced (Noakes, 2004; Johnson & Michaels, 2004).

While the Ryanair gambit may prove to be prescient, for the near term, passenger luggage service will continue to be a cost of doing business for airlines. As such, Gene Alvarez, an analyst with Meta Group, predicts that RFID-based baggage tracking will become standard throughout the airline industry over the next decade (as cited in Brewin, 2004c). By assuring us that it is our black roller bag that ends-up on the luggage carousel at the end of our long journey home, airlines like Delta can seek a competitive advantage through improved baggage service. In the end, we will likely see air carriers, airports, commercial aircraft manufacturers, and national transport and security agencies working cooperatively to smarten baggage handling through RFID tracking through common systems, while looking at other potential applications for automatic identification technology throughout the air transport industry.

REFERENCES

AIM Global. (2004, January). Flying high. *RFID Connections*. Retrieved from http://www.aimglobal.org/technologies/rfid/resources/articles/jan04/0401-bagtag.htm

Anonymous. (2003a). Tag tracking. *Airline Business, 19*(12), 14.

Anonymous. (2003b, July). Luggage tracking trial by Delta Air Lines. *Smart Packaging Journal, 11*, 6.

Anonymous. (2003c, November 10). Wal-Mart lays out RFID roadmap. *RFID Journal*. Retrieved from http://www.rfidjournal.com/article/articleprint/647/-1/9/

Anonymous. (2004a, May 26). Hong Kong airport picks RFID baggage tracking. *Smart Travel News*. Retrieved from http://www.smarttravelnews.com/news/2004/05/hong_kong_airpo.html

Anonymous. (2004b, May 20). Delta rolls out wireless baggage transfer system. *Smart Travel News*. Retrieved from http://www.smarttravelnews.com/news/2004/05/delta_rolls_out.html

Atkinson, H. (2002). The allure of radio frequency. *Journal of Commerce Week, 3*(16), 28-30.

Booth-Thomas, C. (2003). The see-it-all chip: Radio-frequency identification—With track-everything-anywhere capability, all the time—Is about to change your life. *Time, 162*(12), A8-A16.

Bowen, B. D., & Headley, D. E. (2004). *Airline quality ratings 2004.* Retrieved from http://www.unomaha.edu/~unoai/aqr/2004%20synopsis.htm

Brewin, B. (2003a, December 18). Delta has success in RFID baggage tag test, but a wide-scale rollout of RFID technology is being slowed by lack of money. *Computerworld.* Retrieved from http://www.computerworld.com/mobiletopics/mobile/technology/story/0,10801,88390,00.html?f=x10

Brewin, B. (2003b). Delta's RFID trial run has airport predecessors. *Computerworld,* (June 23). Retrieved from http://www.computerworld.com/printthis/2003/0,4814,82381,00.html

Brewin, B. (2004a). Delta to test RFID for parts tracking: Meanwhile, Boeing and Airbus are both pushing for common RFID standards. *Computerworld.* Retrieved from http://www.computerworld.com/mobiletopics/mobile/technology/story/0,10801,93611,00.html?f=x10

Brewin, B. (2004b, April). TSA eyes RFID boarding passes to track airline passengers: Privacy groups view the idea as a "nightmare" for civil liberties. *Computerworld.* Retrieved from http://www.computerweekly.com/articles/article.asp?liArticleID=127364&liFlavourID=1&sp=1

Brewin, B. (2004c, January 6). RFID bag-tag test proves a soaraway success. *Computer Weekly.* Retrieved from http://www.computerweekly.com/articles/article.asp?liArticleID=127364& liFlavourID= 1&sp=1

CNETAsia Staff. (2004, July 17). Japan firms to test radio-tagged luggage. *CNETNews.com.*

Retrieved from http://news.com.com/2102-1009_3-1026860.html?tag=st.util.print

Collins, J. (2004, July 2). Delta plans U.S.-wide RFID system: The airline carrier will spend up to $25 million during the next two years to roll out an RFID baggage-handling system at every U.S. airport it serves. *RFID Journal.* Retrieved from http://www.rfidjournal.com/article/articleview/1013/1/1

Collins, P. (2004). RFID: The next killer app? *Management Services, 48*(5), 20-23.

Coupe, K. (2003, May 15). Customer loyalty initiative being tested by Texas restaurants. *MorningNewsBeat.com.* Retrieved from http://morningnewsbeat.com/archives/2003/05/15.html

Deloitte. (2004). *Tag, trace, and transform: Launching your RFID program.* Retrieved from http://www.deloitte.com/ (registration required).

Dunne, P. M., & Lusch, R. F. (2005). *Retailing* (5th ed.). Mason, OH: South-Western.

Elliff, S. A. (2004, September 6). RFID: Maybe *not* the "next big thing". *The Journal of Commerce, 56*(28), 46.

Fein, A. (2004, September 8). Delta airlines to layoff thousands, outlines plan. *Axcessnews.com.* Retrieved from http://www.axcessnews.com/business_090804a.shtml

Field, D. (2004). Radio waves. *Airline Business, 20*(7), 60-62.

Gonsalves, A. (2004, September 23). Companies adopting RFID despite challenges, survey finds. *Information Week.* Retrieved from http://www.rfidinsights.com/showArticle.jhtml?articleId=47902028&printableArticle=true

Hosenhall, M., & Kuchment, A. (2004, September 6). Crashes: Did "black widows" bring down the planes? *Newsweek, 144*(10), 6.

Johnson, K., & Michaels, D. (2004, July 1). Big worry for no-frills Ryanair: Has it gone as low as it can? *The Wall Street Journal,* A1, A10.

Jones, K. (2004). Are you chipped? *PC Magazine, 23*(15), 21.

Jones, M. I., & Wyld, D. C. (in press). Smart tags + smart professors = smart students. *The Journal of the Association of Marketing Educators.*

Kambil, A., & Brooks, J. D. (2002, September 1). *Auto-ID across the value chain: From dramatic potential to greater efficiency & profit—A white paper from the Auto-ID Center at the Massachusetts Institute of Technology.* Retrieved from http://www.autoidcenter.org/publishedresearch/SLO-AUTOID-BC001.pdf

Kirby, J. (2003). Supply chain challenges: Building relationships—A conversation with Scott Beth, David N. Burt, William Copacino, Chris Gopal, Hau L. Lee, Robert Porter Lynch, and Sandra Morris. *Harvard Business Review, 81*(7), 64-74.

Lacy, S. (2004, August 31). Inching toward the RFID revolution. *Business Week Online.* Retrieved from http://www.businessweek.com/ (subscription required).

McDougall, P. (2004, July 5). No more lost luggage. *InformationWeek, 996,* 14.

McFarlane, D. (2002, May 1). *Auto-ID based control: An overview—A white paper from the Auto-ID Center at the Institute for Manufacturing of the University of Cambridge.* Retrieved from http://www.autoidcenter.org/publishedresearch/ CAM-AUTOID-WH-004.pdf

Morphy, E. (2004, July 2). Delta ups service bar with RFID luggage tracking. *CRM Daily.* Retrieved from http://www.newsfactor.com/story.xhtml?story_id=25711

Murray, C. (2004, July 12). Airline clears RFID luggage tags for takeoff. *EE Times,* Retrieved from http://www.embedded.com/showArticle.jhtml?articleID=22104613

Noakes, G. (2004, July 16). Ryanair bids to banish baggage. *Travel Trade Gazette, 2623,* 20.

Perez, E. (2004, August 16). Flight upgrade: With Delta reeling, chief plans unusual bet on premium routes; as losses mount, Mr. Grinstein pairs cost cuts with extras; leather seats in coach; big risks in crowded skies. *The Wall Street Journal,* A1, A6.

Reed Special Supplement. (2004). RFID: Powering the supply chain. *Logistics Management, 43*(8), R3-R16.

Rothfeder, J. (2004, August 1). What's wrong with RFID? *CIO Insight.* Retrieved from http://www.cioinsight.com/print_article/0,1406,a=133044,00.asp

Schatz, A. (2004, June 8). Airports clash with airlines over Wi-Fi. *The Wall Street Journal,* B1-B2.

Sullivan, L. (2004a, September 20). Retail: General merchandising: Stores look for service with ROI. *Information Week.* Retrieved from http://www.rfidinsights.com/showArticle.jhtml?articleId=47212239&printableArticle=true

Sullivan, L. (2004b, September 29). U.K. grocer expands RFID initiative. *Information Week.* Retrieved from http://www.informationweek.com/story/showArticle.jhtml?articleID=48800020

Tegtmeier, I. A. (2004). RFID knowledge enabled logistics. *Overhaul & Maintenance, 10*(5), 24-28.

Thomas, D. (2004, June 28). Airline industry needs to back IT. *Information World Review.*

Retrieved from http://www.iwr.co.uk/News/1156239

Trebilcock, B. (2003). Ready to fly? *Modern Materials Handling, 58*(3), 40-42.

Tully, S. (2004, June 14). Airlines: Why the big boys won't come back. *Fortune,* 101-102, 104.

U.S. Department of Transportation, Office of Aviation Enforcement and Proceedings, Aviation Consumer Protection Division. (2004). *Air travel consumer report—August 2004.* Retrieved from http://airconsumer.ost.dot.gov/reports/2004/0408atcr.doc

Valentine, L. (2003, September 26). The new wireless supply chain. *CRM Daily.* Retrieved from http://cio-today.newsfactor.com/story.xhtml?story_id=22376

Want, R. (2004). RFID: A key to automating everything. *Scientific American, 290*(1), 56-66.

Weber, H. R. (2004, September 21). Delta pilots agree to recalls. *The [New Orleans] Times-Picayune,* C-1, C-7.

Woods, L. (2004). The fail-proof luggage finder. *Kiplinger's, 58*(10), 34-35.

Wyld, D. C., & Jones, M. I. (in press). A magic pill?: The emergence of radio frequency identification (RFID) technology in the pharmaceutical supply chain. *Journal of Pharmaceutical Marketing & Management.*

Wyld, D. C., Juban, R., & Jones, M. I. (2005). Dude, where's my cow?: The United States Automatic Identification Plan and the future of animal marketing. *Academy of Information and Management Sciences Journal, 8*(1), 107-120.

Chapter LIV
Identified Customer Requirements in Mobile Video Markets—A Pan-European Case

Torsten Brodt
University of St. Gallen, Switzerland

ABSTRACT

Due to a significant cost advantage, mobile multicasting technology bears the potential to achieve extensive diffusion of mobile rich media applications. As weak performance of previous mobile data services suggests, past developments have focused on technology and missed customer preferences. Mobile multicasting represents a radical innovation. Currently, little insight on consumer behaviour exists regarding such services. This chapter presents results of qualitative and quantitative field research conducted in three countries. It provides a continuous customer integration approach that applies established methods of market research to the creation of mobile services. Means-end chain analysis reveals consumers' cognitive reasoning and conjoint analysis drills down to the importance of service attributes. Desire for self confidence and social integration are identified key motivators for consumption of mobile media. Services should aim for technological perfection and deliver actual and entertaining content. Interestingly, consumers appreciate reduced but tailored contents and price appears not to be a superseding criterion.

INTRODUCTION

After its first years of existence, the still emerging mobile telecommunications industry is undergoing a period of fundamental change. Since previously high growth rates of voice revenues started to decrease, the industry is looking for additional sources of revenue, such as mobile data services. However, the development of marketable services proves to be far more challenging than the one of stable, high-quality voice services.

Immature technologies are often blamed to be the reason for bad performances. Undoubtedly, the technological development is dynamic and, in fact, we argue that the intense focus on technology push has been one key factor of the misfortune with mobile data services, as it detracts from customer needs. Furthermore, since vertical integration in the mobile telecommunication industry is low, product development is often organized in cooperative forms (Hagedoorn & Duysters, 2002). Coping with the complexity of innovation network management additionally detaches actors from actual customer needs.

Based on this, we see a need for a thorough understanding of the consumer behaviour side of mobile data services. Numerous studies have addressed issues of adoption and diffusion of mobile data services with the aim to identify diffusion barriers (e.g., Pedersen & Ling, 2003; Pousttchi & Schurig, 2004). However, such research seldom results in operational recommendations for companies on how to align their services with customer needs. We chose to focus on a specific range of services that exploit the investments in larger bandwidths and to develop a thorough understanding of the relations between service characteristics and fulfilment of customer needs and desires.

Since mobile multicasting services are based on a new technology and address a new market, they are termed a radical innovation (Veryzer, 1998). Thus, customer preferences can hardly be drawn from existing resources. By participating in the European "mobile multicasting service development and field trial project" MCAST (www.mcast.info), we were able to conduct the necessary market research.

Within a new product development process, customer integration is best realized after a first internal clarification of product ideas and possibilities, and subsequently after the technical engineering phase before market introduction (Gruner & Homburg, 2000). For this purpose, we integrated qualitative and quantitative methods to explore and formally describe customer needs. In the early stage we aimed to decrease uncertainty by conducting focus groups. We complemented the results by conducting individual laddering interviews following the means-end chain framework (Gutman, 1982). With both methods we were able to obtain a complete set of service characteristics and the underlying cognitive reasoning. In the later stages of development, we conducted a prototype-based adaptive conjoint analysis to quantify relative importance and the preferred levels of service characteristics. These analyses were conducted in Switzerland, Israel, and Greece.

We claim three major contributions to extant research. First, our results provide information on what consumers expect of mobile video services and which reasons drive these expectations. Second, our results quantify the relative importance of service attributes, for example price vs. context dependency. Third, we provide a methodology on how customer needs for break-through mobile service innovations can be obtained. This enables a customer-centric development of radical innovations.

BACKGROUND—MOBILE MULTICASTING

MCAST's multicasting technology enables cellular operators to use shared channel resources for broadcasting video and any other data over 2.5G and 3G networks. MCAST also yields a seamless roaming to WLAN networks. Therefore, MCAST aims at supporting cellular operators to establish affordable flat-fee services for end users and increase operators' revenues per channel resource, allowing economic delivery of media to an unlimited number of cellular and WLAN devices.

Current Technology Constraints

Currently, rich media content can be delivered over cellular networks using unicasting (one-to-one) technology. This has two major short-comings: high delivery cost and limited cell capacity. Delivery cost is high, since each mobile terminal accesses a content server for on-demand content. When users view rich media content, their mobile terminals consume an excessive amount of bandwidth. This results in very high by-the-minute or by-the-packet charges. Due to limited cell capacity, unicasting of rich media can only support a limited number of subscribers at any given time. As the number of online users increases, additional bandwidth is required. Current technology performance, therefore, allows only poor service levels and implies lost revenues.

Challenge, Solution, and Opportunity

Multicasting technology is based on a one-to-many broadcast concept. It enables the delivery of identical content simultaneously to an unlimited number of subscribers. This allows services to scale to almost any number of users while having a manageable and limited impact on available bandwidth per cell. For the end user, multicasting represents a convenient way of accessing rich media content. In this sense, from a user as well as business model perspective, multicasting is believed to be a successful bearer for rich media content over 2.5G and 3G cellular networks.

Since there is currently no competing or ready-to-market technology that can provide multicasting services over 2.5G or 3G cellular networks, the MCAST research project moves on the forefront of technological development (Heitmann, Lenz, & Zimmermann, 2003; Northstream, 2002), and it will contribute to the ongoing standardization process of multicasting in the 3rdGeneration Partnership Project (3GPP). Alternative technologies like DVB-H required substantial investments in new net-

work infrastructure, and others like unicasting have an operational cost disadvantage. With its technological characteristics, multicasting is particularly suitable for rich media content (e.g., video, audio, gaming). Major market research institutions forecast the market potential of video services to nearly double that of audio services (e.g., Müller-Veerse, 2001) and a take up in 2005/06 (e.g., de Lussanet, 2003; Ovum Research, 2002). Based on this, our research focuses primarily on the delivery of video clips to mobile handsets.

EARLY-STAGE IDENTIFICATION OF CUSTOMER REQUIREMENTS

The early and qualitative part of customer integration employs focus groups to determine critical customer requirements as well as individual in-depth interviews following the means-end chain (MEC) methodology (Gutman, 1982) to understand the cognitive structures of decisions and the social motivation for requirements.

Focus Groups and In-Depth MEC Interviews: Background and Methodology

The focus group research was structured according to a theoretical concept for comprehensive and customer-driven product and service design: the OIL product design concept (Schmid, 2002). According to this, an evaluation of product expectations has to consider the levels of organizational design, interaction design, and logic design.

The organizational design level supplies the structural basis for the product design task. It answers the question of *who* and *what* is involved in the product use. Thus, in the case of a customer-oriented design of MCAST services, user groups and content categories must be determined. The interaction design concentrates on the processes and interactions between the relevant elements defined in the organizational level. It thus answers the question of *how* the product will be integrated into everyday life. The logic design examines *why* users use a specific innovation. Based on this understanding of the decision process, the product's language and communication strategy can be designed.

The succeeding means-end chain (MEC) approach (Gutman, 1982) generates an understanding of customers cognitive structure. The MEC concept is partitioning this cognitive structure in three layer—that is, service attributes, needs, and values. In the market research and service design literature, the qualitative MEC analysis has been increasingly an object of scientific debate (e.g., Aschmoneit & Heitmann, 2002; Grunert & Grunert, 1995; Herrmann, 1996a, 1996b; Wansink, 2000). It is based on two assumptions: (1) values, defined as desirable end states of existence, are dominant in the formation of selection structures; and (2) people deal with the variety of services by forming classes to reduce decision complexity.

For the formation of classes, consumers consult perceived and anticipated consequences of their actions or decisions. They associate positive consequences, namely benefits, with certain decisions (Reynolds & Gutman, 1988). Personal values allocate a positive or negative valence to these consequences (Rokeach, 1973). Thus, a correlation between the concrete and abstract characteristics of a service, the functional and psychological consequences, as well as the instrumental and target values is assumed (Gutman, 1997). Since consumers form classes to simplify their decision-making process, relatively few values are connected to a larger number of consequences and attributes. In this hierarchy, the importance of values determines the importance of consequences and attributes (Rosenberg, 1956).

Values represent beliefs about oneself and the reception of oneself by others. They are understood as universal, object-, and situation-independent convictions about desirable end states of life (Schwartz, 1994). The MEC framework is used to reveal the connections between time-stable values and product attributes directly relevant to decision making.

To obtain such results, the laddering technique with individual in-depth interviews is employed (Reynolds & Gutman, 1988). Research has shown that, on average, after 10 to 15 interviews, the number of additionally obtained consumer needs is decreasing radically (Griffin & Hauser, 1993). The technique reveals links between attributes, consequences, and values. The mentioned interactions between the obtained constructs were counted and entered into an implication matrix (not shown), a quantitative, tabular summary of the laddering interviews. This matrix provides the basis for the graphical representation in the form of a hierarchical value map (HVM), which displays the chains between values, benefits, and attributes, and their strengths (Herrmann, 1996a; Reynolds & Gutman, 1988).

Focus Groups and In-Depth MEC Interviews: Results and Implications

In total seven focus groups were conducted, three in Switzerland and four in Israel. Participants were selected from two mobile operators' customer databases according to a screener questionnaire to find high-volume customers with strong interest in innovative services. Each group consisted of five to eight participants; discussions lasted 60 to 90 minutes. The identified issues relate to: (1) relevance and entertainment qualities of content;

Table 1. Exemplary focus group results—key requirements

	Group 1	Group 2
Content	• Good Editing of Content	• Availability/Quality of Content
	• Up-to-Date Content	• Up-to-Date Content
	• Local Content	• Width of Content
	• Fun	• Independence of Content
Technology	• Reliability of the Service	• Battery Consumption
	• Saving Functionality	• Screen Size
	• Picture Quality/Resolution	• Size of Device
	• Sound Quality	• Rapid and Secure Transmission
Service	• Personalization	• Anytime and Any Where
	• No Advertisement	• No Advertisement
	• Forwarding	• Forwarding
	• Easy to Operate	• International Roaming
	• Price	• Price
	• Customizability of Content	• Customizability of Content

Figure 1. Hierarchical value map (HVM)

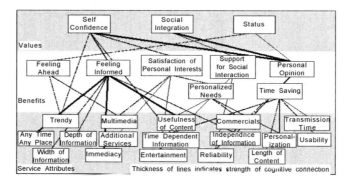

(2) speed, visual quality, and reliability of technology; and (3) customizability of the service. We spare a detailed discussion of the focus group results and provide an exemplary overview of key requirements mentioned in two of the groups in Table 1.

For the MEC analysis, 30 innovators and early adopters were selected in Switzerland. The sample consisted of students and employees between the ages 20 to 40 of companies offering financial and consulting services. Interviews lasted between 30 minutes and one hour.

The obtained constructs complement on the one hand the requirements identified within the focus groups. On the other hand the MEC

approach allows structuring of the cognitive reasoning in an HVM (see Figure 1), summarizing service characteristics at the bottom, service benefits in the middle, and associated personal values at the top layer. For MCAST, two key paths of end user reasoning can be identified. One relates to information and self-confidence, and the other is associated with social integration:

• **Self-Confidence:** Being informed and deriving a personal opinion are among the main benefits associated with the reception of rich media content on a mobile device. That is, end users seek news content, enabling them to feel up to date at any point in time. Three characteristics led to this benefit—the immediacy, the usefulness of the content, and the "any-time-and-any-place" characteristic. A service that follows this reasoning should not only provide updated information, but also ensure the contextual relevance of information.

- **Social Integration:** Consumers feel multicasting services may support them in achieving this goal by providing a basis for social interaction and the development of a personal opinion. While the latter greatly depends on the reliability of the service and the independence of the presented information, the support for social interaction also depends on the entertainment characteristics of the service.

The identified cognitive pathways provide guidance for service development. Immediacy, the relevance of content, and entertainment qualities should especially be taken into consideration. Winning companies will include the benefits of "Feeling Informed," "Support for Social Interaction," and "Forming Personal Opinions" to address the beliefs of consumers.

LATER STAGE CUSTOMER REQUIREMENT ANALYSIS

The preceding analyses show that customers consider a wide range of characteristics when evaluating a mobile multicasting service, which bears still too much complexity for service design. Therefore, we employed an adaptive conjoint analysis (ACA), a sophisticated customer research approach (Green, Krieger, & Wind, 2001; Hauser & Rao, 2002) to determine the weights of characteristics.

Adaptive Conjoint Analysis (ACA): Background and Methodology

The ACA allows identification of the relevance of service attributes and their levels—that is, it reveals the relative importance of different service attributes. The generated database allows the running of price sensitivity analysis for different product scenarios and an estimation of purchase probabilities (Johnson, 1991). Compared to other types of conjoint analyses, the ACA enables a dynamic adoption of a questionnaire according to given answers to preceding questions. This allows generation of robust results also for complex product offerings with a high number of attributes (Huber, Wittink, Johnson, & Miller, 1992; Orme, 1999) and ensures suitability for Web-based survey design (Dahan & Hauser, 2002).

Before implementing the ACA, the attributes under investigation have been reduced in an additional iteration step to 13 attributes. The objective of this step was not only to fulfil the conjoint requirements (of attribute independence, relevance, objective exclusiveness), but also to select attributes in conjunction with technological capabilities and business relevance. Accordingly, defined attribute levels are shown in Table 2.

The ACA was programmed using SSI Web of Sawtooth Software. The ACA questionnaire was hosted online and complemented by supplementary questions on general mobile usage behaviour and content requirements. This survey was conducted in three countries. In Switzerland 125 individuals have been invited from an academic database. Participants were required to be heavy users of mobile services. They were informed about the multicasting service by use of an animated prototype and in-depth information provided with an interactive CD-Rom. Participants were then asked to an-

Table 2. Attribute-level matrix

Attribute	Level 1	Level 2	Level 3
Length of Content	• Max. 30 sec.	• Max. 1 min.	• Max. 2 min.
Number of Clips per Day	• 5	• 10	• 15
Premium Content	• Available	• Not Available	•
Subscription Fee (•)	• 3	• 6	• 9
Forwarding	• Via MMS	• Not Possible	•
Ensured Transmission	• Retransmission	• Clips Lost	•
Supplemental Internet Service	• All Clips Online	• Missed Clips Online	• No Clips Online
Advertisements	• Yes	• No	•
Notification on Missed Clips	• No	• Per SMS	• Per MMS
Number of Content Categories	• 5	• 10	• 15
Number of Clips in MCAST Inbox	• 3	• 5	• 7
International Roaming	• Available	• Not Available	•
Location-Based Content	• Available	• Not Available	•

swer the online survey. After data was cleaned to ensure data robustness, 103 data sets were used for analysis. In Greece and Israel the participants have been recruited from the project partners' databases. The main difference was that users in these countries had the opportunity to use the service in the life network for a duration of four weeks. After data cleaning, the analysis contained 67 participants in Greece and 97 in Israel.

Adaptive Conjoint (ACA): Results and Implications

Questions complementary to the actual ACA asked for the general background of participat-ing individuals (e.g., demographics, telecommunication behaviour). Among others, these questions confirmed the interest in news and location-specific content. Furthermore, video content proves to be the most preferred format.

The following section selectively documents the quantitative conjoint results. Data reveals that in all cases, five service attributes influence almost 50% of the consumer decision (see Figure 2). Not surprisingly, price concerns yield a high score of 13.8% in Israel, 10.6% in Greece, and 11.7% in Switzerland. However, attribute scores are rather evenly distributed, and given an adequate price span and a flat fee pricing model, price appears to be not an over-riding decision criterion for consumers.

Figure 2. Cross-country comparison of attribute importance from ACA

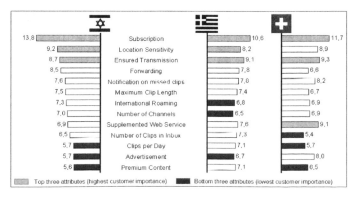

Figure 3. Attribute levels for clip length and number of clips per day (Switzerland)

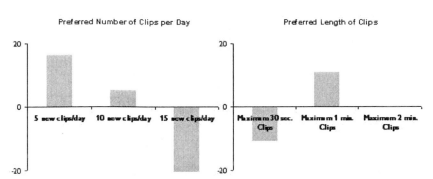

Figure 2 depicts the attribute importance in consumer decision making for the three countries. Certain similarities can be identified—for example, the high weight of the item "Ensured Transmission," which points to the importance of a technically flawless service. Since multicasting services are broadcast to a group of subscribers once and simultaneously, it might happen that a few subscribers do not receive the content due to handset unavailability or interrupted transmission. Users are concerned to lose out on these clips and therefore strongly require the notification and the back-up through supplemental Internet services.

Taking a closer look at single attributes (here we chose the data for Switzerland) and their levels reveals interesting aspects of the willingness to consume mobile data services. As documented in Figure 3, subscribers rather prefer a reduced number of (five) clips per day combined with a maximum clip-length of one minute. This behaviour relates to the concern about content relevance, but also about technical capacity (e.g., transmission speed and memory capacity) mentioned during the preceding qualitative surveys.

Analysing the findings of the three country-specific surveys on an aggregate level reveals four main patterns of consumer behaviour regarding mobile video services. As shown in Figure 4, these patterns relate to (1) the proven existence of a willingness to pay, if the price is controllable, preferably a flat fee. (2) The second pattern describes the users' preference for a reduced but tailored mobile video offer; that is, despite flat pricing, users do not always opt for maximum of outputs. This behaviour is rooted in concerns for relevance, information overload, and technical capacity, as also shown in the MEC-analysis. (3) The very advantage of mobile technology of delivering services "anywhere-anytime" is also a valuable selling point for mobile video services. That is, companies must develop intelligent means to satisfy the need for current and contextualised (personalised/localised) services, without destroying the scale effects of mass-broadcasting. (4) Precision is precious—this pattern represents the users' concerns about technical reliability rooted in past cognitive dissonance and disappointing experiences, and it implies a call for command of technology.

CONCLUSION AND BUSINESS BENEFITS

By reporting insights in terms of methodology and identified customer preferences regarding mobile-rich media services, we address the lack of customer knowledge in marketing practice and research in the mobile media industry.

Figure 4. Evidence for behavioural patterns in observed countries

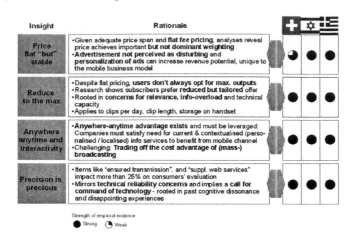

While dealing with the development of a leading-edge multicasting technology, we deployed a set of sophisticated tools for customer integration along the development process. For customer research science, we show a methodology, on how customer needs for breakthrough mobile service innovations can be obtained in a way that generates results, which can be easily communicated within single companies and across innovation networks. With the growing importance of cooperative product development, investigations on the latter, such as a joint customer integration and its qualities, will be an area for future research.

For management, our quantitative empirical results imply precise insights for superior mobile multicasting service design. Additionally, the identified cognitive reasoning of consumers provides input for general communication and marketing strategies. We show that most importantly, management needs to master the doubts on technology performance, and that mobile content must be tailored. The latter point complicates the marketing challenge as it trades off the multicasting cost advantage. For marketing and communication strategy, we have identified that the consumers' desire for self-confidence and social interaction should be addressed.

NOTE

An earlier version of this chapter appeared in: Cunningham, P., & Cunningham, M. (Eds.). (2004). *E-adoption and the knowledge economy: Issues applications, case studies* (Vol. 1, pp. 50-58). Amsterdam: IOS Press.

REFERENCES

Aschmoneit, P., & Heitmann, M. (2002). Customer-centred community application design. *The International Journal on Media Management, 4*(1), 13-21.

Dahan, E., & Hauser, J. R. (2002). Product development—Managing a dispersed process. In R. Wensley (Ed.), *Handbook of marketing* (pp. 179-222). Thousand Oaks, CA: Sage Publications.

de Lussanet, M. (2003). *Limits to growth for new mobile services.* Cambridge, MA: Forrester Research.

Green, P. E., Krieger, A. M., & Wind, Y. (2001). Thirty years of conjoint analysis: Reflections and prospects. *Interfaces, 31*(3), 56-73.

Griffin, A., & Hauser, J.R. (1993). The voice of the customer. *Marketing Science, 12*(1), 1-17.

Gruner, K. E., & Homburg, C. (2000). Does customer interaction enhance new product success? *Journal of Business Research, 49*(1), 1-14.

Grunert, K. G., & Grunert, S. C. (1995). Measuring subjective meaning structures by the laddering method: Theoretical considerations and methodological problems. *International Journal of Research in Marketing, 12*(3), 209-225.

Gutman, J. (1982). A means-end chain model based on consumer categorization processes. *Journal of Marketing, 14*(6), 545-560.

Gutman, J. (1997). Means-end chains as goal hierarchies. *Psychology and Marketing, 14*(6), 545-560.

Hagedoorn, J., & Duysters, G. (2002). External sources of innovative capabilities: The preferences for strategic alliances or mergers and acquisitions. *Journal of Management Studies, 39*(2), 167-188.

Hauser, J. R., & Rao, V.R. (2002). *Conjoint analysis, related modeling, and applications.* Unpublished manuscript, MIT Sloan, USA.

Heitmann, M., Lenz, M., & Zimmermann, H.-D. (2003). *Preliminary user needs analysis for MCAST.* St. Gallen, Switzerland: MCM Institute.

Herrmann, A. (1996a). *Nachfrageorientierte produktgestaltung: Ein ansatz auf basis der "means end"—theorie.* Wiesbaden, Germany.

Herrmann, A. (1996b). Wertorientierte produkt—und werbegestaltung. *Marketing ZFP, 18*(3), 153-163.

Huber, J., Wittink, D. R., Johnson, R. M., & Miller, R. (1992). Learning effects in preference tasks: Choice-based versus standard conjoint. In *Proceedings of the Sawtooth Software Conference,* Ketchum, ID (pp. 232-244).

Johnson, R. (1991). Comment on adaptive conjoint analysis: Some caveats and suggestions. *Journal of Marketing Research, 28,* 223-225.

Müller-Veerse, F. (2001). *UMTS report—an investment perspective [online].* Retrieved August, 2002, from http://www.durlacher.com

Northstream. (2002). *The competitive landscape of mobile video on demand [online].* Retrieved February, 2002, from http://www.northstream.se/21/

Orme, B. (1999). *ACA, CBC, or both?: Effective strategies for conjoint research: Sawtooth software.* Sequim, WA.

Ovum Research. (2002). *Ovum forecast: Global wireless markets.* London.

Pedersen, E., & Ling, R. (2003). Modifying adoption research for Mobile Internet service adoption: Cross-disciplinary interactions. In *Proceedings of the 36th Hawaii International Conference on System Sciences 2003,* Hawaii (pp. 534-544).

Pousttchi, K., & Schurig, M. (2004). Assessment of today's mobile banking applications from the view of consumer requirements. In *Proceedings of the 37th Hawaii International Conference on System Sciences 2004,* Hawaii (pp. 184-191).

Reynolds, T. J., & Gutman, J. (1988). Laddering theory, method, analysis, and interpretation. *Journal of Advertising Research, 28*(1), 11-31.

Rokeach, M. J. (1973). *The nature of human values.* New York: The Free Press.

Rosenberg, M. J. (1956). Cognitive structure and attitudinal affect. *Journal of Abnormal and Social Psychology, 22,* 368-372.

Schmid, B. (2002). *Kommunikations—und medienmanagement.* Unpublished manuscript, St. Gallen, Switzerland.

Schwartz, S. H. (1994). Are there universal aspects in the structure and content of human values? *Journal of Social Issues, 50*(4), 19-45.

Veryzer, R. W. (1998). Discontinuous innovation and the new product development process. *Journal of Product Innovation Management, 15,* 304-321.

Wansink, B. (2000). New techniques to generate key marketing insights. *Marketing Research, 12*(2), 28-36.

Chapter LV
Applying Mobility in the Workforce

Bradley Johnstone
BK Solutions, Australia

Khimji Vaghjiani
BK Solutions, Australia

ABSTRACT

There have been significant advances in mobile technologies in recent years. The euphoric technology void left by the dot-com crash in early 2000 soured many technology users; however mobile computing has provided much needed enthusiasm for both technologists and business users. In this chapter we focus on aspects of mobile technology, from both a business user perspective and a technology view point. Aspects such as total cost of ownership, return on investment and capital investment have been discussed from a financial perspective. Technical aspects of running and maintaining a mobile technology infrastructure have also been explored. The chapter concludes with a review of potential areas of application for mobile technology. The area discussed is mobile technologies in banking; however, many of the aspects covered could easily be applied to any other business vertical. Finally, this chapter is not meant to be a holy grail for mobile computing. It is simply a glimpse of the need to explore the power of this emerging technology.

THE MOBILE WORKER

This chapter discusses the application of mobile technologies in the business processes related to a mobile worker, with specific emphasis on the financial and banking industry. The term mobile worker has been used for many years to describe a worker who travels to various locations in order to conduct their business. This could describe a mechanic going to the car that needs repair, a courier delivering parcels, or even a sales representative who travels in order to showcase and sell his or her products. While these workers have been around for a long time, the term mobile worker has increasingly come to represent a mobile workforce well equipped with mobile technologies and associated devices.

The needs of the aforementioned workforce, coupled with recent advances in telecommunications technologies and infrastructure across the globe, has resulted in abundant opportunities for many businesses to create new mobile-enabled business processes that were hitherto unimaginable. For example, it is now possible to transfer an enormous amount of data at speeds that allow devices and the applications running on these devices to interact in real time with each other irrespective of time and location. The Internet facilitates connection of businesses across geographical and time barriers resulting in increased collaboration demanding innovative and technologically savvy business processes (Unhelkar, 2003). For example, many small business owners are totally eschewing physical offices and, instead, operate their businesses almost entirely on the road by fully utilizing the capacities of their mobile telephones and personal digital assistants (PDAs). Medium to large businesses also have a need to customize those business processes that can incorporate mobile technologies. However, the manner in which businesses and people within businesses utilize mobility depends on the type of industry in which they operate and the specific organization's own methods of doing business. The deployment of a mobile workforce has unique challenges that also depend, to a large extent, on the individuals that comprise the workforce. UK Company Softlab Ltd. (2004) suggests that mobile workers fit into one of three core groups requiring a mobile solution:

- management,
- sales and marketing personnel, or
- customer service representatives.

Managers are able to make key decisions and send and receive important correspondence via laptops and PDAs anywhere within mobile network coverage. Sales and marketing personnel can close deals with their clients and process the order while still at their client's premises. This alone ensures that the level of customer service is higher by allowing the worker to spend more time in front of the customer and not having to wait until they are back at the office to fulfil the order.

Customer service representatives can be armed with many different types of mobile devices designed to enhance the customer experience. An example in the banking sector is the ability to approach a customer waiting in line at a branch and quickly processing their request by connecting wirelessly to the bank's internal network via a tablet PC (Kuykendall, 2004).

The mobile workforce will grow rapidly as the next wave of mobile solutions arrives. We can expect to see a much wider deployment of applications to a greater number of workers from different industries. The applications that mobile workers use will vary depending on the business need. The 2003 findings of UK Company QNB Intelligence, who interviewed IT directors across Europe to gauge what applications they would use in their business if they were to deploy a mobile workforce, showed:

- 88% required general office applications
- 56% required applications to automate the sales processes
- 46% required tools to assist field agents
- 36% required applications to assist delivery and collection of goods
- 24% required management recording tools

Some sample applications within the above categories are explained further.

General office applications include tools such as e-mail and calendar events. A worker can have access to his or her e-mail or schedule via a laptop or other handheld device. This enables him or her to have the latest information on hand at all times, including receiving changes to schedules. This solution requires little investment, as the devices and software are readily available.

Products that automate the sales process along with customer relationship management (CRM) tools are being deployed in enterprises all over the world in order to effectively track customer habits and provide better customer service. It seems only natural that these tools should also extend to the mobile worker. In fact many CRM providers have already built mobile workforce functionality into their systems. The salesperson who has access to all of the clients' data is in a better position to answer any queries a customer may have and also allows for longer face-to-face time with a client.

Tools to assist field agents can also ensure that agents have their schedules arranged and updated throughout the day by the head office. They also ensure agents' access to a company's inventory, allowing them to field customer orders on the spot with up-to-the-minute inventory levels. The agent may use a tablet PC that displays an order form so by the time they are ready to leave the customer, the order is already being processed at the warehouse, allowing for faster delivery.

Delivery and collection of goods improve greatly with the inclusion of mobile technology (Egan, 2003). While delivery may be predetermined based on the parcels a driver may have on board, calls for collections come in at different times. New collection information can be pushed to the driver and placed within the list in the most efficient order. With the use of GPS systems, a mobile device can even give the driver directions to the next port of call.

Management recording tools can be particularly useful for consultants who charge based on effort—they are able to record start and finish times on their handheld devices, which are then relayed back to their office's

billing system for automatic processing (Bailey, Buist, & Vile, 2003).

Of course there are so many more applications available, and the above represents only a mere fraction of the mobile solutions that are already available with these five main categories.

FORMS OF MOBILE WORKER

As previously discussed, a mobile worker is usually considered to be a person who conducts their business by travelling to varied locations. It is important to note that an employee may in fact remain in one fixed location and still be considered a mobile worker. The definition can therefore be extended to include workers who can connect to a company's system with a device that can be easily moved from one place to another. A typical employee will work at a single location with a device that is attached to the company's local area network via a physical cable; this of course limits their ability to perform tasks on the go. Others may require the ability to connect to a network or Internet from any location outside the bounds of physical cables.

In assessing what sort of technology should be deployed, we must consider the employees' roaming habits. It is possible to group mobile technology into four basic categories (Gessel, 2001):

- Stationary at one location
- Stationary at many locations
- Many locations and moving
- Concealed and moving

Stationary at One Location

People who fall under this heading generally work from one location either at or near their desk. These workers may use a laptop which is easily moved or a handheld device such as a PDA or a tablet that accesses the company's wireless network.

Stationary at Many Locations

Employees who find they work in multiple locations may fall under this category. Similarly to the above scenario, these users are in the range of the company's LAN or wireless network generally by being at a premises operated by the business. An example in a banking context would be a financial advisor who operates at different company locations on different days and requires a connection to the network from a device they carry around to all of the locations.

Many Locations and Moving

Many workers find that they perform tasks in many different locations and are subject to many working environment constraints like physical access to a company's network. They may opt to use handheld devices or mobile telephones which allow them to remotely connect to the company's network via a mobile telephone carrier (Broersma, 2004). In a banking context, we might consider a financial advisor visiting a client at their home with the ability to look at the client's online file when and as needed.

Concealed and Moving

There are mobile application scenarios that require no human interaction. A primary example is a warehouse system that tracks stock through using radio frequency identification (RFID) tags. As human involvement is not required, the tasks are generally high speed and non-complex. A typical warehouse RFID system tracks company inventory levels upon request or at pre-determined intervals.

CONSIDERATIONS BEFORE GOING MOBILE

For many years, U.S. courier company FedEx has been using custom-made equipment that allows drivers to record pickups and deliveries on a handheld device. This data is stored until the driver arrives back at base where it can be uploaded to the company's network (Egan, 2003). This legacy technology is being replaced with handheld terminals modeled on the Pocket PC operating system that incorporates General Packet Radio Services (GPRS) that can report changes as they happen. This allows FedEx to accurately track their drivers in order to give better estimated times of arrival to clients.

This mobile solution is returning clear benefits to both the company and their customers, some of which include (Urich, 2002):

- a reduction in paperwork,
- shortening business process cycle times,
- reducing labour requirements, and
- freeing up time to allow for greater productivity.

The benefits of a more efficient and better organized workforce returns real cost savings as well as competitive advantage for the organization.

A business case should justify the investment in new mobile technology. It must be clear that there will be an improvement in the businesses current processes gained from adopting the new technology. Technological solutions have in the past failed because procuring departments thought the technology was "sexy" and they failed to purchase based on a benefits or needs basis. No investment should be made without a strong business case justifying the investment. Preparing a business case for mobile strategies has so far withstood intense scrutiny by company finance departments who demand clear justification on spending (Gold,

2005). A 2003 report by Mobile Competency Founder Bob Egan suggests that when preparing a mobile strategy, management should assess the total financial impact including both *Total Cost of Ownership* and traditional *Return on Investment* analyses.

Total Cost of Ownership (TCO)

When assessing the TCO of a mobile solution, it is important to consider and ensure optimal investment in each of the following four areas:

- Capital Expenditure
- Operations
- Support Services
- Administrative Operations

Capital Expenditure

In deciding where to invest, management should try to standardize the solution wherever possible. This will allow for greater bulk discounts on equipment and services which will in turn lower the ongoing support costs. Deployment costs are also likely to be lower when a standardized solution is implemented.

Operations

How will the implementation of a mobile solution affect the productivity of an end user? Any increase in productivity could potentially reap a large company enormous savings when spread across its entire workforce. The increase in productivity may also equate more face-to-face hours with a sales force and their customers.

Egan (2003) discusses the case of a global bottling group which enabled their sales staff with handheld devices that included a smart selling application. As a result of this new technology, the sales staff close rates dramatically improved. The system provided sales staff with prompts that identified up-selling and cross-selling opportunities, and even helped determine future demand.

Support Services

The success and acceptance of a mobile program by employees may hinge on the available technical support that is provided (Egan, 2003). Standardizing the system should reduce any negative impact by allowing IT support to become familiar with all offerings which will allow them to quickly resolve any issues that staff may have. Through adopting a standardized approach early, change management processes will be easier to control.

Administrative Operations

Administration costs may also improve over time as processes become more efficient. Traditional mobile workers spend their day filling in paperwork only to have a team verify and process the results at a later time. If they used a mobile device, the administration team could have access to their data in real time or some processes could become fully automated.

Return on Investment (ROI)

There are many ways to calculate ROI for new solutions. Some outcomes are predictable and others harder to ascertain without specific use cases. When calculating ROI for a mobile solution, the two benefits below may assist the calculation:

- Efficiency Gains
- Improvements in Staff Productivity and Effectiveness

Efficiency Gains

Speeding up a specific task or application can be a simple way of generating a return. For example, providing financial advisors with the ability to view client data while on site will allow them to keep the customer focused, which may not be the case had the advisor needed to make a phone call back to base which of course will involve another bank employee. This process is likely to keep them in control of the session instead of breaking the momentum of the meeting to retrieve the client's data. The speeding up of this process will allow the advisor to present all facts and better answer questions, which will put them in a better position to get the client's business.

Improvements in Staff Productivity and Effectiveness

Field agents can improve their effectiveness by having access to e-mail, calendar events, and contact data while on the road. Using a laptop connected to a mobile phone for this task is not a new concept, but it is cumbersome. Accessing this information on a handheld device connecting via a mobile telephone network is likely to receive more use. It is also likely to be a more economical solution as the GPRS network is charged on data transferred, whereas the laptop and mobile phone solution will incur charges based on the amount of time the user is connected (Deshpande & Gilbert, 2002).

MANAGEMENT BARRIERS TO MOBILITY

Other than the typical budgetary restraints, there are a number of other barriers whose existence can generally be attributed to a manager's perception of the effectiveness of a mobile workforce or other trust issues.

Many see benefits in adopting mobile technology that assists workers who are on the road, or improves the efficiencies of staff at a company location. There is another group that needs consideration, however: the flexible worker. This is a worker who is granted the opportunity to work from a location other than the office for mainly work life balance reasons or to make savings in office rental costs (Sweeny Research, 2002). The flexible worker tactic may also be used to retain quality staff, and affording them the luxury of flexibility allows them to better handle working pressures.

Some of the key areas that can inhibit a mobile worker are:

- Managers' Lack of Trust in Mobile Workers
- Hostility Between Mobile and Non-Mobile Workers
- Motivation Issues

Most of the above issues could best be addressed by managers setting performance evaluation techniques that would allow them to monitor the workers' output in terms of quality and productivity.

Managers' Lack of Trust in Mobile Workers

A paper by Sweeney Research (2002) revealed that more than 50% of managers surveyed were less likely to trust workers who had a flexible arrangement and could elect to work from home. This attitude is perhaps the greatest inhibitor to realising a widespread flexible workforce.

Hostility between Mobile and Non-Mobile Workers

Another issue which may arise is the mistrust from a non-flexible worker towards the flexible worker. The traditional workers surveyed by Sweeney suggested that they felt that flexible workers did not work as hard as they did.

A common issue was that unsupervised flexible workers were perceived to be more likely to conduct their personal business during working hours and therefore work less productively.

Motivation Issues

Managers surveyed by Sweeney felt that as they did not have traditional control of flexible workers, they would find motivating them to be a very difficult task. A manager could not control the distractions that a flexible worker faced such as watching television or conducting their personal business when the employee should be working.

Before adopting a mobile solution for the workforce, it is crucial that the above areas are considered and a clear strategy is in place to stop these areas from becoming an issue within the organisation. Failure to address this early will lower morale and breed mistrust between workers. This will have a direct impact on quality of customer care, and the bottom line will suffer. With clear ground rules set in stone early, going mobile can produce enormous gains in productivity, and improvements to the workers' quality of life as well as customer satisfaction rates.

SECURITY CONSIDERATIONS

Security is important in all applications, but becomes an essential consideration when applications are used in the field. A mobile product should support a wide variety of authentication and encryption standards to integrate easily into an existing infrastructure. This area should be addressed by managers before deploying a mobile worker, and risks must be identified and mitigated before any investment is considered.

Wireless Policy

Companies are very protective of the data on their networks. Deploying a mobile workforce brings a range of new security challenges that must be addressed (Gallagher, 2004). The solution starts with a top-down policy that addresses why wireless is being used, what the business objectives are, and what policy governs the entire company in this area.

The wireless policy must demand that workers strictly adhere to the standards contained within and also govern how workers will protect the devices within their possession.

Details of what to do to prevent loss of a device or immediate action required in the event that the device is lost should be well known to all users. If the device stores sensitive information, then a high level of security is required. A high level of security should involve having all files on the device password protected and preferably encrypted. There are many different methods that can be used to protect a company's data including encryption, data wipe technology, secure transfer, firewalls, and a worthy protectionist login regime.

Encryption and Secure Data Transfer

The mobile workforce should worry IT security specialists who need to protect company data. Mobile solutions should incorporate security standards that encrypt data between the server and mobile device. Session-based keys should be used with any encryption method, and a secure public key private key exchange should be employed to ensure communications integrity (Hildebrand, 2004).

Processing transactions securely on the Web means that we need to be able to transmit information between the client and server in a manner that makes it difficult for other people to intercept and read. An example of a secure data exchange is through the use of Secure Sockets Layer (SSL) sessions.

SSL sessions work through a combination of programs and encryption decryption routines that exist on the hosting computer and in client browser programs such as Microsoft's Internet Explorer. Below is a high-level description from Intel that explains how an SSL session is negotiated between server and client:

1. Browser checks the certificate to make sure that the site you are connecting to is the real site and not someone intercepting.
2. Determine encryption types that the browser and Web site server can both use to understand each other.
3. Browser and server send each other unique codes to use when scrambling (or encrypting) the information that will be sent.

The browser and server start talking using the encryption, the Web browser shows the encrypting icon, and Web pages are processed secured.

In addition to SSL sessions, it is important to understand that wireless devices can be more of a target for theft or are easily lost. Companies concerned that the device may go missing or fall into the wrong hands should consider the sensitivity of the data on the device and assess that risk. Organizations should have a plan in place to guard against this scenario. Wireless devices can encrypt all data written on its hard drive, rendering it useless to anyone who tries to read the data, whether from the device itself or by any other means (such as mirroring the hard drive).

Encryption Toggling

Depending on the circumstances, encrypting data flow may or may not be critical. When connected to a LAN inside the office, protected by the company firewall, encryption may not be as essential. In addition certain user groups may deal with less sensitive information that does not require encryption thus speeding their communications sessions.

Allowing the system to automatically or manually turn the encryption feature on or off for some applications will ensure that the network is secure and minimizes performance impact according to Intel's 2004 article "Wireless Security Best Practice." This process is known as *encryption toggling* and is designed to enhance the user experience in less secure environments. It is important however to fully assess the risk and sensitivity of data and user permissions to change the settings affecting toggling.

Multiple Authentication Modes

Any new solution should support all current and future company authentication modes. End users should be able to authenticate to the server using existing user IDs and passwords, whether connecting via LAN, WAN, or GPRS.

In addition, adding a new user or a new device by an administrator should require no additional IT intervention.

Sniffer Solutions

Companies are not alone in their fight to secure data over wireless networks (Intel, 2004). Sniffer technologies are available for enterprises that need to lockdown systems on mobile devices. Many of the benefits of wireless technologies also create risk. While the flexibility of being able to operate in any location is the primary attraction of wireless networks, the need for security must also be considered. It is important to have appropriate security to ensure that data is stored properly, travels properly, and is protected from people who should not have access to it, but is still accessible to those who do need it.

Sniffer products function in a wireless environment like LANs or WANs. A wireless sniffer card in a laptop can detect and verify that your encryption protocols are being used and can find rogue access points, enabling you to shut them down if necessary.

MOBILE SOLUTIONS FOR THE FINANCIAL INDUSTRY

Financial service providers have invested heavily in applications with a consumer focus. While investment in this sector will continue, banks are looking at improving internal efficiencies and providing new tools for employees to assist in delivering an exceptional customer service experience.

Financial institutions are in a broadly defined category, and there are many different employees that can benefit from mobile technologies. Some of those who would benefit are financial advisors, fund wholesalers, insurance agents, management, brokers, risk inspectors, investment bankers, analysts, customer service representatives, loan officers, and many other disciplines (Intellisync, 2003). Some of the ways that mobile technology could be used by a bank are detailed as follows.

Branch Applications

The next step in wireless banking is likely to focus on technology used within the bank, including wireless Internet access and access to company information. These types of wireless capabilities can increase employee pro-

ductivity and customer satisfaction. A roving teller can use a number of different devices, like a portable notepad or laptop or even a PDA, and can greet the people as they come in the door. For example, if customers want to find out if a check has cleared, they can find out without needing to wait in line, thereby reducing waiting time for other customers.

Implementing a wireless network within the bank can also prove to be cost effective. Banks looking to expand can benefit from installing a wire-free network. If a bank is going to open up new branches, it will likely be more cost effective to deploy a wireless network vs. a wired network (Ramsaran, 2004). However, banks may not be so quick to move to a wireless network because of security issues and standards protecting customer data. Manufacturers claim that this is because they are not fully informed of the steps taken to secure an internal wireless network. They also claim that there is no more reason to be concerned about that than any other wired network security because it has the same protection.

E-Mail

An obvious first step in extending mobile communications to the mobile workforce is empowering them with handheld devices that can send and receive e-mail. Response times to internal and external customers are greatly improved, as workers can address e-mail as soon as it arrives just like in a fixed location.

The type of device used will depend on the requirements of the worker. If simple text mail messages are being sent, then any device from a mobile phone to a PDA would be fine. In the case of large attachments, anything from a PDA up to a laptop connected via a mobile phone would be required. In any event the type of device deployed must match the business need and be able to perform adequately prior to any investment.

Executive Dashboards

With the financial services moving at a frenetic pace, executives may struggle to keep abreast of what is happening within their organisation. This information can easily be provided on a handheld device. A dashboard shows key performance metrics that will assist the manager in making informed decisions with up-to-date information.

Software Maintenance

IT teams will be able to dial in to the company's mobile device and perform software upgrades remotely. These measures will save time and money since the employee will not have to travel to hand in the device. The list of benefits continue, with more expensive alternatives avoided such as having to burn and send CDs or sending out a field technician to perform the install or update.

Sales Force Automation (SFA)

Account managers with extended SFA applications should boost productivity by speeding up each process along the sales cycle. Allowing anytime access to vital corporate data allows for improved customer responsiveness, better tracking of customer activity, increased customer acquisition, and enhanced communication with the corporate office.

Competitive and Product Information Delivery

Remote and mobile workers benefit greatly from being able to receive frequent updates to competitive analysis reports and regulatory requirements during remote selling opportunities. It is critical for all remote personnel to have the latest product and policy information to avoid making false promises to customers.

Wholesale and Retail CRM Applications

The most important goal for wholesale and retail sales representatives is to cultivate meaningful relationships with clients. To succeed, representatives need easy access to large amounts of contact and sales data, enabling them to make more sales calls and improve the quality of the calls. This data can also assist with up-selling and cross-selling opportunities.

Quoting for Insurance Agents

The abundance of paper-based processes gives the opportunity for a high percentage of submission errors. Automating the task of filling out insurance application forms, and allowing agents to provide easy quoting and sales tracking will improve an organization's close rate.

Commission Reporting and Tracking

The complexity of tracking compensation and incentives for large sales forces is well understood. A mobile application can streamline the process of communicating compensation and progress against goals back to the field. This helps keep representatives motivated and ensures they are spending their time on revenue-producing activity.

Access to the Company Intranet

Mobile workers need to access internal corporate data and documents frequently. Phone lists, vacation requests, expense report forms, travel policy, product information, and other common documents are used daily. Placing a local copy of a company's intranet information on a mobile device helps keep bandwidth costs down, while improving accessibility and providing more rapid access to company procedure. This also allows for offline access to company forms where a network connection is unavailable.

Profiling Tools

The use of profiling tools allows the sales staff to view detailed information about clients and other influencers. Personal interest, purchasing patterns, and responsiveness to past promotions can all be tracked. This allows the company representative to tailor each sales call and personalize the experience for increased effectiveness.

Sales Analysis and Planning

Financial sales operations are flooded with data, both self-created and from third-party data sources. Sales representatives need powerful analytics to help them sift through the information, and develop strategies and plans for boosting sales in their territories. Interactive analysis and planning tools save time and help ensure best practices.

Interactive Selling Tools

Increasingly, sales representatives use slick multimedia interactive selling tools to structure their interactions with clients and provide a richer experience. The materials change frequently and can be automatically updated without significant cost.

Mortgage Origination

Real estate transactions are another data-intensive financial service greatly enhanced with the use of online forms and documents. Mortgage brokers will be able to expedite the loan pre-qualification and approval processes, help-

ing both buyers and sellers get fast and accurate information.

Mobile PIM (Personal Information Manager)

Communications such as flight and hotel confirmation numbers, cancelled appointments, and client emergencies are vital to the remote worker and may need swift action. Workers can stay more productive on the road with mobile access to up-to-date calendar and contact data.

Risk Management

Effective risk managers need to be able to provide swift assessments and deliver detailed recommendations based on those assessments. Automating the processes of risk management—including identifying, analysing, planning, tracking, controlling, and communicating—ultimately leads to less risk exposure for the client.

CONCLUSION

We are clearly at a time of change in the way that we service our customers. Companies around the world see value in a mobile workforce for a variety of reasons. According to The Bulletin's Mark Phibbs (2003), the main drivers will be:

- improvements in staff productivity,
- retaining quality people,
- providing better service to customers with rapid and appropriate response,
- providing tools to cross-promote products, and
- increasing field agents' contact hours with their customers.

These are only some of the reasons tempting companies into adopting a mobile strategy.

There are many factors that will affect how these solutions will be deployed. According to Intel's (2003) *unwired* advertising blitz that the growth of Wi-Fi hot spots will be the tipping point tip for the market, it is only a matter of time before we are all members of a mobile workforce. It would appear that perhaps we are now at, or at least close to the tipping point in terms of mass market appeal. The market seems to moving away from Intel's desired *hotspot* solution and instead investigating deploying applications based on mobile telephone carriers' 2.5G and 3G technologies. GPRS has already gained mass market appeal, and 3G promises more of the same with much richer and more useful content.

There are hurdles when considering how to build a mobile workforce, but all of these seem to be manageable. Strong business cases for mobile access to enterprise systems are very likely to fuel investment in this space (Deshpande & Gilbert, 2002).

The transition to a mobile workforce need not be a painful exercise (Euronet, 2000). Following a set of guidelines, decision makers can make the transformation with ease. Some of the areas that do need consideration are to:

- build a mobile strategy on business use,
- make security a priority,
- have a framework to manage the devices deployed, and
- match the business need with the mobile coverage.

Instead of rolling out a mobile solution "just because it's cool", management should decide on a strategy that determines who will use it and how will it affect the business, what information the field personal need, and how they could access it remotely. Most importantly what is the likely return on investment, and is it a worthwhile proposition?

Security is perhaps the largest inhibitor affecting the decision of a company to invest in wireless technology (Gallagher, 2004). Any mobile solution must incorporate encrypting data, both as it travels in the air and when it is stored on the hard drive. Unscrupulous people will *sniff out* wireless data, and provisions should be put in place to ensure that the data is useless to them. In addition, strong authentication measures will help prevent leaking of valuable information.

The cost of a wireless solution will also depend on the variety of devices within the organisation. Similar products should be sourced to gain better discounts and lessen the technical support and training impact (Egan, 2003).

The choice of device should also be limited to the mobile provider's coverage and capacity available. There is no sense in arming a mobile workforce with tablet PCs so they can download large documents only to find the connection speed is too slow to be beneficial.

The hardware is already available, and GPRS mobile infrastructure is now well and truly in place. It can only be a matter of time before the mobile workforce is a common feature of mainstream business.

REFERENCES

Bailey, R., Buist, C., & Vile, D. (2003). *Corporate wireless data in Europe*. Windsor, UK: QNB Intelligence.

Broersma, M. (2004). Planes and trains: Wireless on the move. *Techworld Magazine*. Retrieved April 20, 2004, from http://www.techworld.com/mobility/features/index.cfm?FeatureID=512

Computer Strategic Innovation. (1998). *Developing a mobile strategy—Preparing for the next discontinuous change.* London: Computer Sciences Corporation.

Deshpande, N., & Gilbert, J. (2002). *GPRS: How does it work and how good is it?* Retrieved from http://www.intel.com/update/departments/wireless/wi10021.pdf

Egan, B. (2003). *Making the case for wireless mobility investment.* North Providence, RI: Mobile Competency.

Euronet Worldwide. (2000). *Keeping a wireless world connected.* Retrieved from http://www.euronetworldwide.com/solutions/biz_lines/software/mobile_banking.asp

Gallagher, H. (2004). *Security still reigns as wireless's weakest link.* Retrieved from http://www.macnewsworld.com/story/32874.html

Gessel, B. (2001). *Editorial comment: Should the telecom industry rethink architecture of its wireless networks?* Retrieved from http://www.mobileinfo.com/Editorial/August15.htm

Gold, A. (2005). *Managing mobility in the enterprise.* Northborough, MA: A. J. Gold Associates.

Hildebrand, C. (2004). *Steps to success: Managing mobile technology.* Retrieved from http://searchcio.techtarget.com/originalContent/0,289142,sid19_gci959674,00.html

Intel Corporation. (2004). *Wireless LAN technology.* Retrieved from http://www.intel.com/business/bss/infrastructure/wireless/solutions/technology.htm

Intel Corporation. (2004). *Wireless security best practice.* Retrieved from http://www.intel.com/business/bss/infrastructure/wireless/security/best_practices

Intellisync. (2003). *Mobile strategies for financial services.* San Jose, CA: Intellisync Corporation.

Kuykendall, L. (2004). *Retail-minded banks shift focus to traffic service.* New York: American Banker.

Phibbs, M. (2003, August). Mobile workforce. *The Bulletin Magazine.* Retrieved from http://www.chamber.org.hk/info/the_bulletin/aug2003/it.asp

Ramsaran, C. (2004). *Tripping over wireless technology.* Retrieved from http://www.banktech.com/news/showArticle.jhtml?articleID=18600398&pgno=3

Softlab Ltd. (2004). *The mobile workforce.* Retrieved from http://www.softlab.co.uk/fm/142/WPThemobileworkforce.pdf

Sweeney Research. (2002). *Mobility and mistrust.* Melbourne, Australia: Sweeney Research.

Synchrologic. (2003). *The CIO's guide to wireless.* Atlanta, GA: Synchrologic.

Urich, K. (2002). *Eight steps to going mobile.* Everett, WA: Intermec Technologies Corporation.

Chapter LVI
Applying Mobile Technologies to Banking Business Processes

Dinesh Arunatileka
University of Western Sydney, Australia

ABSTRACT

This chapter discusses the impact of mobile technologies on service delivery processes in a banking environment. Advances in mobile technologies have opened up numerous possibilities for businesses to expand their reach beyond the traditional Internet-based connectivity and, at the same time, have created unique challenges. Security concerns, as well as hurdles of delivering mobile services "anywhere and anytime" using current mobile devices with their limitations of bandwidth, screen size and battery life are examples of such challenges. Banks are typically affected by these advances as a major part of their business deals with providing services that can benefit immensely by adoption of mobile technologies. As an example case study, this chapter investigates some business processes of a leading Australian bank in the context of application of mobile technologies.

INTRODUCTION

Electronic commerce has become a dynamic force that has changed the way businesses operate on a global scale (Shi & Wright, 2003). Due to increased globalization, individuals, organizations, and governance frameworks have an increasing dependence on communication technologies. The Australian Communication Authority envisions that ubiquity is the "best possible outcome" in terms of the future of business and economy in the country. This ubiquity is based on the elements of technology, market dynamics, users, and rules and guidelines (ACA, 2005). All business organizations in this global context are forced to look at this "best possible outcome" in order to stay competitive. This gives rise to several research questions in the areas of business practices as well as workflow management, and affects the individual and collective social behaviour (Mylonopoulas & Doukidis, 2003). The research areas also focus on the mobile technologies and their application to businesses, with particular emphasis on the method and manner in which services can be delivered using mobile processes. Mobile processes are business processes that are executed with the use of mobile devices such as PDAs (personal digital assistants), mobile phones, or mobile-enabled laptop computers. Thus, mobility, which is the ability to move freely while performing regular business activities, has become an extremely crucial aspect of today's business processes. Furthermore, as per Archer (2004), in order to incorporate mobility, business processes also have to undergo substantial changes themselves to make it essential that the changes are researched and experimented into.

Internet Usage in the Banking Sector

Banks, as primary institutions of service-oriented business, have increasingly leaned towards e-commerce-based operations. Emerging mobile technologies offer "anytime, anywhere" type of banking that results in better customer orientation and provides personalization of services to the customer. The concept of banking using handheld devices, such as PDAs or other mobile devices, is becoming popular as it enhances the Internet connectivity to the fingertips of the customer (Unnithan & Swatman, 2002). The Internet has also provided opportunities for service providers such as PayPal, an online payment processing company founded in 1999, to offer more cost-effective payment-related services similar to banking services to its customers. PayPal, after a mere four years of operation, has become the most used payment system for clearing auction transactions on eBay (Schneider, 2004), competing directly with the traditional banks. Banks thus face a major challenge and are forced to effect substantial cost reductions in order to be more competitive and offer cost-effective services to its customers. Banks aggressively push their customers to use electronic means for most of their banking, as these electronic transactions are far cheaper as compared to over-the-counter or ATM (automated teller machine) transactions. According to a recent study in the U.S., a teller transaction costs the bank U$1.07. as opposed to a telephone transaction costing 54 cents, an ATM transaction costing 27 cents, a software-based PC transaction costing 1.5 cents, and an Internet-based transaction costing a mere 1 cent (Money Central, 2005). Mobile devices enable secure and convenient use of e-banking, payments, brokerage, and other types of transactions which are part and parcel of the banking sector (Herzberg, 2003). Another study reveals that among the Internet-based banking users, there is a positive tendency to use mobile devices to do banking transactions (Coutts, 2002). Hence, factors determining the success or failure of the mobile business and how the corresponding mobile systems and applications are designed, in order to provide banks with cost-effective, flexible, and customer-oriented business processes, are of interest to the banking community. The fact that today's banking customers are more educated, along with increasing demand for state-of-the-art services, also add pressure and push the banks towards mobile technologies.

Global Banking Industry

The educated and technology-savvy customers demanding better service and state-of-the-art technology is a global phenomenon in the banking industry. For example, the banking industry in Europe is undergoing substantial changes as it looks to reduce costs and enhance the utility for customers through new technology. European banks are focused more on their core capabilities while exploring different sourcing options for non-core capabilities. They are disaggregating their value chain into independently operable functional units (Homann, Rill, & Wimmer, 2004). Furthermore, as communication capabilities reach higher levels of performance and reliability, these functional units are combined across corporate borders, providing valuable e-collaborations and flexibility for the organization. The industry sectors are changing the way they do business by using many different collaborations with customers (B2C), service providers (B2B), funding organizations (e-payments), government (B2G), and even competitors (B2B) (Arunatileka & Arunatileka, 2003). The emerging mobile financial applications including both mobile payments and banking services are also being investigated, showing how new financial services could be deployed in mobile networks and also identifying key players in the mobile financing value chain (Mallat, Rossi, & Tuunainen, 2004). Mobile customer relationship management is another area that would personalise business processes, adding value to organisations (Unhelkar & Arunatileka, 2003).

This chapter specifically investigates the implications of service delivery using mobile technologies. Since the author of this chapter has been researching within a well-known Australian bank (name is withheld due to confidentiality issues), this chapter is based on the service-delivery challenges related to incorporation of mobile technologies within the banking environment.

STUDY OF TODAY'S BANKING NEEDS

This section starts with a brief introduction of banking requirements studied in a leading bank (referred to merely as 'the bank') in Sydney. The bank was seeking to create a policy to introduce mobile technology to its banking processes and staff. This policy for mobile services (services using mobile processes) is meant to evolve from a broad framework of existing policies defined for the operations of the entire bank that also encapsulated its values and objectives. The vision of the bank is to inculcate three great values—namely, *teamwork, integrity,* and *performance.* The vision drives the organisation purposefully towards its objectives based on the four foundations of *staff, customers, corporate responsibility,* and *shareholders* (internal documents of the bank). The bank comprises different business units divided based on functionality for management purposes. These various units have trained staff in different disciplines. For example, the financial markets division would have trained investment advisors, the institutional banking division comprising trained relationship managers and customer care personnel, the retail banking division having trained home loan advisors and customer care personnel, whereas the human resources division trained human resource and training personnel. Thus, the bank consists of a multi-disciplined, heterogeneous workforce. The processes and the work methods of separate units could be very different from each other as well. For instance, the corporate and institutional banking divisions work internationally, thus having a 24-hour operation, whereas the retail banking is more likely to be an office hour operation subject to few exceptions. As mobility options that could increase productivity in corporate banking, PDAs could be programmed to notify users of any new e-mails and SMS messages where the employees are notified immediately, and necessary action could be taken depending on the situation. This enables the employees to have more time with their families while still attending to urgent business. In retail banking, loan officers could use mobiles to be more competitive in the field. As parts of an organization, these various divisions have to work together as one entity in achieving the vision and goals of the entire organization.

The mobility policy, once accepted and incorporated, has to address the purpose and objectives underlined by the top management in the facilitation of better service delivery providing higher value to customers. This should fulfil most of the outlined areas by the management by facilitating the staff on better access, wider responsibility, and better tools, motivating them on better service delivery, achieving customer satisfaction, which in turn would fulfil the corporate objectives.

Business Units under Study

There were four business functional units identified by the bank as the initial study areas for introduction of mobile technologies, namely: financial markets, institutional banking, retail banking, and small investor operations. Although the last two units have a high volume of transactions, the first two functional units were given priority by the bank due to their high-value transactions and the need for change by the employees and the customers of these units. Furthermore, financial markets and institutional banking also had a pressing need to change their existing processes due to some existing limitations in their operations. Before the effect of mobility is investigated, a brief summary of how these two units work is described here.

- **Financial Markets:** This is a highly specialised unit, which brings the bank high revenues and very high profitability. Although the number of customers may be lower, the revenues are very high and revenue per customer is very high, resulting in these customers demanding and warranting individual attention. The concerned managers would be international business managers and state managers who are handling time-sensitive corporate accounts. These managers travel a lot, meeting customers and looking after their specific interests.

- **Institutional Banking:** This is also a very highly specialised unit where institutional banking managers generally go to their client organisations in order to serve their needs. Although there are less volumes of transactions, the values could be very high, bringing very high profitability to the bank. Most accounts would belong to large business organisations having diversified needs. Mobile access to the systems would make it much easier for these managers to be in touch with the latest communications, rates, and so forth, which would be very essential in serving business customers in a highly competitive market.

BACKGROUND TO RESEARCH IN THE BANK

The background reasons for researching into the bank's processes in the context of mobility were based on the long-term expectations of the bank. Usually, the expectations of an organisation are summarized in the mission and vision statements. The bank had very specific values and purpose, spelled out in its mission and vision statements. The bank also used the

Figure 1. A balanced score card for the bank

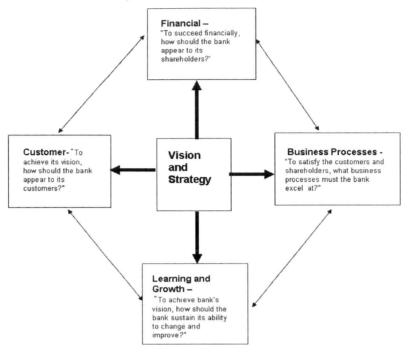

concept of a balanced score card in order to maintain the policy balance and drive its strategies. The balance score card concept (Balanced Scorecard, 2005) is built on four perspectives—financial, internal business processes, customer, and learning and growth, with the vision and strategy in the centre of it.

The bank envisions and makes strategies in the business with respect to these four perspectives. The four perspectives were adapted from www.balancescorecard.org, which describes the perspectives with respect to the organization. All four perspectives are so tightly entwined that one leads to the other. Financials is the starting point, which is the very core of the objectives of a business, simply to pass a value to its owners/shareholders. However, since the banking industry is highly competitive, the customers should be well looked after in order to retain them. Business processes, learning and growth, and customer perspectives all contribute to make the bank more customer oriented. The main focus in this chapter is aimed at

business process perspective, which refers to the business processes of the organisation. The measurement on this perspective allows the managers to know how well the business is running in terms of whether the products and services conform to customer requirements (the mission of the organisation). In addition to strategic management processes, two kinds of processes could be identified—mission-oriented processes and support processes (Balanced Scorecard, 2005). The applications of mobile technologies are looked at in all these processes.

The bank conceptually was looking at two major areas of mobility: mobility at the workplace and mobility in the delivery of service to customers. The mobility at the workplace is in line with a more futuristic plan for a new headquarters building with a smart office where the employees could work anywhere within the building. Mobility in delivery of service is more with the current business processes and how they could be improved to offer better service delivery to customers. In-depth knowledge of the existing work processes, coupled with a detailed plan to transform the existing organization into a mobile enabled (m-enabled) organization without disrupting the day-to-day functions of the organization, is one outcome expected of this exercise. Technology has made progress from earlier setbacks, but the methodologies to fully implement the existing technology into the business operations have to be done carefully in a systematic manner.

The initial expectations of the bank were to look at the business unit of financial markets. The decision was financial and also partly need driven since the financial markets were a big earner and the managers in the unit were very keen to move forward with new technology. Once the focus area was identified, the focus was concentrated on the fundamental questions.

TRANSFORMATION OF BUSINESS PROCESSES

The financial markets area was selected as the first unit for this study since the most pressing demand for change was persistent in this unit. There were several research questions arising with this selection. They were:

- How does the service-based industry change with application of mobile technologies? A generic question focused on the service-based industry as a whole investigating the possibilities to improve efficiency while cutting down on long-term costs.
- What are the changes happening in the banking industry? This is an industry-specific question to look at what is happening elsewhere in the banking industry which is relevant to the current timeframe.

- What should be the bank's response to the change in the service industry and specifically in the banking industry? This probing question is looking inward into the bank's own processes critically to decide how change could be facilitated to improve on the internal processes.
- What is the expected impact after mobile technology is introduced to a selected business unit in the bank? A specific question arising from mobile technology being introduced to the unit which is expected to create a positive impact.
- What would be the direct impact on the customers once the new technology is fully implemented? A probing question to understand the impact on the customers once the change is made.
- Would there be any anticipated problems during the changeover? A question to understand the management of change from the existing to the mobile-enabled organization. It is also important to measure this change in terms of cost factors, time factors, customer satisfaction factors, security issues, and other such factors which the bank thinks important to consider.

Let us look at transforming a simple business process, like the checking of an account balance by a customer as an example. The bank would concentrate on the cost of the process, the satisfaction of the customer in the process, assurance of the process in delivering the right information, security issues in providing the service, and timeliness of the information from the customer's point of view.

If the process is m-transformed from checking the balance through the Internet or at an ATM to use of a mobile phone, how will such a transition be measured in terms of this process? In order to understand the m-transition of the

Figure 2. A business process transformation

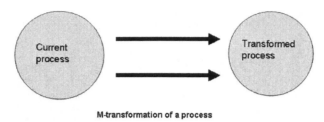

M-transformation of a process

Figure 3. A mobile transition roadmap

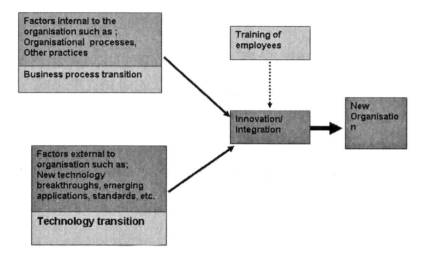

process, it is essential that the process is understood and examined critically. The first step in such a critical evaluation is the creation of a mobile transition roadmap. Such a roadmap is proposed in Figure 3, which shows m-transition applied to a bank focusing on the overall picture while the transition is in progress.

The mobile transition roadmap will capture all the areas that have to be considered in m-transforming an organization in order to become a mobile-enabled organization. The concept of the roadmap for mobile transformation, as shown in Figure 3, has been adopted and evolved from the electronic transformation roadmap (Ginige, Murugesan, & Kazanis, 2001). The mobile transition roadmap is further analysed and investigated using different perspectives in Figure 8, later in the chapter.

The Business Processes in the Financial Markets Division

The Fund Managers and International Business Managers (FM/IBM) are called to service customers at anytime of the day since financial markets are a global business. Different market segments would be working in different time zones. It is important be on top, knowing what is happening all over, all the way from Sydney to New York through the other giants such as Japan, Hong Kong, and Europe. Therefore the managers working in this unit should be dynamically connected. This is one area where mobile technologies could be used very effectively to enhance productivity. The role of an FM/IBM in particular involves visiting clients and understanding their requirements in international business from telegraphic transfers to structuring major import/export deals, to financing solutions for the funding of these transactions to mitigate risks associated.

At present, laptops are being used which contain the data and programs to run the business that is provided online through the system. With the very mobile nature of the worker, being in the field for long periods visiting clients, the access to the system could pose problems. Dial-up access could be time consuming and frustrating. Response time would be one of the major factors for corporate customers in quantifying the service levels of the bank. The customers may be of a captive nature in the short term, and the exit may take some time due to contractual obligations. However, high ser-

vice levels could keep the corporate customers on a long-term basis in line with the bank's objective of "achieving at least a 5% increase in agreed customer satisfaction measures." Thus, the retention and growth would be of great interest to the bank. Expensive acquisition is also valuable in this category of high value customers.

Example 1—Transition of Travel Process of FM/IBM

Travel process is discussed herein where the mobile-transformation was suggested. The process of travel in the financial markets division is undertaken by an FM/IBM. These managers travel often to meet customers all over Australia.

Figure 4. Activity diagram, FM/IBM travel—current process (before m-transition)

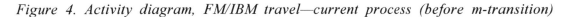

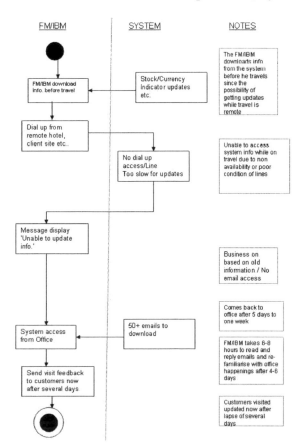

Two activity diagrams are drawn to show the travel process before and after mobile transition. The activity diagram which is a tool from the Unified Modeling Language (UML) is used herein as it is a typical analysis tool. Figure 4 depicts the picture before m-transition and the activities of the FM/IBM involved in the process.

In Figure 4, the current process for travel for the FM/IBM has been depicted. The FM/IBM visits the customers but does not have access to the bank's system most of the time due to bad lines and low line speeds. Therefore the FM/IBM would visit customers and be travelling for about five days every two months with no access to the bank's system most of the time. During this time, the FM/IBM is completely cut off from the news in the bank and e-mails from their own customers as well. Moreover, the customers visited during the trip will not get any feedback for a considerable time until the FM/IBM has access to the bank systems. When he or she finishes travelling and gets back to office, he or she would have to take six hours (approximated, based on current estimates) to read and respond to e-mails before starting on his or her other work such as making proposals for the customers visited.

This process of travel is modelled using an activity diagram (which is like a flowchart, and is derived from www.omg.org) for easy comparison. Note that the customer, the most important person in the process, does not appear as an actor in the current process, since the FM/IBM has to wait until he or she gets back to the office to respond to the customers. Thus, the customer is not in the current process at all active.

Figure 5 depicts the scenario once the existing processes from Figure 4 are transformed, with mobile transition taking place. Each FM/IBM would each his or her laptop and could dial into the bank systems. The information would be initially text based so that speed problems could be overcome at initial stages until mobile technology is mature enough to deliver high bandwidth without problems.

The FM/IBM could download e-mails and relevant figures every night or every morning as and when he or she has free time before or after visiting customers while travelling. This would enable him or her to be in contact with the bank regularly. Customers could get draft proposals via e-mail since the FM/IBM could do the proposals quickly after visiting customers without having to wait until he or she gets back to the bank. It is also time saving since he or she is updated regularly and does not have to spend six long hours reading and responding to e-mails as per current estimates, since all that has already been done during his or her free time while travelling. Also note that the customer is an actor in this process. The customer gets the feedback while on travel.

Thus, mobile transition should save a lot of time for the managers in downloading and reading e-mails, and also should keep them in line with the current rates in the very volatile financial markets area. Also the entire sales process has been expedited, with the customers getting the draft version of their contracts very early. Thus the sales process would be shortened by several days, which could bring substantial income. Customers should also perceive this state-of–the-art process positively, which appears to be very active in comparison to the existing process.

Example 2—Transition of Customer Meeting Process of FM/IBM

The second process discussed herein is that of customer meetings of the FM/IBM. This is also very important for the bank since these meetings could bring very high-value business. Most of these meetings would be happening during the travel process. Therefore, the access to bank computers and systems will only be avail-

Figure 5. Activity diagram, FM/IBM travel—new process (after m-transition)

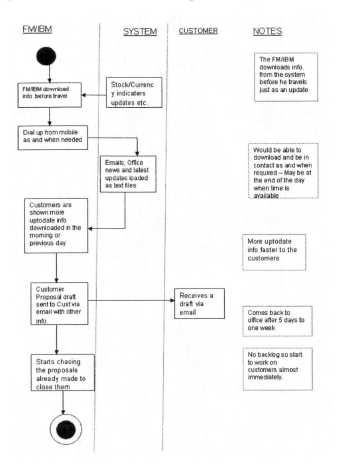

able through dial up. Speeding up the process would be profitable to the bank on the one hand and also would give them an advantage on customer service on the other.

Customer Meeting Process before M-Transition

Figure 6 shows the current process of customer meeting for the FM/IBM. The FM/IBM visits the customers, but does not have any forms in his or her laptop to enter any data at the customer's site. The FM/IBM must get information and then come back to the office and manually enter this data to generate relevant customer documentation. Thus, there is a considerable time gap until the customer receives

final documentation. This could be subject to errors, and due to manual entry without verification, time is needed to get customer feedback after the entry.

Customer Meeting Process—After M-Transition

In Figure 7, the current customer meeting process has been changed with the introduction of mobile technology. The new process anticipated after the m-transition has taken place is shown. The FM/IBM would have his or her laptop preloaded with forms to enter customer data, which could later be loaded into Lotus Notes. Data is entered and verified at the customer's site, saving considerable time. The

Figure 6. Activity diagram, FM/IBM customer meeting—current process (before m-transition)

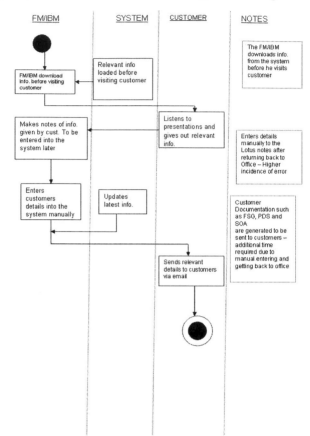

customer could get a draft report until final report (uploaded and updated with latest data) reaches him.

Considerable time savings is apparent with m-transition taking place. This would be crucial in higher productivity for the FM/IBM, while customers perceive higher service levels as well as remain ahead of the competition.

THREE PERSPECTIVES IN M-TRANSITION

To build on Figure 3, where transition of business processes and technology were merged in order to look at new business processes, Figure 8 describes the typical requirements on the bank's side, generalizing all the processes taken into consideration. The diagram considers the processes kept intact, changed, and scrapped, and also looks at the entire change from technology, methodology, and sociology perspectives (Unhelkar, 2003).

These perspectives are useful in managing the entire organizational change due to m-transition.

Technology Perspective

The issues such as what applications should be used, how these applications should be integrated and the networking of these applications, the security issues in the new organization (the mobile transformed organization), and

Figure 7. Activity diagram, FM/IBM customer meeting—current process (after m-transition)

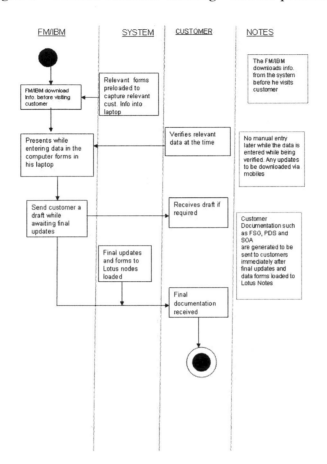

the devices that could be used to facilitate the employees in the service delivery aspects must be decided. In the FM/IBM travel example, the laptops and how to deliver connectivity while travelling has to be looked at from the technology perspective. The aspect of altering certain bank systems' fit enough for the mobiles to download them into a laptop computer has to be investigated. This technology is currently available.

Methodology Perspective

All the procedures that should be followed and adhered to in adopting the new business pro-cesses and the approach to be followed in order to transform into the new organization should be discussed. It is also very important that the employees are trained on the new security measures, devices, and new business processes. In the travel example, the new processes fol-lowed in order to download e-mails, delivery of business proposals, and so forth are the consid-erations in the methodology perspective.

Sociology Perspective

Wider issues include the management of the entire change process along with any legal implications and privacy issues in the new

Figure 8. Mobile transition of business processes in technology, methodology, and sociology perspectives

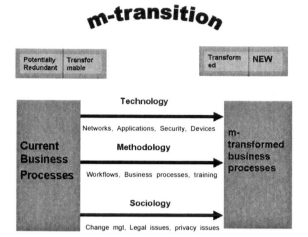

organization. Providing training to employees in these areas is also of utmost importance. In the travel example, the training of the FM/IBM on the new processes, any legal issues falling therein in giving the managers the authority to provide preliminary proposals, and so on will fall within the sociology perspective.

Table 1 shows that considerable time savings could be achieved in m-transition. There are other measures as well that are significant to consider. The customers who are mostly corporate clients would like to get up-to-date information. This falls within the customer perspective of the balanced score card of the bank.

Table 2 shows the comparison of the current and the proposed processes for customer meetings. Considerable time savings appear on delivery of proposals, and so forth. This also leads to better accuracies, as the data is entered at the customer site and verified then and there. Thus considerable monetary savings result due to sped-up process, and also additional revenue results due to customers being signed up earlier than the current system.

There would be additional costs for laptop and software upgrades in the proposed processes. However, benefits would be much more compared to costs involved since large time savings and additional revenue would compensate for one-time costs of upgrades. There will also be intangible benefits such as better customer satisfaction due to timely delivery and better employee satisfaction due to saving of considerable time, which could be used to make more customer visits.

The main concentration was on the business process perspective of the balanced score card. However, the correct implementation of this perspective would lead to improvements on the customer perspective leading to learning and growth and financial perspectives in a rolling effect.

Table 1. Comparison of the current processes with the proposed processes with regard to travel of the FM/IBM

Activity/Attribute	Current Process	Proposed Process
1. Downloading information before travel	Stock/currency indicator updates at office	Stock/currency indicator updates at office
2. Dial up for updates during travel	Too slow and difficult for updates	Mobile dial up for updates as frequently as necessary
3. Liaison with office	Infrequent/almost nil due to bad lines, etc.	Always in touch with frequent periodic updates
4. E-mail contacts	Almost nil	Periodic updates every evening with office
5. Feedback to the customers visited	Several days since the visit	The proposal could be delivered within a day
6. Timeliness of information on hand	Could be several days old	Only few hours since last update

Table 2. Comparison of the current processes with the proposed processes with regard to customer meeting of the FM/IBM

Activity/Attribute	Current Process	Proposed Process
1. Downloading information before visiting customer	Relevant forms are downloaded at office	Relevant form downloaded at office is updated with current information just before visit
2. Verification of customer information	Makes notes to do corrections later while entering	Updates and verifies data at current time before providing a draft to customer at his/her site
3. Re-verification of customer information	May be needed to be done via e-mail due to manual entering	Not required since it is already done
4. Liaison with office	Infrequent/almost nil due to bad lines, etc.	Always in touch with frequent periodic updates
5. E-mail contacts	Almost nil	Periodic updates every evening with office
6. Feedback to the customers visited	Several days since the visit to enter the information and verifications	The proposal could be delivered within a day
7. Timeliness of information on hand	Could be several days old	Only few hours since last update

CONCLUSION AND FUTURE DIRECTIONS

The transformation of the travel process and customer visit process of fund managers and international business managers introducing mobile technology into the bank created a situation wherein the bank stood to gain significantly in terms of savings of time and money. The mobile transition also highlighted potential intangible benefits such as better customer focus, timely proposals boosting customer satisfaction, and significant time savings, leaving the managers with more time to focus on their customers. The outlook created by the intangible benefits should also lead to gain more customers. However, a systematic approach is suggested, introducing the change gradually while educating the employees to be aware of the change. The customers are also being included in this process of gradual change towards an m-enabled organization.

Similar processes in the financial markets division and other divisions are currently under investigation with a view for further m-transformation. Once the other significant processes have also been transformed across all units, to use mobile technology effectively, the chances of the bank achieving a perfect balanced score card will be significantly enhanced.

REFERENCES

Archer, N. (2004). The business case for employee mobility support. In *Proceedings of the IADIS International Conference in E-Commerce,* Lisbon, Portugal.

Arunatileka, S., & Arunatileka, D. (2003, December). E-transformation as a strategic tool for SMEs in developing countries. In *Proceedings of the 1st International Conferences on E-Governance,* New Delhi, India.

Australian Communications Authority. (2005). Vision 20/20: Future scenarios for the communications industry—implications for regulation. *Final Report,* (April).

Balanced Scorecard. (2005). *What is the balanced scorecard?* Retrieved April 17, 2005, from http://www.balance scorecard.org

Coutts, P. (2002). *Banking on the move.* White Paper, Communications Research Forum.

Ginige, A., Murugesan, S., & Kazanis, P. (2001, May). A roadmap for successfully transforming SMEs in to e-businesses. *Cutter IT Journal, 14.*

Herzberg, A. (2003). Payments and banking with mobile personal devices. *Communications of the ACM, 46*(5), 53-58.

Homann, U., Rill, M., & Wimmer, A. (2004). Flexible value structures in banking. *Communications of the ACM, 47*(5), 34-36.

Mallat, N., Rossi, M., & Tuunainen, V. K. (2004). Mobile banking services. *Communications of the ACM, 47*(5), 42-46.

Money Central. (2005). *MsMoney.com—online banking—online fees.*Retrieved April 16, 2005, from http://www.moneycentral.com

Mylonopoulas, N. A., & Doukidis, G. I. (2003). Mobile business: Technological pluralism, social assimilation and growth. *International Journal of Electronic Commerce, 8*(1), 5-21.

Schneider, G. P. (2004). *Electronic commerce: The second wave* (5th ed.). Thomson Course Technology.

Schwiderski-Grosche, S., & Knospe, H. (2002). Secure mobile commerce. *Electronics & Communication Engineering Journal, 14,* 228-238.

Shi, X., & Wright, P. C. (2003). E-commercializing business operations. *Communications of the ACM, 46*(2), 83-87.

Unhelkar, B. (2003). *Process quality assurance of UML-based projects.* Reading, MA: Addison-Wesley.

Unhelkar, B., & Arunatileka, D. (2003, December). Mobile technologies, providing new possibilities in customer relationship management. In *Proceedings of 5th International Information Technology Conference,* Colombo, Sri Lanka (pp. 23-31).

Unnithan, C. R., & Swatman, P. M. C. (2002). Online banking vs. brick and mortar—or a hybrid model? A preliminary investigation of Australian and Indian banks. In *Proceedings of the 7th CollECTeR Conference,* Melbourne, Australia.

Chapter LVII
Mobile GIS—Challenges and Solutions

Pramod Sharma
The University of Queensland, Australia

Devon Nugent
The University of Queensland, Australia

ABSTRACT

This chapter focuses on Mobile GIS (MGIS), which uses wireless networks and small screen mobile devices (such as PDAs and smartphones) to collect or deliver real time, location specific information and services. Such services can be divided into field and consumer (location based services) GIS applications. The use of wireless networks and small screen devices, introduce a series of challenges, not faced by desktop or wired internet GIS applications. This chapter discusses the challenges faced by mobile GIS (e.g. small screen, bandwidth, positioning accuracy, interoperability, etc.) and the various means of overcoming these problems, including the rapid advances in relevant technologies. Despite the challenges, many efficient and effective Mobile GIS applications have been developed, offering a glimpse of the potential market.

INTRODUCTION

A geographic information system (GIS) is a computer-based system designed for the collection, storage, analysis, and visualisation of geographic data. Geographic data includes geographic location as an important attribute. The technology of GIS has undergone rapid development over the past three decades and in the process has transformed itself from mainframe-based systems to Internet-based distributed systems operating on a variety of hardware platforms (see Table 1). During this period, GIS applications have also changed from "the static compilations of the specialist to applications supporting the everyday lives of everyone, everywhere, all the time" (Smyth, 2000). The GIS hardware, software, and services industry was valued at over US$7 billion in 1999 and growing at over 10% per annum—estimated to be over

US$11 billion in 2004 (Longley , Goodchild, Maguire, & Rhind, 2001, p. 13; Daratech, 2005).

Early GISs of the 1970s were static, stand-alone, proprietary systems, focusing on inventory applications (e.g., inventory of natural resources, transportation networks, or utilities infrastructure) and on the automation of existing tasks. The next phase in the evolution of GIS involved the use of the networked client-server model to access remote data servers, and more advanced analysis and modelling capabilities. These systems were still, however, closed, stand-alone systems. They were used to model soil erosion, predict flood risk, model power network outages, and so forth. The GISs of today, however, are open distributed systems utilising the wired and wireless Internet to access distributed GIS services, tools, and spatial information for real-time data management applications (e.g., emergency management systems, location-based services). Not so obvious over the period has been the change in emphasis from "GIS software" to an emphasis on "GIS functionality"—the latter not necessarily delivered via "GIS software".

Traditional GISs are large project, departmental, or enterprise-wide PC- or mainframe-based applications, with full GIS functionality (e.g., natural resource inventory, urban management systems, utilities management systems). Such "legacy" applications continue alongside the newer types of Internet-based GISs—indeed, they are often the core component of the newer applications. Also, while the evolution of technology deserves analysis in its own right, it is the change in the user base ("market") rather than the technology change that is influencing new developments—technology is merely the enabler. Thus Internet GISs tend to have more limited functionality, and can be accessed by clients without GIS software and with little GIS experience/knowledge. The needs of these users are easily met

by simple mapping output to simple queries: Where is x? Where is nearest x? How do I get there? Internet GIS applications do, however, allow a much wider range of people to gain access to GIS tools and data via the wired Internet (e.g., the usage of MapQuest and WhereIs type services).

This "simplification" or "democratisation" of GIS technology finds a natural home in Mobile GIS (MGIS). In MGIS, current technology limitations—both in terms of the wireless communications infrastructure as well as those related to small-screen mobile devices—introduce additional constraints to GIS functionality. MGISs are, however, very well suited to a wide range of field and consumer applications, which do not require a full GIS toolset, processing power, and so on. MGISs complement traditional GISs, extending some GIS functionality into the field (e.g., to collect and update the databases of enterprise GISs).

This chapter focuses on MGIS—where the service coverage is wireless based and the client platforms are small-screen mobile and wireless devices, such as laptop computers, tablet PCs, PDAs, and cellular phones. We exclude laptop computers from any further analysis as, apart from the wireless connection, they essentially mimic the functionality of desktop computers. The chapter begins with a discussion of what MGIS is and the rationale for its introduction. It then examines some common applications, the challenges faced by MGIS, and some of the solutions employed to overcome these problems. It will conclude with a look at the future directions of MGIS services.

MOBILE GIS—THE INNOVATION

What is MGIS? MGIS refers to the access and use of GIS data and functionality through mobile and wireless devices such as mobile laptop

Table 1. Evolution from static to Mobile GIS

Stand-Alone GIS 1970s-1980s	Early Network GIS Mid-1980s-1990s	Complex Network GIS Late 1990s -	Mobile GIS 2000s -
Large, project applications, GIS software, and data reside on fixed computers; full GIS functionality; proprietary data formats; closed, stand-alone systems; applications concentrate on the automation of existing tasks; static inventory applications (e.g., natural resource inventories, early urban management systems, tax assessment); computer cartography	Data can be accessed on a server via LAN or WAN; allows centralised storage and remote access to data; software must be installed on fixed computers; full GIS functionality; proprietary data formats; closed, stand-alone systems; more advanced analysis and modelling applications (e.g., modelling flood risk, power network outages, transportation networks)	Distributed GIS; software, and data reside on one or more remote servers; clients can access GIS tools, services, and data through Web browsers; existing Internet applications are customised GIS applications with limited functionality (simple analysis and mapping); utilises open standards (e.g., OpenWMS), many desktop or professional GISs also allow access to remote data via the Web	Infrastructure features similar to complex network GIS; data and GIS software tend to reside on the server; some software can be installed on the mobile device (e.g., ArcPad); utilises wireless networks; a limited but specialised set of tools for fieldworkers and consumers; highly customised applications; utilises mobile devices; standards based; real-time data management applications (e.g., utilities maintenance, disaster management); location-based services

computers, tablet PCs, PDAs (such as Palm Pilots and pocket PC devices), and Web-accessible smartphones. They differ from traditional GISs and Internet GISs in that they utilise wireless networks and small-screen mobile devices. MGIS applications development requires a complete rethink of existing GIS models, although they are essentially a "repackaging" of Internet GISs for mobile devices, offering functionality similar to Internet GIS applications that utilise the thin client model (Peng & Tsou, 2003, p. 479). However, it is not necessary to have specialised MGIS software in order to offer such applications (e.g., such services can be provided using J2ME Location API, Oracle Spatial, and Oracle Locator). Major specialised MGIS software products include: ArcPAD from ESRI, MapXtend from MapInfo, IntelliWhere from Intergraph, and OnSite from Autodesk.[1]

Why MGIS? As GIS has evolved from essentially static systems to a technology which is available to "everyone, everywhere, all the time" (see Table 1), it is able to address needs which have resulted in the creation of MGIS. These needs include (Hassin, 2003):

- access to geodata in the field where it is often needed the most,
- ability to capture data in the field and in real time,
- ability to append positional information to data capture, and
- GIS functionality where it is often needed the most.

Given these needs, it is not surprising that the driving force behind the development of MGIS has been the needs of fieldworkers, as well as meeting the management's objectives of "field force automation" (i.e., to improve the work process of field service or field sales employees). MGISs have revolutionised field data collection through the many advantages they offer to fieldworkers and organisations: improved accuracy of field data collection and update; improved access to relevant and up-to-date data; improved efficiency in data collection, validation, and analysis; improved timeliness of the decision-making process; and so forth (ESBI Computing, 2005a; ESRI, 2005).

MGISs have also enabled users with small-screen mobile devices to access GIS applications and spatial information that was previously restricted to fixed desktop or workstation applications (even Internet GIS applications were only available for large-screen devices). The widespread use of mobile phones, especially smartphones, and the increasing use of PDAs for managing day-to-day activities have also led to the extension of MGIS applications for consumers. Indeed, for many users the main use of GIS services is when they are mobile (e.g., users of various directories, locator or navigation services).

What is the basic MGIS architecture? MGISs have unique requirements and considerations that set them apart from other types of GISs. MGISs require the following components:

- a mobile device that is capable of being detected by some position-determining technology,
- position-determining technology,
- a GIS content provider (Web server, GIS server, data server), and
- a wireless network provider (including gateway server).

They differ from wired Internet GISs in three respects: (i) the use of wireless networks to connect the mobile devices, (ii) the use of small-screen mobile devices, and (iii) a gateway server is required to interface the Web server and wireless network (Peng & Tsou, 2003, pp. 477-479). The first two components have major impacts on the applications that can

Figure 1. Components of an MGIS

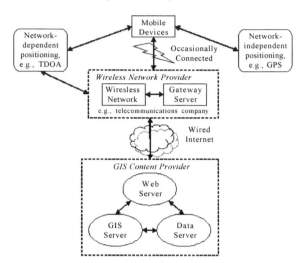

be developed, as they impose a set of significant constraints. On the positive side, they do allow the development of a new range of services that were not previously possible.

The wireless network provides a means of uploading or downloading information to or from a server to the workers in the field. Positioning technologies, particularly GPSs, enable the recording of the location of the mobile device and so can be used to record the location of features and objects at a known level of accuracy. Hence, MGIS has demonstrated the capability for real-time data collection and delivery in the field for a wide range of purposes (see Table 2).

APPLICATION CASE STUDIES

The major users of MGISs are field workers (data collection + service delivery) and consumers who need real-time geographic information and services (Peng & Tsou, 2003, p. 492). Hence, applications can be divided into Field GISs (e.g., utilities maintenance, pavement management, responding to outages, re-

Table 2. An overview of potential MGIS services[2]

	Applications	Example
Field GIS	Data collection	Recording soil sample data
	Data update and validation	Updating information on crop conditions
	Utility and facility maintenance (fieldwork/repairs)	Pavement management; telecommunications repairs
	Asset inventory and maintenance	Collect information on assets and their condition, e.g., street lights
	Customer repair services	Managing telecommunications repair crews
	Site analysis/inspections	Recording the presence of fire ants
	Incident reporting	Reporting the location of a car accident
Location-Based Services	Tour guide	Pre-defined tour around Brisbane City
	Routing	Best route from the hotel to the Queensland Museum
	In-vehicle navigation	Best route from home to the CBD
	Traffic information	Updates on traffic accidents, congestion, and so on in the nearby area
	Fleet tracking	Tracking parcel delivery vehicles
	Emergency response	Transmits the location of an accident to emergency services; finds the best route to an accident site; displays real-time information of the spread of a bushfire
	Concierge services	Where is the nearest Italian restaurant?
	Location-sensitive billing	Subscribers charged according to their location
	Marketing	Information on specials at a specific retailer
	Child safety	Monitoring where children are and if they move outside certain approved areas
	Criminal (release) tracking	Ankle bracelet that home release prisoners wear

Table 3. Fieldwork GIS case study 1 (Geological Survey of Western Australia, 2005)

Application Area	Asset inventory and maintenance
Summary	The Geological Survey of Western Australia (GSWA) is using Mobile GIS and GPS to map the location of abandoned mine sites (mines that closed before 1990). The aim of the project was to locate and document abandoned mine sites, identify safety and environmental hazards, and assess their state of preservation.
Push/Pull Services	Collect data on the location and condition of abandoned mine sites
Geographic Coverage	Western Australia
Positioning	LinksPoint's GlobalPoint GPS receiver
Mobile Devices	Symbol PPT 2800 mobile computers; now use an integrated PDA/GPS
Platform	ArcPad
Communications Network	Downloaded to laptop in the field and database updated later
Comments	Allows for effective and accurate data collection, and improved productivity

Table 4. Fieldwork GIS case study 2 (North London Strategic Alliance, 2005)

Application Area 2	Incident reporting
Summary	North London Strategic Alliance (NLSA) is using Mobile GIS in order to combat crimes "such as graffiti, fly-tipping, anti-social behaviour and vehicle abandonment and to improve environmental conditions that create the opportunity for these activities."
Push/Pull Services	Report the location and details of crimes in real time; allow immediate action to be taken
Geographic Coverage	North London, UK
Positioning	GPS
Mobile Devices	-
Platform	ArcPad, ArcSDE database hosted by ESRI; database can be accessed on a secure Web site hosted by ESRI (using ArcIMS)
Communications Network	GPRS
Comments	Mobile GIS is seen as "improving the end-to-end cycle of reporting an incident to taking corrective action." It improves efficiency and productivity. The recording of such information also allows for spatial analysis, the identification of trends, and hotspots.

cording accident locations) and Consumer GISs or location-based services (LBSs). This classification highlights a fundamental divide in the MGIS industry, and as such it is also a useful framework for discussing almost any aspect of the industry.

The main advantages offered by MGISs for fieldwork are the ability to collect, enter data, validate, and manage geographic data in real time at the relevant location. The use of this technology eliminates delays in data collection, data entry, validation, and database updates (Peng & Tsou, 2003, p. 453). Hence, the immediate advantage for field workers is apparent. Fieldworkers can also utilise MGISs in order to navigate to clients or facilities needing repairs, to manage repair crews and jobs in real time, and to access information in order to carry out specific tasks such as telecommunications repairs (Peng & Tsou, 2003, pp. 493-495). The use of Mobile GIS improves the efficiency, accuracy, and productivity of fieldwork. At the same time the system also permits managers to manage their field workers. Consequently, it is easy for enterprises to justify their investment in MGIS.

The Consumer submarket of MGIS is essentially what is referred to as location-based services, although some LBS applications (e.g., vehicle tracking systems) do not sit comfortably under the Consumer MGIS label. Both terms refer to a suite of applications which utilise the location of a mobile device in order to deliver real-time, location-specific, value-added, personalised services. The worldwide location services market is expected to increase from US$.7 billion in 2003 to US$9.9 billion by 2010 (Giaglis, Kourouthanassis, & Tsamakos, 2002, p. 65). LBSs encompass a wide range of services from in-car navigation (e.g., Navman GPS) to mobile tourist guides, concierge services, emergency services, location-specific marketing, and "find a friend" services. These services can be categorised into 'push' and 'pull' services. Pull services are those services elicited by the mobile client (e.g., where is the nearest Chinese restaurant?). Push services are those initiated by the service providers (e.g., location-specific marketing). These services support the client on the move and would have to be monitored to avoid becoming a nuisance (e.g., a client may not wish to be

Table 5. Consumer GIS case study 1 (AT&T Wireless, 2005)

Application Area	Friend Finder
Summary	AT&T Wireless Find People Nearby
Push/Pull Services	Find friends nearby; add and delete people on list; directions to restaurants, bookstores, etc.; locate people, call or SMS friends; hide your location from friends
Geographic Coverage	North America
Positioning	Network based (location of cellular transmission tower most recently contacted)
Mobile Devices	GSM/GPRS mobile phone
Platform	-
Communications Network	Need a phone with an mMode plan and within AT&T Wireless GSM/GPRS network coverage
Comments	Privacy is a major issue

Table 6. Consumer GIS case study 2 (PARAMOUNT, n.d.; Lohnert, Wittmann, Pielmeier, & Sayda, 2001; Lohnert, Mundle, Wittmann, & Heinrichs, 2004)

Application Area	Tour Guide
Summary	PARAMOUNT Public Safety & Commercial Info-Mobility Applications & Services in the Mountains. The purpose of PARAMOUNT is to improve information access, navigation, and safety for mountaineers/hikers. It will help mountaineers to locate, orient, and navigate in unfamiliar terrain, and improve safety and emergency response services.
Push/Pull Services	INFOTOUR delivers information on points of interest, navigation; access to weather forecasts, etc. SAFETOUR delivers safety information on weather, fire warnings, etc.; provides avalanche forecasting; tracks the user's location, sends emergency calls, routes rescue teams to aid mountaineers DATATOUR tracks willing users, updates trail network information, captures information on points of interest or trail difficulty, etc.
Geographic Coverage	Mountainous terrain, outdoors; Pyrenees and Alps
Positioning	GPS
Mobile Devices	Smartphones and PDAs
Platform	-
Communications network	GSM/GPRS, UMTS (in the future)
Comments	Problem of network coverage in mountainous regions; problem of GPS availability in rugged terrain

contacted every time he or she walks past Starbucks or McDonalds). The information delivered to the client may be in the form of text, audio, maps/graphics, or multimedia information. Before the implementation of LBSs, there was no support for the mobile user, except in printed form.

LBSs maintain three common data types: (i) base data on a region (e.g., street network, aerial photographs); (ii) point of interest (POI) data (e.g., museums, cinemas, restaurants, ATMs, etc.); and (iii) the location of the mobile user. They also involve a much longer value chain, including data providers, network providers, GIS software vendors, application developers, and so forth (McKee, 2004, p. 152). Such services rely on the delivery of location-based information, according to user profiles, preferences, and characteristics. As they continuously record the location of the mobile user and already have their personal details, privacy and security are a major consideration for location-based services—they are, after all, the closest manifestation of 'big brother is watching' concerns. Many services address the privacy issue by allowing the user to opt-in or opt-out of different services and levels of service at any time. However, LBS applications store vast amounts of personal information, and there is need to provide security (encryption) during transmission. The user must also be assured that service providers will not use this information in any other way or pass it onto a third party (the EU has specific legislation covering this; see also Vodaphone UK Privacy Policy).

CHALLENGES

The value and benefits of MGISs are apparent from the applications listed above. Yet despite the advantages and the general availability of the technology by the year 2000, they are not yet widely deployed. Why? A range of technical and non-technical issues are involved. These include: integrating MGIS architecture, hardware issues with mobile devices, issues with positioning technologies, issues related to the wireless communications network, and mapping issues.

Integrating MGIS Architecture

MGIS applications depend on a relatively complex set of components. The first challenge is to ensure that the various components that make up such applications can communicate. This is a significant problem given that different mobile devices, networks, GIS software, data servers, and so forth utilise different interfaces and protocols (Chen, 2004). The second challenge is to ensure that MGISs can be integrated with existing (non-mobile) systems (e.g., billing systems) (McKee, 2004; Spinney, 2003)—hence,

the need for standard APIs and protocols (e.g., XML, GML, Java, OSA/Parlay API, CORBA). Such standards ensure that each MGIS application can access distributed and heterogeneous data sources, distributed tools, and services, regardless of the underlying platform, hardware, vendor, network, positioning technology, data formats, and so forth (Zadorozhny & Chrysanthis, 2004, p. 147). Standards also allow for MGISs to be easily implemented with minimal cost and effort (McKee, 2004). MGIS interoperability in its various forms is an *essential* requirement for success—it is not optional.

Hardware Issues with Mobile Devices

Mobile devices are typically regarded as underperforming and overpriced compared to a desktop PC or even a laptop. While the small screen (including its poor resolution and limited colour palette) and limited memory are often cited as the chief failings of these devices, experience also suggests that poor battery life is a major problem—a device with only three hours of continuous use capability is not likely to endear itself to consumers (especially when the battery is not easily replaceable by the user). Some solutions, however, are beginning to emerge (e.g., PDA Trip Project, 2005). Added to these limitations is the fact that these screens can be easily damaged and the devices are easily lost.

The small screen size means that special consideration must be given to the interface design of MGISs. Buttons, tabs, menus, lists, maps, text, images, and so on must generally fit into a much smaller area. Standard Web pages and Windows applications are inappropriate for these devices. The information on such devices must be readable under all conditions (day and night, indoors and outdoors) and so requires enhanced colour contrast.

Table 7. Characteristics of different types of mobile devices[3]

	Resolution	Colour	Processor	Storage	Memory
WAP phone	96 x 32, 24 x 84, 60 x 80 – 120 x 160, 132 x 176, 176 x 220	Monochrome, greyscale; 256 colour; 4096 colour; 65k 16 bit TFT LCD screens	?	? Limited	Limited; 1, 2, 3, 5, 7 MB
Smart phone	Up to 208 x 320	3375-65536 colours; 64k; 16 bit 65k TFT reflective display	33, 130, 144 MHz	Greater storage; 32; 64 MB Flash ROM; 2GB memory stick	1.7, 4, 5, 8, 10, 16, 32, 64, 128 MB
PDA	160 x 160, 240 x 320, 320 x 320, 320 x 480, VGA resolution 480 x 640	Monochrome; 65k 16 bit colour; 65k colour transflective TFT screens	126, 2oo, 300, 400, 624 MHz	7.2, 10, 20, 32, 48, 128, 256, 512 MB, 1GB, 4GB	8, 25, 32, 64, 92, 128, 152 MB

Nor do small-screen mobile devices support the same data input methods as large-screen devices such as laptops and desktop computers. While some PDAs have QWERTY keyboards, many do not, relying instead on handwriting recognition or digital on-screen keyboards for entering text. They have touch screens and utilise pointing devices instead of a mouse. Some also support voice entry (Peng & Tsou, 2003, p.456). Smartphones also utilise touch screens, pointing devices, and on-screen keyboards, as well as having a phone keypad for text entry. WAP phones generally do not have touch screens, on-screen keyboards, or utilise pointing devices. They have much more restricted data entry capabilities. Hence, different devices require different interface designs and support different methods of user interaction.

The fact that there are many different devices—each with different characteristics and capabilities, and supporting different operating systems, mark-up languages, and protocols—is another significant issue (see Table 7; Dao, Rizos, & Wang, 2002). These variations impact negatively on the development of the industry, as they slow down application development,

and multiple platform-specific versions of software prevent scale economies from being realised.

Mobile GIS must be able to support a wide range of devices with different capabilities. It should also be noted that despite the limitations of current devices, there are many high-end mobile phones and PDAs that can adequately support such applications. Hence, extensive 'proof of concept' type hardware development is not required. Furthermore, the capabilities of such devices are improving rapidly, although (as it is the defining feature of the device type) small screen size will continue to be an issue. This requires that applications must be designed in such a way that if a device cannot support all the features of an application, the application must fail gracefully (e.g., if a device can be locationally aware, but cannot display maps, it still must operate on text only and not crash).

Issues with Positioning Technologies

A wide variety of positioning technologies are available for MGIS development, ranging from

Table 8. Positioning technologies and their limitations (Readman & Mojarrabi, 2004; Tanner, 2002)

Positioning Technology	Accuracy	Limitations
Cell ID	50m-20km	Greatest in rural areas; often around 200m in urban areas
Cell Triangulation, e.g., EOTD	50-500m	Within a cell sector; often 100m
GPS	10m	Not for indoor use; problem of urban canyons
AGPS	5m	For high-end users
DGPS	<1m	Not for indoor use; problem of urban canyons
Galileo (2008-)	1m	Not for indoor use; problem of urban canyons
Bluetooth	3m	Indoor only, short range

cell-based positioning (e.g., Cell ID, AOA, TDOA, EOTD) and satellite positioning (e.g., GPS), to wireless networks and radio frequency ID (RFID). Hence, applications must support a variety of positioning technologies. The choice is mainly cost driven. Most field-work applications rely on GPS-based positioning, as positional accuracy is important and the cost of the device is not usually a deterrent. In these applications the mobile device establishes its own position. In contrast, for LBS the position is usually fixed by the network, as consumers are generally not expected to be willing to pay the higher price for a mobile device which would be capable of fixing its own position.

Each positioning technique has its own limitations (see Table 8). For example, while the use of GPS is a much more accurate technique than cell-based positioning, they have limitations in indoor and urban environments. However, as GPS receivers become more compact and affordable, they are more likely to be incorporated into a wider range of mobile devices.

The positional accuracy of these different techniques is also quite varied, with some methods such as Cell ID being coarse to the point where the location of a user may be quite misleading (e.g., a map showing the location of the user may place the user in the wrong street or in a building when they are actually driving on the road). Positional accuracy is also influ-

enced by map scale, being quite significant on a 1:10,000 map (Gartner & Uhlirz, 2001; Uhlirz, 2001). The widely used method of symbolising the location of the user (e.g., the use of cross hairs) also implies a false accuracy (Gartner & Uhlirz, 2001; Pospischil, Umlauft, & Michlmayr, 2002, p. 146; Uhlirz, 2001).

Different applications have different accuracy requirements, and there are applications for which coarse positional accuracy is adequate (see Table 9). Supplementary services can sometimes be used to improve locational accuracy, as is often done within car navigation kits, using the odometer and steering angle if the vehicle goes into an urban canyon. This is used to estimate the position until another fix is found—some heuristic logic would be involved as well, in that the car will only drive on roads.

As many of these positioning techniques are based on cellular phone networks, they cannot be used in areas where coverage is not available. Hence, GPS-based positioning is regarded as superior because of its inherent greater

Table 9. Positional accuracy requirements[1]

Applications	Required Accuracy (m)
Concierge services	100
Outdoor navigation	25
Indoor navigation	<10
Tour guide	30
Emergency services	100
Incident investigation	<10
Facility maintenance	<10

accuracy and better availability in rural areas (Mountain & Raper, 2002, pp. 5-6). In order to overcome this problem of the coverage of mobile networks and to improve the accuracy of cell-based techniques (as well as to combat the limitations of GPS), many applications utilise hybrid positioning techniques (e.g., Cell ID and GPS) (Mountain & Raper, 2002).

Issues Related to the Wireless/ Communications Network

Existing wireless environments for MGIS tend to have the following characteristics (Peng & Tsou, 2003, p. 467):

- **Less Bandwidth:** Bandwidth is the amount of content that can be passed down the narrowest point of a network at a given time.
- **More Latency:** Latency refers to the length of time it takes the content to flow a certain distance. It is a measure of the amount of time it takes for a request to make it from the client and back ("ping").
- **Less Connection Stability:** The mobile user in motion has to handoff from cell to cell, which could cause information drops during the handoff process. While this should be handled by the connection protocols, it means that the information has to be resent a large number of times—effectively reducing available bandwidth.
- **Less Predictable Availability.**

The limited bandwidth of wireless networks, compared to wired networks, is a major issue for MGISs. Given the large volumes of geographic information and the need for real-time response, MGISs require fast transmission and rendering of spatial information. 2G networks such as GSM typically have a bandwidth of 10kbps and 2.5G networks (e.g., GPRS) up to

115kbps. 3G networks currently being implemented offer vastly improved rates of up to 2Mbps (Peng & Tsou, 2003, pp. 458-467). 2.5G and 3G networks also have the advantages of streaming, being always on, and supporting the transmission of multimedia information (Gartner & Uhlirz, 2001; Nissen, Hvas, Münster-Swendsen, & Brodersen, 2003; Uhlirz, 2001). Caching of surrounding data while the user is looking at current data, or when they are in range of a larger capacity network (such as when a worker goes to a Wi-Fi hotspot and caches a lot of data and then updates and refreshes the data using GPRS) goes some way to addressing this problem.

Coverage is another major issue with cellular networks, as without coverage, no data or network-based positioning information can be received or transmitted. In this case, the user must rely on the information stored locally in the mobile device. Given the limited memory and storage of many mobile devices, this has been an issue. Technological improvements, however, are significant (e.g., availability of 4gb+ memory cards). Lack of network coverage also reduces the real-time data updating benefit of field applications of MGIS. The client-server model used has important implications here, and the most suitable model depends on the types of mobile devices, network characteristics, and required functionality (Mountain & Raper, 2002, pp. 5-7).

Mapping Issues

The change from 17" to 3" screens and the poor bandwidth in wireless communications create major problems for GIS developers. While acknowledging that there are several significant mapping issues (discussed below), it should not be assumed that traditional GISs and MGISs always have the same cartographic objectives or requirements; clearly, the application requirements are very different—simple maps or even text-only output are quite acceptable for many MGIS queries (e.g., routing instructions might look 'nicer' in a colour map, but plain text on a monochrome screen is quite acceptable). With the continuing advances in 'text to speech' applications, audio may be more practical in some circumstances as the driver is not required to take his/her eyes off the road.

Maps form a major component of MGISs and must not only fit within the small screen, but also within the application interface, with its menus, buttons, and so forth. Maps in Mobile GISs need to be simple, uncluttered, and easy to read at arm's length (Hjelm, 2002, p. 247). In order to combat the bandwidth problem, many applications have used predefined, static, scanned raster maps (Pospischil et al., 2002). However, while scanned raster maps result in lower storage, transmission, and rendering costs, they allow for little manipulation or interactivity (Abowd et al., 1997; Hjelm, 2002, p. 257; Reichenbacher 2001a, 2001b, 2003, p. 1317). Rescaling also results in information loss unless data pyramids are used—adding to the storage overhead. Raster data layers also require high storage and transmission costs.

With the improvements offered by 2.5G and 3G networks, there has been an increasing focus on the use of vector data, which allow for lower storage and transmission costs, easier manipulation of data, a wider range of applications, links to a database, and adaptive, dynamic, and interactive mapping (Abowd et al., 1997; Long, Kooper, Abowd, & Atkeson 1996, p. 104; Reichenbacher, 2001a, 2001b). Traditional vector data formats, however, require much greater processing for map display and are not suited to mobile mapping; new data types and formats are required (De Vita, Piras, & Sanna, 2003; Hjelm, 2002, p. 257; Reichenbacher, 2001a, 2001b).

Adaptive mapping refers to the situation where maps are created on the fly according to the user's context, preferences, and characteristics (age, language, culture, preferences, etc.). Traditionally, context is regarded as the user's location, but other attributes such as weather, time, bandwidth, and mobile device can be included (Reichenbacher, 2001a, 2001b; Zipf, 2002). It is the context and user profiles that determine the content, level of detail, and degree of generalisation of the maps in adaptive mapping (Reichenbacher, 2003).

Other important cartographic issues include the need for common rules and standards for mobile cartography. Symbology needs to be automatically assigned according to widely understood colours and symbols, taking into consideration cultural conventions and the need to maintain high contrast (Nissen et al., 2003; Zipf 2002). Labels, varying according to the map scale and level of detail, must also be provided in the preferred language of the user. The small screen taxes any automated name placement routine for clutter and overlap (Reichenbacher, 2003).

TOWARDS SOLUTIONS

Open standards are required for interoperability to ensure that MGISs can operate across different wireless networks, platforms, devices, positioning technologies, data types, and so forth. They also allow for the various components of an application to communicate and to be easily integrated with existing systems. There are a number of standards bodies working on standards for MGISs, the most significant of which are the Organisation for Standards (ISO), Open Mobile Alliance (OMA), and Open GIS Consortium (OGC). For example, the Location Interoperability Forum (LIF), which is now part of the OMA, developed the mobile location protocol (MLP). The MLP is an open interface for determining the position of a mobile device, regardless of positioning method and network (i.e., it interfaces the location server and application server) (Zadorozhny & Chrysanthis, 2004, pp. 151-153). The OGC OpenLS GeoMobility Server (GMS) defines open interfaces for core spatial services (McKee, 2004, pp. 160-168; OpenLS, 2004).

In order to address the issue of small screen size, some applications only support particular devices or devices above a certain resolution. For example, the GiMoDig project only supports devices with a resolution of 180 x 180 or above, Java ME capability, and suitable memory (Nissen et al., 2003). It is also recommended that applications should reduce the length of their pages and utilise hypertext links. Shorter pages and a deeper hierarchy are recommended for the wireless Web (Kacin, n.d.). Using a hierarchy of maps, interactive mapping, and multimedia information also allows for optimization of content for small screens (Gartner & Uhlirz, 2001; Pospischil et al., 2002; Uhlirz, 2001). Maps must also be simple, uncluttered, and have minimal detail. Hence, many maps are predefined and static (Gartner & Uhlirz, 2001). While screen size will continue to be a problem, technological improvements in mobile devices will result in better resolution, colour range, battery life, and so forth. Owing to the difficulty of text entry on many mobile devices, voice entry remains an important, but as yet underdeveloped, means of input.

Improvements in positioning technologies will result in greater positional accuracy, as will the use of more precise positioning techniques (e.g., Galileo). Accuracy can further be improved through the use of hybrid positioning techniques, user input, the use of passive landmarks and active sensors (e.g., Bluetooth), as well as video and augmented reality (Gartner, 2004; Gartner & Uhlirz, 2001; Uhlirz, 2001).

Furthermore, it is important to acknowledge the uncertainty of the user's location—the symbology utilised should indicate the accuracy of the location or error term (e.g., using a circle of varying size to indicate the error term). In addition, rather than showing a precise location, it is possible to indicate the region or street within which the user is located (Gartner & Uhlirz, 2001; Pospischil et al., 2002; Uhlirz, 2001). The symbology used should also indicate when real-time positional information is not available.

The communications bandwidth problem is partially solved by the use of 2.5G and particularly 3G technology. They also allow for a larger range of client-server models (Mountain & Raper, 2002, p. 4). Adaptive mapping and generalisation have reduced the bandwidth problem further, by reducing the volume of information transmitted (Gartner & Uhlirz, 2001). The improved caching ability of current devices also reduces network traffic and the problem of network coverage.

Furthermore, the client-server model chosen has an impact on the network traffic. Server-side applications place a greater load on the network, and require high transfer rates and streaming mode. However, they also support a wider range of devices, particularly lower end devices, and have greater functionality, with support for interactive and dynamic mapping. Client-side applications, while reducing network traffic and allowing offline processing, require high-end devices; the software on these devices has limited functionality and they favour prepared, static maps (Mountain & Raper, 2002, p. 6; Reichenbacher, 2001b; Gartner & Uhlirz, 2001). It is relatively simple to have a raster background with a vector layer overtop so that most of the data is static, but the 'relative' information is in vector format, so that all the vector advantages are available.

Vector data formats offer greater potential to MGISs than raster data formats, and so a number of vector data formats have been developed especially for mobile devices; examples include Compact GML (cGML), SVG Basic (for PDAs), and SVG Tiny (for mobile phones) (De Vita et al., 2003; Reichenbacher, 2003). Currently, most LBSs utilise data in GML and then convert it to SVG for display (e.g., PARAMOUNT; GiMoDig).

The mobile mapping challenges have also led to the development of new forms of representation such as focus maps, bird's eye perspectives, floor plans, and the use of multimedia information (Gartner, 2004). Focus maps reduce transmission costs by providing a detailed representation of features within a certain distance of the user and a coarse level of detail for more distant features (Zipf & Richter, 2002). Adaptive mapping also reduces transmission and rendering overheads.

With regard to label placement, one way to mitigate the problem of cluttered and overlapping labels is to provide few details on overview maps and to require the user to zoom into the maps for more information (Heidmann, Hermann, & Peissner, 2003). This issue can also be addressed through the use of labelling algorithms, but is most frequently solved through the use of tool tips and hot spots. This greatly reduces the number of labels required, and the user can still obtain information on features by passing the cursor over the feature or clicking on it (Gartner & Uhlirz, 2001; Uhlirz, 2001).

FUTURE DIRECTIONS

With improvements in technology (software and hardware), many of the challenges mentioned above will be reduced or eliminated in the near future. For example, with the development of mobile devices with greater memory, better battery life (including user-replaceable batteries), more powerful processors, and screens with higher resolution, many of problems of mobile devices will no longer be applicable. While devices are constantly improving, however, small screen size will remain an issue. A reduction in the cost of such devices will also allow for greater adoption of this technology and better user acceptance.

The implementation of 3G and better networks will also alleviate bandwidth problems and support a greater range of client-server models. The use of more precise positioning technologies (e.g., DGPS, Galileo), and improvements in the availability and costs of such technologies will also improve the locational accuracy of applications. Most significant here will be the increasing availability of mobile devices with built-in GPS receivers.

These technology improvements will make Mobile GISs of even greater importance to enterprises in achieving operational efficiencies in field operations generally and with significant gains in 'field force automation' in particular. However, while technology improvements are a prerequisite, more widespread consumer adoption of this technology will be hindered unless applications of greater utility to the consumer (as well as more "compelling content") become available.

REFERENCES

Abowd, G., Atkeson, C., Hong, J., Long, S., Kooper, R., & Pinkerton, M. (1997). Cyberguide: A mobile context-aware tour guide. *Wireless Networks, 3,* 421-433.

AT&T Wireless. (2005). *Find people nearby.* Retrieved March 29, 2005, from http://www.attwireless.com/personal/features/organisation/findfriends.jhtml

Chen, A. (2004). *Open standards will evolve location-based services.* Retrieved November 30, 2004, from http://www.eweek.com/print_article2/0,2533,a=131075,00.asp

Conolly, N. (2001). Software for mobile mapping. *GIS User 47.* Retrieved March 29, 2005, from http://www.gisuser.com.au/GU/content/2001/GU47/gu47_frame.html

Dao, D., Rizos, C., & Wang, J. (2002). Location-based services: Technical and business issues. *GPS Solutions, 6,* 169-178.

Daratech. (2005). *Leading manufacturers validate PLM though confusion remains.* Retrieved from http://www.daratech.com/press/releases/2005/050228.html

De Vita, E., Piras, A., & Sanna, S. (2003). *Using compact GML to deploy interactive maps in mobile devices.* Retrieved from http://www20003.org/cdrom/papers/poster/p051/p51-devita.html

ESBI Computing. (2005a). *Mobile GIS: Benefits.* Retrieved March 17, 2005, from www.esbic.ie/geobusiness/mobile_gis/GIS_Benefits.htm

ESBI Computing. (2005b). *Mobile GIS: Application to industry.* Retrieved March 17, 2005, from www.esbic.ie/geobusiness/mobile_gis/GIS_Industries.htm

ESRI. (2005). *Mobile GIS.* Retrieved March 17, 2005, from www.esri.com/software/arcgis/about/mobile.html

Gartner, G. (2004). Location-based mobile pedestrian navigation services—The role of multimedia cartography. In *Proceedings of ICA UPIMap 2004,* Tokyo. Retrieved from http://ubimap.net/upimap2004/html/papers/UPIMap04-B-03-Gartner.pdf

Gartner, G., & Uhlirz, S. (2001). Cartographic concepts for realising a location based UMTS service: Vienna city guide "LOL@". *Proceedings of the 20th International Cartographic Conference* (vol. III, S.3229-3239), Beijing. Retrieved from http://lola.ftw.at/homepage/content/a40material/Vienna_City_Guide_LoLa.pdf

Geological Survey of Western Australia. (2005). Retrieved March 15, 2005, from http://www.linkspoint.com/docs/GSWA_CS.pdf

Giaglis, G., Kourouthanassis, P., & Tsamakos, A. (2002). Towards a classification for mobile location services. In B. Mennecke, & T. Strader (Eds.), *Mobile commerce: Technology, theory, and applications* (pp. 64-81). Hershey, PA: Idea Group Publishing.

Hassin, B. (2003). *Mobile GIS: How to get there from here.* Retrieved March 29, 2005, from http://gis.esri.com/library/userconf/proc03/p0988.pdf

Heidmann, F., Hermann, F., & Peissner, M. (2003). Interactive maps on mobile, location-based systems: Design solutions and usability testing. In *Proceedings of the 21st International Cartographic Conference* (ICC) (pp. 1299-1305), Durban, South Africa.

Hjelm, J. (2002). *Creating location services for the wireless Web: Professional developer's guide.* New York: John Wiley & Sons.

Kacin, M. (n.d.). *Optimizing Web content for handheld devices.* Retrieved from http://www.wirelessdevnet.com/channels/pda/features/handheldcontent.html

Karimi, H. A., & Hammad, A. (2004). *Telegeoinformatics: Location-based computing and services.* New York: CRC Press.

Lohnert, E., Mundle, H., Wittmann, E., & Heinrichs, G. (2004). Wireless in the Alps: An LBS prototype for mountain hikers. *GPS World, 15*(3), 30-37.

Lohnert, E., Wittmann, E., Pielmeier, J., & Sayda, F. (2001, September 11-14). PARAMOUNT—Public Safety & Commercial Info-Mobility Applications & Services in the Mountains. In *Proceedings of the 14th International Technical Meeting of the Satellite Division of the Institute of Navigation* (ION GPS 2001), Salt Lake City, UT (pp. 319-325).

Long, S., Kooper, R., Abowd, G., & Atkeson, C. (1996, November). Rapid prototyping of mobile context-aware applications: The Cyberguide case study. In *Proceedings of the 2nd ACM International Conference on Mobile Computing and Networking (MobiCom '96)*, Rye, NY (pp. 97-107). New York: ACM Press.

Longley, P., Goodchild, M., Maguire, D., & Rhind, D. (2001). *Geographic information systems and science.* New York: John Wiley & Sons.

McKee, L. (2004). *LBS interoperability through standards.* In J. Schiller & A. Voisard (Eds.), *Location-based services* (pp. 149-171). Amsterdam: Elsevier.

Mountain, D., & Raper, J. (2002, September 18). Location-based services in remote areas. In *Proceedings of the Association of Geographical Information* (Paper B5.1, pp. 1-9). Retrieved from http://www.soi.city.ac.uk/~dmm/research/pubs/B05.3.pdf

Nissen, F., Hvas, A., Münster-Swendsen, J., & Brodersen, L. (2003). *KMS, National Survey and Cadastre–Denmark: Small-display cartography.* Retrieved November 30, 2004, from http://gimodig.fgi.fi/pub_deliverables/D3_1_1.pdf

North London Strategic Alliance. (2005). Retrieved from http://www.publictechnology.nct/print.php?sid=861&POSTNUKESID=1a6d6da2c722c833fb65eddf8ff072e3

Open LS. (2004). *Open GIS location services.* Retrieved December 15, 2004, from http://www.opengeospatial.org/specs/?page=specs

PARAMOUNT. (n.d.). Retrieved from http://www.paramount-tours.com

PDA Trip Project. (2005). Retrieved March 29, 2005, from http://www.7volts.com/travel.htm

Peng, Z., & Tsou, M. (2003). *Internet GIS: Distributed geographic information services for the Internet and wireless networks.* New York: John Wiley & Sons.

Pospischil, G., Umlauft, M., & Michlmayr, E. (2002). Designing LOL@, a mobile tourist guide for UMTS. *Lecture Notes in Computer Science, 2411,* 140-154.

Readman, D., & Mojarrabi, B. (2004). *Location based and communication systems.* Unpublished manuscript, University of Queensland, Australia.

Reichenbacher, T. (2001a). Adaptive concepts for a mobile cartography. *Supplement Journal of Geographical Sciences, 11,* 43-53.

Reichenbacher, T. (2001b). The world in your pocket—Towards a mobile cartography. In *Proceedings of the 20th International Cartographic Conference,* Beijing, China. Retrieved from http://citeseer.ist.psu.edu/cache/papers/cs/23234/http:zSzzSzwww.lrz-muenchen.dezSz~t583101zSzWWWzSzpublicationszSzreichenbacherzSzICC2001_Paper.pdf/reichenbacher01world.pdf

Reichenbacher, T. (2003, August 10-16). Adaptive methods for mobile cartography. In *Proceedings of the 21st International Cartographic Conference* (ICC) (pp. 1311-1322), Durban, South Africa. Retrieved from http://www.carto.net/geog234/readings/

reichenbacher_mobile_cartography_durban_2003.pdf

Smyth, C. (2000). *Mobile geographic information services: Turning GIS inside out.* Retrieved from http://www.giscience.org/GIScience2000/invited/Smyth.pdf

Spinney, J. (2003). *A brief history of LBS and how OpenLS fits into the new value chain.* Retrieved from http://www.esbic.ie/geobusiness/Mobile_GIS/overview.htm

Tanner, J. (2002). Location-based services: Where it's at. *Wireless Asia,* (November), 22-24.

Uhlirz, S. (2001). Cartographic concepts for UMTS-location based services. *Proceedings of the 3rd Workshop on Mobile Mapping Technology,* Cairo, Egypt. Retrieved from http://lola.ftw.at/homepage/content/a40material/Cartographic_Concepts_for_UMTS_Location_based_Services.pdf

Zadorozhny, V., & Chrysanthis, P. (2004). Location-based computing. In H. Karimi & A. Hammad (Eds.), *Telegeoinformatics: Location-based computing and services* (pp. 145-170). Boca Raton, FL: CRC Press.

Zipf, A. (2002). User-adaptive maps for location-based services (LBS) for tourism. In K. Woeber, A. Frew, & M. Hitz (Eds.), In *Proceedings of the 9th International Conference for Information and Communication Technologies in Tourism* (ENTER 2002), Innsbruck, Austria. Retrieved from http://www.eml-development.de/english/homes/zipf/ENTER2002.pdf

Zipf, A., & Richter, K. F. (2002). Using focus maps to ease map reading. developing smart applications for mobile devices. *Artificial Intelligence* (special issue on spatial cognition). Retrieved from http://www2.geoinform.fh-mainz.de/~zipf/ki-04.2002-zipf.pdf

ENDNOTES

[1] For more information on Mobile GIS software, see Peng & Tsou (2003, pp. 480-492) and Conolly (2001).

[2] Information for this table has been adapted from Peng & Tsou (2003, pp. 493-495). For a broader range of applications, see ESBI Computing (2005b).

[3] Information for this table was derived from Internet research on currently available mobile phones and PDAs.

[4] Some figures were derived from Gartner (2004).

Chapter LVIII
Mobile Technologies and Tourism

Pramod Sharma
The University of Queensland, Australia

Devon Nugent
The University of Queensland, Australia

ABSTRACT

This chapter examines the potential of mobile technologies for the tourism industry. Mobile technologies have the capacity to address not only the pre- and post- tour requirements of the tourist, but also to support the tourist on the move. It is this phase of the tourist activity upon which mobile technologies can be expected to have the greatest impact. The development of applications for the mobile tourist will allow for the creation of a new range of personalised, location and time specific, value added services that were not previously possible. Before such applications can be widely deployed, however, some fundamental technical and business challenges need to be addressed. Despite these challenges, mobile technologies have the potential to revolutionise the tourist experience, delivering context specific services to tourists on the move.

INTRODUCTION

Tourism is one of the world's largest and most rapidly expanding industries, contributing over 10% to global GDP (WTTC, 2003). Information and communication technology (ICT) has played a critical role in its development, as evidenced for example by the development of massive global distribution systems (GDSs) and airline computerised reservation systems (CRSs), as well as enterprise systems, such as FIDELIO for the hospitality industry. Until relatively recently most ICT applications dealt with enterprise operations of the B2B type. The Web has had a "liberating" effect by making possible the introduction of B2C applications (e.g., using a Web site for the distribution of product information, destination promotion, online bookings, and e-commerce in general).

The B2C applications have been relatively "static" systems offering support in the pre- and, to some extent, post-tour phase of a trip.

For example, they support information retrieval on places to visit, online reservations, personal Web pages, and so forth. A key gap in ICT applications, however, has been the lack of support for the tourist on the move (e.g., "Where is the nearest hotel?," "How do I get to the museum from here?," etc.). Solutions such as kiosks, Internet cafes, and the increasing availability of Internet connections in hotel rooms address only part of the problem, as they still require the tourist to be wired ("tethered"). The answer lies in developing applications which address the mobility issue *directly* by delivering applications on wireless-enabled platforms such as mobile phones and PDAs to tourists while they are on the move. Herein lies the challenge: at its simplest the issue is one of providing information-rich, "bandwidth guzzling" content (colour photos, graphical and audio-visual content) via "capacity-challenged" hardware platforms and communications infrastructure. Thus, exhortations to develop applications that provide "personalised" services "in context" are not surprising, both because they are desirable (CSTB, 2003) and also because for the foreseeable future, hardware limitations will dictate it.

Despite the challenges, the vision is clear enough:

... the opportunity for providing location dependent information and reservation is critical for tourism[,] and the constantly moving consumer ... will support a whole new way of communicating, accessing information, conducting business, learning and being entertained while on the move...With access to any service anywhere, anytime from one terminal, the old boundaries between communication, information, media and entertainment will gradually disappear, offering convergence between technologies and tourism services. (Buhalis, 2003, p. 323)

In a sense the issue is transitional, for it is not difficult to foresee that, in the longer term, wireless forms of communication will be the dominant technology, and that technology and business challenges will be addressed. However, regardless of the duration of the transitional period, certain fundamental issues need to be addressed: Why are mobile technologies of interest to tourism? What are the key components of a mobile system? What kind of services are required by the mobile tourist? What are the issues involved in applications development for mobile services? Case studies of existing applications in tourism will be used to illustrate some of these points. This chapter will answer these questions and evaluate the potential impact of mobile technologies on tourism.

INFORMATION REQUIREMENTS FOR MOBILE TOURISTS

Using "old" and "new" tourism models (O'Looney, 2004) has highlighted some aspects of information use by tourists. In the "old Tourism Model", tourists used guidebooks, paper maps, and printed media. They were inundated with cluttered and broad-based advertising, and lacked the ability to focus or follow-up on a topic. Furthermore, much of this information was out of date soon after printing. Guidebooks can also be hard to follow: it is not easy to link locations on a map with information in guidebooks, one must read through lots of information in order to get what one wants, and such guides are designed for the "general" tourist (Brown & Chalmers, 2003). Web information searches, when available, essentially mimicked the traditional library service, providing vast amounts of unstructured data. In contrast, the 'new tourism model' uses mobile phones, PDAs, laptops, and electronic maps; it provides a personalised, information-rich, and location-specific experience; multiple media, content-linked advertising; and the new ability to focus or follow-up on a topic. The user can decide how much information they want. However, while this vision is accurate in terms of technology, it is far too "rosy in terms of user acceptance; *much* remains to be done both in terms of enabling technology as well as tourism information content to convert the "new" tourism model into reality.[1]

What are the limitations of current tourist information systems? A recent paper (Watson, Akselsen, Monod, & Pitt, 2004, pp. 315-316) identifies three fundamental problems:

1. tourists are overwhelmed by the variety and the volume of information on the Web, and experience information overload;
2. there is little ICT support during the touring phase; and
3. experiences gained during a trip are not easily shared, and reminiscing is rarely supported.

While all of these matters are of research interest, our focus is mainly on issues related to ICT support during the touring phase.

The tourist activity is frequently envisaged as having three stages or phases—each has different information requirements. The first of these is the pre-tour or planning phase, during which the tourist is actively researching products and destinations. The Web, despite its shortcomings, is increasingly becoming the preferred source for information (Epstein, Garcia, & Fiore, 2003, p. 29) on which the decision to travel or not to travel is made. The proliferation of travel and tourism Web sites has meant that a huge volume of variable quality information is available on almost any tourism topic—ranging from destination sightseeing information to information on types of plugs for electrical appliances.[2] If this frequently results in "information overload" for the consumer in a typical Internet environment (a desktop PC with a dialup modem, if not broadband, connection), it is not hard to imagine the "system overload" it will lead to in a hardware- and infrastructure-challenged mobile environment.

The third stage is post-tour or reminiscing when there is little need to research information, but there is a need to record and share the tourist's own information and experiences with family and friends. This reminiscing or recording and sharing of information also has other uses—it is invaluable information for other tourists. Thus it should be noted that the Web permits 'word of mouth' dissemination through the various "user review" or "feedback" facilities; available evidence suggests that tourists rate the experiences of other tourists as equal to, if not better than, those of 'experts' (Brown & Chalmers, 2003). An additional aspect of interest to applications developers is the use of this information as input into travel recommender systems (TRSs).

It is our view that it is in the second stage—the touring itself (i.e., when the tourist is on the move)—that ICT in general and mobile technologies in particular can make a significant contribution to the tourism industry by meeting the information needs of the tourist. These information needs take on a certain urgency when we realise how stressful lack of information can be when the tourist may be in unfamiliar ("threatening") surroundings and also when plans may need changing in light of changing circumstances. The need for flexibility is an important aspect of the tourist experience, as tourists enjoy being able to change their plans—over planning can produce unhappy experiences (Brown & Chalmers, 2003). A number of promising applications, such as LOL@, Webpark, and The Electronic Guidebook, address all three phases of tourist activity (see Tables 3 and 4).

What kinds of information does the tourist need during the touring phase? Two broad categories can be envisaged. The first category consists of 'administrative' information regarding the tour itself. These include confirmation of or changes to travel arrangements and bookings (e.g., delayed or cancelled flights, changes to accommodation arrangements, or other changes to itineraries). While e-mails are clearly superior for their information content, SMS has been embraced more widely, as such service can be sent to even the very basic handsets, which generally cannot handle e-mails. This service is already being used by the various airlines. Mobile technologies have clearly represented a great advance in keeping the tourist updated—industry developments such as e-ticketing reinforce this further.

The second, a broader category, usually containing a large volume of information, is made up of destination information. Traditionally the content has been the justification for the creation and maintenance of Destination Management Systems (DMSs) managed by both semi-governmental, as well as private tourism authorities (e.g., national, state, regional, or district tourism boards). Destination information for the mobile tourist is made up of three categories (Eriksson, 2002; for a very ambitious proposal involving a richer content, see O'Looney, 2004):

- information about tourism products (about restaurants, hotels, museums, amusement parks), and various directory ("yellow pages") services;
- information about transport networks and traffic; about transport network infrastructure (e.g., roads, railways and airports, timetables, etc.); and
- information about locations and positions (to be able to provide answers to: Where is the nearest x? How do I get there?),

information on points of interest (POIs), navigation/route guidance, tracking friends/other members of the party.

To be of value, this information has to be current, context sensitive, personalised, and preferably in the chosen language. Currently, GIS service and content providers can meet most of these requirements on fixed/wired/non-mobile systems; however, it is a big task to deliver these services on mobile systems. The load on the system also varies with the behaviour of the tourist. Thus a tourist who has undertaken all research and downloads off a fixed Internet connection prior to starting the tour and merely accesses the Internet during the tour for a "top-up" or updates, imposes a significantly lower load than the tourist who starts the process during the tour. In the latter case, a series of factors (the learning process, longer connect times, more information downloads, and the limitations of mobile hardware and the communications infrastructure) combine to make it a potentially difficult experience.

Are mobile tourism information systems/services different from fixed Internet applications? At first glance it is reasonable to assume the mobile tourism services are merely the "light" versions of standard Internet tourism applications and should pose no particularly difficult or unique problems. Unfortunately this is not the case, for both the communications network (wired *vs.* wireless systems), as well as client terminals (large screen, mains powered *vs.* small screen battery powered), pose major problems for the development of mobile applications for tourists. While the problems relating to the communications network may be eliminated with the advent of universal 3G or better phone systems, the hardware limitations of mobile devices (especially the small screen constraint) are not likely to be eliminated in the foreseeable future. Furthermore, mobile tourism information systems allow new kinds of services and fulfil user requirements that are not met via fixed networks.

EMERGING SOLUTIONS FOR THE MOBILE TOURIST

Despite the challenges, significant applications do exist either as prototypes or as operational systems. In order to provide a context for an examination and evaluation of these applications, we need to address some issues: What are the ways of getting information and services to the tourist? What is the basic architecture of these systems?

How do we get the information and services to the mobile tourist? Two situations can be envisaged (see Table 1):

1. **The Unconnected Mobile Tourist:** This is the situation where the tourist is not connected to a communications network and is using either a "dumb" or a "smart" electronic guidebook type of product—dumb or smart refers to whether the mobile device is context aware in terms of location (i.e., is able to determine its own geographic location). Dumb devices are units which allow access to "softcopy" versions of travel guides, books, and so forth; smart devices are the various GPS-enabled PDAs (e.g., NavMan units, Garmin IQUE3600) or GPS-equipped phones.

2. **The Connected Mobile Tourist:** This is when the device is connected to a network either by wireless (e.g., Wi-FI networks (Wireless Internet Network 802.11b) to surf the Internet, check e-mail, and fix position (Epstein et al., 2003)) or to a mobile phone system—the connection to a communications network permits data transfer in both directions. The device may be smart (i.e., it can establish its own position), but even if it is dumb, it may be possible to "fix" its position by the communications network.

While it is the connected mobile tourist who is the focus of this chapter, a brief comment is necessary to illustrate that ICT support of other types (e.g., electronic guides) are possible. As the unconnected dumb mobile electronic guides

Table 1. Types of mobile tourist guides

Type of Mobile Tourist	Location Aware	Description
Unconnected to mobile network	Dumb	No positional information; no data updates (e.g., e-book, digital guidebook)
	Smart	Handset-based positioning (GPS); location-specific information delivery; no data updates (e.g., NavMan)
Connected to mobile network	Dumb	Network positioning; real-time, location-specific information delivery (e.g., Sony Ericsson 910)
	Smart	Handset-based positioning; real-time, location-specific information delivery (e.g., WheriFone)

Figure 1. Components of mobile tourism information systems

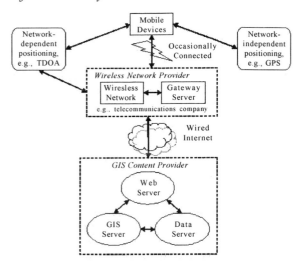

have the limitations of mobile devices and do not allow for information updates, they offer few advantages over printed guides and maps (i.e., users have to deal with all the difficulties of using small-screen devices without the full benefits of mobility). In contrast, unconnected smart mobile guides offer the advantage of locating the user in real time and are able to deliver information relevant to that location (although the information is only as current as the last date at which the data on the mobile device was updated).

What is the basic system architecture of a mobile tourism application? Mobile tourism applications are developed around the following components:

- A mobile device (the client terminal), which is capable of being detected by some position-determining technology. The mobile devices tend to be mobile phones (especially those with better, often colour screens), smartphones, and wireless capable PDAs.

- Position-determining technology—the key distinction here is whether the position is determined by the mobile device (e.g., using GPS) or by the communications network (by using one of several triangulation methods).

- GIS content provider (Web server, GIS server, data server).

- Wireless network provider—these are required to communicate with the mobile device.

While superficially similar to standard Internet tourist applications, closer inspection reveals that wireless connected mobile tourist applications are radically different. These applications have the ability to extend the traditional user experience, providing real-time, location-specific, value-added, personalised services that were not previously possible. They are able to support all three phases of the tourist activity, especially the tourist on the move. While touring, the user can access a wide range of additional material and can interact with this information (Semper & Spasojevic, 2002); they can interact with dynamic maps, multimedia content, and text; explore information on a point of interest (POI); record information on places visited (in a travel diary); ask questions about what features are nearby; access the recommendations of other tourists; get directions; and so forth. Most importantly with access to dynamic and up-to-date information, they can ask new kinds of questions such as, "Where is the nearest *open* museum?" (Cheverst, Davies, Mitchell, Friday, & Efstratiou, 2000, p. 18; Watson et al., 2004, p. 317). They combine the advantages of maps and guidebooks with those of dynamic and interactive digital content, which is accessible "anywhere, anytime". Tourism applications of this kind not only provide a new means of delivering information available on the Web, in guides, and so forth, but also provide new "innovative" services (Eriksson, 2002, p. 17).

Such applications utilise three common data types: (1) base data (street network, satellite imagery, etc.); (2) point of interest (POI) data (e.g., restaurants, ATMs, cinemas, museums, etc.); and (3) the location of the user. While maps are the most common form of output, data is delivered to the user in a variety of forms— voice, text, images, video, maps, and so on— according to the user's context (location, device, network, the weather, etc.) and profile (age, interests, language, disability, etc.) (Reichenbacher, 2001a, 2001b, 2003). For example, for navigation when driving, it is more appropriate to deliver audio directions, as the user may not be able to read text or a map while driving. In addition, the user can access a wide variety of textual and multimedia information relating to POIs. The depth to which data is explored is essentially an exchange between content availability and the user profile.

These mobile tour guides can offer a wide variety of services to business and leisure travellers (see Table 2).

Mobile tourists have different requirements from those of tourists searching the Internet. They are often very busy/distracted—looking around, taking in the scenery, navigating, and so forth. They want maps that are easy to read on the small screen, at arm's length, with easily legible text and high colour contrast for outdoor use (Hjelm, 2002, p. 247). Nor do they have time to sift through pages of text. Mobile tourists want short, directed text, and the ability to find out further information if they wish; simple interfaces and personalised information delivery; applications that work the first time ('out of the box') and every time; and they must offer significant advantages over traditional printed tour guides and Internet applications (Umlauft, Pospischil, Niklfeld, & Michlmayr, 2003). For excellent 'proof of concept' examples, see Tables 3 and 4.

Other examples of mobile tourist guides include:

- PARAMOUNT, which was developed to improve information access, navigation, and safety for mountaineers in the Pyrenees and Alps (Lohnert, Mundle, Wittmann, & Heinrichs, 2004; Lohnert, Wittmann, Pielmeier, & Sayda, 2001);
- LOVEUS, which aimed to "provide European citizens with ubiquitous services for personalised, tourism-oriented multimedia information related to the location and orientation within cultural sites or urban settings" (Karagiozidis, Zacharopoulos, Xenakis, Demiris, & Ioannidis, 2003);
- TellMaris, which developed TellMarisOnBoard and TellMarisGuide for leisure boat tourists of the Baltic Sea Region; the main purpose was to explore

Table 2. Some services offered by mobile tour guides

Service	Example
Routing	The best route for visiting the British Museum, National Galley, and Tate Gallery in a single day.
Navigation	Driving directions from the Marriot Hotel in Brisbane to Toowoomba.
Tour Guide	A pre-defined, one-day tour around Paris.
Concierge Services	Where is the nearest Italian restaurant?
Travel Diary	Record a photo of the Eiffel Tower and comment on the visit.
Travel Recommendations	Access other travellers' opinions on a particular cycling tour of the Swiss Alps.
Emergency Response	Call for assistance when your hired car breaks down.
Marketing	Receive information on special tour offers.
Tracking	Tracking the location of a tourist along a walking track in a national park.

Table 3. LOL@ mobile tourist guide (Umlauft et al., 2003; Pospischil, Umlauft, & Michlmayr, 2002; Annegg, Kunczier, Michmayr, Pospischil, & Umlauft, 2002)

APPLICATION AREA: Mobile Tourist Guide	
B2B/B2C: B2C	**IMPLEMENTED:** On laptop
SUMMARY: LOL@ is a prototype of a mobile electronic interactive tour guide for the city of Vienna. The aim of this project was to develop a location-based multimedia service for UMTS. It supports three kinds of services: pre-tour planning (maps, text, and multimedia information on tours and POIs, filtered according to the user's profile/behaviour); support for the tourist on the move; access to a tour diary (stored on the server) via the Web or PC—the user can record text and multimedia information while on tour, friends can also follow the user's activity.	
PUSH/PULL SERVICES PROVIDED: Push—predefined or user-defined tours; records visited POIs, route segments travelled, and so forth; information (text and multimedia data) on POIs/landmarks; route guidance and navigation; adding to and accessing the electronic tour diary. Pull—records information on visited POIs in tour diary, user location, and orientation.	
GEOGRAPHIC COVERAGE/ENVIRONMENT: Outdoor urban environment, Vienna's First District	
POSITIONING TECHNIQUES: GPS; A-GPS; Cell ID; radio signal propagation; user input; small active or passive senders (e.g., Bluetooth) in the future	
MOBILE DEVICES SUPPORTED: Smart phones, PDAs, and laptops; demonstration version uses a laptop, as PDAs do not provide for full functionality of the map viewer applet or speech recognition software (Pospischil et al., 2002, p. 152); for PDA-like phones 120 x 320 pixels, with Java MExE, colour LCD display, and pen input (Annegg et al., 2002); platform comparable to Windows Pocket PC 2002 with Internet browser, 32 or 64 MB memory, virtual keyboard, text recognition, and simple GUI (Umlauft et al., 2003)	
COMMUNICATIONS INFRASTRUCTURE: UMTS or GPRS (always on)	
DESIGN INTERFACE: Uses the map metaphor and browser metaphor; consistent design, multi-modal interaction; hierarchy of maps (overview maps in raster and detailed maps in vector); high contrast for outdoor use; POIs grouped into regions to avoid clutter on overview maps; speech commands; users views, not application modes; multi-lingual (English, German, French); graphical, text, and voice routing	
BUSINESS/REVENUE MODEL: Charged through phone bill; additional revenue for tourist agencies from access to tour diary and production of diary on CD/DVD (Umlauft et al., 2003)	
PERFORMANCE/PROBLEMS/ISSUES: Network and device constraints; user in a strange environment; must work 'out of the box'; connection loss addressed through seamless restart; network-initiated push mechanism to reduce network traffic, part of business logic resides in mobile device; locational accuracy and scale; automated labelling issues (use of tool tips and hotspots); use of 3D silhouettes and stretching the map to improve user orientation; interactive and dynamic mapping; assume low-level devices (limited hard keys, uses soft keys); data compression; data integration; symbology (simple, self-explanatory)	

the use of 3D maps on mobile devices (Laakso, Gjesdal, & Sulebak, 2003);

- The Guide, which aimed to develop a context-aware tourist guide for Lancaster (England), which overcame many of the limitations of existing information and navigation tools (Cheverst et al., 2000);
- The Electronic Guidebook, which was developed to show how wireless technology could be used to enhance all three phases of a visitor's museum experience (Semper & Spasojevic, 2002).

The development of such tourism applications can be expected to have a significant impact on the tourism industry, as it will create a greater awareness of the tourism industry for business and leisure activities, as well as a greater appreciation of the history, culture, and environment of places. It will improve the knowledge-acquisition process of tourists, and this increased knowledge can then help conservation and preservation efforts. This in turn can be expected to lead to an increase in tourism. Hence, the revenue and economic development generated from such applications could be considerable (O'Looney, 2004, p. 14). In this way such services could be of great benefit to the tourism industry, and revenue generated from mobile tourism could be fed back into the industry.

Table 4. Webpark mobile tourist guide (Krug, Mountain, & Phan, 2003; Edwardes, Burghardt, & Weibel, 2003)

APPLICATION AREA: Rural and Recreation Tourist Guide	
B2B/B2C: B2B and B2C	**IMPLEMENTED:** Prototypes trialled
SUMMARY: Webpark aims to create a platform for LBS for tourists in rural, recreation, and protected areas across Europe (e.g., National Parks). Utilising existing information, Webpark will create new value-added services (and new value chains), providing location-specific, timely, and personalised information (maps, text, multimedia information from several databases) and services to cyclists, hikers, and so forth to aid their choices. Such services are enhanced through data mining (e.g., to predict travel times or accessibility). Webpark can also be used by administrators to not only collect and manage information, but also to aid in the protection of natural resources, educate, and influence attitudes towards conservation (www.webparkservices.info/pages/project.html)	
PUSH/PULL SERVICES PROVIDED: Push—safety, weather, and ecological alerts; spatial and temporal queries about flora and fauna (e.g., where are eagles found, which species are found close by at this time of year); species identification; reconnaissance; route guidance, tracking, and route profiles; information on POIs, hotels, restaurants (e.g., visible POIs filtered according to user preferences); record personal information on places visited (notes, photos, etc.) Pull—emergency services; visitor tracking (providing information to administrators on user preferences, where visitors go and when, travel times according to different modes of transport, etc.); the user is requested to opt in for visitor tracking	
GEOGRAPHIC COVERAGE/ENVIRONMENT: Developed for outdoor rural and recreation environments of Europe; trialled in the Swiss National Park, Dartmoor National Park, and Wadden Sea National Park	
POSITIONING TECHNIQUES: GPS (not restricted by network coverage); also utilises mobile communications networks; able to utilise any positioning technology	
MOBILE DEVICES SUPPORTED: Smart phones and PDAs; initial trials used a Compaq PDA (iPAQ and Navman GPS) and a Nokia GPRS phone	
COMMUNICATIONS INFRASTRUCTURE: GPRS capabilities tested in Swiss National Park trials	
DESIGN INTERFACE: Initial trials used ArcPad, which was replaced by their own platform due to problems, such as too many buttons and menus, and instability of the application (Edwardes et al., 2003, p. 1013); Webpark utilises three groups of controls—a query interface (lists and maps), explore answers interface, and display answers interface (lists, maps, multimedia information); for map interfaces, commonly used controls such as pan, zoom, and so forth are available	
BUSINESS/REVENUE MODEL: Free and pay-per-use services; micro-payments can be used from an Internet wallet account; premium services require user subscription (for additional personalisation and push data, such as weather alerts)	
PERFORMANCE/PROBLEMS/ISSUES: User needs, information availability, delivery mechanisms, interoperability; knowledge discovery, the use of intelligent agents; dynamic visualisation on small displays (e.g., scale and generalisation issues); limited network coverage in rural areas; the need for spatial and temporal metadata for value-added personalised services; data integration; privacy; security; pricing	

Such applications will also fundamentally alter the experience of the tourist, allowing the user to spend more time exploring sites and attractions, and less time finding information, making bookings, and so forth. This is an important consideration when a significant component of the tourism market is dominated by "time poor" individuals who take frequent but shorter holidays. The implementation of mobile tourist guides will also result in the creation of new and longer value chains, including network providers, a number of content providers, application developers, and so on. They will also lead to increased specialisation (e.g., different content providers specialising in different types of information) (Peng & Tsou, 2003, pp. xxx-xxxi). Another possible benefit is improved safety for tourists.

ADDRESSING APPLICATION DEVELOPMENT CHALLENGES

While prototypes and applications have been developed and they clearly show the potential and benefits of mobile technology for the mobile tourist, there are a number of technical and business issues which must be addressed before such applications will be widely adopted. The technology problems must be solved first (i.e., the applications have to "work" before cost and other issues can be tackled). Yet studies have shown that customers are flexible, provided there are significant benefits and the costs of the new technology are lower (e.g., SMS) (Eriksson, 2002, p. 15). It should also be remembered that we are dealing with B2C applications here and that consumers require them to be easy to use, low cost, and of better value than existing products/solutions.

Technical Challenges

As mobile tourism applications depend on a relatively complex set of components, the first challenge is to ensure that the various components that make up such applications can communicate with each other. This is a significant problem given that different mobile devices, networks, GIS software, data servers, and so forth utilise different interfaces and protocols (Chen, 2004). The second challenge is to ensure that mobile tourism applications can be integrated with existing (non-mobile) systems (e.g., billing systems) (McKee, 2004; Spinney, 2003)—hence, the need for standard APIs and protocols (e.g., XML, GML, Java, OSA/Parlay API, CORBA). Such standards ensure that each application can access distributed and heterogeneous data sources, distributed tools, and services, regardless of the underlying platform, hardware, vendor, network, positioning technology, data formats, and so forth (Zadorozhny & Chrysanthis, 2004, p. 147). Standards also allow for mobile tourism applications to be easily implemented with minimal cost and effort (McKee, 2004). A number of standards have been developed for LBSs, the most significant of which are the Open GIS

Consortium's (OGC) Open Location Service (OpenLS) GeoMobility Server (GMS) and the mobile location protocol (MLP) of the Location Interoperability Forum (LIF).

The methods of user interaction also differ from those for Internet- or PC-based applications. Small-screen mobile devices allow fewer means of user input. PDAs and smartphones utilise a touch screen and pointing device rather than a mouse; some PDAs have a qwerty keyboard, but most rely on onscreen keyboards or handwriting recognition software; they also have scrolling keys, and some support voice entry. Smartphones also utilise onscreen keyboards and scrolling keys. WAP phones, on the other hand, tend to have even more restricted means of user input, being limited to phone keypad entry and scrolling keys (Peng & Tsou, 2003, p. 456) (see Table 5). Hence, PDAs and smartphones are the preferred devices.

Such devices are also limited by their small screen size, so the user interface must fit within a very small area. Menus, button, tabs, lists, maps, and so on must all be arranged so as to optimise the limited space available. Each application must be customised in order to provide a clear, uncluttered, easy-to-read, and easy-to-use user interface. Most applications adopt the Web metaphor for their user interfaces. The standardisation of interfaces (e.g., icons and symbols) (Eriksson, 2002) for the wireless Web would be a significant advance in the development of mobile tourism applications.

Maps and graphics should also be designed, especially for such small-screen devices—they should be simple, uncluttered, have high contrasting colours (so they can be read outdoors in bright light), and be easy to read at arm's length (Hjelm, 2002, p. 247). Maps should also take into account cultural conventions and utilise widely understood symbology, so that they can be understood regardless of language (Nissen, Hvas, Münster-Swendsen, & Brodersen, 2003).

Table 5. Characteristics of mobile devices (largely derived from Mountain & Raper, 2002, pp. 3-7)

Device	Description	Client-Server Model	Communication	User Interaction
WAP Phone	Largest number of users, small screen, poor resolution, sometimes monochrome but now many colour displays; narrow colour palette; mainly supports text (SMS); little processing power, closed platform; pre-installed applications, little customisation; can only operate as a connected mobile tour guide	Server side processing	Poor caching ability means can only be used online; reliant on network coverage	Phone keypad; scrolling keys
Smartphone	Larger screen size, better resolution, greater storage and memory, better colour palette and processor; can download applications, utilises common OS, e.g., Symbian; supports text, maps, e-mails, word processing, etc.; allows customisation; can operate as unconnected or connected mobile tour guides; greater functionality; fewer users	Server-side and client-side processing	Better caching capability means can be used online, as well as off-line when network coverage is unavailable	On-screen keyboard, touch screen and pointer; scrolling keys; phone keypad
PDA	Can operate as unconnected or connected mobile guides; they have the advantage of an even larger screen, higher resolution, larger colour palette, higher processing power, larger storage and memory; customisation, can download applications; Palm OS, Windows CE, etc.; supports greater client-side processing; greater functionality; higher-end applications; fewer users; supports a wider range of positioning techniques	Support the widest range of client-server models	Even better caching capability means can be used online, as well as off-line when network coverage is unavailable	Qwerty keyboard or onscreen keyboard and pointer; hand writing recognition, scrolling keys; voice input

Labelling is also a problem on such small-screen devices, in that automated labelling can result in cluttered and overlapping labels. In order to address this issue, labels should be scale dependent, and tool tips and hotspots should be used (Gartner & Uhlirz, 2001; Uhlirz, 2001).

Other limitations related to the use of mobile devices include poor screen resolution, limited colour palette, short battery life, limited storage and memory, and slow processing power. Many of these limitations, however, are already less of an issue given the pace at which hardware is improving. Nevertheless, small screen size will continue to be an issue, regardless of other technological improvements.

Applications must be smart enough to be able to deliver content to the user, taking into account the user device and its capabilities, as well as the characteristics of the wireless network. For example, owing to the limitations of WAP phones, it is preferable to deliver driving directions to these devices as text rather than as maps and multimedia information; the same query on a PDA may result in the delivery of maps and audio to the driver.

Web sites developed for the wired Internet are generally unsuitable for small-screen mobile devices and must be redesigned for these mobile devices—shorter pages, a deeper hierarchy, and so forth (Kacin, n.d.). Furthermore, the information delivered to such devices must be in a form the mobile devices can read (e.g., WML, WAP, HDML, C-HTML, XHTML) (Peng & Tsou, 2003, p. 468). The format of spatial data delivered to the Internet is also inappropriate to the wireless Web (e.g., GML, SVG) (De Vita, Piras, & Sanna, 2003; Reichenbacher 2003). Hence, data formats suitable for wireless devices have been developed. Most applications using vector data use GML to store the data on the server and convert to it to SVG Tiny (for mobile phones) or SVG Basic (for PDAs) for the display of vector data (e.g., Reichenbacher, 2001a, 2001b, 2003; Nissen et al., 2003).

Many applications also utilise predefined, static, scanned raster maps, as they result in lower storage, transmission, and rendering costs (Pospischil et al., 2002). The problem is, however, that they allow for little manipulation, interactive, or adaptive mapping, and rescaling results in information loss (Reichenbacher, 2001a, 2001b, 2003; Hjelm, 2002, p. 257). Raster data formats also result in high storage and transmission costs. Vector data formats, on the other hand, result in lower storage and transmission costs, but greater rendering costs (Hjelm, 2002, p. 257). However, they do support dynamic, interactive, and context-adaptive mapping, connection to a database, and a wider range of applications (Abowd et al., 1997; Long, Kooper, Abowd, & Atkeson, 1996; Reichenbacher 2001a, 2001b).

The use of wireless networks also poses a challenge for mobile tourism. The limited bandwidth of wireless networks is a particular issue for the delivery of spatial data and multimedia content. As applications for mobile tourists require real-time delivery of information, users will not take up these services if they have to wait (i.e., the issue of latency). Hence, such applications require reasonably fast transmission and rendering of spatial data. Other issues particularly relevant to telecommunications networks are connection stability and availability (Peng & Tsou, 2003, p. 467). Coverage is also a major issue for wireless networks, as without access to the network, no information can be exchanged with the server. In this case, the applications must rely on locally cached data. Given the limited memory and storage of many mobile devices, this is a problem. Some wireless networks, such as WLAN and Bluetooth, also only cover small geographical areas.

Given that mobile phones are the most common form of mobile devices for leisure tourists, another important issue is that of international roaming and whether the mobile device can connect to the "local" mobile network. For example, a GPRS phone, operating in a GSM network environment, can only operate in GSM mode. If a GSM phone is being used in North America where the network is AMPS, then the phone cannot connect to the local network. With the introduction of GSM networks in North America, this is becoming less of a problem, although tourists from GSM countries will need to own a phone with the ability to access the GSM1900 band (these units are popularly referred to as "tri-band" phones; "quad-band" phones, which can also access GSM850, are available now).

Another major issue for mobile tourism is that of locating the user at the required accuracy. Information delivered to the tourist is location dependent, so this is an important aspect of such applications. Not only are there a wide variety of positioning technologies available (mobile network based and handset based), but they each have different limitations and accuracy. GPS, a handset-based technique, is generally regarded as superior, as it provides global coverage and is more accurate (around 10m) than network-based techniques, but GPS has operational limitations in urban environments and indoors.

Mobile network-based positioning does not require any direct expenditure by the user, as existing mobile phone networks already have the ability to identify the user's location by Cell ID. More precise positioning techniques, however, require upgrades to the network infrastructure (e.g., Time of Arrival—TOA). Some, such as Enhanced Time Difference of Arrival (EOTD), also require more intelligent handsets. As such techniques require no additional outlay by the user (other than the cost of their mobile device) and can be utilised by even the simplest devices, they are the easiest positioning techniques to use for consumer applications. However, network-based techniques, such as Cell ID, TOA, and so on, have a very coarse accuracy (50-500m in urban areas), which can be a major problem for tourists trying to navigate around a city.

GPS positioning is the better choice for mobile tourism applications, but requires the mobile device to have a built-in GPS receiver or to be able to connect to a GPS receiver—hence, more cost for the tourist. While PDAs with built-in GPS receivers are on the market, they are costly and few mobile phones today come with in-built GPS receivers. Nor are mobile tourists likely to want to carry around two devices (a GPS receiver and a mobile device).

Business Challenges

Revenue/business models for mobile tourism will vary according to the application/services provided (Giaglis, Kourouthanassis, & Tsamakos, 2002, p. 76). They can, however, be quite complex, owing to the longer value chain involved in such services. Hence, while it may be that the network providers may charge the users directly, other stakeholders, such as content providers, will want a share of the revenue (Giaglis et al., 2002, p. 74). Thus the stakeholders will have to agree on costs, as well as on issues such as roaming and privacy protocols (Eriksson, 2003, p. 17). A centralised billing system in which network providers collect fees and then forward them on to other stakeholders is easiest for both tourists and providers alike (Watson et al., 2004, p. 324).

There are a number of business/revenue models adopted by existing applications with the basic distinction being whether the service is free of charge or the user is charged (Giaglis et al., 2002). In some cases only certain services are free, while charges are incurred for higher-level services (e.g., Webpark). Charging may be by subscription or on a pay-per-use basis. If the pay-per-use model is used, there must be a system in use for micropayments. Giaglis et al. (2002, pp. 76-78) offer a framework for matching services/applications, technology, and business models, in order to assist in the development of such applications.

Fundamentally, there are three sets of charges: (1) cost of handsets (purchase or rental); (2) mobile telephone infrastructure costs (rental of the sites the base stations are situated on, the cost of the base station equipment, ongoing maintenance costs, access costs to the sites, roaming charges); and (3) charging for value-added services. O'Looney regards syndication as an effective business model for such services, owing to the number of stakeholders involved. Syndication, however, requires standardisation of information and transaction procedures (O'Looney, 2004, p. 4).

The costs of mobile devices are also an issue, as are the costs of the services themselves. Until the cost of suitable mobile devices is more affordable to leisure travellers, mobile tourism applications will not be widely adopted (although the current popularity of "multi–function" phones—which include some PDA functionality, camera, and MP3 players—suggests that consumer resistance may be overstated). At the same time, tourists have to be willing to pay for such services. In order to gain acceptance, mobile tourism applications must offer real-time, value-added, accurate, location-specific services that provide significant advantages over traditional modes of tourism. The availability of suitable and accurate content (data quality) is another factor inhibiting the widespread adoption of such services—some argue that suitable content will only become available when there is a critical mass of mobile tourism applications in the marketplace. The debate can be summarised simply as one of whether "applications availability drives content" or "content availability drives application development." In order to ensure the widespread development of mobile tourism, Eriksson (2002, p. 17) states that the tourism industry needs to: store all information digitally and make it accessible via digital devices; all services must be able to be paid for digitally; information must be made available in a variety of languages; and all facilities must be able to be plotted on maps.

A major factor limiting consumer acceptance is the issue of privacy/security. These applications store vast amounts of personal information on users (e.g., user profiles, user location, etc.), and service providers need to protect this sensitive information by implementing clear security and privacy protocols; secure encryption during transmission is a minimum requirement. Consumers need to know that service providers will not use this information in any other way or pass on their personal details to any third party. Many applications address this issue by requiring the user to opt-in or opt-out of services (e.g., user tracking). This also solves the problem of some services becoming a nuisance (e.g., marketing), as they can be turned off. Another way to ensure trust is to make sure control of the location information stays in the hands of the consumer (Giaglis et al., 2002, p. 78). Until consumers trust the service providers, such applications will not be widely adopted.

FUTURE DIRECTIONS

The adoption and implementation of standards for mobile tourism will assure "plug and play" capability of LBS applications on different devices or different telecommunications networks and across different platforms. With technological improvements in hardware and wireless networks (especially with the adoption of 3G or better networks), many of the technical challenges mentioned above will no longer be an issue. Along with these improvements, other features will become more prominent, such as 3D visualisation, virtual reality, and animations. Applications are already being developed with 3D visualisation, virtual tours, and animations as their focus (e.g., TellMaris end DeepMap).

The rapid rate at which mobile technology is improving and the decreasing costs of mobile devices will mean that suitable and cost-effective devices will soon be available to a wider range of consumers. Currently the cost of fully featured smartphones and PDAs is too high to lead to widespread use among leisure travellers, although there is widespread use of these devices among business tourists.

Technological improvements in positioning techniques will also result in improved positional accuracy (e.g., as a result of improvements in networks—3G networks, Galileo, etc.). The decreasing costs and shrinking size of GPS receivers and the further development of mobile phones with inbuilt GPS receivers (early versions of these units are available) will also mean that GPS positioning will be more widely available. GPS provides more accurate positioning (than those based on telecommunications networks), as well as global coverage. Hence, it is more appropriate for providing accurate and value-added services to the mobile tourist. Even when mobile networks are unavailable, real-time positioning is still possible. While more precise non-GPS-based positioning is available using WLAN, Bluetooth, and so forth, they have very restricted local coverage.

A significant restriction to the widespread adoption of mobile technologies for tourism is the unavailability of suitable content. As more applications are developed and more content providers become aware of the potential of mobile tourism, more content will become available. This will also result in increased specialisation of the content providers.

CONCLUSION

Mobile tourism has the potential to enhance the experience of the mobile traveller and to deliver new kinds of information to the user—animations, video, audio. The most significant advantage to the tourist is to not only support the pre- and post-touring phases of activity, but also to support the mobile tourist on the move. By providing value-added, location-specific, personalised services to mobile tourists, mobile technologies will not only offer superior services to the travellers (services not provided by existing Web-based applications), but also fulfil a requirement of the mobile user.

Services that improve the efficiency and experience of the tourist are needed more than ever before, in an environment where people have less leisure time and shorter holidays. Such services will, however, not be adopted unless they provide significant advantages at reduced costs. A number of technical and business challenges must also be addressed before applications for the mobile tourist will be widely adopted. With the rapid speed at which technology is improving, a major market is likely to emerge in the next few years.

ENDNOTES

[1] Such applications fall under the banner of location-based services (LBSs), which refer to a suite of applications which utilise the location of a mobile device in order to deliver real-time, location-specific, value-added, personalised services.

[2] The tourism industry could certainly benefit from a standard data model or interface from which to access structured data, so enhancing user satisfaction with the planning stage of the tourism activity (Watson et al., 2004, p. 316).

REFERENCES

Abowd, G., Atkeson, C., Hong, J., Long, S., Kooper, R., & Pinkerton, M. (1997). Cyberguide: A mobile context-aware tour guide. *Wireless Networks, 3,* 421-433.

Annegg, H., Kunczier, H., Michlmayr, E., Pospischil, G., & Umlauft, M. (2002). LOL@: Designing a location based UMTS application. *Elektrotechnik und Informationstechnik, 119*(2), 48-51.

Brown, B., & Chalmers, M. (2003, September 14-18). Tourism and mobile technology. In K. Kuutti, E. Karsten, G. Fitzpatrick, P. Dourish, & K. Schmidt (Eds.), *Proceedings of the 8ᵗʰ European Conference on Computer Supported Cooperative Work,* Helsinki, Finland (pp. 335-355). Dordrecht: Kluwer Academic Press.

Buhalis, D. (2003). *E-tourism: Information technology for tourism management.* London: Prentice-Hall.

Chen, A. (2004). *Open standards will evolve location-based services.* Retrieved November 30, 2004, from http://www.eweek.com/print_article2/0,2533,a=131075,00.asp

Cheverst, K., Davies, N., Mitchell, K., Friday, A., & Efstratiou C. (2000). Developing a context-aware electronic tourist guide: Some issues and experiences. In *Proceedings of CHI2000, the Conference on Human Factors in Computing Systems* (pp. 17-24). New York: ACM Press.

CSTB (Computer Science and Telecommunications Board). (2003). *IT roadmap to a geospatial future.* Washington, DC: The National Academies Press.

De Vita, E., Piras, A., & Sanna, S. (2003). *Using compact GML to deploy interactive maps in mobile devices.* Retrieved from http://www20003.org/cdrom/papers/poster/p051/p51-devita.html

Edwardes, A., Burghardt, D., & Weibel, R. (2003, August 10-16). Webpark—location based services for species search in recreation area. In *Proceedings of the 21ˢᵗ International Cartographic Conference, "Cartographic Renaissance",* Durban, South Africa.

Epstein, M., Garcia, C., & dal Fiore, P. (2003). *History unwired: Venice frontiers—mobile technology for intelligent tourism and citizenship.* Cambridge, MA: MIT. Retrieved from http://web.mit.edu/frontiers

Eriksson, O. (2002). Location based destination information for the mobile tourist. In *Information and Communication Technologies in Tourism—2002* (pp. 255-264). New York: Springer.

Gartner, G., & Uhlirz, S. (2001). Cartographic concepts for realising a location based UMTS service: Vienna city Guide "LOL@." In *Proceedings of the 20ᵗʰ International Cartographic Conference* (vol. III, S.3229-3239), Beijing. Retrieved from http://lola.ftw.at/homepage/content/a40material/Vienna_City_Guide_LoLa.pdf

Giaglis, G., Kourouthanassis, P., & Tsamakos, A. (2002). Towards a classification for mobile location services. In B. Mennecke, & T. Strader (Eds.), *Mobile commerce: Technology, theory, and applications* (pp. 64-81). Hershey, PA: Idea Group Publishing.

Hassin, B. (2003). *Mobile GIS: How to get there from here.* Retrieved March 29, 2005, from http://gis.esri.com/library/userconf/proc03/p0988.pdf

Hjelm, J. (2002). *Creating location services for the wireless Web: Professional developer's guide.* New York: John Wiley & Sons.

Kacin, M. (n.d.). *Optimizing Web content for handheld devices.* Retrieved from http://www.wirelessdevnet.com/channels/pda/features/handheldcontent.html

Karagiozidis, M., Zacharopoulos, I., Xenakis, D., Demiris, A., & Ioannidis, N. (2003). *Location aware visually enhanced ubiquitous services.* Retrieved from http://loveus.intranet.gr/docs/LoVEUS_TechPaper.pdf

Karimi, H., & Hammad, A. (2004). *Telegeoinformatics: Location-based computing and services.* New York: CRC Press.

Krug, K., Mountain, D., & Phan, D. (2003, March). Webpark. Location-based services for mobile users in protected areas. *GeoInformatics, 6*(3), 26-29.

Laakso, K., Gjesdal, O., & Sulebak, J. (2003, September 8-11). Tourist information and navigation support using 3D maps displayed on mobile devices. In B. Schmidt-Belz & K. Cheverst (Eds.), *Proceedings of the Workshop on Mobile Guides, Mobile HCI 2003 Symposium,* Udine, Italy (pp. 34-39).

Lohnert, E., Mundle, H., Wittmann, E., & Heinrichs, G. (2004). Wireless in the Alps: An LBS prototype for mountain hikers. *GPS World, 15*(3), 30-37.

Lohnert, E., Wittmann, E., Pielmeier, J., & Sayda, F. (2001, September 11-14). PARAMOUNT: Public Safety & Commercial Info-Mobility Applications & Services in the Mountains. In *Proceedings of the 14th International Technical Meeting of the Satellite Division of the Institute of Navigation* (ION GPS 2001), Salt Lake City, UT (pp. 319-325).

LOL@ Local Location Assistant. (2004). Retrieved November 25, 2004, from http://lola.ftw.at/homepage/

Long, S., Kooper, R., Abowd, G., & Atkeson, C. (1996, November). Rapid prototyping of mobile context-aware applications: The Cyberguide case study. In *Proceedings of the 2nd ACM International Conference on Mobile Computing and Networking* (MobiCom '96), Rye, NY (pp. 97-107). New York: ACM Press.

LOVEUS. (2004). Retrieved December 5, 2004, from http://loveus.intranet.gr/

McKee, L. (2004). LBS interoperability through standards. In J. Schiller & A. Voisard (Eds.), *Location-based services* (pp. 149-171). Amsterdam: Elsevier.

Mountain, D., & Raper, J. (2002, September 18). Location-based services in remote areas. In *Proceedings of the Association of Geographical Information* (Paper B5.1, pp. 1-9). Retrieved from http://www.soi.city.ac.uk/~dmm/research/pubs/B05.3.pdf

Nissen, F., Hvas, A., Münster-Swendsen, J., & Brodersen, L. (2003). *KMS, National Survey and Cadastre–Denmark: Small-display cartography.* Retrieved November 30, 2004, from http://gimodig.fgi.fi/pub_deliverables/D3_1_1.pdf

O'Looney, J. (2004). GIS and enlightened location-based tourism: An innovation whose time has come. In *Proceedings of the ESRI User Conference,* San Diego, CA.

Open LS. (2004). *Open GIS location services.* Retrieved December 15, 2004, from http://www.opengcospatial.org/specs/?page=specs

PARAMOUNT. (n.d.). Retrieved from http://www.paramount-tours.com

Peng, Z., & Tsou, M. (2003). *Internet GIS: Distributed geographic information services for the Internet and wireless networks.* New York: John Wiley & Sons.

Pospischil, G., Umlauft, M., & Michlmayr, E. (2002). Designing LOL@, a mobile tourist guide for UMTS. *Lecture Notes in Computer Science, 2411,* 140-154.

Readman, D., & Mojarrabi, B. (2004). *Location based and communication systems.* Unpublished manuscript, University of Queensland, Australia.

Reichenbacher, T. (2001a). Adaptive concepts for a mobile cartography. *Supplement Journal of Geographical Sciences, 11*, 43-53.

Reichenbacher, T. (2001b). The world in your pocket—Towards a mobile cartography. In *Proceedings of the 20th International Cartographic Conference,* Beijing, China. Retrieved from http://citeseer.ist.psu.edu/cache/papers/cs/23234/http:zSzzSzwww.lrz-muenchen.dezSz~t583101zSzWWWzSzpublicationszSzreichenbacherzSzICC2001_Paper.pdf/reichenbacher01world.pdf

Reichenbacher, T. (2003, August 10-16). Adaptive methods for mobile cartography. In *Proceedings of the 21st International Cartographic Conference* (ICC) (pp. 1311-1322), Durban, South Africa. Retrieved from http://www.carto.net/geog234/readings/reichenbacher_mobile_cartography_durban_2003.pdf

Semper, R., & Spasojevic, M. (2002, April). The electronic guidebook: Using portable devices and a wireless Web-based network to extend the museum experience. In *Proceedings of the Museums and the Web Conference.* Retrieved from http://www.archimuse.com/mw2002/papers/semper/semper.html

Spinney, J. (2003). *A brief history of LBS and how OpenLS fits into the new value chain.* Retrieved from http://www.esbic.ie/geo business/Mobile_GIS/overview.htm

TellMaris. (2004). Retrieved December 10, 2004, from http://www.tellmaris.com/

The Electronic Guidebook. (2005). Retrieved April 4, 2005, from http://www.exploratorium.edu/guidebook/

Uhlirz, S. (2001). Cartographic concepts for UMTS-location based services. In *Proceedings of the 3rd Workshop on Mobile Mapping Technology,* Cairo, Egypt. Retrieved from http://lola.ftw.at/homepage/content/a40 material/Cartographic_Concepts_ for_UMTS_Location_based_Services.pdf

Umlauft, M. Pospischil, G., Niklfeld, G., & Michlmayr, E. (2003). LOL@, a mobile tourist guide for UMTS. In H. Werthner, & E. Veit (Eds.), *Journal of Information Technology & Tourism, 5*(3), 151-64.

Watson, R., Akselsen, S., Monod, E., & Pitt, L. (2004). The Open Tourism Consortium: Laying the foundations for the future of tourism. *European Management Journal, 22*(3), 315-326.

Webpark Mobile Tourist Guide. (2004). Retrieved December 10, 2004, from http://www.webparkservices.info/

Wherify. (2005). Retrieved March 24, 2005, from http://www.wherifywireless.com/univLoc.asp

WTTC. (2003). *Travel and tourism: A world of opportunity.* Retrieved from http://www.wttc.org/measure/PDF/Executive% 20 Summary.pdf

Zadorozhny, V., & Chrysanthis, P. (2004). Location-based computing. In H. Karimi & A. Hammad (Eds.), *Telegeoinformatics: Location-based computing and services* (pp. 145-170). Boca Raton, FL: CRC Press.

Chapter LIX
Mobile Computing—An Enabler in International Financial Services

N. Raghavendra Rao
SSN School of Management & Computer Applications, India

ABSTRACT

Information and telecommunication technologies are the major stimulus for changes in trade and commerce. Recent convergence of the above technologies has become possible due to the rapid advancements made in the respective technology. This convergence is termed as information and communication technology (ICT) and considered as a new discipline. The new discipline has made cross border commerce in the present globalization scenario a reality. This chapter talks about a model for financial services sector in international market under the new discipline. The model explains the creation of knowledge based financial services system incorporating the sophisticated concepts of information technology. Further, it provides an access to the system with devices which can be used under wireless communication environment, across the globe.

INTRODUCTION

The effects of the convergence of telecommunication and information technology are being felt in the present global corporate world. This new discipline has made economics across the globe closely interconnected and integrated. Business processes are constantly changing at an exponential rate. The new discipline is also advancing by delivering exponential increase in computing power and communication capability. The result of this advancement has created a new generation of computers working on wireless technology, cell phones having the features of portable computers, and notebooks offering similar performance of desktop computers by using the same software. Portable computers and cell phones are no longer just for globetrotting executives. Innovations and radical changes are taking place in these products.

The approach of the makers of these products is to provide fast and unwired connections in their products, enabling their clients to make use of the rich resources of their organizations located across the globe.

The policy of globalization followed by many countries is changing the world's financial markets. In this context, Buckley (2003) observes that the world economy is internationalizing and, further, firms may engage in the international business by undertaking portfolio investment (p. 35).

This has led to deregulation. This is also providing opportunities to many financial institutions across the globe who are rendering investment advisory services. Accordingly every country is rapidly adapting itself to the new global changing vistas in the financial market. It is high time the investment advisory service providers take advantage of the benefits from the new discipline. A model is suggested to help investment advisors who are involved in the international financial market analyzing data and information for investment. Further it provides information to their team members who are located at various locations across the globe for providing services.

Business Process

The international financial market mainly comprises the corporate securities, Forex, metals, and commodity segments. Investment decision and advice in these segments need vast information. Information is required for corporate companies regarding the industry, natural resources such as metals, commodities, and the country level of each segment. The types of databases which can hold a high volume of data and information are required for this model. Sophisticated software tools are also needed for analyzing the data and information from these databases.

Investment financial analysts often explore an incredible amount of data about instruments, markets, and the corporate sector. They analyze the different market segments, price movements, economic forecasts, and news events. They react on the basis of market information, price trends, historical data, and their own experience. In this process, they can make many observations from the data and information available. They can try to determine the patterns from their observations.

Case Study for International Financial Services

A London-based investment consultancy organization, which has been operating in securities trading at the London Exchange market, has decided to go global. The organization decided to add other activities such as securities related to companies in different countries, Forex, metals, and commodities as their core services under its umbrella. It also changed its name to the Global Finance Services Advisory Group (GFSAG). GFSAG hired domain experts located in different countries under its business process outsourcing strategy. The group decided to follow the concept of virtual office for its operations in different countries. Domain experts and their team members can operate from anyplace of convenience. Their approach for virtual offices is to save the cost of infrastructure and to avail the benefits under the new discipline. The respective domain expert groups are expected to monitor, guide, and assist their counterparts and team members at different locations across the globe. The corporate office in London provides services for all activities to existing clients and prospective clients through the executives located at various locations across the world. In case of additional information and clarification, the executives are permitted to be in touch with the

Table 1. Activities of GFSAG

Country	Location	Activities	Controlled By
USA	New York	Corporate Securities	Domain Experts
Australia	Sydney	Foreign Exchange	Domain Experts
Middle East	Bahrain	Metals	Domain Experts
Japan	Tokyo	Commodities	Domain Experts
UK	London	All the Above Activities	Corporate Group

respective domain experts while they are at their clients' offices. The activities of GFSAG are summarized in Table 1. The places of operations are assumed for the purpose of case study.

These domain experts can analyze the information from the knowledge-based system for forecasting and identifying the risks associated with the operations in investments in the international financial market. On the basis of their analysis, inferences can be drawn and solutions can be suggested by them. These solutions are stored in an application database in a mobile computing server. The executives at the respective locations of their offices across the globe will be guiding their clients by having access to this server.

MODEL FOR GLOBAL FINANCIAL SERVICES ADVISORY GROUP

The business process explained in the case study will be the base for creation of knowledge-based international financial services systems under a wireless communication environment. This model will be referred as the GFSAG model. The GFSAG model has the following four stages:

- **Step 1:** Creation of Knowledge-Based System in GFSAG Model
- **Step 2:** Simulation and Forecasting for the Probable Risks in Financial Market

- **Step 3:** Mobile Computing Function in GFSAG Model
- **Step 4:** Requirements for GFSAG Model

Knowledge-Based System

Step 1: Creation of Knowledge-Based System in GFSAG Model

A core team of GFSAG will consist of the domain experts, hardcore software professionals, and telecommunication experts. The importance of the services of hardcore software professionals and telecommunications experts cannot be underestimated because they are the backbone of the knowledge-based and core team. The macro-level design of an investment knowledge-based system is described in Figure 1.

The inputs received from the respective domain teams and other related information are stored in the centralized legacy system transferring to text database, and data warehouse depends on the type of data and information required for the creation of an investment knowledge repository system. The importance of a data warehouse in the financial sector is best described by Humphries, Hawkins, and Dy (1999):

A data warehouse contains data extracted from the many operational systems of the enterprise, possibly supplemented by external data. For example, a typical banking data warehouse will require the integration of

Figure 1. Knowledge base in GFSAG model

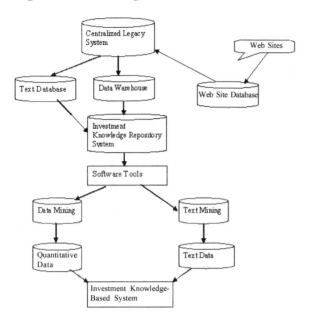

data drawn from the deposit systems, loan systems and the general ledger, just to name three. (p. 34)

The significance of a data warehouse is also highlighted by Adriaans and Zantinge (1999) when they say:

In order to perform any trend analysis you must have access to all the information needed to support you and this information stored in large data bases. The easier way to gain access to this data and facilitate effective decision making is to set up a data warehouse. (p. 25)

Text Database

This database will contain business practices, procedures, policies, culture, legal, taxation, accounting standards, political environment, various organizations profiles, information pertaining to natural resources of metals and commodities, and views and opinions of domain experts at each country and global level.

Data Warehouse

This will contain the quantitative data related to corporate securities, foreign exchange, metals, and commodities at each country and global level.

Web Site Database

This will contain the downloaded relevant information from the various sites in respect to the financial services sector at each country and global level.

Knowledge Repository

The data and information in the text database and data warehouse are to be grouped and stored as per the segments of the business activities of GFSAG.

Software Tools

The analysis of quantitative data stored in the investment knowledge repository system is carried through data mining. This tool helps one to know the relationship and patterns between data elements. The analysis of textual data is carried through text mining like data mining; it helps to identify relationships among the vast amount of text data. Pujari (2002) also states that text mining corresponds to the extension of the data mining approach to textual data (p. 239).

The investment knowledge-based system is created after the analysis by domain experts. This will contain how financial markets react to an event and the behavior of the market in the recent times.

Step 2: Simulation and Forecasting for the Probable Risks in the Financial Market

Virtual reality is one of the concepts among the number of other concepts provided by information technology. Simulation is the basic element in virtual reality applications. The time dividing between simulated tasks and their real-world counterparts is very thin.

The synergy between real-world and simulated facts yields a surprising amount of effectiveness. The possible financial risks can be envisaged through simulations from the investment knowledge-based system. The simulated information can be provided through a mobile computing system. The use of features of virtual reality for simulation and forecasting for probable risks in financial markets are illustrated in Figure 2. Chorafas and Steinmann (1995) observe that each financial institution has a different way of looking at the market and business opportunity. The strategic approach must be mapped into the machine and then interactively visualized. Not surprisingly, some banks are very advanced and are leaving their competitors in the dust (pp. 174-175).

The harsh realities of risks are well known and understood by domain experts and their team members. The sophisticated tools help financial analysts form their views and opinions. It must be remembered that these tools are useful for keeping the unpleasant surprises to a minimum. The culture of the country influences the percentage of risks one takes.

Wireless Environment: Concept of Mobile Computing

Mobile computing can be defined as a computing environment over a physical mobility. Schiller (2004) rightly says that GSM (Global System for Mobile communication) is the most successful digital mobile telecommunication system in the world today (p. 96).

The main features of a GSM system are indicated in Figure 3. A GSM has three subsystems: RSS (Radio SubSystem), NSS (Network and Switching Subsystems), and OSS (Operation SubSystem).

Support for Mobility

A mobile computing network becomes more useful when it supports business applications on its network. Now it is becoming possible by adding components such as file systems, databases, and security in mobile and wireless communication. The Web has been designed for conventional computers and fixed networks. Several new system architectures offer the opportunity to change the phase in telecommunication technology. Mobile communication is being influenced by merging telecommunication with computer networks. The present trend

Figure 2. Virtual reality concepts in GFSAG model

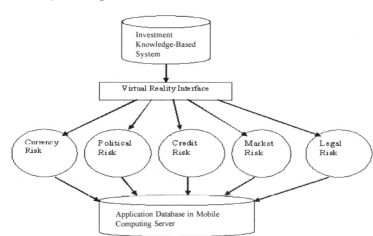

Figure 3. Overview of GSM system (adapted from Schiller, 2004)

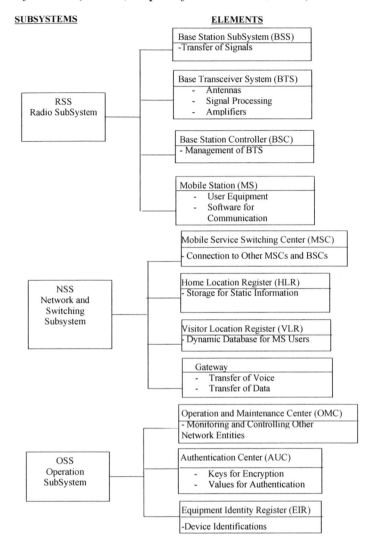

SUBSYSTEMS   ELEMENTS

in the mobile phone market is cell phones being designed that take care of some features of computers besides voice calls. It would be apt to refer to the present mobile phone as a "mobile device" because these have additional features besides the conventional cell phones. It is interesting to note the observation of Giussani (2001) on mobile phones (pp. 227-247). He classifies it into four categories of devices: (1) dedicated devices, (2) integrated devices, (3) modular devices, and (4) federated devices. The essential features of the four categories of devices are mentioned in the Table 2.

Present Scenario

Recently, handset makers have been queuing up to launch a new generation of smart phones which provide many features that have been special to pocket PC and PDAs, showing it is a buyer's market. Dood (2003) rightly points out that competition has benefited customers by

Table 2. Categories and the features of devices

Categories of Devices	Features of Devices
Dedicated	Designed for a particular functionalities
Integrated Devices	Integration of functions of different devices
Modular Devices	Bringing devices into one shell
Federated	Connecting different parts of devices

triggering price decreases and wider availability of service (p. 387).

Makers of pocket PCs are enhancing many new features in their new models. Here it would be apt to quote Dornan (2001), who states:

The hype surrounding mobile data is ultimately founded on one thing: The Internet, Vendors and Operators alike use slogan such as 'Internet anywhere' and 'Internet in your pocket', promising to cut the Internet free from its PC-based roots. (p. 190)

In the case of smart phones, the additional features indicated are sending and receiving mail, Excel spreadsheets, PowerPoint presentations, and PDF files. PDAs allow the users to go online even while enjoying the usual office tools like Word, Excel, and Internet browser. A camera phone is used to capture visual information such as a phone number on a billboard instead of looking for a paper or pen to jot down the number. In the work environment some workers take pictures of finished projects to secure a visual record of completed work in case management requests such a record. On the same lines of handsets, the manufactures of notebooks and laptop computers are launching their products with wireless technology. These products have convenient mobility available with modem, integrated LAN, or wireless connections with desktop power. Now it has become a necessity to establish synergy between mobile computing and knowledge-based business systems through these sophisticated devices. Taulkder (2002) confirms this view by saying:

Mobile computing not only offers instant information to a mobile worker; to a mobile worker it is indeed a productivity tool. Further the list of possible mobile applications can never be complete. (p. 18)

Security

The international financial services knowledge-based system is more sensitive and critical. Encryption is a solution that ensures the data content is not altered during the transmission between originator and recipient in a wireless environment. This is elaborated by Minoli and Minoli (1999), who observe:

Cipher technique[s] lend themselves more readily to automated. Theses technique[s] are uses [sic] in contemporary security tools and there are three kinds of cryptographic functions, such as Hash functions, Secret key and Public key. (p. 215)

Figure 4. Components of asymmetric crypto system

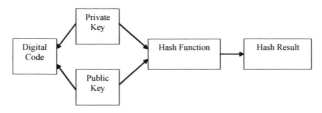

Figure 5. Mobile computing functions in GFSAG model

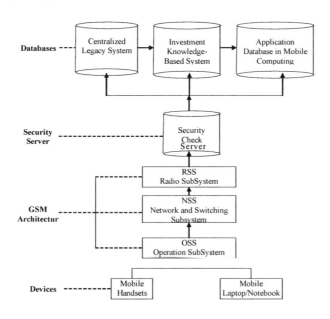

The concept of an asymmetric crypto system for encryption of data may be used in a wireless environment. An asymmetric crypto system is a system of secure key pair consisting of private key for creating a digital code and public key to verify the same. Hash function in this system means obtaining "hash result" by applying a predefined logic, control, or arithmetical process. Hash result means that every time the predefined procedure is applied, it should give the same result. The components in an asymmetric crypto system, taking the results obtained from the hash results, are explained in Figure 4.

With the advancement of information technology, the concept of cryptography used earlier by kings for secret communication is becoming popular in global commercial applications.

Step 3: Mobile Computing Function in GFSAG Model

The integration of the functions in mobile computing with a GFSAG model are illustrated in Figure 5.

Step 4: Requirements for GFSAG Model

The user groups for the GFSAG model at the respective country level and end users across the globe will be making use of it. The requirements of hardware, software, and mobile devices for the GFSAG model are mentioned in Table 3.

CONCLUSION

The disintegration of barriers in previously protected and insulated markets has created a new era of competition in the global economic environment. The present challenge for global players in international financial markets is how they should take advantage of the opportunities from the severe competition and survive in the market. The rewards for the opportunities are always are accompanied by risks. Assessing risks and incorporating the same in the final decision is an integral part of the decision making. The new discipline plays an important role for acquiring and processing information for analysis and decision making. The GFSAG model offer an idea for using the services of domain experts across the globe and for minimizing the risks in financial market. The word "international" prefixed to "financial market" will become redundant once the concepts of new discipline are taken advantage of by the corporate world.

FUTURE STUDY

Many more financial markets are opening up and becoming integrated with global markets. Many virtual financial services organizations will make their presence felt in the global

Table 3. Requirements for GFSAG model (hardware, software, and mobile devices)

Particulars		Purpose
Hardware	Server	System and application programs
	Server	Exclusively for encryption and recognition of users
	Server	Storage of data and information of various activities of financial services
	Desktop/Workstation	Development of programs, and updating and retrieving data
Software	Data warehouse	Quantitative data of various segments of the financial services market
	Text Data warehouse	Text data of various segments of the financial services market
	Data mining	Analyzing data from data warehouse
	Text Mining	Analyzing text data from text data warehouse
	Virtual Reality Interface	Visualizing through simulation and applying forecasting techniques
	Other Related Software	Software for supporting the system and routine software for business purpose
Mobile Devices	Laptop/Notebook	Interaction with domain experts and corporate office Viewing the selected information from the knowledge base system Sending reports from the marketplace to corporate office and domain experts Downloading the analyzed data for understanding
	Handsets	Discreetly informing some specific information from client's place Capturing important data by using camera features in handsets, and transmitting to corporate and domain experts offices Browsing the knowledge base system for specific purpose Sending latest short news from the financial market

market. There will be an increase in reliable forecasting by using real-time data and financial modeling. Many more different activities related to financial services will be required to be added to GFSAG model.

RECOMMENDATION

The concept of virtual organization is the key to GRID computing. All the virtual organizations will share the common resources for computing power and accessing data across the globe. Grid and mobile computing concepts will be required to be integrated, once many financial markets are interconnected with each other in the global market under the concept of virtual organization. Referring to GRID computing in the financial sector, Joseph and Tein (2004) state that grid computing provides the financial analysis and services industry sector with advanced systems delivering all the competitive solutions in grid computing. These solutions exemplify the infrastructure and business agility necessary to meet and exceed the uniqueness that the financial analysis and services industry sector requires. This particular value statement is accomplished by the fact that many of these solutions in this industry are dependent upon providing increased access to massive amounts of data, real-time modeling, and faster execution by using the grid job scheduling and data access features (p. 14).

REFERENCES

Adriaans, P., & Zantinge, D. (1999). *Data mining and data warehouse data mining* (pp. 25-36). Harlow, UK: Addison Wesley Longman.

Buckley, A. (2003). *The internationalization process, multinational finance* (pp. 35-46). New Delhi: Prentice-Hall.

Chorafas, D. N., & Steinmann, H. (1995). Implementing virtual reality in financial institutions. In *Virtual reality: Practical applications in business and industry* (pp. 161-179). Englewood Cliffs, NJ: Prentice-Hall.

Dodd, A. Z. (2003). *Wireless services, the essential guide to telecommunications* (pp. 371-408). New Delhi: Pearson Education Asia.

Dornan, A. (2001). *Inside a mobile network, the essential guide to wireless communications applications* (pp. 175-195). New Delhi: Pearson Education Asia.

Giussani, B. (2001). *The intimate utility: Roam making sense of the wireless Internet* (pp. 227-247). London: Random House Business Books.

Humphries, M., Hawkins, M. W., & Dy, M. C. (1999). *Data warehouse concepts, data warehousing architecture and implementation* (pp. 31-48). Englewood Cliffs, NJ: Prentice-Hall.

Joseph, J., & Fallenstein, C. (2004). *Introduction, the grid computing anatomy* (pp. 12-14, 47-57). New Dehli: Grid Computing, Pearson Education.

Minoli, D., & Minoli, E. (1999). *Encryption, Web commerce technology handbook* (pp. 213-225). New Delhi: Tata McGraw-Hill.

Pujari, A. K. (2002). *Text mining, data mining techniques* (pp. 239-250). Hyderabad: Universities Press (India).

Schiller, J. (2004). *Telecommunication systems, mobile communications* (pp. 93-130). New Delhi: Pearson Education.

Talukder, A. K. (2002). Mobile computing—impact in our life. In C. R. Chakravarthy, L. M. Patnaik, T. Sabapathy, & M. L. Ravi (Eds.), *Harnessing and managing knowledge* (pp. 12-24). New Delhi: Tata McGraw-Hill.

ADDITIONAL READING

Cudworth, R. (2003). *The demand for continuous information: The source online business* (pp. 12-17). London: Kogan Page.

Haugen, R. A. (2002). *Securities and markets: Modern investment theory* (pp. 6-31). New Delhi: Prentice-Hall.

Lasserre, P. (2003). *Global financial management, global strategic management* (pp. 335-351). Hampshire, UK: Palgrave MacMillan.

Lumby, S. (1998). *Foreign exchange risk management, investment appraisal and financial decisions* (pp. 579-596). London: International Thomson Publishing.

Chapter LX
Mobile Computing:
An Australian Case Study

Paul Hawking
Victoria University, Australia

Gina Reyes
Victoria University, Australia

Stephen Paull
Victoria University, Australia

ABSTRACT

Companies are investigating how they can extend existing business process through the implementation of mobile computing solutions. Deloitte has developed a model which can describe the evolution of mobile solutions within a corporate setting. This chapter adopts a case study approach to investigate the adoption of a mobile solution within an Australian company and classifies the implementation as per the Deloitte model.

INTRODUCTION

If we can make a $500 handheld device to do the same thing as a $4,000 laptop, we've saved a ton of money. (Billy Wang, Business Development Manager, Coca Cola Corporation, Mobile Planet, 2004)

Mobility, as used in the context of technology, can be described as the ability of users, systems, or data to perform or participate in information-processing tasks without being constrained to a fixed location. Although the possible applications incorporating mobile technology have been well documented, the actual realisation of these applications has only been a recent phenomenon. There have been a number of key enablers which have facilitated this realisation (Dedo, 2004). Paavilainen (2001) suggests that the specific characteristics of mobile devices contribute to the ease of use of these devices and the subsequent expansion of mobile markets. Convenience, instant connectivity, ability to personalise a device, and the independence of time and location are cited as the characteristics of these mobile computing units. It was been predicted that due to the rapid expansion of high-speed mobile services, by 2007, 60% of the U.S. population will receive mobile data, an increase from 2% in 2001. The Cellular Telecommunications and Internet Association (CTIA) expects that the most popular Internet access devices will be mobile and wireless technologies, surpassing PCs (Strategis Group, 2001).

Another contributing factor is the advent of enterprise systems, in particular enterprise resource planning (ERP) systems. These systems have provided the necessary infrastructure for companies to move towards "best business practice" while at the same time providing real-time access to information. This access, originally only available internally via desktop PCs, has now been extended to Web-based applications and to mobile computing devices. Many of these ERP systems have incorporated technology and scenarios to assist with the interaction with mobile devices.

IMPACT ON ORGANISATIONS

A number of industries already feel the impact of mobile computing devices (Varshney, Vetter, & Kalakota, 2000). In universities, mobile technology infrastructures are being implemented so that students can access academic databases from any campus location (Willard, 2000). In government, police and criminal justice organisations use mobile computing technologies, as they need mobile access to information for law enforcement. In police organisations in the U.S., mobile computing terminals are used for access to federal, state, and county records in order to facilitate auto registrations, summons, and warrants of arrest (Seaskate, 1997). These are also used for online offence reporting. In a recent study of the impact of these devices on the organisation, these mobile computing terminals enabled better communication among officers, increased the availability of information, and have been found to have a significant positive impact on officers' job satisfaction (Agrawal, Rao, & Sanders, 2003). Gartner (Casonato, 2001) found in a survey of 212 respondents who had implemented mobile technology to support "business-to-employee" scenarios that the main benefits were increased employee productivity, followed by cost reduction and cost management, new information channel, and experimentation.

In the service industries that involve product delivery, for example, it is expected that mobile inventory management systems used to track the location of goods help improve delivery times and customer service. United Parcel Services has been an early adopter of wireless technology, using radio transmitter technology in trucks to send package delivery data back to the central UPS network, so customers can track package delivery in real time. It recently invested $100 million to upgrade and consolidate its wireless network to Bluetooth technology in order to reduce operating costs. In a recent survey conducted among business-technology professionals, improved mobile technologies, business applications, and lower prices are factors that drive the use of mobile technology in business. The benefits cited by those surveyed were: increased employee communication and data sharing, increased employee productivity, improved customer service and satisfaction, easier collaboration with business partners, and increased access to corporate data for decision making (Ewalt, 2004).

IMPACT OF MOBILE COMPUTING

The benefits and impact of such technologies are factors that can be considered in the success of these mobile computing systems. In analysing the success of information systems, DeLone and McLean (1992) suggest that six dimensions of the system need to be considered: system quality, information quality, use, user satisfaction, individual impact, and organisational impact. These dimensions relate to each other in the following manner:

System Quality and Information Quality singularly and jointly affect both Use and User Satisfaction. Additionally, the amount of Use can affect the degree of User Satisfaction—positively or negatively—as well as the reverse being true. Use and User Satisfaction are direct antecedents of Individual Impact; and lastly, this Impact on individual performance should eventually have some Organisational Impact. (DeLone & McLean, 1992, pp. 83-87 as quoted in Myers, Kappelman, & Prybutok, 1998, p. 102)

This framework can be used in analysing the overall success of mobile computing systems. For example, in the case of Agrawal et al.'s (2003) study on the use of mobile computing systems in police organisations, it can be argued that the officers using the mobile technology gained mobile access to important law enforcement information. The infrastructure of the system provided a level of system quality and information quality as officers felt they were able to reliably communicate and share important information. With remote and mobile access, this promoted high use and resulted in satisfaction in using the system. The individual impact of this was felt in job satisfaction. Because of the speed of licence plate checks on the system, this helped deter criminal activ-

ity. Thus, it can be argued from this point of view that this mobile computing system made some organisational impact in that it helped in increasing organisational effectiveness.

Deloitte Touche (2001), in its management briefing on mobile technology, identified a barrier to the uptake of mobile technology being the limitations of the available mobile devices in that users often required access to data and voice simultaneously. The recent convergence of wireless technology has overcome this problem. Many single-purpose devices such as bar code readers, pagers, mobile communicators, and personal digital assistants (PDA) have now evolved and converged into single devices offering a broad range of functionality. This increased functionality and processing power has encouraged the development of various mobile application solutions. However, even with this improved functionality, the applications would be limited without the introduction of high-speed wireless networks. The increased coverage of the GSM/GPRS network has enabled companies to transfer large volumes of data in real time. Internally companies have the choice of wireless technologies such as 802.11b to interface to the local area network or Bluetooth to interface with other technologies. But the advent of improved communication technologies is presently geographically constrained, limiting the range of mobile applications while at the same time necessitating a range of communication technologies to be implemented in companies that operate in diverse geographic locations.

Even though many of the existing barriers have been addressed, there are still limitations associated with security and application development (Deloitte Touche, 2001). In terms of security, many companies are still coming to grips with issues within their companies rather than those caused by extending the boundary of the company onto developing technologies such as mobile computing devices. Presently most mobile applications are merely extensions of existing enterprise applications, and therefore human computer interface issues pertinent to the mobile device are often overlooked.

IMPLEMENTATION CONSIDERATIONS

Companies have implemented mobile technologies to address a variety of business scenarios (ESRI, 2002; Symbol, 2002; Gedda, 2004). Deloitte Research (2002) developed a model which classifies the maturity or evolution of mobile applications into three different stages (see Figure 1). In the first stage, "Mobile Enablement," the existing applications interface is extended onto a mobile device. In this stage there is limited new functionality other than portability, but the device acts as an alternate input device. In the second stage, "Mobile Reinvention," probably due to the familiarity with the mobile solution, further efficiencies can be gained using a business process reengineering approach whereby new functionality can be achieved due to the nature of the mobile device. The final stage, "Mobile Discontinuity," is where new innovative busi-

ness scenarios can be developed which transform the organisation.

Up until now, very little research has been conducted in the area of mobile solutions, and their development and implementation in Australian companies. Even though many companies have or are in the process of implementing such solutions, the documentation of benefits and issues is lacking. This chapter attempts to document such a solution in an endeavour to provide a foundation for future case studies which can assist Australian industry with examples of the application of mobile technology.

RESEARCH METHODOLOGY

This chapter adopts a case study approach to examine the impacts and issues associated with replacing a paper-based system with a mobile computing solution. Yin (1994, p. 35) emphasises the importance of asking "what" when analysing information systems. Yin goes further and emphasises the need to study contemporary phenomena within real-life contexts. Walsham (2000, p. 204) supports case study methodology and sees a need for a move away from traditional information systems research methods, such as surveys, toward more interpretative case studies, ethnographies, and action research projects. Several works (Chan & Roseman, 2001; Lee, 1989; Benbasat, Goldstein, & Mead, 1987) have used case studies in presenting information systems case-study research. Cavaye (1996) used case study research to analyse inter-organisational systems and the complexity of information systems.

The data-collection process for the present research included:

* examination of existing documentation,
* interview of actors, and
* direct observations.

Figure 1. Deloitte Touche mobile application maturity

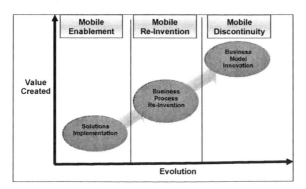

CASE STUDY BACKGROUND

The Water Corporation has been in operation for more than 100 years, and is responsible for the provision of water and removal of wastewater, servicing more than 1.8 million people in the state of Western Australia. The corporation services both urban and rural communities spread across the 2.5 million square kilometres of the state. As part of this service, the corporation is responsible for the establishment and maintenance of the necessary infrastructure. This includes:

- 250 water treatment plants,
- 110 dams and reservoirs,
- 715 bores in 107 bore fields,
- 30,538 kilometres of water mains,
- 12,579 kilometres of sewers, and
- 2,782 kilometres of drains.

The corporation employs more than 2,000 staff, with the majority based in the state's capital. An important component of the information systems infrastructure utilised by the corporation is its ERP system. The corporation had been a user of SAP's R/3 ERP system since 1998 when they installed version 3.1H. In 2001 they upgraded to the latest version available (4.6c). The scope of this implementation included financial, human resource, plant maintenance, and project management modules, as well as SAP's data warehouse solution. The SAP system is used at 50 sites by more than 1,400 users, with 350 of these being concurrent at any one time.

A major responsibility of the corporation is the maintenance of the extensive range of infrastructure responsible for the delivery of water and removal of wastewater both in the rural and metropolitan areas. The workforce responsible for this is made up of employees from within the company and external contractors provided by a partner company. The exter-nal contractors predominately worked within the metropolitan area. Traditionally, each day the workers would be allocated work orders for reported faults and new installations. This to-talled more than 300,000 orders annually, with approximately 800 orders per day communi-cated via mobile telephones. The workers would collect the appropriate materials and maps for each job before travelling to the various work locations. Getting to the correct location was often difficult due to the vast network of pipes and the remote locations. The corporation had developed a geographic information system (GIS) that recorded the location of infrastruc-ture items down the mains and manhole level of detail. Workers were provided with a printout of the necessary map for each work order.

Once field staff arrived at a location, they would assess the task and order any additional materials that may be required. This was done by telephone or radio. When the work was completed, employees were required to com-plete a range of paperwork. This included worker details, duration and type of job, mate-rials used, and work completed. This paper-work was required for each allocated task.

On returning to their base office, the paper-work was collected and the data was entered into the SAP system. This was then used for necessary time calculations including payroll and job costing. The materials used were en-tered into the materials management module, and were used to assist with inventory of components and costing of work completed, which satisfied the plant maintenance require-ments. The possible efficiencies from the cap-turing of this information were hindered due to the difficulties associated with capturing of this information in the field in a paper-based format. Due to the nature of the plumbing work and the less-than-suitable nature for recording infor-mation on paper in the field, often the paper-work was incomplete, inelEgible, or lost.

In 2001 it was decided to investigate how mobile computing devices could possibly overcome some of the issues associated with paper-based forms. A business case was developed with the objectives of:

- providing support for business objectives, and enhancing the efficiency and effectiveness of the key activities performed by field service delivery staff;
- enabling timely and resource-efficient scheduling of work;
- facilitating timely feedback to customers as to the status of work; and
- creating the capacity and flexibility for field support staff to meet the evolving business requirements of the corporation and its customers (Water Corporation, 2002).

The specific benefits identified included a reduction in data entry staff, reduction in errors resulting in an improvement in data integrity, improved scheduling resulting in a reduction of overtime hours, and reduction in field staff and vehicles.

It was expected that the project would enable efficient linkages, through mobile devices, to the corporate systems. The project would support 420 field crews. Of these, 150 service the Perth metropolitan region, while the remaining 270 service outlying rural regions. It was intended that the metropolitan employees would interact with the corporate systems in "real time" utilising GPRS technology, while the regional employees would synchronise with the corporate system in batch mode via the corporation's wide area network or via dial-up connection. The overall project budget was expected to be $AUD3.5 million, which included the typical implementation costs as well as costs associated with making staff redundant.

The main issues associated with the business case were application development, mobile device cost and functionality, infrastructure stability, transformation of business processes to mobile technology, and ongoing support. It was believed that due to the complexity of the solution, it was preferable to award the contract to a consortia of implementation partners who best addressed the above issues rather than awarding individual components to separate partners. This was believed to alleviate different partners, laying blame on other partners involved.

A request for proposal was distributed to interested parties and was eventually awarded to a consortia composed of Deloitte Consulting and Telispark in early 2003. They were expected to supply the mobile computing application licences and implement the mobile computing solution as well as providing ongoing support.

The Mobile Solution

The project commenced in April 2003 and was completed in December 2003. The solution enabled field staff to log on to their mobile computing device and display a list of work orders. They could then *"drill down"* to get more details of the order. The corporation's PC-based GIS system had been further developed to operate in a mobile computing environment. Now instead of having paper-based maps for each work order, the solution enabled workers to enter a location into the device and then display the appropriate map. The maps included addresses, streets, suburbs, and infrastructure locations. The workers had the ability to zoom in or out depending on the amount of detail they required.

When they were about to leave for the site, they would click a button in the device to indicate that they had started travelling, and once they arrived they clicked another button which then calculated travel time for each job. This also indicated that work had commenced. Once the job had been assessed, they would select the type of work from a predetermined list and materials required. When the job was completed or work had been halted for some other reason, this was also recorded in the device. The field worker would then display the next work order.

The information was synchronised between the mobile device and the SAP system using a variety of technologies depending on the type of mobile device and the communication technologies available in the area. In the metropolitan area, GPRS/GSM technologies were utilised which provided real-time synchronisation, while in outlying areas where communication technologies were limited the data was synchronised in batch mode using dial-up facilities when staff returned to their offices.

Technology

As mentioned previously a component of the project was the selection of appropriate mobile computing devices and the creation of the technological infrastructure to support the integration of these devices.

After a review of the available mobile computing devices suitable for the identified tasks, it was decided to implement two distinctly different devices. The first was a device made for rugged conditions (70%) which used the traditional telephone network (PSTN) for communication. The second was a sleeker and stylish device produced by Oxygen (XDA) which accounted for the remaining 30%. The XDA was a combination of a mobile phone with GPRS/GSM capabilities as well as the traditional PDA functionality.

In terms of the technological infrastructure, the complexity was considerable, with 21 components from a variety of vendors. With so many interdependent components, an analysis was undertaken to identify single points of failure, whereby if a particular component failed, the solution would cease to operate. The analysis identified 12 separate components on which the solution was reliant, with the worst case scenario of a 70-hour recovery time for one particular component.

Implementation

From the commencement of the project, the importance of change management and training was identified. There needed to be a considerable change in culture to gain successful adoption of the solution. As part of the change management program, a short video was produced to explain the purpose of these devices and possible benefits they could provide. It also attempted to allay fears associated with increased scrutiny by management. Training courses were then conducted with staff at numerous locations, and they were then supplied with their mobile computing device.

One unexpected factor which contributed to the adoption of the devices was the selection of the XDA device. This device uses the Microsoft Pocket PC operating system which was supplied with a range of software including Microsoft Outlook used for contacts, appointments, and so forth, as well as entertainment software including games. When staff were issued their mobile devices, the GIS solution was implemented but the work order solution was not completed and therefore not implemented on the device. Initially it was decided to remove any unnecessary software other than those related to work tasks. But this did not happen and probably partly contributed to the success of the implementation. It encouraged workers to become competent with their device, as it could be used for a range of non-work-related tasks. For many of the workers, the XDA had a novelty attraction, as very few of them had used such a device before and were impressed with its capabilities. The increased usage of the device facilitated the uptake of the work order solution when it was rolled out.

Interestingly the solution was more readily adopted by employees of the corporation compared to external contractors. It was felt that the contractors saw this solution as a means to better manage them and reduce their payments. Another unexpected factor is that the more "ruggedised" devices were not required, as minimal damage has been reported with the XDA devices.

In terms of business benefits at the time of the interviews, it was too early to assess. The adoption of the devices had been successful and preliminary results have indicated improved data integrity and plant maintenance reporting. Staff involved in managing the project have investigated other mobile solutions adopted by various utility companies throughout Australia; they believe that based on these examples, the corporation will achieve a good return on investment.

Already the corporation is considering extending the project to include further functionality. They expect to rollout PDAs in the future which incorporate global positioning systems (GPSs) to facilitate workers' navigation to specific locations. An optional extra in the RFP document was automated vehicle location functionality. This would enable the corporation to identify which vehicle was nearest any particular fault, resulting in better response times and scheduling.

CONCLUSION

In terms of the mobile application maturity as per the Deloitte Touche model (Figure 1), the Water Corporation's mobile solution would fall into the "mobile enablement" category whereby an existing application or business process has been extended via mobile technology. The mobile solution is only in the initial stage of rollout and needs further investigation to assess the overall success and impact on the organisation. It is intended to revisit the organisation later in the year to conduct follow-up research. The issues that will be investigated will be the uptake of the various mobile devices, the robustness of the technical infrastructure, the tangible and intangible benefits, and future mobile solutions.

REFERENCES

Agrawal, M., Rao, H. R., & Sanders, G. L. (2003). Impact of mobile computing terminals in police work. *Journal of Organisational Computing and Electronic Commerce, 13*(2), 73-89.

Benbasat, I., Goldstein, D., & Mead, M. (1987). The case research strategy in studies of information systems. *MIS Quarterly, 11*(3), 215-218.

Casonato, R. (2001). *Mobility and business to employee applications.* Retrieved May 2004 from http://www3.gartner.com/pages/story.php.id.2545.s.8.jsp

Cavaye, A. (1996). Case study research: A multi-faceted approach for IS. *Information Systems Journal, 6*(3), 227-242.

Chan, R., & Roseman, M. (2001, December 4-7). Integrating knowledge into process models—A case study. In G. Finnie, D. Cecez-Kecmanovic, & B. Lo (Eds.), *Proceedings of the 12th Australasian Conference on Information Systems,* Coffs Harbour, Australia (pp. 113-120).

Dedo, A. (2004). *The return on your mobile investment.* Retrieved June 2004 from http://

download.microsoft.com/download/1/a/5/1a572c42-10b5-469d-9acb-cedd2e634985/MobileDevices_ROI.doc

Deloitte Touche. (2001). *Management briefing: Mobile technology.* Dublin: Deloitte Touche.

Deloitte Research. (2002). *Mobilising the machine.* London: Deloitte Touche Tohmatsu.

DeLone, W. H., & McLean, E. R. (1992). Information systems success: The quest for the dependent variable. *Information Systems Research, 3*(1), 60-95.

ESRI. (2002). *Boulder County, Colorado, sign maintenance receives big dividends from ArcPad.* Retrieved June 2004 from http://www.esri.com/news/arcnews/winter0102 articles/boulder-cnty.html

Ewalt, D. (2004, June 7). The pros and cons of wireless connectivity. *InformationWeek.* Retrieved from http://www.informationweek.com/showArticle.jhtml;jsessionid=CCYMSQZTK4WR0QSNDBCSK0CJUMEKJVN?articleID=21401672

Gedda, R. (2004, June). Mobiles make the grade. *Computerworld.* Retrieved from http://www.computerworld.com.au/index.php/id;1903328574;relcomp;1

Lee, A. (1989). Case studies as natural experiments. *Human Relations, 42*(2), 117-137.

Mobile Planet. (2004). *Mobile manufacturing: Capturing critical job tracking and inventory data.* Retrieved May 2004 from http://www.mobileplanet.com/askexperts/solutions/syware_case10.asp

Myers, B. L., Kappelman, L. A., & Prybutok, V. R. (1998). A comprehensive model for assessing the quality and productivity of the information systems function: Toward a theory of information systems assessment. In E. J.

Garrity, & G. L. Sanders (Eds.), *Information systems success measurement.* Hershey, PA: Idea Group Publishing.

Nelson, M. G. (2001). Wireless delivers for UPS overhaul. *Informationweek,* 109-113.

Paavilainen, J. (2001). *Mobile business strategies: Understanding the technologies and opportunities.* London: Wireless Press.

Seaskate. (1997). *The evolution and development of police technology.* National Institute of Justice. Retrieved from http://www.nlectc.org/virlib/InfoDetail.asp?intInfoID=169

Strategis Group. (2001). *Mobile data penetration to be nearly 60% by 2007 says the Strategis Group.* Purchase, NY: Strategis Group.

Varshney, U., Vetter, R. V., & Kalakota, R. (2000). Mobile commerce: A new frontier. *IEEE Computer,* 32-38.

Walsham, G. (2000, June 9-11). Globalisation and IT: Agenda for research. In R. Baskerville, J. Stage, & J. DeGross (Eds.), *Organisational and social perspectives on information technology,* Aalborg, Denmark (pp. 195-210). Boston: Kluwer Academic.

Water Corporation. (2002). *Request for proposal document for the provision and implementation of a mobile computing system.* Perth, Western Australia: Water Corporation.

Water Corporation. (2003). *Annual report.* Retrieved from http://www.watercorporation.com.au/annrep/index.cfm

Willard, C. (2000). High wireless act. *Computerworld, 34*(31), 49.

Yin, R. (1994). *Case study research, design and methods* (2nd ed.). Newbury Park, CA: Sage.

Chapter LXI
Introducing Mobile Technology into an Australian City Council:
Experiences and Lessons Learned

Joanne Marie Curry
University of Western Sydney, Australia

ABSTRACT

In an ongoing bid to provide high quality local government services, Penrith City Council partnered with the University of Western Sydney to derive a mobile strategy for the development of a range of handheld systems for use in the field. Several R&D projects aimed at determining the viability of using mobile technology for the conduct of off-site health, building and development and sewerage inspections and the allocation of parking and waste management infringements were conducted over a two-year period. Some significant issues relating specifically to the implementation of mobile technologies in a large Australian city council were encountered including: release hype vs. the implementation realities of mobile technology, technological options for the introduction of mobility, user acceptance of new technologies, management of client expectations, and local government standards and guidelines and their impact on development directions. The experiences and lessons that were learned from these projects can be of assistance to other local government agencies and similar organisations employing a heterogeneous workforce that is restrained by external legislation and policy.

INTRODUCTION

Penrith is a city on the western fringe of the Sydney metropolitan area, in east central New South Wales, Australia. Located at the foot of the Blue Mountains just 55 kilometres from the Sydney CBD, the City of Penrith covers an area of some 407 square kilometres. With an estimated resident population of approximately 178,000 (as at June 2004), there has been a 15% growth rate since 1991 (*Australian Bureau of Statistics*, 2004). With such rapid expansion and approximately 4,500 businesses operating in the local government area, Penrith is a fast-growing metropolitan region, and in just a generation has grown from a rural town to the major regional service centre for the outer western Sydney area. In the bid to continue to provide high-quality local government services, the IT Department of Penrith City Council, working in partnership with the University of Western Sydney, undertook several projects aimed at determining the viability of using mobile technology for Council operations such as the conduct of off-site health, building and development, and sewerage inspections, and the allocation of parking and waste management infringements.

The introduction of the Penrith City Council Mobile Strategy evolved over a two-year period and provided some interesting experiences for the development teams and contributed significantly to their learning in many areas, including: release hype vs. the implementation realities of mobile technology, technological options for the introduction of mobility, user acceptance of new technologies, management of client expectations, and local government standards and guidelines and their impact on development directions (Bryan, Holdsworth, Sharply, Curry, & McGregor, 2002; Curry & Lan, 2004). This chapter provides highlights of the development work performed by "high-achieving" student groups, and discusses some of the resultant experiences and the detailed lessons learned from this work of incorporating mobility within the Council's business processes.

PROJECT BACKGROUND

In the mid-1990s Penrith City Council (PCC) entered into a partnership with the University of Western Sydney (UWS) for engaging final-year computing students for Research and Development and Prototyping projects. This chapter concentrates and expands on a few select "high-achieving" student projects dealing with incorporation of mobility within the Council's business processes that were conducted between March 2002 and December 2004. These are invaluable projects for both their academic importance as well as their practical importance to the Council, wherein their overall aim is to reengineer existing processes for PCC field officers required to complete official documentation off-site. Field officer work ranged from inspections for development and building applications, food and hygiene surveillance reports, and issuing of parking and general counsel by-law infringement notices.

Penrith City Council is required to make many inspections in areas such as health, building and development, and sewerage. The management of the data collection and data entry for inspections is becoming increasingly cluttered and time consuming using the existing manual system.

The current system requires a field officer (inspector) to attend an inspection site equipped with the necessary paper forms for collecting data. These forms are completed in duplicate on site: one copy for the subject of the form (e.g., the owner of the premises being inspected), and a duplicate copy for Council's records. Once all inspections have been completed for the day, the inspector returns to the office where the data for each inspection is typed manually into the database, either by the field officer or an administration clerk.

This resulted in double handling of the data, thus increasing the general workload for Council administrative staff, as well as making the duplicate form redundant once the data was entered into the PCC database. Data entry errors were also common, leading to corrupt data residing on the Council database leading to poor quality management reporting.

One of the key PCC goals in the achievement of this reengineering initiative was the introduction of mobile technology to the new process environment. The justification for this approach, based on Gronlund (2002) and Garson (1999), can be given as follows:

- **Improved Security of Documentation Completed Off-Site:** Paperwork would not be misplaced as it would be securely stored in the mobile device.
- **Easier Off-Site Information Retrieval:** Information could be easily retrieved and updated if errors were identified while the field officer was still off-site.
- **Reduced Volumes of Manual Data Entry:** Automatic download of off-site data will dramatically reduce the time need for manual data entry. PCC had estimated the tangible benefits for reduced data entry at over $15,000.
- **Reduced Data Entry Errors:** Results of off-site work could be transferred to the PCC database automatically, thus significantly reducing the amount of incorrect data held.
- **Reduction of Physical Paper Volumes Requiring Storage:** Hard copies of reports are only printed as needed.
- **Improved Management Reporting:** Due to reduced data entry errors, the quality of resulting management reports would increase.

Several general environmental constraints related to the processes used by the Council's field officers were identified including:

- Field officers are required to be extremely mobile and therefore cannot be burdened by heavy and/or large equipment.
- Field officers are out in the elements everyday, therefore the solution must be unaffected by rain, heat, and other normal working conditions an officer experiences.

- There may or may not be access to a continuous power source, therefore the solution must be self-dependant for a substantial length of time.
- Due to the nature of the business (government agency), the solution must also be very secure with regards to data.

In order to address the Council's problems, the system needed to include the following capabilities:

- Ability to download lists of inspection appointments to be completed by the field officers, from the Council database to a handheld device.
- Capture an electronic version of data currently being recorded on paper forms.
- Automatically store captured electronic data into the relevant fields of the applicable Council database.
- Allow for a paper copy of the data to be produced for issue to the client upon request.

During the period March 2002 to December 2004, six selected student project teams worked on six phases of the PCC Mobile Strategy (see references for the project reports from these teams). Each team was required to follow a formal Systems Development Lifecycle, including production of system documentation for ongoing system development and maintenance by PCC IT staff (Curry & Stanford, 2005). The remainder of this chapter will detail the work completed in each of the six phases, and will highlight experiences and lessons that were learned that are relevant to the introduction of mobile technology into a local government environment.

PHASE 1: FOOD AND HYGIENE SURVEILLANCE REPORTING VIA A REMOTE INTEGRATED SYSTEMS TERMINAL (MARCH 2002)

The scope for Phase 1 was to develop a mobile systems architecture that catered to the development of an automated data entry system to be used for building and development inspections and food and sewerage inspections, using handheld devices. The first module to be developed dealt with inspections for food production premises and collected data relating to food production and hygiene standards.

The new system was to be designed to suit the unique needs of Penrith City Council. This involved data being input directly to a PDA, with the resulting files being transferred across the palm top operating system to the Council system, with the information being downloaded directly into the Council database. It was proposed that for the first-stage prototype, Universal Serial Bus (USB) connections to upload the data from the handheld device would be used. Bluetooth or similar technology would be researched at a later date to enable the uploads to be performed remotely. The initial application used Microsoft's eMbedded Visual Basic 3.0 as the development platform with a Microsoft Access database. ActiveSync was used to manage the USB connection and synchronise the data transfers.

The Remote Integrated Systems Terminal (R.I.S.T.) handheld system was designed on the Compaq IPAQ 3870 using Microsoft's Pocket PC version 3.0.11171 (Build 11178). The device ran the Arm SA1110 CPU, as well as being Bluetooth enabled (for the future).

The primary purpose of the first user interfaces was to make them look like the original paper-based form the field officers were used to. Only minor cosmetic changes were made where necessary.

Project Progress

The initial phase saw the development of the underlying architecture for the use of handheld devices for the PCC field officers with comprehensive development of the Food and Hygiene Surveillance Report application. The R.I.S.T. application was developed to a Beta software level, with the high-priority modules readied for implementation. These included the recording of the inspection data, and the synchronisation between the database on the Pocket PC and the database on the host PC (Team S02106A, 2002).

PHASE 2: ENHANCEMENT TO PHASE 1 AND DEVELOPMENT OF THE SEWERAGE INSPECTION MODULE (AUGUST 2002)

The following semester the R.I.S.T. project entered its second phase. The R.I.S.T. beta software application (Phase 1) had been deemed successful by the IT Manager and his staff as an R&D activity, and they were now prepared to enhance Phase 1 and develop another "inspection" module ready for release to the field officers.

The second phase of the development focused on turning the beta software into a commercial quality application, and adding extra functionality to the existing application. Once this was complete, the team was to begin work on the next inspection module—sewerage.

Initially the existing MS Access database needed to be reviewed due to some intermittent data entry errors, which were resulting in the R.I.S.T. application not being able to synchronise with the host PC for every upload. A new transition table was created that was only accessed by the Pocket PC. The new table was then used in the data upload to the PCC database.

The initial version of the Food and Hygiene Surveillance Report application was only of beta quality. This meant that the software contained the basic functions that it required but still needed some finetuning to make it of implementable quality. Testing of the beta software had yet to start in the field, and it was envisaged that further modifications might be needed if the field officers were not completely satisfied.

Slight modifications were made to the beta software, with extra functionality being added to the interface. Modifications to the report formats were also made.

Once the Phase 1 enhancements outlined above were complete, development work began on the second inspection module: Sewerage Inspection. Development work followed the initial architecture and basic functionality of Phase 1. Field officer testing began and an issue was identified with re-inspection visits, which had not been previously catered for. As re-inspections dealt with an existing inspection report and only a minority of items needed updating, all inspection details from previous inspections needed to be retrieved to the PDA prior to the field officer attending the inspection site. A new database table was added to the existing database, and the data from four other tables were integrated into the new table. Following the inspection, updated data from the new table was segmented as needed and updated at the original source.

Project Progress

The Sewerage Inspection module was completed and thorough testing by the inspectors commenced. Some minor bugs were reported and, once these were overcome, PCC management directed that the new PDA-based inspection system be implemented in the field (Team P02203A, 2002).

PHASE 3: BUILDING INSPECTION MODULE DEVELOPMENT (MARCH 2003)

Following the success of Phases 1 and 2, PCC decided to expand the new mobile architecture into the larger and more complex area of building and development applications (BA/DA). PCC already had an advanced online submission system for building and development applications, and was looking to further its technological leadership by enabling the building inspectors to carry out mobile inspections. With the increased growth in industry and population in the Penrith Valley area, the number of building and development application inspections had amplified significantly and now averaged 30-40 per day. This had bought with it similar problems for the manual system as outlined above. Most significantly it had been estimated by PCC that the time that could be saved by automating the transfer of building inspection data to the Council database could save in excess of $15,000.

The purpose of Phase 3 was to create a mobile system that leveraged the existing systems and equipment into which Penrith City Council had already invested time and money. Development progressed smoothly and the solution was handed over for field-testing as scheduled.

Project Progress

During early testing of the Building Inspection module by the building inspectors, the inspectors expressed some initial concerns about the technical complexity of the solution, the small screen size of the PDA, and the resulting difficulty they had in reading and completing the online forms. The building inspectors expressed a view that they would be much happier completing a hardcopy form. They asked if

Table 1. System architecture at the completion of Phase 3

Hardware	Software	Development Environment
• Pocket PC iPAQ 3900	• ActiveSync 3	• Pocket PC 2000 Software Development Kit

• Windows Desktop PC	• MS Access 2000	• eMbedded Visual Tool 3.0
		   o eMbedded Visual Basic
		   o eMbedded Visual C++

there was some way that this form could be transmitted to the PDA automatically and then uploaded to the Council database as originally planned. This required the project team to complete some additional research into other technical options that could accommodate these needs and that were more acceptable to the current user group (Team S03101A, 2003).

The system architecture at the completion of Phase 3 is shown in Table 1.

PHASE 4: PARKING INFRINGEMENT MANAGEMENT MODULE

Phase 4 of the PCC Mobile Strategy involved development of a Parking Infringement module for the handheld device. The added complexity for this module revolved around the fact that a hard copy of the parking infringement was required to be left on the windshield of the offending vehicle.

Initial investigations had found that only one other council in Australia was currently using PDA technology for the issue of parking infringements, but that the system was still in its infancy and not ready for commercialisation.

Following significant research, the development team recommended a solution that used the existing iPAQ with a Thermal Micro Printer. The printer unit from Unique Micro Design's PP-50 range was chosen because of its slimline and ergonomic design. The Datec PP-50 is a portable cradle type printer for a variety of palm and pocket PC handheld-type computers (Unique Micro Designs, 2005).

The unit can also act as a Hot Sync cradle and a recharge station for rechargeable devices. Software utilities, Printer Manager, Font Manager, Screen Pad, Logo Manager, and Print Emulator were also included, and Software Developers Kits (SDKs) were available for MS-embedded VD, C/C++, and .NET Compact Framework.

The unit features included: ergonomic handheld design; fast, direct thermal printing; data, barcodes, and graphics printing; long-life rechargeable battery; and 32- or 42-column printing.

Project Progress

The resulting solution delivered the new Parking Infringement module with the added facility of printing a copy of the parking infringement for attachment to the offending vehicle. Additional features included a desktop content management system for the management of parking officer staff details and infringement types.

Figure 1. Example of a Thermal Micro Printer with iPAQ inserted (paper printout also shown)

The solution was considered functional, light-weight, compact, and relatively user friendly. It also provided PCC with a strong business potential for on-selling the resulting system to other interested parties such as councils, national parks and wildlife bodies, universities, shopping centre car parks, and any other major public venue where parking restrictions existed. The project team highlighted that the mobile parking infringement market was virtually untouched in Australia, and any available systems were based on overseas designs and therefore were not suited to Australian regulations and legal practice. PCC accepted this recommendation under advisement and proceeded to investigate the full commercialisation of the newly developed system (Team S03204A, 2003).

PHASE 5: NEW WASTE MANAGEMENT INFRINGEMENT MODULE, AN INTEGRATED GENERAL INFRINGEMENT MANAGEMENT SYSTEM AND REMOTE UPDATING RESEARCH (MARCH 2004)

The initial work in Phase 5 leveraged on the existing PDA development work and saw the construction of a new Waste Management module. This new module was then integrated with the existing Parking Infringement module to create a General Infringement Management System for the creation and management of all off-site infringement notices issued by PCC field officers. Storage of field officer signatures and the ability to include the issuing officer's signature on an infringement notice was also incorporated.

The second stage of this phase involved research into a remote updating capability, with the aim of further reducing the need for the field officers to personally attend the Council offices. As part of this research, PCC also asked the student team to look for options that could overcome the issue of not being able to leave a hardcopy of the completed building inspection form with the client onsite.

Project Progress

The project team successfully implemented the new Waste Management module and integrated it with the existing Parking Infringement module to create the PCC General Infringement Management System for handheld devices. Research into the remote updating capability and production of an onsite hardcopy of the completed building inspection form led to the

Figure 2. Example of User Interface Design for Parking Infringement module

presentation of three alternatives and the recommendation that the Council look at the purchase of SmartPad devices combined with new GPRS-enabled PDAs (Team S04104A, 2004).

PHASE 6: REENGINEERING OF THE BUILDING APPLICATION INSPECTION MODULE AND IMPLEMENTATION OF REMOTE UPDATING (JULY 2004)

During testing of the initial PDA-based Building Inspection module, the building inspectors had expressed a certain degree of dissatisfaction. Functionally they felt that it was vital that a copy of the inspection result be left onsite with the client, and usability wise they were uncomfortable with the PDA technology and the size of the information they were required to work with (see above, and "Experiences and Lessons Learned below). This necessitated reengineering of the existing application to better accommodate these new requirements. PCC also took this opportunity to enhance the Building Application module to include the remote updating facility researched in the previous phase.

After reviewing the research outcomes of Phase 5 and confirming whether any other options provided further advantages, the team suggested that PCC take a different technological direction and utilise the facilities offered by a SmartPad system with a GPRS-enabled PDA (Seiko Instruments, 2005). Although this appeared to be a backward step technologically, it provided a simple solution to both of the highlighted issues.

The new process required the building inspectors to record the inspection results through the SmartPad interface. This was done by placing the SmartPad's digitising tablet sheet under the Inspection results form and clipping the form directly onto the SmartPad's tablet. The inspectors then completed the form with an ink pen, just as they normally would during the course of the building inspection. Once the inspection is complete, the inspector removes both the inspection results form and the digitising tablet sheet—leaving the inspection results onsite, thus removing the need to print an extra copy for the client onsite.

The inspector then removes the digitising tablet, and changes the mode of the pen to stylus—which enables him/her to write directly onto the SmartPad. The inspector saves the recorded results and sends them (via infrared) to the accompanying PDA.

The formatting and transmission of the data to the PDA and the subsequent transmission of data to the PCC database did raise some interesting technical issues however. At the time of this project, the SmartPad was limited to saving the written results from the tablet into Graphics Interchange Format (GIF). The PDA in conjunction with handwriting recognition software then translated the .gif file, and prepared it for transmission. When ready, the inspector sent the text-based results back to Penrith City Council via GPRS. Once the results were received, OCR (Optical Character Recognition) software converted them into a readable format and saved the results into the database.

More specifically this involved:

- The results actually being e-mailed via GPRS in a .bmp format to a dedicated e-mail address at PCC's office.
- A new module R.I.S.T. 2 was installed on the Council's PC. This is a timer-based software that performs hourly polling, looking for any new building inspection results received in the inbox.
- R.I.S.T. 2 then extracts the .bmp files from the Microsoft Outlook inbox into a specified directory.
- The .bmp-formatted files are then converted into .tiff format.
- The iReadForms software is activated.

Figure 3. Example of SmartPad and GPRS-enabled PDA environment

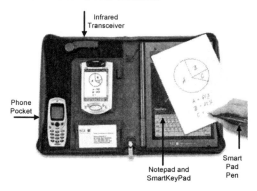

- iReadForms then translates the results and creates a .txt file.
- This file is then transformed into an MS Access readable format and creates a temporary database table.
- The results from the temporary table are then uploaded to the PCC database.

Once the R.I.S.T. 2 system had completed the data transfer and manipulation, the information is then ready for processing as per normal Council guidelines.

Figure 3 shows an example of the mobile office environment offered by the SmartPad solution.

Project Progress

The requirements were fulfilled (not without some angst on the student's side!), and the new system approach was accepted by both the IT and building inspection staff (Team S04208A, 2004).

The resulting architecture now included:

- Active Sync Version 3.5
- Microsoft .NET Framework 1.1
- iReadForms Evaluation 4.1 Build 166
- iMate Personal Digital Assistant (PDA) with In-Built GPRS Capabilities

- InkNote Manager Version 3.2
- R.I.S.T Version 2.0
- Seiko SmartPad2 (Model SP582)

EXPERIENCES AND LESSONS LEARNED

Over the two years of these development projects, Penrith City Council, along with the students and UWS academics, all learned a great deal about the nuances of implementing mobility into a local government environment. Specific areas of this process of implementing mobility include: release hype vs. the implementation realities of mobile technology, technological options for the introduction of mobility, user acceptance of new technologies, management of client expectations, and local government standards and guidelines and their impact on development directions.

Release Hype vs. the Implementation Realities of Mobile Technology

The students were very excited to be working with mobile technology because it was new and leading edge, and they were keen to present the client with options that were at the forefront of the mobility movement. As with most new technologies, however, there is a time lag in the software and hardware features publicised by the vendors in their press releases and the implementation support available for these new features. The students, and their client, learned that trying to be on the "bleeding edge" of a new technology brings with it specific development issues and a certain degree of rework. This is evidenced in Phase 6 and the lack of support for form-based input. This point also impacted the development client's expectations (see below).

Technological Options for the Introduction of Mobility

In conjunction with the point above, there were always several options available for implementing the different features of mobility required by PCC. As evidenced in the introduction of the SmartPad option, what appeared to be an ideal solution on the surface required quite a number of "work-arounds" to actually achieve the client requirements.

User Acceptance of New Technologies

Although IT management at PCC were very supportive of the mobile strategy, they did not fully investigate the actual users' acceptance of mobile technology into their working environments. The primary driver in the reaction of the different user groups was "age". The field officers involved in issuing parking infringements were quite a bit younger than the building inspectors and had no problems using the PDA with the thermal printer for their duties. The building inspectors were quite a deal older, and their reactions were not as positive due to the size of the font they were required to work with and the perception that the new technology was too complex. As with any technology, it is vital to assess the users' acceptance of the new environment and develop specific training to overcome this, or as in PCC's case, realign the development direction to accommodate the users' concerns.

Management of Client Expectations

The IT staff at PCC were very active in identifying potential new technology that would give them a competitive advantage and deliver their clients (the residents of the Penrith Valley area) the highest quality services. At times this led to excessive client expectations of what

was possible, and the students and UWS academic staff spent a good deal of time discussing how some of these expectations would impact the delivery schedule if they were pursued. The students also learned that all requirements identified as high risk (new technological area) needed to be fully investigated before development commitment was given to the client.

Local Government Standards and Guidelines and Their Impact on Development Directions

For most of the students, this was their first 'real-world' development experience. In addition to the normal learning curve for such situations, they were also faced with having to accommodate specific conditions that were imposed on them in relation to local government policy and bylaws. This affected the development work in several ways. In most cases the teams were not permitted to change the input forms that the field officers completed, as the Local Government Association defined these and they were standard across the state. This affected the user interface design on the PDAs on many occasions. The teams learned that in certain situations, external factors that were out of their control significantly impacted some of the potential quality improvements that would have been possible.

CONCLUSION AND FUTURE DIRECTION

The work undertaken between Penrith City Council and the University of Western Sydney was very rewarding for both parties. PCC was able to advance its mobility R&D efforts significantly by using low-cost resources (students). The Council was able to implement its mobile strategy considerably earlier than originally planned, thus delivering advanced high-quality services to local communities. The students from UWS participated in projects at the leading edge of the mobility movement and gained invaluable experience for their future careers. Academic staff were also able to integrate the experiences and lessons learned into their teaching material for other students. The mobile strategy work is continuing, with PCC currently investigating the possibilities for commercialising the Parking Infringement module for on-selling to other councils and venues with existing parking restrictions.

REFERENCES

Australian Bureau of Statistics. (2004). *Census data*. Retrieved April 20, 2005, from www.abs.gov.au

Bryan, G. M., Holdsworth, D., Sharply, R., Curry, J., & McGregor, C. (2002, January 7-10). Using XML to facilitate information management across multiple local government agencies. In *Proceedings of the 35th Hawaii International Conference on Systems Sciences* (HICSS'35), Big Island, HI (pp. 119-128).

Curry, J. M., & Lan, Y. (2004, May 23-26). Web-enabled business operations: Development of the Semantic Web evolution model. In *Proceedings of the 15th Information Resources Management Association International Conference* (IRMA2004), New Orleans, LA (pp. 1123-1125).

Curry, J. M., & Stanford, P. (2005). *Practical systems development—A project-based approach*. Sydney: Pearson Education Australia.

Garson, G. D. (1999). *Information technology and computer applications in public administration*. Hershey, PA: Idea Group Inc.

Gronlund, A. (2002). *Electronic government, design, applications and management*. Hershey, PA: Idea Group Inc.

Seiko Instruments. (2005). *Product description, SmartPad2*. Retrieved April 2, 2005, from http://www.siibusinessproducts.com/products/sp582.html

Team P02203A. (2002). *System Development Lifecycle documentation, Computing Projects Archive, UWS Library*. Retrieved from http://library.uws.edu.au/

Team S02106A. (2002). *System Development Lifecycle documentation, Computing Projects Archive, UWS Library*. Retrieved from http://library.uws.edu.au/

Team S03101A. (2003). *System Development Lifecycle documentation, Computing Projects Archive, UWS Library*. Retrieved from http://library.uws.edu.au/

Team S03204A. (2003). *System Development Lifecycle documentation, Computing Projects Archive, UWS Library*. Retrieved from http://library.uws.edu.au/

Team S04104A. (2004). *System Development Lifecycle documentation, Computing Projects Archive, UWS Library*. Retrieved from http://library.uws.edu.au/

Team S04208A. (2004). *System Development Lifecycle documentation, Computing Projects Archive, UWS Library*. Retrieved from http://library.uws.edu.au/

Unique Micro Designs. (2005). *Product description, Thermal Microprinter*. Retrieved April 5, 2005, from http://www.umd.com.au/itd/products/datecs_pp50series.html

Chapter LXII

Emerging Mobile Technology and Supply Chain Integration:
Using RFID to Streamline the Integrated Supply Chain

Richard Schilhavy
University of North Carolina at Greensboro, USA

A. F. Salam
University of North Carolina at Greensboro, USA

ABSTRACT

This chapter explores how a mobile tracking technology is able to further streamline the integrated supply chain. Previous technologies which have attempted to integrate suppliers, manufactures, distributors and retailers have lacked the flexibility and efficiency necessary to justify the prohibiting costs. Radio frequency identification (RFID) technology however enables various organizations along the supply chain to share information regarding specific products and easily remotely manage internal inventory levels. These applications are only a sample of what RFID is able to accomplish for the integrated supply chain, and this chapter seeks to explore those applications.

INTRODUCTION

This chapter sets forth to provide a holistic view of how a recently adopted wireless identification technology, specifically radio frequency identification (RFID) tags, could potentially revolutionize the integrated supply chain. Companies are able to become more flexible and efficient by using a combination of mobile technologies and RFID to provide for remote inventory control and real-time, information-rich tracking of shipments in the distribution channel (Lapide, 2004). Although this technology has several hindrances currently blocking it from mass usage (Thompson, 2003), recent advancements in the technology have increased the viability of RFID for widespread organizational use, increasing the capacity and strength while decreasing the size and cost. RFID now rests in a unique position wherein large organizations are strongly considering its viability in a variety of applications to streamline the supply chain.

Organizations have already begun considering its application in the realm of supply chain management, attempting to further streamline the process. However, while many authors have discussed the benefits of RFID tags for parts of the supply chain, this insight has only focused on a localized level, such as inventory management in retail outlets (Atkinson, 2004; Lapide, 2004; Kinsella, 2003; Schindler, 2003). Much of this discussion is centered on reducing costs for those isolated parts of the supply chain. For example, several large manufacturers are pushing the technology by actively conducting trials in manufacturing, distribution, and even retail. These companies include Proctor and Gamble, Gillette, Unilever, and retail giant, Wal-Mart (Kinsella, 2003). These RFID trials have been limited to single stages of the supply chain, focusing on the reduction of costs as the ultimate goal. Although cost reduction is commendable, true improvements in value for industry and consumers come through a unified effort to improve the entire supply chain network, reducing costs and improving accuracy and efficiency for all companies integrated into the network.

This chapter will first provide an overview of the technical aspects of RFID. Following this, an analysis of two perspectives of the integrated supply chain will be framed in light of the current and possible future applications of RFID in each area and the relationships between those areas. Finally, RFID will be framed in a holistic view of both the integrated supply chain as well as the demand chain, addressing some inter-organizational issues.

BACKGROUND

The RFID Tag and Reader

The core of RFID technology consists of two components, the identification tag and the tag reader. The identification tag itself is composed of a small antenna and a microchip, which stores a small amount of information pertinent to the object tagged (Rappold, 2003). Although the information stored may take a wide variety of forms, for many objects a simple code would be sufficient to identify the item. Asif and Mandviwalla (2005) identified five types of RFID tags in their RFID Applications Framework, including active, semi-passive, passive, chipless, and sensor. Tag readers may also be stationary or mobile, depending on the application. Of those tags which contain chips, RFID tags may either be active, passive, or a combination of the two. Active tags are powered by some external power source, such as a small battery. Passive tags, on the other hand, have no individual power source and receive power from the electromagnetic waves the tag reader uses to access the information from the tag. Some tags may use a combination of the two strategies, where an active tag containing a battery is recharged by the transmission used to read it. Chipless tags have the lowest power consumption, range, as well as cost of all the types of RFID tags since they do not contain either a battery or a silicon chip. Information storage is also significantly less, often only enough to store a simple product code.

The tag reader uses electromagnetic waves in the radio frequency band to transmit the data stored on the identification tags to the reader and, in some cases, power the identification tags. The reader may be a mobile or stationary unit depending on the application, and an organization could easily employ both. Mobile readers naturally benefit from being able to change location; however, power limitations have a severe impact on the range and may even become an issue if passive RFID tags are used extensively. The effective range of a tag reader is a function of the frequency the tag reader is operating on and the power output available. At lower frequencies, range is severely diminished; however, power output is minimal. At higher frequencies, identification tags are able to be accessed from further away, but require significantly more power. Here, active RFID tags may need to be utilized to increase the range along with increasing the power output of the tag reader. Finally, sensor tags combine a small sensor targeted at a particular purpose, such as measuring temperature, viscosity, movement, and so forth, with an antenna, chip, and battery to store and transmit information from the sensor to a reader or network.

Not unlike any other wireless technology, RFID comes with a few limitations or issues affecting communication reliability (Angeles, 2005). The use of radio frequencies becomes a significant issue since these bands are often open to a multitude of other devices, such as wireless phones, computes, radios, and other office equipment. These common workplace devices may cause interference with reading a RFID tag and should not be overlooked when problems arise. Another problem common among mobile technologies is the increased collision of packets when more senders and receivers (in this case, more tags) are present. Since the reader is not limited to line of sight, a reader may pick up a multitude of tags in any direction of the reader, and if a large amount is present surrounding the reader, the transmissions between identification tag and reader may become interrupted due to such collisions. Anticollision technologies are currently being developed to confront this problem with collisions (Angeles, 2005).

The RFID Network

Thanks to the Auto-ID Center at MIT, recent developments in RFID technologies have expanded RFID technology to create a holistic product identification system that consists of four components (Rappold, 2003; Smith & Konsynski, 2003; Asif & Mandviwalla, 2005). The identification tag and tag reader, again, transmit and read the information stored on the identification tag attached to the particular physical object. Stored on the identification tag is an electronic product code (EPC), which identifies the particular object or the state of that object. The EPC is a 96-bit identification code similar to conventional bar codes which uniquely identifies a product. The object name server (ONS) is a local or remote server that acts as a directory service, mapping the EPC to additional information about the physical object. This additional information in Physical Markup Language (PML) provides a standard format for describing products and storing other information about them (Smith & Konsynski, 2003). For example, the PML documents may contain information about the product manufacturer, source, and destination, or simply more detailed information about the product itself (Rappold, 2003). By mapping the EPC code to the PML documents containing information about the product, the tag may be significantly smaller since all the information is not required to be stored on the tag itself.

RFID Applications

Companies are able to become more flexible and efficient by using a combination of mobile technologies and RFID to provide for remote inventory control and real-time, information-rich tracking of shipments in the distribution channel (Lapide, 2004). Although this technology has several hindrances currently blocking it from mass usage (Thompson, 2003), the potential long-term benefits are astounding for both the integrated supply chain and other mobile technology applications as well. RFID technology is able to provide item- and product-specific information which remains with the physical object. Since no line-of-sight is required to read a tag and multiple tags may be read simultaneously, inventories may be tabulated quickly with little manual labor and items may be tracked regardless of their location in the range of the readers. Rich information can be stored on the tags themselves, or simply mapped to the tags via an EPC, allowing this new technology to be easily mapped into current systems. PML provides additional information through a standardized markup language, providing additional interoperability between existing systems and the systems of other organizations. The technology is small, flexible, and relatively inexpensive. In the following sections, we will look more closely at the applications of RFID—specifically in the supply chain—and analyze the relative costs and benefits for each.

RFID AND THE SUPPLY CHAIN

One of the problems of current implementations of RFID tags and readers in the supply chain is that they have largely been efforts of a single company operating independently in their area of the supply chain. Technology improvements in the supply chain which are isolated to a single stage, such as manufacturing or retail, are limited to minor improvements in costs. To further illustrate this point, the following paragraphs will explain potential implementations of RFID in each stage of the supply chain considered entirely in isolation. Therefore, suppliers will be considered apart from manufacturers, retailers apart from distributors and consumers, and so on. Relative costs and benefits will be weighed with each implementation, as well as the possible risks and rewards in undertaking the endeavor.

Suppliers

A common theme among many of the stages of the supply chain is inventory management, even when considered in isolation. Technology improvements in inventory management allow for significant improvements in labor and capital costs, and accuracy over traditional inventory systems which require manual operation. Suppliers are no different, requiring the maintenance of large amounts of raw and processed materials in various forms. However, there are unique considerations for each stage of the supply chain in regards to inventory management which needs to be addressed. In particular to suppliers, some materials and parts require significant specialization and complexity, such as composition requirements or well-defined specifications, which can be maintained through the information stored in RFID tags or using EPCs to map the product to the information stored in a database.

Materials which require constant monitoring of temperature, viscosity, or other physical qualities could also benefit. This also applies to those materials which are heavily time dependent in regards to time-to-disposal, time-to-shipment, and so forth. RFID tags could wirelessly transmit updated information of the state of the material in real time, without human interaction or the costs of installing and/or maintaining an infrastructure based upon a physical connection. While the wireless monitoring devices would require maintenance in case of failure, the overall complexity of the system and of the maintenance would significantly decrease.

Naturally, there are significant costs involved with such implementations of RFID technology in the supplier's world. Compared

Table 1.

	Costs	*Benefits*
Supplier	Moderate.	Improved inventory management and control.
	High for complex systems, such as monitoring devices.	Improve demand forecasting.
Manufacturer	Moderate, increasing with the complexity of the product or equipment.	Improved inventory management and control.
		Improve demand forecasting.
Distributor	Moderate. Package-level tagging.	More accurate tracking information.
	High costs when integrating with tracking systems.	En-route location tracking of packages.
Retailer	High. Item-level tagging required for pervasive implementation.	Improved inventory management and control.
		Reduce stock-outs.
Consumer	Variable. Low for luxury items to extremely high for low priced goods.	Improved shopping experience.
		Reducing in price.

with manufacturing or distribution, the information necessary to store is uniquely different. The cost of identification tags and tag readers are a common theme among all of the stages in the supply chain—an unavoidable cost. The supplier does have a slight advantage in this regard, in that relatively few tags are necessary in comparison to the other stages in the supply chain. However, if the materials required highly precise specifications or other physical properties, the complexity of the system required to maintain the information could increase the cost exponentially. Monitoring the state of the materials poses even more problems, requiring specialized identification tags attached to sensors. Additionally, simple EPC codes may no longer be sufficient to monitor the possible states of the materials and therefore require a more specialized system for each individual monitoring tag. In this regard, semantic mark-up languages similar to PML may become incredibly useful in such applications.

Manufacturers

Similar to suppliers, manufactures have much to gain in regards to simplifying inventory management and increasing the robustness and richness of the inventory system on the whole. However, what is unique to manufactures is the need to maintain large amounts of data on highly specialized or customized parts or products. RFID tags will provide item-level information unique to the particular part or product, which remains with the individual part of product. Manufacturers will find a significant reduction in the effort necessary to manage the inventory of parts or products on hand and find an increase in the accuracy of that inventory.

Manufacturers often require many complex pieces of equipment for specific applications which, in a large bustling factory, may become lost or simply difficult to find. TransAlta found that tagging pieces of equipment across the 600-foot plants made finding and maintaining the equipment easier and more flexible (Malykhina, 2005). Using Wi-Fi and Bluetooth wireless technologies in conjunction with RFID to blanket the entire facility, TransAlta was able to locate equipment regardless of its location in the company's large facilities. Active RFID tags were used to eliminate the need for manual operation and provide real-time information about the location of the equipment and specific metrics from temperature gauges, vibration probes, and a variety of other peripherals (Malykhina, 2005).

If suppliers could have a problem due to complexities in inventory management, manufacturers have an overabundance of them. Implementing RFID throughout the manufacturing process requires that the individual raw materials and parts from other manufacturers be tagged, and the ultimate product to be shipped out the door also be tagged, either individually or as a package. The system becomes exorbitantly more complex when the manufacturing process requires multiple steps where information of the part or product at each stage is required.

Here, at the manufacturing stage, managers will be first posed with a difficult question when considering implementing RFID technology at their site. The question is whether package level or item level will be more economical for the specific application. For larger products or specialized equipment, tagging individual items is an economical choice. However, if the factory produces millions of widgets per month with little or no variation in those widgets, tagging them on an individual level would be a foolish choice. In most cases, manufacturers have little need for individual tagging of parts and products that come off the assembly lines, leaving package-level tagging a more prudent choice.

Distribution

In isolation, distributors are able to see significant benefits. One of the most significant proposed implementations of the technology in distribution channels is the ability to locate in real time each individual shipment, regardless of its location, and to provide information about its shipment location, destination, content, and so forth. This can be accomplished through combining several other mobile technologies. One implementation suggested the use of GPS technology to locate and identify individual

vehicles, then remotely transmit the information obtained by scanning the individual shipment tags, thus providing remote access to the current contents of the vehicle, regardless of its location. Such a system could provide some value to the business and consumers alike. Richer tracking information allows interested parties to know exactly where a highly important package was last scanned, even if the package or shipment is "en route."

Wireless mobile technologies are used throughout distribution channels for varieties of business benefits, such as geographic positioning systems (GPS), for tracking and monitoring vehicles in distribution channels (Faber, 2001; Schindler, 2003). However, one issue distributors have with current mobile tracking technologies is that en-route information of vehicle contents is almost impossible to monitor, which is where RFID technology has significant promise.

However, several obstacles hinder the practical feasibility of such a system. First, current tracking systems are easily able to provide similar information, but nevertheless lack the remote, real-time tracking such a system would offer. The practical issue comes with the fact that the cost involved with implementing such a system still surpasses any benefit, even considering how the cost of RFID per tag and per reader has become more reasonable.

Retail and Consumers

Similar to suppliers, if the product retail is selling is dependent upon time—that is, perishable with a limited shelf life—those must be sold, moved, or discarded by a particular date. By outfitting each individual item sold within the store, single, outdated items will not find themselves sitting on then shelves for extended periods of time. A simple system could automatically read the tags and inform the employees which items have or will soon pass the date in question.

There have been several technologies over the years which have been considered as replacement for the aging UPC standard. However, few have provided sufficient benefits over and above UPC labels. Electronic tags were considered as a replacement, but the benefits over the current standard were so minimal that the small cost associated with each tag was too great to justify the switch. On the other hand, RFID tags provide richer information about the product over and above the current UPC standard, which simply identifies the product. Since the RFID tag can be read via a wireless connection, many items can be scanned and identified in seconds, whereas the present UPC standard requires line-of-sight reading through an optical scanner. This application could further be extended into the often-dreamed automatic checkout machines.

Current theft-deterrent technologies are an independent system from the limited UPC standard. The technology is often unreliable, resulting in countless false positives, and requires that the electronic tag passes through a small area before detection. RFID, on the other hand, would be able to provide for both the UPC label functionality discussed previously, as well as a rudimentary theft deterrent system similar to the ones currently used in retail outlets. Through RFID, retailers would be able to know exactly what products are entering and leaving the store, and which have been purchased and which have not. However, cost barriers are significant in this application since item-level tagging is necessary for it to be effective. Until the cost of individual RFID tags drops substantially in comparison to the price or quantity of the product, in the realm of fractions of a cent, it is not likely we will see widespread, item-level application of RFID. However, some retailers have begun tagging larger, higher priced items

for this purpose. Both Wal-Mart and Woolworth's in the United Kingdom have begun tagging items considered high risk, such as CDs, mobile phones, computer accessories, and other electronic goods (Smith & Konsynski, 2003).

However, retail has a particularly difficult time when isolated from the rest of the supply chain in the implementation of RFID technology. Although the obstacles from the distribution side of retail are not insurmountable—requiring the contents of palettes and packages to be tagged at the distribution centers, similar to package-level distribution or manufacturing inventory, before being sent to individual retail stores—on the consumer side however, implementation becomes a much more difficult task. To be effective, individual items must be tagged. While other stages could have survived with a small amount of identification tags and tag readers, retailers cannot avoid the substantial costs associated with the thousands upon thousands of tags to cover individual items in some RFID applications. Inventory management and automatic checkouts require each individual item in the entire store to be appropriately tagged, with no assurance that the tags could be reused. Effectively, the costs of tagging individual items would be directly added to the wholesale cost of each item, and at the current five cent-mark—a substantial portion of many items' wholesale costs—further reducing the already meager margins in retail. Some organizations have implemented theft-prevention systems utilizing RFID technologies, but it is a small fraction of merchandise, particularly high-risk or high-cost items. However, retail systems are particularly simple in comparison to those found in the other stages. Since all items already possess UPC labels, mapping EPC to individual items via RFID tags may not be a daunting task and would provide some technological redundancy in case of system failure.

STREAMLINING THE SUPPLY CHAIN

Implementing technologies in the supply chain ultimately creates value when each organization at each stage of the supply chain vertically integrates, standardizing on the single technology. RFID is no different. Using the same technology from the manufacturing of a product to the sale of the product throughout the entire supply chain substantially reduces costs and provides business value for everyone (Poirier, 1999). However, this does not always occur, for example, when an integral part of the supply chain chooses not to implement or share information, or organizations force their supply or demand chain to implement a particular technology, largely at the cost of those implementing (Kinsella, 2003). In the following section, how RFID can provide value to a more integrated supply chain through the implementation of RFID technologies will be discussed, ultimately culminating on a view of the entire supply chain. While many of the applications, costs, and benefits have been covered in previous sections, how their applications tie together in the supply chain and provide value will be the focus.

Suppliers to Manufacturers

The real power and value of RFID technology in supply chain management sadly comes later in the supply chain, although many benefits can be realized between suppliers and manufacturers. Materials are cultivated, packaged, distributed, received, and processed by the respective manufacturers. Throughout this entire process, RFID is able to track the inventories of materials, the source of the materials (the supplier), and the destination (the manufacturer). Manufacturers then benefit by knowing what inventory they have on hand of pre-production materials and from what vendors those materials originated. This improves the production process at manufacturing facilities greatly if such information is already accounted for by the source of the materials. However, because of the transforming nature of the manufacturer, the same RFID tags are not valuable to the remainder of the supply chain. Once the materials are transformed into products and finally head downstream to distributors and retailers, they must be retagged. In fact, after every transformation of the product, the nature of the product changes and tags must be reapplied.

Manufacturers to Distributors

Here, at the manufacturing phase, we begin to see the real potential of implementing RFID technologies and the scope of their effect on the integrated supply chain. Package-level tagging at the manufacturing level before distribution occurs both helps maintain inventories at the manufacturing sites and aids distribution and other inventory management and control systems to be equally as streamlined. Incoming materials from suppliers are able to be logged, and inventories updated and maintained throughout the manufacturing process. However, the beauty occurs when item-level tagging is implemented. As the product is produced, the item may be tagged with manufacturing information and other specifications particular to the product. Other organizations down the supply chain will be able to access this information even when the package it was contained in was dismantled and the contents strewn across retail stores and the consumer population. Additionally, the benefits from improved forecasting comes from information downstream, at the retail level, where manufactures are able to determine which products are sold, at what locations, in specific quantities. For retailers to provide this information, with the RFID network previously discussed in place, would be far from an insurmountable feat, and the value coming downstream from the manufacturers would be to their benefit. While manufacturers ultimately may bear much of the implementation cost, they will receive equally in benefits, with even more significant benefits for organizations downstream from the manufacturer.

Distributors to Retailers

Leaving the distribution centers are countless cases and palettes of merchandise heading to different retail stores with varying quantities of thousands of different products. Managing what products are leaving or being received and where they are going can become a daunting task, even with some of today's technologies. For distributors, RFID technologies provide some of the more impressive benefits, even in isolation with relatively smaller increases in cost. However, having package-level inventories tagged by suppliers and manufacturers before entering the distribution channels improves the efficiency of the logistics systems for both those parties. As the tagged packages move through the distribution channels, retailers ultimately will benefit as well as the packages move through their receiving centers, actively managing the incoming inventory at individual stores. However, between manufacturing and retail, distributors must retag if the packages themselves are repacked. Luckily, if this is not the case, the RFID network system provides separate semantic information for each of the EPCs associated with the packages. Again, the beauty is at the item level. Since each individual product is now tagged, even repackaging the products does not require new tags to be placed on them. In fact, distributors need not retag any items whatsoever, only change the information associated with the

corresponding EPC, which is unique for all the items entering and leaving the distribution channel, regardless of the location.

Tesco, a large UK retailer, recently implemented an RFID system of significant size, totaling 20,000 identification tags for their stores and distribution centers, with 4,000 tag readers and 16,000 antennas to receive the identification tag signals (Sullivan, 2005). RFID tags have been installed in order to track the merchandise cases and palettes which grace the docks at the distribution centers and receiving doors at retail stores. Unlike the retail giant Wal-Mart, however, Tesco made the investment in RFID themselves independent of their suppliers, in hopes that they perceive similar cost benefits as Tesco.

In some situations, logistics systems may need to make a sudden re-route of product or material in case of a sudden stock-out. When a product is in demand, having no inventory of a product on hand means lost sales for retailers or lost production time for manufacturers. In conjunction with GPS and cellular technologies (Schindler, 2003), distributors now may locate items en-route between destinations and calculate precise inventories of those vehicles. If a reroute is economically reasonable, the vehicle is able to be informed of where the reroute is located and what products the reroute are for. All in all the distribution system becomes more flexible and capable of providing for the retail and customer base.

Retailers to Consumers

One of the significant benefits from implementing RFID technology at the retail level is the reduction of labor costs from managing inventory, which now can be accounted for and monitored with little or no manual operation. Reduction in labor costs provides for two potential outcomes that benefit consumers. First, the additional labor capacity can be used to improve customer service of the retail establishment for customers. Or, the additional savings in labor costs not rerouted to another activity could be brought directly to the consumer in the form of lower prices, ultimately an increased value for the customers.

Merchandise Security

RFID technologies also move the retail and consumer relationship to the Holy Grail, the market of one. Prada of New York is one of the first retailers to use RFID technologies to revolutionize the shopping experience of its customers. Each item sold in the store is tagged with an RFID tag. Naturally, this provides additional security in the case of stolen merchandise; however, the interesting aspect of the implementation occurs when the customer re-enters the store. Whether carrying or even wearing the previously purchased garment, the tag readers at the entrance to the store scan for an RFID tag; if found, information pertinent to that garment appear on large flat-panel displays around the store. For example, items matching that garment may appear, in reasonable sizes to the item purchases, directing the customer. In addition, richer marking information is obtained through this system, as item purchases are now tied together with how frequently the customer visits the store, what items are purchased in combination or sequences, and so forth. However, consumer privacy concerns have already arisen in regards to this implementation.

From the information gathered here, at the retail and consumer level, suppliers and manufacturers are now better able to forecast demand and control inventories, sending the business value back upstream (Lapide, 2004). The value invested earlier in the supply chain to tag either packages or items leaving manufacturing facilities returns to them through this improved ability to forecast demand.

CONCLUSION AND FUTURE DIRECTION

RFID technology is a fairly simple wireless technology, composed of a small antenna and microchip, and able to streamline the mobile supply chain. Technologies surrounding the RFID technology, such as EPC and PML, improve the interoperability, transparency, and flexibility of implementing RFID systems with current inventory management and distribution systems. The mobile nature of the technology incorporates additional advantages only found with more complex, higher cost systems. However, important cost considerations must be given, as choosing between the costs and benefits for package level and item level becomes an important decision. While it provides substantial value for organizations downstream, it requires significant investment upstream. As additional implementations appear throughout the supply chain, the cost of the technology will fall and the relative benefits will increase. If standardized on RFID technology, regardless of package- or item-level implementation, the entire supply chain benefits from a standard mechanism to identify objects moving up and down the supply chain, through distribution channels, and off the shelves at retail stores. RFID is poised to revolutionize the supply chain by streamlining operations, providing flexible, transparent communication between organizations.

REFERENCES

Asif, Z., & Mandviwalla, M. (2005). Integrating the supply chain with RFID: A technical and business analysis. *Communications of the AIS, 15,* 393-426.

Angeles, R. (2005). RFID technologies: Supply-chain applications and implementation issues. *Information Systems Management, 22*(1), 51-65.

Anonymous. (2003). Supply chain technologies—At Woolworth's. *Work Study, 52,* 44-46.

Atkinson, W. (2004). Tagged: The risks and rewards of RFID technology. *Risk Management, 51,* 12-19.

Kinsella, B. (2003). The Wal-Mart factor. *Industrial Engineer, 35,* 32-36.

Lapide, L. (2004). RFID: What's in it for the forecaster. *Journal of Business Forecasting Methods and Systems, 32*(2), 16-19.

Leary, D. E. O. (2000). Supply chain processes and relationships for electronic commerce. In M. Shaw, R. Blanning, T. Stradder, & A. Whinston (Eds.), *Handbook on electronic commerce* (pp. 431-444). Berlin: Springer-Verlag.

Malykhina, E. (2005). Active RFID meets Wi-Fi to ease asset tracking. *Information Week, 1022,* p. 38.

Poirier, C. C. (1999). *Advanced supply chain management.* San Francisco: Berrett-Koehler.

Rappold, J. (2003). The risks of RFID. *Industrial Engineer, 35,* 37-38.

Schindler, E. (2003). Business: The 8th layer: Location, location, location. *netWorker, 7*(2), 11-14.

Smith, H., & Konsynski, B. (2003). Developments in Practice X: Radio frequency identification (RFID)—An Internet for physical objects. *Communications of the AIS, 12,* 301-311.

Sullivan, L. (2005). UK retailer goes in RIFD shopping spree. *Information Week, 1022,* p. 36.

Yang, B. R. (2000). Supply chain management: Developing visible design rules across organizations. In M. Shaw, R. Blanning, T. Stradder, & A. Whinston (Eds.), *Handbook on electronic commerce* (pp. 445-456). Berlin: Springer-Verlag.

Chapter LXIII
Mobile Batch Tracking—A Breakthrough in Supply Chain Management

Walter Hürster
T-Systems International, Germany

Hartmut Feuchtmüller
T-Systems International, Germany

Thomas Fischer
T-Systems International, Germany

ABSTRACT

Globalization and expanding markets has invariably led to increasingly higher loads of goods traffic. This has resulted, amongst other things, in challenges to supply chain management in terms of cost pressure and demands for short-term availability of the goods. Considering that an increasing number of goods will be "on the road" (on rails, on ship, in the air) for an appreciable percentage of the life-cycle, there is an urgent need to bridge the information gap between the automated systems at the factory sites and the storage control systems at the destination sites. This chapter reports on a system solution that has been developed by T-Systems' Solution and Service Center Ulm / Germany, within the Service Offering Portfolio "Embedded Functions". The system solution has been gained as a synergy effect of connecting mobile communication solutions with Auto ID Services. It is presented here and discussed in the context of online surveillance during transportation, providing both downstream batch tracking, as well as upstream traceability.

INTRODUCTION AND BACKGROUND

Traditional problems of managing resources and the flow of material appear to have been solved by enterprise resource planning (ERP) systems as well supply chain management (SCM). This is true of the stationary case of an isolated factory and of the goods that form part of its inventory. However, with the increasing movement of goods, a new dimension of problems has arisen that makes it inevitable to consider transport status itself—particularly to improve the supply chain planning and the execution process. This chapter is an attempt to cope with the new challenges that result from a higher degree of mobility, a higher percentage of the mobility phase with respect to the total lifecycle, and a higher flexibility with respect to transport media and changes of the transport mode within one single transaction, such as conveying a pallet from A to B (where A and B may be located anywhere on the surface of the earth, thus indicating that also increasing distances have to be bridged).

Goods, spare parts, and assembly components are no longer kept in storage for long periods of time, but are fed in when needed. This is the effect of the popular just-in-time (JIT) approach to inventory management. Thus, managing the supply chain effectively means managing more and more of the transportation chain.

Successful attempts have been made to manage the internal transport at a factory site by means of new technologies, such as radio frequency identification (RFID) tagging or other auto ID technologies (ten Hompel & Lange, 2005). Within this context, a new class of middleware is emerging, acting as a platform for managing the data and routing them between tag readers and enterprise systems (Leaver, 2004). However, a huge gap of information exists for the increasing time of external transportation—either between two factory sites for a semi-product or between factory site and end user location for a final product.

THE CORE CHALLENGE IN SUPPLY CHAIN MANAGEMENT

In order to obtain an exact overview at any time, it is essential to track the flow of goods on batch level at least, if not on item level (for larger items). This requires acquiring knowledge about the geographical position whenever needed plus detailed information about the goods—that is, batch identification and batch description, including information about origin and destination, plus all intermediate agents involved in the process. Regulation (EC) No. 178/2002 of the European Parliament and of the Council of January 28, 2002, as an example, is laying down the general principles and requirements of food law and at the same time the procedures in matters of food safety. This includes strong implications with respect to downstream trackability (from origin to destination), as well as upstream traceability (from end user back to the production site). In the case of non-preservative food, it is of essential importance to monitor and to record the environmental data of the transport—for example, to ensure that the refrigerating chain has not been interrupted (or only for a very short period of time and within a certain temperature range). The big challenge therefore consists of getting all the required information while the goods are on their way on a transport medium in motion.

THE SOLUTION TO THE CHALLENGE

The requirements mentioned above directly lead to the way of finding an appropriate solution by a decomposition of the system into its two basic components:

a. Subsystem to determine the geographical position of the transport medium (container, lorry, trailer, wagon, ship, aircraft, etc.).

b. Subsystem to gain information about the goods transported by that medium—that is, batch identification and batch description (plus additional environmental parameters).

Figure 1. Trailer equipment

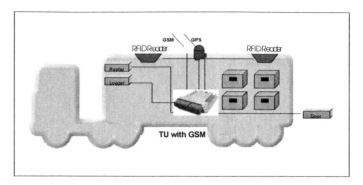

The first subsystem (a) preferably consists of a GPS antenna and a GPS receiver to obtain the geographical position. For the second subsystem (b), an advanced approach would be to use RFID technology—that is, RFID tags affixed to the packaging units and RFID readers installed on the transport medium to read the tags. An example is shown in Figure 1 (for the case of a trailer/lorry configuration).

The trailer contains the GPS equipment plus RFID reader(s) to identify and to read the tags which are fixed at the package units. A great advantage of RFID vs. other auto ID technologies is due to the fact that no direct geometrical line-of-sight between tag and reader is required—that is, the packages may be oriented in any arbitrary way and do not have to be aligned or rotated in a specific manner. Addi-

tional environmental parameters, such as temperature, acceleration (shock), door status, and so on (including intrusion alarm) are polled from adequate sensors by a so-called reefer and are stored locally on a data logger. All data are collected online by a telematic unit (TU) and are transmitted instantaneously or at given time intervals to a Transport Tracking Center (TTC), preferably by means of GSM or by using satellite communication (depending on the coverage and the location of the transport medium on its route). This Transport Tracking Center, thus defining the third subsystem (c), collects the batch data from all connected transport units and makes them available to all subscribers and stakeholders being entitled to use them. The TTC itself will consist of a computer cluster with distributed tasks for I/O handling, central

Figure 2. Network of information

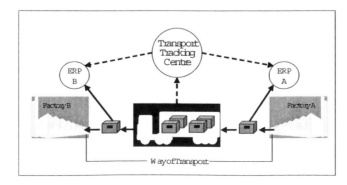

Figure 3. Intermediate waypoints

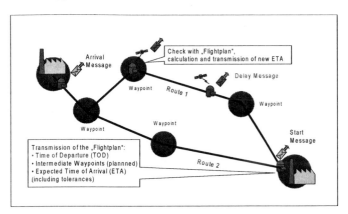

data storage, archiving, and data retrieval. Figure 2 shows an example for this network of information, again for the lorry/trailer configuration mentioned above:

Whenever a batch is leaving Factory A, a stationary tag reader identifies the batch, and the batch data are transmitted by an appropriate middleware to the ERP system of Factory A. The same procedure will take place when the batch arrives at Factory B (or at the site of final destination). The full information is available at any time by connecting the stationary ERP system to the Transport Tracking Center by means of a proprietary telecommunication link or via Internet. By those means, it is possible to obtain a more reliable estimate of the time of arrival for a specific good—thus allowing for a rearrangement of the production line at the destination site within due time (if necessary).

The eminent advantages of such a system become increasingly obvious if the transport is not a single point-to-point connection, but if a number of intermediate waypoints have to be covered, including unloading of some batches and loading of new batches, as illustrated in Figure 3, and whenever the transport medium is changed (e.g., from lorry to train) and a new batch configuration has to be assembled.

The system described here allows the user to gain an exact overview at any time and to track the flow of goods from the origin to the destination online and in real time. On the other hand, by means of archiving and retrieval, it allows for backward (upstream) traceability at batch level (or item level). In other words, by closing the mobility gap, this system is covering the full supply chain without any interruptions.

T-Systems has implemented such a system (called "eCargo") for RAILION, Europe's largest international logistics enterprise for railway-based transports, as described by Epple and Feuchtmüller (2005). More than 100,000 individual RAILION transports per day are crossing all over the continent, carrying a huge variety of goods. About 13,000 wagons are equipped with GPS and GSM devices at least, plus environmental sensors, reefers, and data loggers.

For a pilot installation and for operational use, a similar system (called "iTM"®—intelligent Tracking Management) has been developed by T-Systems for Schmitz CargoBull, one of Europe's leading trailer manufacturers. Reliability, safety, and security are top priority requirements for both systems.

Table 1 provides a summary of the benefits and advantages of the described system solution.

Table 1. Benefits and advantages of mobile batch tracking systems

Commercial Aspects	Security Aspects
• Fulfilling the requirements with respect to batch tracking	• Secure authentication by full time coverage and uninterrupted data history
• Timely implementation of EU Regulation 178/2002 with respect to food batches	• Basis for certification according to IFS (International Food Standard)
• Online trackability ("downstream")	• High security by "closed-door" principle (reliable content management)
• Data retrieval and traceability ("upstream")	• Documentation of the grower, producer, or manufacturer
• High degree of automation by using RFID technology	• Documentation of the receiver/user and of the intermediate agents
• Minimization of damage in case of recall actions	• Documentation of the wares, the raw materials used, and of all relevant time stamps
• Interface to customer ERP systems	• Documentation of the environmental conditions during transportation
• High degree of intermodal flexibility	• Documentation about storage and status of semi-products and intermediate processing stages

LIMITATIONS AND CHALLENGES TO THE SOLUTION

Basically, there are no other limitations to a worldwide use of the system than those imposed by physics. Perturbations of the radio frequency (RF) may have to be faced if RFID is used in a ferro-metallic environment, or Faraday screening may prevent readers from identifying tags if they are "hidden" by a metallic foil. A further electromagnetic threat is encountered if the system is operated in environments with spark discharges or in cases of other events causing high-voltage electromagnetic pulses (EMPs). In those cases, the transponder chip may be completely damaged.

Besides these physical challenges and threats, a global use is rather endangered by incompatible or even competing systems with respect to the performance of readers and transponders, and with respect to the RF used. While in the U.S. the UHF range between 868 and 915 MHz is favored, many developments in

Germany prefer a frequency of 13.56 MHz (ten Hompel & Lange, 2005).

A different limitation to an increasing use of RFID-based systems may be given by a more commercial point of view. As for all new technologies in the beginning, the unit prices are relatively high (approximately 0.60 USD per transponder tag, depending on the storage capacity, ranging from a few bytes to several Kbytes). Drastic price reductions can be expected for the time to come when large numbers of tags will be produced.

Finally, like in many other cases, it has to be considered that the system is subject to some security risks and to the possibility of criminal attacks. Removing or destroying tags by brute force is the simplest way, followed by more sophisticated acts such as unauthorized reading of the tags, cloning of tags by means of electronic devices, and emulation of tags with any desired content (Oertel et al., 2005).

Here again, further development of the technology and international security standards will help to reduce the inherent risks.

CONCLUSION AND FUTURE DIRECTION

This chapter has shown that a system solution for mobile batch tracking is feasible that allows for online batch tracking during downstream transportation, as well as for upstream traceability. The system presented here bridges the information gap between the automated systems at the factory sites and the storage control systems at the destination sites. By using finest technology according to the state of the art, this mobility system can be considered to represent a breakthrough in supply chain management—especially when taking into account that an increasing number of goods will be "on the road" (on rails, on ship, in the air) for an appreciable percentage of the lifecycle, thus resulting in an urgent need to cover this mobility phase.

Nevertheless, a number of problems and difficulties still persist. Due to the international nature of the system, it is quite obvious that full functionality across borders will require international agreements, legal regulations, and standards. Technological standards will have to deal with reserved frequency ranges for the RFID equipment, the transmission speed, coding, protocols, and anti-collision procedures. Data standards will have to take care of a scheme for unique numbering (e.g., according to ISO/IEC Standard 15963), and application standards will have to consider new coding standards, such as the Electronic Product Code (EPC) replacing the UPC Barcode Standard (ten Hompel & Lange, 2005). Unique identification will require a well-elaborated coding standard based on a worldwide agreement, especially when thinking in terms of progressing from unit- and pallet-tagging down to item-tagging.

In parallel, sophisticated security measures will have to be developed in order to overcome the criminal risks inherent to each new technology.

Provided that these prerequisites are given, there is no doubt that mobile batch tracking systems based on RFID technology—like the one presented here—will result in a tremendous improvement of supply chain management.

REFERENCES

Epple, M., & Feuchtmüller, H. (2005, January). Weichen für die transportsicherheit (Points for the safety of transportation). *Europäische Sicherheit (European Safety)*, (1), 56-57.

Leaver, S. (2004, August 13). *Evaluating RFID middleware* (Company Research Report), Forrester Tech Choices, Forrester Research, Inc., Cambridge, MA, USA.

Oertel, B., Wölk, M., Hilty, L., Kelter, H., Ullmann, M., & Wittmann, S. (2005). Der gläserne kunde (The glassy customer). *DoQ, H&T Verlag*, (1), 53-55.

ten Hompel, M., & Lange, V. (2005). Barcode geknackt (Barcode cracked). *DoQ, H&T Verlag*, (1), 48-50. München, Germany.

T-Systems. (2005). Retrieved April 12, 2005, from www.t-systems.com

About the Authors

Bhuvan Unhelkar has 24 years of strategic as well as hands-on professional experience in information and communication technology. Founder of MethodScience.com, he also has notable consulting and training expertise in software engineering (modeling, processes, and quality), enterprise globalization, Web services, and mobile technologies. He earned his doctorate in the area of "object orientation" from the University of Technology, Sydney. In his academic role at the University of Western Sydney, he leads the Mobile Internet Research and Applications Group (MIRAG), has authored/edited eight books, and has extensively presented and published research papers and case studies. He is a sought-after orator, a fellow of the Australian Computer Society, a Rotarian, and a previous TiE mentor.

*   *   *

Christopher Abood has worked in ICT for over 20 years with a Master of Commerce degree, majoring in Information Systems, from the University of New South Wales. He is an active member of the Australian Computer Society; during 2004, he led the development of the Mobile Camera Phone policy for the Australian Computer Society, and has done a number of newspaper and radio interviews on the inappropriate use of mobile camera phones.

Harpreet Alag is currently working as a senior business process consultant with Agilisys Limited, UK. He has a number of years of experience in process change and business analysis for developing and implementing business systems in multiple domains and business areas. In the last few years, Mr. Alag has been involved in major change programs, implementing CRM and ERP applications along with business integration. The focus of his work has been business process redesigning and implementation of application packages, and he has also been involved in change and transition management.

Chantal Ammi is a professor at the National Institute of Telecommunications, Paris, France. Specializing in marketing and strategy in high-tech sectors, she has published several books in different fields, has worked on different international contracts of research, and has developed different products and services.

Dinesh Arunatileka is researching for his PhD at the University of Western Sydney, Australia, in the area of mobile technologies and their application to business processes. He has also been a teaching fellow at the same university for the past two-and-a-half years. He earned his BSc in Computer Science from the University of Colombo and his MBA from the University of Sri Jayewardenepura, Sri Lanka. He has over nine years of experience in business development in the computing and telecommunications industry. He has published and presented conference papers in the area of methodologies to introduce mobile technology into business practices.

Achraf Ayadi is PhD student in Management Science at the National Institute of Telecommunications, Paris, France. He has published several articles on banking technologies management, electronic and mobile financial services, entrepreneurship, and Science Parks creation. He is currently the chairman of the PhD graduates and PhD students association of INT (Doc'INT) and associate general-secretary of the Tunisian association of high-schools graduates (ATUGE).

R. M. Banakar received a BE degree in electronics and communication engineering from Karnatak University, India, in 1984 and an MTech in Digital Communication from Regional Engineering College, Surathkal, Karnataka. She has a couple of years experience with the Indian Space Research Organization (ISRO). She completed her PhD in the area of low-power, application-specific design methodology from IIT Delhi in 2004. Presently she is working as an assistant professor at the engineering college, Hubli Karnataka. She is the member of the ISTE, IETE, MIE, and IEEE scientific and professional societies. Her current areas of research include SOC, VLSI architecture, and WCDMA. She can be reached via e-mail at banakar@bvb.edu.

Franck Barbier is a professor of software engineering at the University of Pau, France. Previously, he was the director of the Computer Science Research Institute of the University of Pau from October 2000 to October 2004. His main research focuses are software component modeling, evaluation, distribution, and mobility in the context of UML. He was the scientific consultant of Reich Technologies, a French company among the 17 companies that built UML 1.1 at the Object Management Group in 1997.

Francisco Barcelo earned a degree in telecommunications engineering and a PhD from the Technical University of Catalonia (UPC) in 1986 and 1997 respectively. In 1987 he joined the School of Telecommunications Engineering of Barcelona at UPC, where he has been teaching design and planning of communication networks. After graduation, he did research in the areas of digital network synchronization and switching. Since 1997 he has been an associate professor at UPC. He also serves as a consultant to the telecommunications industry and operators in Spain, and is currently involved with several research projects supported by the Spanish Government (Plan Nacional de I+D) and the European Commission (IST 5th Framework Program). His current research interests lie in the study of the evaluation and planning of the capacity of wireless networks and in the area of location technologies for cellular and WLAN networks.

Joseph Barjis is currently working as an assistant professor in the department of IT at Georgia Southern University, USA. He is also chairman of the research committee in this department and a member of the Faculty Research Committee at the university level. He earned BSc, MSc, and PhD degrees in computer science. He is actively conducting research on modeling and simulation of business processes, information systems design, and related topics. He has published more than 70 conference and journal papers, four book chapters, and one edited proceedings.

Raghunadh K. Bhattar received his BSc (Engg) from Regional Engineering College, Kurukshetra, India, and his MSc (Engg) from Indian Institute Science (IISc), Bangalore, India, in 1985 and 2000, respectively. Presently he is pursuing a PhD degree in Electrical Engineering IISc. He joined the Space Applications Centre, ISRO in 1990, where he is actively involved in the design and development of Turbo codec on FPGA for ground-based SATCOM systems and contributed to MPEG2 video codec, JPEG, and JPEG2000 systems. His current research interests are digital signal processing, image and video coding, multimedia, wavelets, channel coding, and wireless technologies.

Mohamed Boulmalf received his BSc in Communications from the National Institute of Telecommunications in 1987, Rabat, Morocco. He worked for five years as a network engineer at ONCF Company, Rabat, Morocco. In August 1992, Dr. Boulmalf moved to Canada to pursue his graduate studies at the National Institute for Scientific Research, Montreal, Canada. He received MSc and PhD degrees, both in wireless communications and networking, in 1994 and 2001, respectively. From September 1994 to December 1998, he was with INRS-Telecom as a radio communications research engineer. In January 1999, he joined ETS, Quebec University where he worked as a lecturer. In September of the same year, Dr. Boulmalf moved to Microcell Telecommunications, GSM Operator in Canada, where he worked as a senior network engineer. From 2000 to 2002, he worked as a principal engineer in the Multi-Vendor Integration Department at Ericsson, Montreal, Canada. In February 2002, he joined the College of Information Technology at the United Arab Emirates University, Abu Dhabi–Al-Ain, where he is now an assistant professor.

Matthias Brantner studied Information Systems at the University of Mannheim, Germany, from 1999 until 2004. He is currently employed and working on his doctorate at the Chair of Practical Information Technology III in Mannheim. The topic of his doctorate is query evaluation of XML query and transformation languages.

Torsten Brodt joined the Mcminstitute at the University of St. Gallen as a PhD researcher in 2003 specializing in mobile communication and media-related research. He is leading international research projects with MNOs, media, and high-tech companies. From 2001 to 2003 he worked as a consultant at A.T. Kearney management consultants and gained substantial experience in the European mobile communication industry. Mr. Torsten holds master's degrees in management science from the University of Mannheim, Germany, and Trinity College, Dublin, Ireland. His research interests include the market-orientation of innovation networks, and user-centric product and service development.

Narottam Chand received a BTech degree from the National Institute of Technology, Hamirpur, India, and an MTech degree from the Indian Institute of Technology, Delhi, in Computer Science & Engineering. He is currently a research scholar in the Electronics & Computer Engineering Department at the Indian Institute of Technology, Roorkee. His research interests include mobile computing and ad hoc networks. His PhD research work continues in the subject of data dissemination and caching techniques in mobile computing environments.

Chris Chatwin, under the auspices of the Industrial Informatics and Manufacturing Systems Research Centre (IIMS), has successfully completed a number of EPSRC, European, DTI, DFT, and industrial research projects in e-commerce, electro-optics, optical computing, control and systems integration, digital electronics, and image processing. Professor Chatwin has published two research level books: one on new numerical methods for simulation of laser materials processing, and the other on hybrid optical/digital computing. He has also published more than 200 international papers.

Xiao Chen holds a Bachelor's of Computing (E-Commerce) degree from the University of Western Sydney from their joint course run at Nanjing University of Chinese Medicine (NUCM), China. He is currently studying for his master's degree as an international student at Monash University, Australia. He has had special training and work experience in enterprises such as Siemens AD and Gori in China, enabling him to form a good understanding of mobile technologies in China.

Andrew P. Ciganek is approaching the final stages of his PhD program at the University of Wisconsin-Milwaukee. His research focus is on the adoption, diffusion, and implementation of Web services and other Net-enabling initiatives, mobile computing devices, knowledge management systems, and the role that the social context of an organization has in information systems research. He has published in a number of refereed conference proceedings and has manuscripts under various stages of review with scholarly IS journals.

Fred Claret-Tournier received his engineering degree from the School of Electrical Engineers, Grenoble, France, in 1996. In 1997 he received a master of science degree from the University of Sussex, Brighton, UK. In 1998 he joined the Department of Engineering, University of Sussex, as a research fellow. His research interests are in digital image processing, digital signal processing, embedded systems, manufacturing systems, and interface programming.

Joanne Marie Curry joined the University of Western Sydney (UWS) in 1996 specializing in the teaching of information systems. She has been the unit coordinator of the final-year Computing Project units for the last five years. In 2004, she was recognized for her contributions to practical project work, being the winner of the UWS 2004 Vice Chancellors Award for Excellence in the area of Regional and Community Engagement. Ms. Curry established the "XML and the Semantic Web¾Implications and Applications" mini-track at HICSS in 2001, and is also on the Organizing Committees for the 2005 Web Engineering Conference and the 2006 International Conference on Information Management and Business. She is currently researching for her PhD studies.

Ritanjan Das is a lecturer in information systems in the University of Portsmouth Business School. His research interests include e-business and e-commerce/m-commerce, information systems failure, and information systems security management.

Sipra Das(bit) received her BE in electronics and telecommunication and ME degree in electronics and telecommunication (specialization in computer science) in 1984 and 1986, respectively, and her PhD from the Department of Computer Science & Engineering in 1997. Since 1988 she has been with Department of Computer Science & Technology, Bengal Engineering and Science University (formerly Bengal Engineering College), where currently she is an assistant professor. She is also the recipient of the Career Award for Young Teachers from the All India Council of Technical Education.

K. S. Dasgupta received bachelor's and postgraduate degrees in computer science, with Honors, from Jadavpur University, India, in 1972 and 1973, respectively. He received his PhD from the Indian Institute of Technology, Mumbai, India, in 1990. He joined the Space Applications Centre, ISRO, India, in 1974, and since then has contributed significantly in the field of image processing and satellite communications. As group director of the Advanced Digital Communication Technology Group (ADCTG), he was instrumental in the design and development of a PC-based multimedia system for satellite-based distance education. His current areas of research interest address digital signal processing, digital image processing, computer architecture, and digital communications. He is a senior member of IEEE and a fellow of IETE India.

Samir El-Masri earned an engineering degree in electronics from the Lebanese University in 1993. Dr. El-Masri holds a master's degree (1994) and a PhD in Speech Production from the Institut National Polytechnique de Grenoble (INPG), France. He worked on building a real talking robot at Hokkaido University and NTT in Japan from 1998 to 2001. From 2001-2003, He was a lecturer at Central Queensland University and the University of Southern Queensland. Currently Dr. El-Masri is a senior lecturer at the University of Western Sydney, Australia. His current research interests are Mobile Web Services.

Francesco Falcone started as a project manager in 1986, developing ICT solutions for Banks. After a short move to the Fiat Group in 1989, in 1991 he was with the U.S. system integrator, EDS, and led the development of Internet applications. In 1999 he was named CTO of Digital Business, leader in the development of VAS applications based on new emerging mobile Internet technologies, partnering with Nokia and Openwave. Leading the Research & Development Department, he managed important projects with Vodafone and Telecom in the mobile VAS, messaging, and IP multimedia environment. He currently is a technology manager at Nok Services Ltd., London.

Hartmut Feuchtmüller (Dipl.-Ing. technical computer science) is a senior consultant for manufacturing services and intelligent tracking management systems at T-Systems International. He is experienced in software project management and consulting for telematics projects and has expertise in business processes for commercial and railway telematic systems, tracking and tracing systems, auto-identification, supply chain event management, and RFID technologies.

Thomas Fischer (Dipl.-Ing. aerospace) is currently head of manufacturing services at T-Systems International. He is experienced in software project management and executive management, and he has gained expertise in navigation systems and avionic systems, tracking and tracing systems, rail control systems, auto-identification, and supply chain event management.

Jason Gan specializes in the software design and development of Web front-ends to integrated business systems. He has 10+ years of experience as a layout artist and currently works as a Web designer and programmer in the industry, developing Web front-ends against Microsoft Great Plains ERP, Microsoft CRM, and Microsoft SQL Server. Mr. Gan graduated from Macquarie University in 1994 with a major in English Literature (Literary Craftsmanship). This was followed by a master's degree in computing science from UTS, Sydney, and he is currently working to produce "Great Plains C#" developer documentation.

Marco Garito was born in Milan, Italy, in 1967. He graduated with a law degree from the University of Milan and started his professional development in a professional and telecommunication services company before entering the Internet world with an incubator, where he stayed until late 2001. He then moved to Sydney to study for a master's degree at the University of Technology. After returning to Europe, he worked with IBM in Scotland (2004) and is currently with Cisco in Northern Ireland dealing with marketing and CRM.

Abbass Ghanbary holds a bachelor's degree in applied science, Honors, and is currently undertaking PhD research at the University of Western Sydney (UWS), Australia. His specific research focus includes the issues and challenges in incorporating Web services in businesses with the aid of mobile technologies and the subsequent issues of business process reengineering. Mr. Ghanbary earned a scholarship from the University of Western Sydney to undertake his research on the effects of and how to improve Web services with the aid of mobility. He also teaches and tutors at UWS, and is a full member of the Australian Computer Society.

Nina Godbole has more than 12 years of experience in the IT industry in system analysis, business development, support services, quality management, and training. She holds a key managerial position at IBM Global Services (India) Ltd., wherein she has successfully driven organization initiatives such as the P-CMM, CMM-I, BS 7799, and participated in the security/IT audits, as well as researching in them. She is author of a software quality assurance professional book (published in 2004 by NAROSA in India¾www.narosa.com¾and distributed in the U.S. and UK by Alpha Science). Ms. Godbole has also presented a number of papers on this topic in national and international conferences. She holds an MS degree from India's reputed Indian Institute of Technology (IIT-Bombay) and also earned an MEngg (computer science) from Newport University, USA. Her professional certifications include CQA, CISA, PMP, and CSTE.

John Goh is a PhD student at the School of Business Systems, Faculty of Information Technology, Monash University, Australia. He is conducting research in the field of mobile data mining, which focuses on extracting interesting patterns and knowledge from raw data collected from mobile users, including mobile phones and personal digital assistants. Since his enrolment into a research degree in late 2003, he has published a number of papers in the area of mobile data mining.

C. Gomathy acquired a BE (Honors) in electronics and communication engineering from the Government College of Engineering, Tirunelveli, in the year 1986, and an MS in electronics and control engineering from the Birla Institute of Science and Technology, Pilani, in 1992. She also obtained an MS (by research) from Anna University in 2001. She is currently pursuing her PhD in the Department of Electronics and Communication Engineering, College of Engineering, Anna University, Chennai, India. She has published more than 18 research papers in national and international conferences and journals. Her areas of interest include mobile ad hoc networks, high-speed networks, and digital communication.

Sheng-Uei Guan received his MSc and PhD from the University of North Carolina at Chapel Hill, USA. He is a chair professor with the School of Engineering and Design at Brunel University. Dr. Guan has worked in a prestigious R&D organization for several years, serving as a design engineer, project leader, and manager. After leaving the industry, he joined Yuan-Ze University in Taiwan for three-and-a-half years. He served as deputy director for the Computing Center and the chairman for the Department of Information & Communication Technology.

A. Hameurlain is a professor with Paul Sabatier University; his main research interests include: parallel databases, mobile databases, mobility, distributed systems, and query response time optimization.

Robert Harmon is professor of marketing and technology management at Portland State University, USA. His research includes value-based marketing, new product development, e-business, mobile commerce, strategic pricing, and all phases of the strategic market planning process. He has published in such journals as the *Journal of Marketing Research, Journal of Marketing, Journal of Advertising, Journal of Advertising Research,* and *Decision Sciences,* among others. He is currently a principal investigator for a $250,000 National Science Foundation grant focused on value-based software engineering for wireless thin-client applications.

Paul Hawking is a senior lecturer in the School of Information Systems at Victoria University, Melbourne, Australia. He is the SAP academic program director for the Faculty of Business and Law, and is responsible for the facilitation of ERP education and research across the university. Accordingly he is coordinator of the university's ERP Research Group (www.businesandlaw.vu.edu.au/sap/research.html). He is past chairman of the SAP Australian User Group, and is now responsible for education and research for this group. This has provided him with strong links with SAP customers. Professor Hawking is a leading researcher in the area of ERP systems and has produced many research publications, including a number commissioned by the SAP user community.

Sven Helmer studied computer science at the University of Karlsruhe in Germany from 1989 until 1995. Following that, he acquired a PhD in 2000 doing research in the area of database performance at the University of Mannheim, Germany. He stayed there until 2005, working in the area of native XML database systems as an assistant professor. In October 2005 he joined Birkbeck College, London, as a lecturer. He has published more than 30 papers in various journals, conference proceedings, and books. Furthermore, he served as a reviewer for different journals and as a member of several program committees.

Wen-Chen Hu earned BE, ME, MS, and PhD degrees, all in computer science, from Tamkang University, Taiwan, National Central University, Taiwan, the University of Iowa, USA, and the University of Florida, Gainesville, USA, in 1984, 1986, 1993, and 1998, respectively. He is currently an assistant professor in the Department of Computer Science, University of North Dakota, USA. Dr. Hu has published more than 30 articles in refereed journals, conferences, books, and encyclopedias, and one book, titled *Advances in Security and Payment Methods for Mobile Commerce*. His current research interests are in electronic and mobile commerce, Web technologies, and databases. He is a member of the IEEE Computer Society, ACM, and the Information Resources Management Association.

Walter Hürster holds a PhD in physics and is a principal consultant at T-Systems International; he has been involved in research work in nuclear and particle physics (Germany and Switzerland). He has extensive experience in software project management (large projects) and executive management, expertise in radar systems and avionic systems, radiation protection, coastal protection, remote monitoring of nuclear power plants, early warning and risk management systems, tracking systems, and air traffic control. He has presented numerous papers at international conferences and has contributed to several books on environmental informatics. Dr. Huerster has been called on several times by the EU Commission as an expert for risk management and early warning systems. He is also a member of several organizations and national working groups.

Mohammad Mahfuzul Islam received his BEngg (Honors) and MEngg degrees in computer science and engineering (CSE) from the Bangladesh University of Engineering and Technology (BUET) in 1997 and 2000, respectively. He is currently an assistant professor of CSE at BUET, and is close to completing his PhD at the Gippsland School of Computing and IT, Monash University, Australia. His research interests are mobility support resource management for cellular multimedia networks, genetic algorithms, neural networks, and fuzzy systems. He has published 18 journal and peer-reviewed research publications.

Bradley Johnstone spent many years working as an account executive in the advertising industry in Australia and the United Kingdom. After completing a computer science degree, he changed careers and is now working in the banking and financial industry. His research interests include application of mobility to banking business processes.

Matthew R. Jones lectures in information management in the Department of Engineering and the Judge Institute of Management at the University of Cambridge, UK. His research interests concern the relationship between information systems and social and organizational change. He has published widely in this area, including several studies in the health care domain.

R. C. Joshi received the BE degree from Allahabad University, India, and ME and PhD degree in electronics & computer engineering from the Indian Institute of Technology, Roorkee. Currently he is a professor in the Department of Electronics & Computer Engineering, Indian Institute of Technology, Roorkee. He has served as chair for various international conferences, including Expert System and Robotics at Pittsburgh (1994), Parallel Processing at St. Charles (1994), and ADCOM at IIT Roorkee (1999). His current research interests include database systems, data mining and knowledge discovery, mobile and distributed computing, sensor networks, and security.

Carl-Christian Kanne studied computer science at the Rheinisch-Westfälische Technische Hochschule (RWTH) Aachen in Germany from 1992 until 1998. In 2003, he acquired a PhD from the University of Mannheim, based on his research on native XML database systems. He is a founder of "data ex machina GmbH," a German company whose products include the Natix XML DBMS. He has published more than a dozen papers in various journals and conference proceedings.

Heikki Karjaluoto is a research professor at the Faculty of Economics and Business Administration, University of Oulu, Finland. His research interests concern electronic business in general, and mobile business and mobile commerce in particular. He has published extensively on electronic business in marketing and information system journals, and has collaborated with several researchers in Finland and abroad and with Finnish high-tech companies in common research projects.

Anand Kuppuswami is a research associate pursuing his PhD in the School of Computing and IT at the University of Western Sydney. He received his MS (Honors) from the University of Western Sydney and his BE from the University of Madras. His primary research interest lies in the application of neural networks and agent technology in the field of pattern recognition and intelligent migration.

Yi-chen Lan is a senior lecturer in the School of Computing and IT, the University of Western Sydney. Dr. Lan holds a Bachelor's of Commerce—Computing and Information Systems (Honors) degree and a PhD from the University of Western Sydney. He teaches information systems and management courses at both the undergraduate and graduate levels. Prior to his current academic work, Dr. Lan served industry for five years, wherein he held senior management responsibilities in the areas of information systems and quality assurance programs in a multinational organization. His main areas of research include global transition process, global information systems management issues, globalization framework development, integrated supply chain development, and health-related information systems development and management.

Sandra Synthia Lazarus is currently completing her PhD in Medical Informatics at the University of Sydney in which she hopes to evaluate the impact of information technology on clinical processes. As part of her research, she has evaluated wireless technology in a hospital environment. Her pervious contributions include research in biochemistry, inhabiting fungal actives, and agile methodologies in computer science. She has completed a bachelor of science degree, a master's in computer science, and hopes to continue her education, while working with various e-health companies to develop e-health solutions.

Thomas Leary is a doctoral student in the Information Systems and Operations Management (ISOM) Department at the Bryan School of Business and Economics at the University of North Carolina at Greensboro, USA. His research interests include mobile technologies, Semantic Web, mobile e-business and strategy, service descriptions and composition, and intelligent multi-agent architectures.

Chean Lee holds a master's in information technology from the University of Western Sydney and a Bachelor's of Commerce from Griffith University. He has been working throughout e-business, Internet, Web development, and ERP application industries. His current research interests include application of mobile technologies to Customer Relationship Management applications.

Maria Ruey-Yuan Lee is the department head and an associate professor in the Information Management Department, Shih Chien University, Taipei, Taiwan. Her current research interests include ontology, wireless Internet applications, and mobile commerce. She has published in a number of international conferences and journals, and has been referee, panelist, program committee member, and organizer at many international conferences. She worked at CSIRO Mathematical and Information Sciences, Sydney, Australia, for more than 12 years before she joined academics in Taiwan. She led a group conducting research and development in applying artificial intelligence technologies to electronic business applications while at CSIRO.

Pouwan Lei is a lecturer at the University of Bradford, UK. Previously she was a lecturer at the University of Macao. Her research interests include electronic commerce, mobile commerce, semantic Web, ubiquitous computing, and intelligent agents. She holds a Doctor's of Philosophy degree from the University of Sussex, UK.

Chye-Huang Leow is a lecturer with the School of Business at the Singapore Polytechnic. She has lectured on subjects including retail environment and technology, organizational management, management and organizational behavior, and services marketing. Ms. Leow has assisted with several key projects for retailers in Singapore in the area of successful business strategies. Her key area of research is business-related and technology applications to education institutions and businesses. Prior to joining the Polytechnic, Ms. Leow was in marketing and retailing practice with local and international companies for many years, providing technical and professional advice to retailers and commercial companies. She is also a certified Casetrust auditor in Singapore.

Matti Leppäniemi is a researcher with the Faculty of Economics and Business Administration, University of Oulu, Finland. He is interested in the areas of mobile marketing and integrated marketing communications.

Feng Li is chair of E-Business Development at the University of Newcastle upon Tyne, UK. His research centrally focuses on the interactions between information systems and emerging strategies, business models, and organizational designs. He has worked closely with organizations in banking, telecommunications, manufacturing, retailing, and electronics, as well as the public sectors. He is the chair of the E-Business & E-Government Special Interest Group (SIG) in BAM and a member of the BAM Council. His work on Internet banking and on telecom value networks and pricing models has been extensively reported by the media. He can be via e-mail at Feng.li@ncl.ac.uk.

Wei Liu holds a Bachelor's of Computing (E-Commerce) degree from the University of Western Sydney from their joint course run at Nanjing University of Chinese Medicine (NUCM), China. Having studied information-centric technologies, he started researching Internet technology, mobile technology, and supply chain management during his final year at NUCM. His research interests include the application of mobile technologies to business processes.

M. Mammeri is a professor at the Paul Sabatier University. His main research interests include: real-time systems, distributed applications, wireless networks, and mobile environments.

Ioakim (Makis) Marmaridis holds a bachelor's in business computing and information management, as well as an honors degree in computing with First Class Honors. He is also the recipient of many prestigious awards for academic excellence and outstanding academic performance. Mr. Marmaridis is an MCP and has been working in the IT sector for several years, gathering very substantial experience in systems administration and networking, information systems security, and high availability systems including Web farms and DB clusters. He is currently undertaking his PhD in computing and also enjoys golf, Toastmasters writing, and blogging on his Web site at http://marmaridis.org/.

N. Marsit is a PhD student at the Paul Sabatier University. His research work focuses on mobile database queries.

Israel Martin-Escalona earned a degree in telecommunications engineering from the Technical University of Catalonia (UPC) in 1999 and is currently an assistant professor at the Department of Telematic Engineering. Since 2001 he is involved in the IST Emily Project, funded by the European Commission and dealing with location services and technologies for GSM/GPRS networks. In 2002 he joined the project PPT, funded by Telefónica, on location services for the UMTS network. In 2004 he joined the IST Liaison Project on location services for WLAN and ad-hoc networks.

Henrique M. G. Martins is a medical doctor who also holds a master's in management Studies and is currently finishing his PhD at the University of Cambridge, working within the Information Management area on the topic: "The Use of Mobile Information and Communication Technologies in Clinical Settings." His interests are primarily in the interfaces between medicine and management as facilitated by mobile clinical information systems, as well as management education to medical students and junior doctors. He has presented work at several conferences and has forthcoming articles published in these areas.

Carolyn McGregor holds a PhD in computing science, wherein she developed new ways to use intelligent decision support systems to assist organizations in the monitoring of business performance through the use of workflow audit logs. Furthermore, she has also provided strategic guidance in the areas of decision support, data warehousing, and data mining to some of Australia's leading corporations. In 2001, Dr. McGregor was the first Australian PhD student to be awarded a three-month IBM PhD Research Internship at the IBM TJ Watson Research Center in Yorktown, New York. Currently, she is active within the CASE Research Center at University of Western Sydney, where her research focus has been on development of an intelligent decision support system for business process performance management, known as the solution manager service, which she has refined and applied to problem domains such as health and medicine.

Shailendra Mishra earned his ME in computer science and engineering from MLNREC, (now MNIT Allahbad), Allahbad, and his master's degree in electronics in 1992 from the University of Allahbad, India. From August 1994 to February 2000, he led the Department of Computer Science and Electronics at ADC, University of Allahbad. From February 2000 to February 2001, he was with RG Engineering College, Meerut, affiliated with the UP Technical University, Lucknow, as assistant professor in the Department of Computer Science and Engineering. Presently, he is with the DehraDun Institute of Technology (DIT), affiliated with UP Technical University, as an assistant professor in the Computer Science & Engineering Department, and he is a joint faculty member of IGNOU, Delhi, India. His recent research has been in the field of mobile communication. He has also been conducting research on communication systems and networks with performance evaluation, and design of multiple access protocol for mobile communication networks. He is the author of more than 15 technical papers in international and national journals, and conference proceedings.

Manoj Misra received a BTech degree from HBTI Kanpur, India; an MTech degree from the Indian Institute of Technology, Roorkee; and a PhD from Newcastle upon Tyne, UK, in Computer Science & Engineering. He is currently an associate professor in the Electronics & Computer Engineering Department, Indian Institute of Technology, Roorkee. Dr. Misra's current research interests include mobile computing, performance evaluation, and distributed computing. He is also serving as co-coordinator of the Information Superhighway Center, IIT Roorkee.

Sulata Mitra received a BE in Electronics and Telecommunication and an ME in Power Electronics in 1986 and 1996, respectively. She is pursuing her PhD. Since 2000 she has been with the Department of Computer Science & Technology, Bengal Engineering and Science University (formerly Bengal Engineering College), where currently she is a lecturer. Her area of interest is mobile computing.

Guido Moerkotte studied computer science at the Universities of Dortmund, Massachusetts, and Karlsruhe from 1981 to 1987. The University of Karlsruhe awarded him a Diploma (1987), a doctorate (1989), and a postdoctoral lecture qualification (1994). In 1994 he became an associate professor at RWTH Aachen. Since 1996 he holds a full professor position at the University of Mannheim, Germany, where he heads the database research group. His research interests include databases and their applications, query optimization, and XML databases. Dr. Moerkotte authored/co-authored more than 100 publications and three books.

F. Morvan is an associate professor at the Paul Sabatier University. His main research interests include parallel databases, distributed systems, and query response time optimization.

Manzur Murshed received his BEngg (*Honors*) in computer science and engineering from Bangladesh University of Engineering and Technology (BUET) in 1994, and his PhD in computer science from the Australian National University in 1999. He is currently the director of research and a senior lecturer at the Gippsland School of Computing and Information Technology, Monash University, Australia, where his major research interests are in the fields of multimedia communications, wireless communications, video coding and transcoding, video indexing and retrieval, video-on-demand, image processing, parallel and distributed computing, grid computing, simulation, complexity analysis, multilingual systems, algorithms, digital watermarking, and distributed coding. He has published more than 70 journal and peer-reviewed research publications and two refereed book chapters. He is the recipient of numerous academic awards including the *University Gold Medal* from BUET.

Sashi Nand holds BComm, MAcc, and PhD degrees, and provides administrative and accounting services to an engineering computer consultancy through Nandtech Pty Ltd. She has been a guest lecturer at the University of Technology Sydney Faculty of Engineering on the topic of Emerging New Technologies. Dr. Nand has published and presented papers at various international conferences including in Egypt and Perth. She is the winner of the PriceWaterhouseCoopers Prize in Advanced Accounting Theory, a member of the Australian Institute of Company Directors, as well as a member and ambassador for the Australian Computer Society. She was recently awarded as Toastmaster of the Year 2004/05 by Hornsby Toastmasters, and was designated Compctent Leader and Competent Toastmaster.

Devon Nugent is a PhD candidate at the University of Queensland. Her research interests include Geographical Information Systems and mapping in location-based services.

V.S. Palsule received an ME in communication systems (Spread Spectrum Systems) from Jabalpur University, India, in 1979 and then joined the Space Applications Centre, ISRO, India. Presently he is heading the Advanced Communication Technology Division (ACTD) and is project director for EDUSAT, a satellite dedicated to education. He handled many prestigious projects of national interest, which includes spread spectrum systems for satellite communications, small communication terminal (SCOT), GPS, and MSS.

Amol Patel is founder and president of ConvergeLabs Corporation. Prior to that he was director of new ventures at ADC Telecommunications, a provider of integrated voice, video, and data solutions for the last mile of the communications network. His responsibilities included strategic planning, international mergers and acquisitions, and electronic commerce. Prior to ADC, he held several positions in marketing and engineering at Intel Corporation, Cirrus Logic, and Sun Microsystems. He holds a BS (Honors) in Electrical Engineering and Computer Science from the University of California at Berkeley, an MS in Electrical Engineering from Stanford University, and an MBA from the Kellogg School of Management, Northwestern University.

Keyurkumar J. Patel received his First Class BEngg from the Bangalore University, India, and his MEngg (Robotics) from Swinburne University of Technology, Australia, in 1997 and 2000, respectively. He is currently pursuing his PhD. He is a Microsoft Certified Systems Engineer (MCSE), Microsoft Certified Trainer (MCT), Cisco Certified Academy Instructor (CCNA/CCAI), and Certified Novell Engineer (CNE), Novell Academic Instructor (NAI). He is currently employed with the Centre for Computer Technology and Super Cisco Academy Training Centre (SuperCATC) at the Box Hill Institute of TAFE, Australia, as a leader of ICT Higher Education Programs. He was a founder of the Emerging Communications Technologies (ECT) Research Cluster in his previous position at the University of Ballarat, Australia. His research interests include a good mixture of IT and engineering, including communication technologies, application implementation, image processing, systems design, health informatics, and advanced manufacturing techniques. He has successfully executed research and commercial projects including for major IT companies like Cisco and Microsoft. To date he has published more than 40 research studies in international refereed conferences and journals. He is an IASTED Technical Committee member for the term 2004-2007, a member of the Australian Computer Society (MACS), and a member of IEEE (Communication and Computer Societies).

Stephen Paull has more than 30 years of experience working in software development, project management, and the design, implementation, training, and support of manufacturing-based ERP systems. In addition, for the last 16 years, he has been a senior lecturer in the School of Information Systems at the Victoria University of Technology. He is currently involved in the teaching of the SAP software package as part of the curriculum for undergraduate and postgraduate courses, specifically in the areas of manufacturing, project management, and programming (ABAP).

Priyatamkumar received BE and MTech degrees in electronics and communication from the Karnataka University and National Institute of Technology, Karnataka, India, in 1989 and 2004, respectively. He joined the BVB College of Engineering and Technology, Hubli Karnataka, India, in 1989, where he was engaged in teaching antennas and advanced communication systems. He is currently a senior faculty member doing research in WCDMA and mobile communication. He is a member of the Institute of Electronics and a member of the Indian Society of Technical Teachers (MISTE). He can be reached via e-mail at priyatam@bvb.edu.

Jon Tong-Seng Quah is an associate professor of electrical and electronic engineering, Nanyang Technological University. He lectures in both undergrad as well as graduate courses such as *Software Development Methodology, Software Quality Assurance and Project Management, Object-Oriented System Analysis and Design*, and *Software Engineering*. His research interests include financial market modeling using neural networks, software reliability, e-commerce, as well as Web and WAP technologies and applications. Other than academic services, Dr. Quah has undertaken joint projects with major companies in banking and airline industries, as well as statutory boards of the government body. Prior to his academic pursuit, Dr. Quah was a director of a local company dealing with industrial chemicals.

Mahesh S. Raisinghani is an associate professor at TWU School of Management's Executive MBA program. He is also a certified e-commerce consultant and a project management professional (PMP). Dr. Raisinghani was the recipient of the 1999 UD Presidential Award and the 2001 King Haggar Award for excellence in teaching, research and service. His previous publications have appeared in *Information and Management, Information Resources Management Journal, Journal of Global IT Management, Journal of E-Commerce Research, Information Strategy: An Executive's Journal, Journal of IT Theory and Applications, Enterprise Systems Journal, Journal of Computer Information Systems,* and *International Journal of Information Management,* among others. He serves as an associate editor and on the editorial review board of leading information systems/e-commerce journals and on the board of directors of Sequoia, Inc. Dr. Raisinghani is included in the millennium edition of *Who's Who in the World, Who's Who Among America's Teachers,* and *Who's Who in Information Technology.*

S. Rajeev is with the Department of Electronics & Communication Engineering, PSG College of Technology, Coimbatore, India. He has more than 13 years of industrial and academic experience. His research interests are policy provisioning systems and distributed computing and systems. He has authored three books in the area of computer communication.

K. R. Ramakrishnan received BE, ME, and PhD degrees in electrical engineering from the Indian Institute of Science, Bangalore, India, in 1974, 1976, and 1983, respectively. He is currently a professor with the Department of Electrical Engineering, Indian Institute of Science. His research interests include image processing, computer vision, medical imaging, water marking, and multimedia communication.

K. Ramamurthy is the Roger L. Fitzsimonds scholar and professor of MIS at the University of Wisconsin-Milwaukee. He has a bachelor's degree in mechanical engineering, a graduate diploma in Statistical Quality Control and Operations Research, and an MBA. He also earned his PhD in management information systems from the University of Pittsburgh. He has nearly 20 years of industry experience and has held several senior technical and executive positions prior to entering academia. His current research interests include electronic commerce including Internet; adoption, implementation, and diffusion of modern information technologies; supply chain management; strategic IS planning; self-directed teams; business process reengineering; and management of computer-integrated manufacturing technologies. He has published more than 30 articles in major scholarly journals like *MIS Quarterly* (where he is an associate editor) and *IEEE.* He is a charter member of the AIS, and was elected to the Beta Gamma and Sigma honor society.

Anne-Marie Ranft has a BS in computing science from the University of Technology, Sydney, and has seven years of commercial IT experience in Internet/e-business applications. She is currently completing her MS in computing. Her research interests include application of mobile technologies to globalization.

N. Raghavendra Rao is a professor at the SSN School of Management & Computer Applications, Madras. He holds a PhD in Finance from the University of Poona. His teaching experience in the disciplines of information technology and financial management spans 10 years. He has more than two decades of experience in the development of application software related to manufacturing, service-oriented organizations, financial institutions, and business enterprises. He regularly contributes IT articles to mainstream newspapers and journals.

Gina Reyes is a lecturer in the School of Information Systems at Victoria University. Her main teaching areas are in systems analysis, systems implementation, project management, and management of IT. She received her PhD from RMIT University with her thesis on the impact of organizational context on IS performance. Her current research interests include mobile computing applications and the effective design of e-learning systems.

Fabien Roméo is a graduate student in computer science (MSc). He is currently preparing his thesis on the administration of wireless software components.

A. F. Salam is an assistant professor in the ISOM Department at the University of North Carolina-Greensboro. He earned both his PhD and MBA degrees from the School of Management at SUNY Buffalo. His research has been published or is forthcoming in *Communications of the ACM, IEEE Transactions on Systems, Man and Cybernetics, Information & Management, Communications of the AIS, Information Systems Journal,* and *Information Systems Management.* He has also been a co-editor of a special issue of the *Journal of Electronic Commerce Research* and co-guest editor of special sections of the *CACM on Internet and Marketing,* and the *Semantic E-Business Vision.*

Jari Salo is a researcher with the Faculty of Economics and Business Administration, University of Oulu, Finland. His present research interests include digitization of business relationships and networks, electronic commerce, and mobile marketing.

Richard Schilhavy is a PhD student in the Department of Information Systems and Operations Management (ISOM) at the Bryan School of Business and Economics at the University of North Carolina-Greensboro. His research interests are in technology innovations and their influence on business functions and implications for supply/value chain operations. He has published papers in the proceedings of INFORMS and DSI conferences which cover issues related to queuing problems, information assurance, and software development.

B. Shankaranand received his BE in electronic and communication engineering from Mysore University, India, in 1973, and his MSc (Engg) in Microwave Engineering from the Kerala, India, in 1977. He joined the National Institute of Technology, Karnatak, India, in 1979, where he has been engaged in teaching and research and development of broadband communication systems, focusing on communication applications, especially in satellite, optical, and microwave applications. He is currently a professor at NITK India, and is a member of ISTE, IETE, and MIE. He can be reached via e-mail at bsnanda2000@yahoo.co.uk.

S. Shanmugavel graduated from the Madras Institute of Technology in Electronics and Communication Engineering in 1978. He obtained his PhD in the area of Coded Communication and Spread Spectrum Techniques from the Indian Institute of Technology (IIT), Kharagpur, in 1989. He joined the faculty of the Department of Electronics and Communication Engineering at IIT as a lecturer in 1987 and became an assistant professor in 1991. Presently, he is a professor in the Department of Electronics and Communication Engineering, College of Engineering, Anna University, Chennai, India. He has published more than 68 research papers in national and international conferences, and 15 research papers in journals. He was awarded the IETE-CDIL Award in September 2000 for his research efforts. His areas of interest include mobile ad hoc networks, ATM networks, and CDMA engineering.

Pramod Sharma is the director of the IT Program at the Cooperative Research Center for Sustainable Tourism and is based at the University of Queensland in Brisbane, Australia. His research interests include Geographical Information Systems, and the development of information and communications technology applications in tourism.

Khaled Shuaib holds a PhD in electrical engineering from the City University of New York (1999). While doing his graduate studies, he worked for two years as a senior member of the technical staff at GTE Labs (currently Verizon Technology Center) in Waltham, Massachusetts. In 1999, Dr. Shuaib joined Ascend Communications (currently Lucent Core Switching Division) in Westford, Massachusetts, as a principle performance engineer, where he worked on performance evaluation, reliability, and scalability of IP and ATM networks of ISPs. Since September 2002, he has been with the College of Information Technology at the UAE University as an assistant professor. Currently, he is the program coordinator of the Network Engineering Track and the director of the UAEU Cisco Regional Networking Academy. His research interests are in the areas of network design and performance, QoS IP networks, and protocols for ad-hoc and wireless networks.

Nipur Singh received her PhD from Gurukul Kangari University and her MCA from Banasthli Vidaypith, India. She headed the Department of computer science, K.G. second campus at Gurukul Kangari University. She has authored numerous technical papers in national and international journals and conference proceedings, which include her recent area of research in the field of mobile communication and mobile computing.

Rahul Singh is an assistant professor in the ISOM Department at the Bryan School of Business and Economics at the University of North Carolina-Greensboro. His research interests include intelligent multi-agent architectures, Semantic Web, emerging technologies, and agents in supply chain. His published research includes articles in the *Communications of the ACM, E-Services Journal,* and *International Journal of Production Economics.* He has also served as co-chair of the mini-track on Semantic E-Business at AMCIS 2004. He also co-guest edited the special section on "Semantic E-Business Vision" in the December 2005 issue of *Communications of the ACM.* He earned his PhD from Virginia Commonwealth University.

Jaakko Sinisalo is a researcher with the Faculty of Economics and Business Administration, University of Oulu, Finland. His research interests lie at the intersection of mobile marketing and customer relationship management.

S. N. Sivanandam heads the Department of Computer Science and Engineering, PSG College of Technology, Coimbatore, India. He has more than 25 years of professional teaching experience. His research interests include control systems, networking, distributed architectures, and soft computing. He has over 250 publications to his credit, and he has authored five books in the areas of control systems and soft computing.

K.V. Sreenaath is a student doing research in the Department of Information Technology, PSG College of Technology, Coimbatore, India. His research is in cryptography, service-level provisioning, and automated systems.

Hsiang-Ju Su is a graduate student from the Institute of Enterprise Innovation and Development, Department of Information Management, Shih Chien University. Su's main research activity addresses consumer behaviors in mobile phone services.

David Taniar received his bachelor's (Honors), master's, and PhD degrees in computer science, with a particular specialty in databases. His research now extends to data mining and mobile databases. He has published research papers in these fields extensively including a recent book *Object-Oriented Oracle*. Dr. Taniar now works at Monash University, Australia. He is an editor-in-chief of several international journals, including *Data Warehousing and Mining, Business Intelligence and Data Mining, Web Information Systems, Web and Grid Services, Mobile Multimedia,* and *Mobile Information Systems.* He is a fellow of the Institute for Management Information Systems.

Ramaprasad Unni is an assistant professor of marketing at Portland State University. His research interests are in understanding the role of technology (including the Internet) in marketing. He has also conducted several projects that have examined differences in consumer behavior in online and off-line environments, and has presented his research at several national and international conferences.

Khimji Vaghjiani has spent 16 years in the IT industry, covering manufacturing, telecommunications, and banking and finance. He has extensive experience advising organizations in the field of emerging technologies, with particular focus on the user needs with mobile technologies. He is currently working on his PhD at the University of Technology, Sydney, in the area of innovation and its application in management and transitions.

Ketan Vanjara is a lead program manager with Microsoft in its Global Delivery Center, India (GDCI). Previously, he held several senior positions in the software industry and has led large software enterprise software development teams on diverse domains and technology platforms. During his 17+ years in the software industry, he has worked with various delivery models on multiple platforms. He specializes in the creation of application software for various industry verticals like health care, knowledge services, and manufacturing. He has in-depth knowledge and experience of working on various software development models, and implementing quality processes based on frameworks like CMM and ISO. He has published in numerous industrial as well as research journals and has contributed to chapters in edited books.

Amrish Vyas is in the process of completing his PhD at the University of Maryland-Baltimore County. Prior to joining the PhD program, he earned his MBA in Business Computer Information Systems from the Zarb School of Business, Hofstra University, USA, and his LLB from the Maharaja Sayajirao University, India. His research interests encompass expert (intelligent) systems, including intelligent agent systems and their application to an array of business problems, including location-aware applications in mobile services domain.

Fu Lee Wang received BEngg (Honors) in computer engineering and MPhil in computer science and information systems degrees from the University of Hong Kong, in 1997 and 1999, respectively. He earned his PhD in systems engineering and engineering management from the Chinese University of Hong Kong in 2003. He is currently an instructor in the Department of Computer Science at the City University of Hong Kong. He is a member of ACM and IEEE. His research interests include document summarization, digital library, and information retrieval.

Jia Jia Wang received an MSc with distinction in personal, mobile, and satellite communications from the University of Bradford, UK, in 2003, and a First Class BSc in electrical and electronic engineering from the HeiLongJiang University in China in 2002. In 2003, she joined the Mobile, Satellite, and Communication Research Center at the University of Bradford, where she participated in the EC's IST FIFTH project. She is currently in the second year of her PhD study; her research interests include mobile commerce, security, wireless mobility, and network protocol.

Tom Wiggen received his PhD from the Louisiana State University in 1973. He is an associate professor in the Department of Computer Science, University of North Dakota. His research interests include intelligent systems, modeling and simulation, and computer security.

David C. Wyld serves as the Mayfield professor of Management at Southeastern Louisiana University, where he also directs the Strategic E-Commerce Initiative. Dr. Wyld is a widely published author and invited presenter in the area of RFID technology. He is currently working on a report on RFID in the public sector for the IBM Center for the Business of Government and writing a book on RFID for Cambridge University Press.

Christopher C. Yang is an associate professor in the Department of Systems Engineering and Engineering Management at the Chinese University of Hong Kong. He received his BS, MS, and PhD degrees in electrical and computer engineering from the University of Arizona. Before he joined the Chinese University of Hong Kong, he was an assistant professor in the Department of Computer Science and Information Systems and the associate director of the Authorized Academic Java Campus at the University of Hong Kong. He has also been a research scientist in the Artificial Intelligence Laboratory in the Department of Management Information Systems at the University of Arizona. His recent research interests include cross-lingual information retrieval, multimedia information retrieval, digital library, information visualization, Internet searching, automatic summarization, information behavior, and electronic commerce. He has published more than 100 refereed journal and conference papers including in the *Journal of the American Society for Information Science and Technology* and *IEEE*.

Hung-Jen Yang received his PhD from the Iowa State University in 1991. He is an associate professor in the Department of Industrial Technology Education, National Kaohsiung Normal University, Taiwan. His research interests include computer networks, automation, and technology education.

Victoria Yoon is an associate professor in the Department of Information Systems at the University of Maryland-Baltimore County. Before that, she was an associate professor in the Department of Information Systems at Virginia Commonwealth University. She received her MS from the University of Pittsburgh and her PhD from the University of Texas at Arlington. Her primary research interests are the application of expert systems to business decision making in organizations and the managerial issues of such systems.

Rupert Young is currently a reader at the University of Sussex; he graduated with a degree in Engineering from Glasgow University in 1984. Subsequently, he was employed as a research assistant and research fellow at Glasgow, during which time he gained wide experience in image and signal processing techniques. He participated in two Brite/EuRam collaborative European programs during that time, the second with Glasgow as the consortium leaders. His research led to the award of the PhD degree in 1994. In April 1995 he took up his present post as a lecturer in the School of Engineering, University of Sussex. He was promoted to senior lecturer in 1997 and then reader in 1999. He has disseminated his work to the research community by publication of more than 80 refereed journal and conference papers. He is a member of the Institute of Electronic and Electrical Engineers.

Index

Symbols

2.5G Technology 146
2G (see second generation)
3G (see third generation)
3GPP (see third-generation partnership project)
4G (see fourth generation)
8-PSK (8-Phase Shift Keying) 147

A

a posteriori probabilities (APPs) 188
a priori 188
abstract data type 275
  model 276
abstract location 60
ACA (see adaptive conjoint analysis)
access cost 615
accessibility 697
accuracy 39
ad hoc 308
adaptation 609
adaptive 8
  conjoint analysis (ACA) 759

adaptive modulation and coding (AMC) 152,
  153
additive white Gaussian noise (AWGN) 186
administrative operations 769
advanced
  communication age 719
  encryption standard (AES) 396, 496
  medical priority dispatch 106
  mobile phone system (AMPS) 144
advertising 607, 608
AES (see advanced encryption standard)
affordability 505
after-purchase evaluation 634
agent 3, 367
  activation 373
  freeze 371, 373
  ontology 511
  pre-activation 372
  receptionist 370
  transport 372, 373
agent's
  fitness function 532
  GUI 534